AF541426

Pvt. Ltd.

BUFFALO PRODUCTION

P.N. Bhat

Centre for Integrated Animal Husbandry Dairy Development,
Flat No. 205, Block No. F – 64/C9,
Sector – 40, Noida – 201 301

2010

Studium Press (India) Pvt. Ltd.

BUFFALO PRODUCTION

ISBN: 978-93-80012-03-2

Published by:

Studium Press (India) Pvt. Ltd.
4735/22, 2nd Floor, Prakash Deep Building
(Near Delhi Medical Association),
Ansari Road, Darya Ganj, New Delhi-110 002
Tel.: 23240257, 65150447; Fax: 91-11-23240273;
jngovil@gmail.com; jngovil@hotmail.com

Printed at:

Salasar Imaging Systems
Delhi-110035 (India)

Series ISBN: 978-93-80012-00-1

ABOUT THE SERIES

According to the 2003 Census data, the country had 485 million (M) livestock and 489 M poultry, having the second highest number of cattle 185 M, the highest number of buffaloes 97 M, the third highest number of sheep 61 M, the second highest number of goats 124 M, the sixth highest number of camels 632 M, the fifth highest number of chickens 489 M and the fourth highest number of ducks 33 M in the world.

Livestock Sector has been playing an important role in Indian economy and is an important sub-sector of Indian agriculture. The contribution of livestock to GDP was 4.36% in 2004–05 at current prices. According to CSO estimates, gross domestic product from livestock sector at current prices was about Rs 935 billion during 1990–2000, (about 22.51% of agriculture and allied GDP). This rose to Rs1239 billion during 2004–05 with 24.72% share in agriculture and allied GDP. But the share of livestock sector in the plan allocation hovered at around 7% of the agricultural out lay.

This sector plays an important and vital role in providing nutritive food, rich in animal protein to the general public and in supplementing family incomes and generating gainful employment in the rural India, particularly among the small, marginal farmers, land less labourers and women. Distribution of livestock wealth in India is more egalitarian, compared to land. Hence, from the equity and livelihood perspectives, it is an important component in poverty alleviation programmes. This fact however has not been appreciated by Policy planners and implementers.

The development of animal husbandry has been envisaged as an integral part of system of diversified agriculture. With its large livestock population, India has vast potential for meeting the growing need of millions, in respect of livestock products such as milk, eggs, meat and wool. This sector has the greatest potential in creating new self sustaining jobs in villages, if the knowledge base in veterinary and animal husbandry technology is improved and is used in transforming India by creating entrepreneurships, small and big, poverty can be banished from India in five years..

Livestock production systems are based on low cost agro by-products as nutritional inputs, using current day technologies. The spectacular growth of livestock products especially milk, meat, eggs and poultry meat is attributable to the several initiatives taken by Government and the organized private sector, which has primarily been driven by horizontal increase in numbers. It has been observed that with increasing income, demand for cereals is decreasing, which is causing a demand driven livestock revolution.

With the livestock sector assuming an important role in the national economy, there is a requirement to improve the present state of knowledge gathering and information dissemination. Although considerable resources have been directed towards collecting and disseminating information on basic crops, little attention has been given to collecting, analyzing and disseminating information on livestock.

It is necessary that livestock units are made financially viable through generating a service provider industry which becomes a technology catalyser in a low educated farming community. This requires several initiatives, one of the major initiative is to upgrade the knowledge base and make the information available to students, teachers, planners and farmers.

The **Studium Press (India) Pvt. Ltd.** has decided in association with **Centre for Integrated Animal Husbandry and Dairy Development (CIAH&DD)** to bring out a series of books for under graduate and post graduate scholars in Animal and Veterinary Sciences in several volumes under the chief editorship of Professor (Dr) P.N. Bhat, Former Vice Chancellor and Director of Indian Veterinary Research Institute, Izatnagar–243122 (U.P) and former Animal Husbandry Commissioner of India and Deputy Director General (Animal Science) of Indian Council of Agricultural Research, Ministry of Agriculture, Krishi Bhawan, New Delhi. The basic idea of the series is to provide first rate text books to students and scholars in developing countries based on the experiences of developing countries themselves with special focus on Southern Asia in conformity with standards laid out by regulatory agencies in India (VCI, ICAR, UGC, AICTE) and similar agencies in other developing countries.

The titles to be brought out in present series are given below.

1. Goat Production
2. Dairy co-operatives in India
3. Sheep Production
4. Buffalo Production
5. Dairy cattle production
6. Pig Production
7. Poultry Production

8. Cross breeding of cattle for improved milk production in tropics
9. Camel Production
10. Yak Production
11. Mithun Production
12. Rabbit Production
13. Laboratory Animal Production
14. Dog Management, Breeding and Health
15. Animal Biodiversity
16. Livestock Statistics
17. Livestock Economics
18. Livestock Extension
19. Breeding and Health of Equine
20. Animal Nutrition
21. Animal Physiology

It is hoped that in the next two years, these books will be available for the benefit of student, teachers and professionals in the area and fill the gap which is currently wide.

Prof (Dr) Pushkar Nath Bhat
Chairman- World Buffalo Trust
And
Centre for Integrated Animal Husbandry & Dairy Development
and
Chief Editor

ABOUT THE AUTHORS

Prof (Dr) P.N. Bhat got his Ph.D in population genetics from Institute of Population Genetics Purdue University, West Laffayette, Indiana, USA followed by several post doctoral assignments and visiting professorship in genetics, biotechnology, livestock production systems. He returned to India and joined Punjab Agricultural University at Hisar–Ludhiana followed by several professional assignments in India and abroad. He joined as Project co-ordinator (Animal Breeding) and subsequently joined as Head, Division of Animal Genetics at Indian Veterinary Research Institute in 1971. He was responsible for establishing coordinated projects on cattle, buffalo, sheep, goat, pigs and poultry during 1970–74. He was founder Director of Central Institute for Research on Goats. In 1984 he became Vice-Chancellor and Director of IVRI. He joined as Deputy Director General (Animal Science) in May, 1992 and Animal Husbandry Commissioner in December, 1992. He is fellow of several National and International Science Academies.

Chapter 13 and 20 have been contributed by Dr. M.C. Sharma, Director, Central Institute for Research on Goats, Makhdoom, Farah, Mathura, Dr. Mahesh Kumar, Prof. and Head, Veterinary Preventive Medicine and Epidemiologist, College of Veterinary Sciences, G B. Pant University of Agriculture and Technology, Pantnagar and Dr. Umesh Dimri, Principal Scientist, Veteinary Medicine IVRI. Izatnagar.

Chapter 16 has been contributed by Dr. A. J. Pandya, Prof. and Head, Department of Dairy Processing and Operations, Faculty of Dairy Science, Anand Agricultural University, Anand, Gujarat.

PREFACE

The total population of buffaloes in the world during 2005 was 174 million heads. Major concentration of buffaloes is in India (98 million), Pakistan (26 million) and China (23 million). India possesses largest number of buffaloes (56.4%), followed by Pakistan (15.1%), China (13.1%), Nepal and Egypt (2.3%) each, the Philippines (1.9%). Thailand, Indonesia, Vietnam, Mayanmar, Turkey, Sri Lanka, Iraq, Iran, Brazil and other countries account for the remaining 8.9% of the population (FAOSTAT–Website year, 2006).

The river buffalo constitutes around 65% of the total world buffalo population and accounts for 92% of the total milk produced. India alone contributes 42% of total milk produced by buffaloes. The largest concentration of swamp buffaloes is found in rice growing countries of Asia.

The distribution of the buffalo is conspicuous by its being confined principally to the areas where animal husbandry is poorly developed and badly organized. By and large, buffaloes are owned only in small numbers by resource-poor farmers. In fairness to buffalo stock owners in these regions, it may be stated that they maintain buffaloes not from ignorance of the potentialities of other large ruminants but because they find that in the prevailing agricultural situation no other domestic animal will thrive like the buffalo and at the same time be so useful and economical.

Water buffaloes have been responsible for more than 10% of world milk production for several years, but the potential of these animals has seldom been appreciated or recognized. One of the main reasons for this is that those who have a stake in buffalo rearing are generally poor and underprivileged, and not able to project the impact this animal has on their livelihood and well-being.

The water buffalo is an important beast of burden in Asian farming. It is widely used to plough, level land, plant crops, puddle rice fields, cultivate field crops, pump water, haul carts, sleds and shallow-draft boats. It is also used to carry people, thrash grain, press sugarcane, haul logs and more. Buffaloes have an advantage over other draught animals in wet or muddy areas, with their large hooves. Their legs can withstand wet conditions better than cattle, although they are not as fast as cattle, horses or mules.

Buffaloes have been used as draught animals for centuries. This has lead to exceptional muscular development: some animals can weigh more than 1000 kg. Though buffalo is a major source of meat, they have not been used solely for meat production until recently. Most buffalo meat is derived from old animals so not surprisingly the meat is considered to be of poor quality. However, this is not true of meat from younger animals. Buffalo meat from animals properly reared and fed, is tender and palatable. Buffaloes are lean animals. In general, a buffalo carcass has a higher proportion of muscle and a lower ratio of bone and fat than a cattle carcass.

Buffaloes are managed under very different conditions over the world. These depend both on the geographic situations and for what purpose the buffaloes are used; any system from multi-purpose animals kept in backyards to high yielding milk producers in advanced farms may exist. Over time buffalo rearing has shifted from the backyard to commercial farms and large business enterprises. The immense popularity of buffalo milk and meat products has ensured that buffalo production follow the path of the dairy cattle industry. However, for this species to perform optimally under the pressure of intensive production system, the breeds have to be improved, with clear focus on the desired output. This has not yet happened. Buffalo, although potentially excellent for both milk and meat production, still languishes in obscure conditions of poor nutrition, breeding, management and welfare. Changes in breeding, feeding and management can bring about notable improvements in the productive and reproductive performance of buffalo.

The Indian Council of Agricultural Research has been in the forefront of buffalo development. Besides launching a major All India Coordinated Project on Buffalo breeding, it has established the Central Institute of Research on Buffaloes (CIRB) at Hisar, the breeding tract of Murrah breed (the jewel among buffalo breeds) with a sub-centre at Naba for Nili-Ravi breed.

The first draft of the manuscript prepared by the author has been revised by Dr. Stella Esther and Dr. A. Bandyopadhyay, both worked very studiously and carefully on the draft. It was edited by Mrs. Aruna T. Kumar, ICAR, New Delhi. I am grateful to her for carefully going through the manuscript and preparation of index and for making several new suggestions which have improved the text. The advantage of having outstanding colleagues and friends like Dr R.M. Acharya and Dr N.K. Bhattacharyya for referral discussion is acknowledged.

Dr J.N. Govil, Publishing-Director and Managing Editor, Researchco Books & Periodicals Pvt. Ltd., Daryaganj, New Delhi, who is the brain behind this initiative deserves special thanks for making it possible to see that this volume is brought out in time and to the expected standards. Mr Anil Jain and Mr Shrey Jain, proprietors of Studium Press (India) Pvt. Ltd. need to be complemented for sustained support. The hard work put up by Dr Stella Esther and Dr. A. Bandyopadhyay for proof reading the manuscript is gratefully acknowledged. The coordination work of Mr G.P. Gangadharan Pillai, Executive Assistant to Chairman, in preparation of the draft manuscript and typing by Mr Pius Joseph and Lal Babu Singh is gratefully acknowledged.

New Delhi
8th August, 2009

Prof (Dr) Pushkar Nath Bhat
Author

CONTENTS

LIST OF COLOUR FIGURES

LIST OF TABLES

CHAPTER 1

INTRODUCTION

Buffaloes belong to the Bovinae sub-family. Their relationship to other Bovinae such as cattle and bison is shown in Fig 1.1. Although the American bison (*Bison bison*) is sometimes referred to as a buffalo, this is a misnomer that is apparently due to a past misconception. The Fig 1.1 bison belongs to a separate sub-genus of the Bovinae.

There are four wild species of buffalo (Fig 1.1), but all existing types of domestic buffalo appear to have been derived from *Bubalus arni*, the wild buffalo of mainland Asia.

It is customary to divide all types of domestic buffalo (*Bubalus bubalis*) into two groups–the river and the swamp buffaloes. As the name implies, the water buffalo, whether of the river or the swamp type, has an inherent predilection for water and loves wallowing in water or mud pools. Like cattle, buffaloes are good swimmers.

Swamp buffaloes are lighter in weight (males approximately 700 kg and females 500 kg mature weight) and have a lower milk production capacity, *i.e.* 430 to 620 kg of milk/lactation (Webster and Wilson, 1980). These buffaloes are used mainly for draught power in rice cultivation in the paddy fields of southeast Asia. The swamp buffalo is a semi-aquatic animal, spending the hottest part of the day (10.00 to 16.00 hr) partly submerged in natural swamps or self-made wallows (Macgregor, 1941). The riverine buffalo, on the other hand, is heavier (males approximately 1100 kg and females 550 kg mature weight). Their milk production capacity is far higher than that of swamp buffaloes, ranging from 1000 to 2000 kg/lactation and varying among countries and strains. The river buffalo, as a rule, shows preference for clean running water.

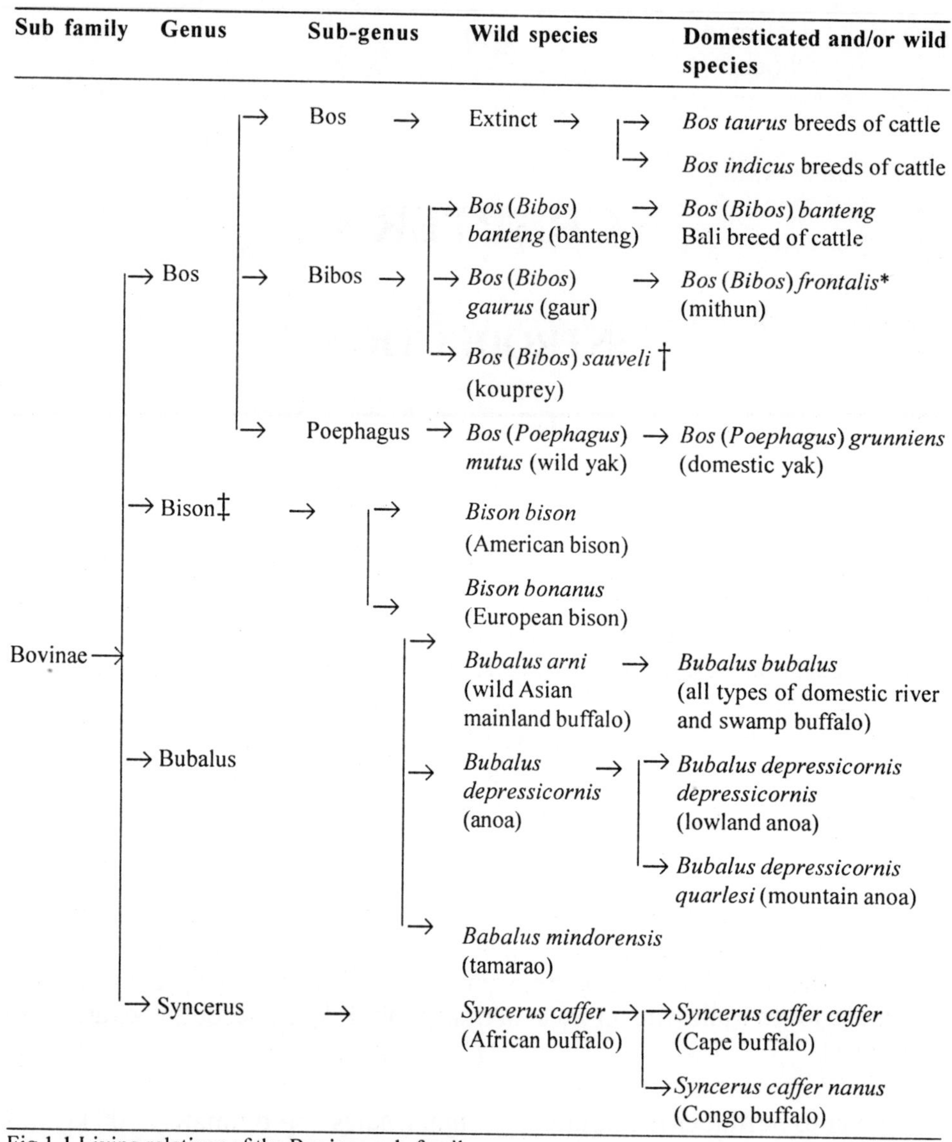

Fig 1.1 Living relatives of the Bovinae sub-family

* Listed by some authorities as a separate species

† May be the result of crossbreeding between *Bos* (*Bibos*) *banteng* and *B.* (*Bibos*) *gaurus*.

‡ Linnaeus includes this genus under *Bos*.

The water buffalo *Bubalus bubalis* is the domestic form of the wild progenitor *Bubalus arni* of India, southern Asia and possibly the wet areas of western Asia (Brock, 1989). This domesticated buffalo has been little exposed to human interference such as artificial breeding because of its perfect adaptation to the harsh environment in which it has to perform. The water buffalo is potentially the most important tropical bovine species, especially in very hot areas where rivers and swamps abound. It competes very successfully with other bovine species and if fed properly, it excels them.

The water buffalo is ranked highly in the tropics and subtropics since it thrives in hot conditions. In particular, the genus *Bubalis* surpasses the cattle genus *Bos* in its ability to adapt to the hot, humid areas of muddy and swampy lands; therefore, water buffaloes have special importance in the swamps of southeast Asia, the marshes of southern Iraq and the valleys of flooding rivers in the Indian subcontinent, as well as near the river Nile in Egypt.

The buffalo is well suited to hot and humid climates and muddy terrain due to its morphological and anatomical characteristics. Buffalo skin is covered with a thick epidermis, the basal cells of which contain many melanin particles that give the skin surface its characteristic black colour (Shafie, 1985). The melanin particles trap the ultraviolet rays and prevent them from penetrating through the dermis of the skin to the lower tissue. These rays are abundant in solar radiation in the tropics and subtropics, and excessive exposure of animal tissue could be detrimental, perhaps even resulting in skin tumours. Buffalo skin is almost devoid of hair, particularly in older animals, because of its exposure to water and mud. This beneficial characteristic is reinforced by its well-developed sebaceous glands. With greater secretary activity than in cattle (Shafie and Abou El-Khair, 1970), these glands secrete a fatty substance called sebum, which emerges on the skin's surface and covers it with a lubricant, making it slippery for water and mud. This greasy sebum, along with the thick cornified top layer of skin, prevents water and the solutes in it from being absorbed into the skin. In this way, the animal is protected from the harmful effects of any deleterious chemical compounds in the water. Moreover, the sebum layer melts during hot weather and becomes glossier to reflect many of the heat rays, thus relieving the animal from the excessive external heat load. With this physical adaptation to the tropical and subtropical hot humid conditions, the water buffalo has acquired a reproductive-productive pattern that conforms to the sequential seasonal changes in climate, terrain and vegetation conditions.

Water buffaloes have been responsible for more than 10% of world milk production for several years, but the potential of these animals has seldom been appreciated or recognized. One of the main reasons for this is that those who have a stake in buffalo rearing are generally poor and underprivileged, and not able to project the impact this beast has on their livelihood and well-being.

World milk production has doubled in the last few decades and it is noteworthy that in the last few years, buffaloes have supplied about 12% of the total world milk production. In 2004, according to statistics from the United Nations' Food and Agriculture Organization (FAO), the world production of buffalo milk was 75.8 million tonnes (Mt). Trends in world milk production over the five years to 2004 indicate that the volume of buffalo milk is increasing steadily at about 3% per year. India produces 60% of the world's buffalo milk (FAO, 2004).

The water buffalo is an important beast of burden in Asian farming. It is widely used to plough, level land, plant crops, puddle rice fields, cultivate field crops, pump water, haul carts, sleds and shallow-draft boats. It is also used to carry people, thrash grain, press sugar cane, haul logs and more. Buffaloes have an advantage over other draught animals in wet or muddy areas, with their large hooves. Their legs can withstand wet conditions better than cattle, although they are not as fast as cattle, horses or mules.

Buffaloes have been used as draught animals for centuries. This has lead to exceptional muscular development: some animals can weigh more than 1000 kg. Though buffalo is a major source of meat, they have not been used solely for meat production until recently. Most buffalo meat is derived from old animals so not surprisingly the meat is considered to be of poor quality. However, this is not true of meat from younger animals. Buffalo meat from animals properly reared and fed, is tender and palatable. Buffaloes are lean animals. In general, a buffalo carcass has a higher proportion of muscle and a lower ratio of bone and fat than a cattle carcass.

Buffaloes are managed under very different conditions over the world. These depend both on the geographic situations and for what purpose the buffaloes are used; any system from multi-purpose animals kept in backyards to high yielding milk producers in advanced farms may exist. Over time buffalo rearing has shifted from the backyard to commercial farms and large business enterprises. The immense popularity of buffalo milk and meat products has ensured that buffalo production follow the path of the dairy cattle industry. However, for this species to perform optimally under the pressure of intensive production system, buffalo breeds have to be improved, with clear focus on the desired output. This has not yet happened. Buffalo, although potentially excellent for both milk and meat production, still languishes in obscure conditions of poor nutrition, breeding, management and welfare. Changes in breeding, feeding and management can bring about notable improvements in the productive and reproductive performance of buffalo (Sastry, 1983).

The Indian Council of Agricultural Research has been in the forefront of buffalo development. Besides launching a major All India Coordinated Project on Buffalo breeding, it has established the Central Institute of Research on Buffaloes (CIRB) at Hisar, the breeding tract of Murrah breed (the jewel among buffalo breeds) with a sub-centre at Naba for Nili-Ravi breed.

In the following chapters some practical aspects related to improving buffalo production are discussed for the benefit of students, researches and farmers.

CHAPTER 2

HISTORY, ORIGIN AND DOMESTICATION

2.1 Classification

The domestic or water buffalo (*Bubalus bubalis*) belongs to the family Bovidae, sub-family Bovinae, genus *Bubalus* and species *arni*. The world's buffaloes are classified into two groups the African and the Asian. These two groups have been assigned different generic names: *Syncerus* and *Bubalis* respectively. The African buffalo *Syncerus,* consists of a single species, *Syncerus caffer* (Sparrman), with a small number of sub-species. The Asian buffalo, *Bubalus*, includes three species– the anoa of Celebes, the tamarao of Mindoro and the arni or the Indian wild buffalo. Of the African and Asian buffalo species, the Indian wild buffalo, *Bubalus arni* (Kher) is the only one to be domesticated and has since been given the name *bubalis*.

2.2 Taxonomy

The essence of any modern classification scheme is to group all those organizms that are related to one another because of morphology, anatomy, habit, reproduction, etc. The buffaloes have close zoological relationship with the cattle. Both belong to the same subfamily, Bovini, under the family, Bovidae. Its relationship with other Bovinae such as cattle and bison is shown in Fig 1.1 (Chapter 1).

2.3 Origin

Although American bison (*Bison bison*) is sometimes referred to as buffalo, this is a misnomer. The bison belongs to a separate subgenus of the Bovinae. There are four wild species of buffalo but all existing types of domestic buffaloes appear to have been derived from *Bubalus arni*, the wild buffalo of mainland Asia.

Information on origin and the precise period of domestication of the buffalo is lost in antiquity. However, the crescent horns, coarse skin, wide muzzle and low carried heads of buffaloes represented on seal struck 5000 years ago in the Indus valley suggest that the buffalo had already been domesticated in the Indian subcontinent around that time. The domestication of swamp buffalo took place independently in China about 1000 years later. The movement of buffaloes to other countries, both east-wards and west-wards, is thought to have occurred from these two countries.

Buffaloes were unknown in Egypt during the time of Pharaohs. Their movement to Egypt took place *via* Iran and Iraq when Arabs took these animals from India during the first invasion in the 9th Century. Water buffaloes were later introduced into Europe by pilgrims and crusaders returning from the Holy land in the middle ages.

Importation of buffaloes to other countries in southeast Asia, western Asia, Europe, Australia and South America followed, the spread being slow and gradual. However, attempts at the introduction of these animals into some countries did not meet with success. The Earl of Cornwall, a brother of Henry III, brought buffaloes to England, but the animals did not thrive (Cockrill, 1967). Introduction of the buffalo to African countries south of the Sahara, has not so far been very successful. At present attempts are being made in Australia to redomesticate the feral buffaloes.The results achieved so far are encouraging.

All types of domestic buffaloes (*Bubalus bubalis*) are generally grouped as the river and swamp buffaloes. As the name implies, the water buffalo, whether of the river or the swamp type, has an inherent predilection for water and loves wallowing in water or mudpools. The river buffalo, as a rule, shows preference for clean running water, whereas the swamp buffalo likes to wallow in mudholes, swamps and stagnant pools. Buffaloes are good swimmers.

Buffaloes are generally easily domesticated. The African buffalo (*Syncerus caffer*) which was believed to be difficult to domesticate could be tamed easily when scientific efforts at domestication were made (Rouse, 1970). On the other hand domestic buffaloes, both of swamp and river types, can easily revert back to semi-wild or wild state if let lose to fend for themselves, as has happened in northern Australia.

Buffaloes do not cross with cattle or vice versa.

Buffaloes have been introduced into more than 20 countries in Latin America. The importations have primarily been from India, Italy, Bulgaria, Indochina and Australia, the latter contributing rather small numbers. The first introductions are believed to have been made in 1890, but subsequent importations have mostly been of the breeds from the Indian subcontinent. Initially they were introduced for work and later for meat. They were primarily managed on ranches on extensive system, but presently milk production is in high focus and the dairy type breeds are stall-fed and properly housed in barns and kept under intensive management. The breeds of Indian origin are maintained as pure breeds, but they have also been interbred leading to new breeds like Brazil Mediterraneo and Brazil Murrah.

2.4 History of Domestication

Studies of human settlement down the ages show that domestication of animals is intrinsic to any progressive civilisation. Archaeological findings and historical data point to the fact that buffaloes were first domesticated around 2500 BC, in the Indus valley: present day India and Pakistan (Chantalakhana and Falvey, 1999). It has been contended that the habitat of *arni* in pre-historic time extended across north India, west to Mesopotamia and east to China (Epstein, 1969). Thus it seems possible that domestication took place in all the three regions independently. However, systematic development of well established breeds of distinctly different qualities and characteristics occurred in the Indian subcontinent. Around 600 AD, Arab traders brought water buffalo from Mesopotamia towards the near East: modern day Syria, Israel and Turkey.

Spread of buffalo in the western direction took place after 7th century AD, particularly into Russia by way of Persia and Egypt from Mesopotamia. The buffalo apparently reached Europe in 13th century. Buffaloes are now found in Italy, Hungary, Romania, some Balkan countries, Greece and Bulgaria. The domestic water buffalo has also been introduced into South America, the United States and Australia (BSTID, 1981).

2.5 Global Distribution of Domesticated Buffalo

Dairy buffalo production has been a tradition in parts of the world like the Caucasian countries, Asia and Egypt, where fresh buffalo milk, *dahi* (cultured sour milk), *ghee* (butter oil) and yoghurt are popular. In Italy, the dairy buffalo industry is flourishing, thanks to the popularity of buffalo mozzarella cheese. Because of the market for mozzarella, buffalo farming is a profitable enterprise and is carried out in an organised manner with modern equipment. In south American countries like Brazil and Argentina, buffaloes are reared for both milk and meat. In recent years, buffalo milk and milk products, especially mozzarella cheese, have become immensely popular and dairy buffalo production has found its way into non-traditional areas with the number of buffalo farms mushrooming even in the UK and USA . In India as well as Pakistan, in the vicinity of all the major cities like Mumbai, Kolkata and Karachi, one can find a large number of buffalo farms of varying herd sizes. In Mumbai alone, there are more than 2 lakh buffaloes in downtown districts and probably another 1 lakh in suburban areas. Some of these farms have herd sizes of more than 1000 buffaloes and on an average these herds have more than 100 buffaloes. Large-scale dairy buffalo production is a greater reality in India and Pakistan than anywhere else in the world–even though it represents less than 2% of the buffalo farms in these two countries.

CHAPTER 3

WORLD POPULATION AND PRODUCTION

The total population of buffaloes in the world during 2005 was 174 million heads. Major concentration of buffaloes is in India (98 million), Pakistan (26 million) and China (23 million). India possesses largest number of buffaloes (56.4%), followed by Pakistan (15.1%), China (13.1%), Nepal and Egypt (2.3%) each, the Philippines (1.9%). Thailand, Indonesia, Vietnam, Mayanmar, Turkey, Sri Lanka, Iraq, Iran and other countries account for the remaining 8.9% of the population (FAOSTAT–Website year, 2006).

The river buffalo constitutes around 65% of the total world buffalo population and accounts for 92% of the total milk produced. India alone contributes 42% of total milk produced by buffaloes. The largest concentration of swamp buffaloes is found in rice growing countries of Asia.

The distribution of the buffalo is conspicuous by its being confined principally to the areas where animal husbandry is poorly developed and badly organized. By and large, buffaloes are owned only in small numbers by resource-poor farmers. In fairness to buffalo stock owners in these regions, it may be stated that they maintain buffaloes not from ignorance of the potentialities of other large ruminants but because they find that in the prevailing agricultural situation no other domestic animal will thrive like the buffalo and at the same time be so useful and economical.

3.1 Recent Buffalo Population and Production in Asia

The Asian buffalo and the smallholder farmer are almost synonymous in many respects and have been in tandem for centuries, generating food and agricultural products in the mixed farming system dominant in developing countries of Asia.

The number of animals per household range from 1 to 5 in East and southeast Asia where the buffaloes are of the swamp type and mainly used for draught and meat. A similar situation can be seen in south and southwest Asia where riverine type buffaloes abound and the primary purpose is for milk, work and meat. While majority of these animals are reared in a mixed farming system where cut and carry, tethering and grazing system on communal pasture land is practised. Maximizing the utilization of crop residues and farm by-products is a traditional practise. Semi-commercial and commercial size dairy operations around the peri-urban areas are emerging where buffaloes are reared in total confinement system and are fed forages and feed materials raised in commercials. Dairy cows are kept while in lactation and thereafter are sold to be replenished by pregnant animals derived from village sources. The recent production output in terms of milk, meat, draught and hide, and the growth in population are summarized in Table 3.1 together with an enumeration of the currently existing buffalo breeds.

Asian buffaloes dominate the world population, representing 96.36% of the worldwide population of 172.2 million as of 2004. Within the Asian region, about 90.7% of buffaloes are in South and southwest Asia (77.39%) and east Asia (13.38%) (Table 3.1).

Table 3.1 Regional Distribution and Average Annual Growth Rate of Buffalo Population in Asia (2004)

Region	Population ('000 head)	Percentage of Asia	Percentage of world	Growth rate (2001–2004)%
Southeast Asia	15331	0.20	8.80	2.68
South and southwest Asia	128844	77.30	74.70	1.55
Central Asia	–	0.005	0.005	0.0
East Asia	22287	13.38	12. 30	–0.7
Asia	166472	100.00	6.63	1.34
WORLD	172263	–	100.00	1.34

Source: FAO Production Yearbook, 2006.

During the period 2001 to 2004, the Asian buffalo population registered an average annual growth rate of 1.34%, with the highest growth occurring in Southeast Asian countries at an average annual growth rate of 2.68%. This recorded positive growth in buffalo population in SEA is interesting considering that this region experienced dramatic decline in inventory in the 1990's with the massive introduction in farm mechanization and intensive irrigation in rice-producing areas where the water buffaloes are utilized primarily as source of draught power.

3.2 Buffalo Population in Indian States

Uttar Pradesh has the major concentration of buffalos (22.9 million) followed by Andhra Pradesh (10.6 million) and Rajasthan (10.4 million). Buffalo population in various States in India is given in Table 3.2.

Table 3.2 Total Number of Buffaloes 1997 and 2003 State-wise (in thousands)

State/U.Ts	Buffaloes						Annual growth rate (%) 1997–03	
	Male		Female		Total			
	1997	2003	1997	2003	1997	2003	Female	Male
Andhra Pradesh	1376	1638	8282	8991	9658	10630	1.38	1.61
Arunachal Pradesh	4	4	8	8	12	11	–0.37	–0.73
Assam	323	307	404	370	728	678	–1.45	–1.19
Bihar	1619	1032	4260	4711	5879	5743	1.69	–0.39
Chhatisgarh	1395	1078	546	520	1941	1598	–0.80	–3.19
Goa	13	11	28	27	40	37	–0.66	–1.11
Gujarat	544	667	5740	6473	6285	7140	2.02	2.15
Haryana	1063	870	3760	5164	4823	6035	5.43	3.81
Himachal Pradesh	35	47	713	727	748	774	0.31	0.56
J and K	148	131	639	909	787	1039	6.06	4.75
Jharkhand	–	710	–	634	–	1343	–	–
Karnataka	632	464	3736	3527	4367	3991	–0.95	–1.49
Kerala	57	24	54	40	111	65	–4.93	–8.62
Madhya Pradesh	1344	1195	5304	6381	6648	7575	3.13	2.20
Maharashtra	1065	870	5008	5274	6073	6145	0.87	0.20
Manipur	46	34	49	43	95	77	–2.12	–3.45
Mehalaya	9	10	8	8	17	18	0.20	0.95
Mizoram	2	3	3	3	5	6	–0.86	2.25
Nagaland	19	19	18	15	36	34	–2.45	–1.07
Orissa	748	753	640	641	1388	1394	0.03	0.07
Punjab	753	597	5417	5398	6171	5995	–0.06	–0.48
Rajasthan	1182	1126	8589	9288	9770	10414	1.31	1.07
Sikkim	1	1	1	1	2	2	0.26	0.96
Tamilnadu	506	268	2235	1390	2741	1658	–7.61	–8.03
Tripura	8	6	9	8	18	14	–2.19	–3.60
Uttar Pradesh	4887	5163	14109	17751	18996	22914	3.90	3.17
Uttaranchal	155	134	939	1094	1094	1228	2.58	1.95
West Bangal	654	693	579	393	1233	1086	–6.27	–2.09

Table 3.2 (*Contd.*)

State/U.Ts	Buffaloes						Annual growth rate (%)	
	Male		Female		Total		1997–03	
	1997	2003	1997	2003	1997	2003	Female	Male
Union Territory								
A and N Island	7	7	8	9	14	16	1.83	2.47
Chandigarh	1	1	21	21	23	23	0.17	–0.24
D and N Haveli	1	1	3	3	4	4	–3.07	–0.57
D and Due	0	0	1	1	1	1	–0.51	–0.52
Delhi	26	24	177	207	203	231	2.58	2.14
Lakshadweep	–	–	–	–	–	–	–	–
Pondicherry	0	1	4	3	4	4	–0.94	–0.48
All India	**18624**	**17888**	**71293**	**80034**	**89915**	**97922**	**1.95**	**1.43**

Note: 1. Totals may not tally due to rounding up of figures. 2. 0 is less than thousand figure. 3. '–' not applicable.

Source: Directorate of Economics and Statistics, M/O Agriculture and Department of Animal Husbandry, Dairying and Fisheries.

3.3 Buffalo Population in the World

Distribution of buffalo population in top 10 countries is given in Table 3.3

Table 3.3 Distribution of Buffalo Population in Selected Countries

Country	% share in global population in 2005
India	56.4
Pakistan	15.1
China	13.1
Nepal	2.3
Egypt	2.3
Philippines	1.9
Vietnam	1.7
Myanmar	1.6
Indonesia	1.4
Thailand	1.0

Source: FAOSTAT–Website year, 2006.

Nearly 97% of world population is concentrated in Asia, 2% in Africa–mainly in Egypt, 0.2% in Europe–mainly in Italy (FAO, 2004).

The more recent data on world buffalo population by countries provided by FAOSTAT is given in Table 3.4 and world buffalo population by continents is given in Table 3.5.

The buffalo renders invaluable service to mankind by way of work and the supply of milk and manure while alive, and meat, hide, horns, hooves, bones etc. after death. Buffalo contributes 72 million tonnes (Mt) of milk and 3 Mt of meat annually to world food. There are large areas where human survival is dependent on this animal. In spite of these facts the study of these animals has remained neglected. The benefits of science and technology have hardly impinged on the husbandry of the buffalo population to enable a fuller expression of its production potentialities to be achieved. The greatest handicap to its development lies in very large gaps in our knowledge of the physiology of the animal, disease prevalence in the species and the husbandry practises most suitable for animals located in widely varying environmental conditions. Of late, scientific workers in India, Egypt, Pakistan, Thailand, etc., have been studying the animal in depth with a view to evolve efficient buffalo husbandry techniques, and as a consequence considerable information is now accumulating. However, available published material is still very scanty and what is published is often fragmentary and incomplete and sometimes contradictory.

The buffaloes of India and Pakistan, as well as those of Europe, the near East, formerly USSR, and south America and the Caribbean are mostly of the riverine type, usually larger in size and noted for their milk production. The buffaloes of southeast Asia and China are of the swamp type, mainly used for work but becoming increasingly important for meat as well.

The production of milk and meat from buffaloes in Asian countries over the last decades has shown a varying pattern: in India, Sri Lanka, Pakistan and China, the milk yield per animal has increased by 2.44%, 1%, 1.45% and 1.55%, respectively, while there has been either no change or only a negligible change in milk production in Bangladesh, Myanmar, Nepal and Vietnam. In some regions of east and south-east Asia, there has been a negative growth. Meat production from buffaloes has shown a growth of 1.43% only in Pakistan, while in other countries there was no change or a decline. At the Asia level, although buffalo milk production increased by 2.26%, meat production marginally declined (Dhanda, 2004).

Table 3.4 Buffalo Population in the World by Countries

Countries	1998	1999	2000	2001	2002	2003	2004	2005	2006	2007
Albania	120	120	120	120	120	120	120	120	120 120	
Armenia	70	67	67	67	538	473	392	402	399	440
Azerbaijan	293300	292144	296981	298649	303559	306356	308551	302887	299121	295400
Bangladesh	820000	828000	830000	850000	850000	850000	850000	850000	850000	850000
Bhutan	2700	2400	2200	2000	1800	1900	2000	2000	2000	2000
Brazil	1017246	1068059	1102551	1118823	1114720	1148808	1133622	1173629	1156870	1180000
Brunei-Darussaiam	5899	5812	5150	5748	5565	5903	4675	4790	4685	4580
Bulgaria	10553	10365	9277	7790	6528	7489	7875	7973	8198	8247
Cambodia	693651	653850	693631	626016	625912	660493	650572	676646	724378	775000
China	22553806	22674439	22595017	22764781	22689620	22729162	22287212	22366080	22499950	22722010
China (mainland)	22545000	22665000	22587000	22758000	22684000	22724000	22282000	22361000	22495000	22717000
Egypt	3149429	3329700	3379410	3532244	3550000	3777000	3845000	3920000	3920000	3950000
Georgia	32836	32780	34923	33220	36487	37000	32700	33200	33200	34500
Greece	788	865	877	975	1000	1024	1110	1270	1237	2593
Guam	65	65	65	65	65	65	65	65	65	70
India	91252000	92586000	93920000	95254000	96588000	97922000	98175000	98875000	98805000	98700000
Indonesia	2829291	2503788	2405277	2333429	2402990	2459434	2403300	2428190	2166606	2246017
Iran	474000	474000	490600	506800	523500	540000	560000	580000	600000	620000
Iraq	80000	90000	115000	109000	120000	115000	111000	213000	120000	120000
Italy	162000	200000	182000	182000	193774	185000	223393	210000	205000	231000
Jordan	100	100	100	100	100	100	100	100	100	100
Kazakhstan	9000	9000	9000	9000	9000	9000	9000	9000	10000	10000
Lao	1092700	1008000	1028000	1051400	1089400	1111000	1125000	1096000	1108000	1120000
Malaysia	160413	149544	142043	140000	131000	133000	138000	137000	137000	138000
Mauritius	25	25	25	25	25	25	25	25	25	25

Table 3.4 (*Contd.*)

Countries	1998	1999	2000	2001	2002	2003	2004	2005	2006	2007
Micronesia federated states of	130	130	130	130	130	130	130	130	130	140
Myanmar	2337574	2391018	2441240	2502000	2552020	2598044	2648739	2704540	2769659	2820000
Nepal	3419150	3470600	3525952	3624027	3700864	3840013	3952654	4081463	4202886	43668113
Pakistan	21422000	22032000	22669000	23335000	24030000	24800000	25500000	26300000	28400000	2730000
Philippines	3013000	3005989	3024403	3065812	3122026	3179536	3269980	3327000	3357956	3365000
Russian federation	16586	16141	16500	17370	14004	15863	17943	17288	14362	15654
Serbia and Montenegro	16000	21000	23000	25900	28700	28700	28700	28700	–	–
Sri Lanka	316400	319500	304500	290300	282087	280480	301500	307750	314080	318920
Suriname	540	650	425	395	395	395	395	395	395	395
Syrian Arab Republic	1280	2803	2824	2477	2794	3500	4000	4470	5000	5200
Tajikistan	14566	14486	13189	15000	16272	15952	14367	14550	–	–
Thailand	2286417	1911518	1711573	1523627	1612534	1689762	1737698	1770625	1763444	1743546
The former Yugoslav Republic of Macedonia	908	899	725	725	600	566	565	725	750	640
Timor-Leste	80754	90000	100000	103723	105000	107000	108000	110000	110000	110000
Trinidad and Tobago	5400	5400	5450	5500	5600	5650	5700	5700	5700	5700
Turkey	194000	176000	165000	146000	138000	121000	113356	103900	104965	100516
Vietnam	2951400	2955700	2897200	2807900	2814452	2834886	2869802	2922155	2921000	29211000
World	160714957	162332827	164143294	166292008	168669051	171521699	172442111	174586658	176635281	202386626

Source: http://faostat.fao.org.

Table 3.5 Number of Buffalo by Continents

Countries	1998	1999	2000	2001	2002	2003	2004	2005	2006	2007
Africa	3149454	3329725	3379435	3532269	3550025	3777025	3845025	3920025	3920025	3950025
Eastern Africa	25	25	25	25	25	25	25	25	25	25
Northern Africa	3149429	3329700	3379410	3532244	3550000	3777000	3845000	3920000	3920000	3950000
Americas	1023186	1074109	1108426	1124718	1120715	1154853	1139717	1179724	1162965	1186095
Caribbean	5400	5400	5450	5500	5600	5650	5700	5700	5700	5700
South America	1017786	1068709	1102976	1119218	1115115	1149203	1134017	1174024	1157265	1180395
Asia	156335297	157679538	159422869	161400076	163753520	166350994	167177598	169220748	171309429	196979042
Central Asia	23566	23486	22189	24000	25272	24952	23367	23550	10000	10000
Eastern Asia	22553806	22674439	22595017	22764781	22689620	22729162	22287212	22366080	22499950	22722010
Southern Asia	117706250	119712500	121742252	123862127	125976251	128234393	129341154	130996213	133173966	132157733
South eastern Asia	15450089	14675219	14448516	14159655	14460899	14779058	14955766	15176946	15062728	41533143
Western Asia	601586	593894	614895	589513	601478	583429	570099	657959	562785	556156
Europe	206955	249390	232499	234880	244726	238762	279706	266096	242667	271254
Eastern Europe	27139	26506	25777	25160	20532	23352	25818	25261	22560	23901
Southern Europe	179816	222884	206722	209720	224194	215410	253888	240835	220107	247353
Oceania	65	65	65	65	65	65	65	65	195	210
Australia and New Zealand	0	0	0	0	0	0	0	0	0	0
Micronesia	65	65	65	65	65	65	65	65	195	210

Source: http.//faostat.fao.org.

CHAPTER 4

GENERAL CHARACTERISTICS

The buffalo is a slow, ponderous and docile animal. It is semi-aquatic in its natural habiat and is exceedingly fond of wallowing. It has remarkable capacity to adapt itself in the most adverse climatic conditions. Buffaloes have a higher tolerance to cold temperatures than domestic cattle, and therefore exhibit greater winter hardiness.The best milch animals are found in areas where temperature goes as high as 40 °C. Such areas are Haryana, Punjab, Uttar Pradesh, western parts of Rajasthan and Gujarat. The non-descript buffaloes are found in areas of high rainfall and humidity. They have small size and have heavy draught capacity for working in paddy fields. These are, therefore, called as "A living tractor of the east". Because of the buffalo's superior digestion of low quality feeds, it may be better suited for production on marginal rangelands.

Buffaloes are herd animals. In the wild, bulls will form separate herds or bachelor groups and join with the cow herd only during the breeding season. However, buffalo raised domestically will not separate by gender and will remain in the same herd throughout the year. Buffaloes have a highly structured social "pecking" order determined by seniority in the herd, size and age. The pecking order is much more rigorously enforced than in domestic cattle, and ignoring the social status can result in serious injury.

4.1 General Body Conformation

Buffaloes are big-boned, rather massive animals with body set low on strong legs with large hooves. Dewlap and hump are absent. The body frame of a good milk-type river buffalo conforms to that of dairy-type cattle, and of the swamp buffalo to that of draught breeds of zebu. Horns are generally more massive than those of cattle. They are broad, flat and almost rectangular in cross-section near the base and with prominent ridges across the long axis. Swamp and several breeds of river and nondescript buffaloes have backswept horns. Two of the milk breeds of river buffaloes have very tightly curled horns. Some buffaloes have fantastically long horns that may measure more than three-quarters of the total length of the body.

At birth and during early calf hood, buffaloes have a good coat of soft hair but the hair becomes sparser and coarser as the animal grows, depending on the breed, season and housing practises. In some adult buffaloes, body is practically bare. The hair is black, dun, creamy yellow, dark or light grey or white. Some breeds show white markings in the form of stripes and some others are conspicuous in having a piebald coat.

Star, blaze or socks and a white switch at the end of the tail are frequently found in river buffaloes. Among swamp buffaloes, white animals are not uncommon. The incidence varies in different areas and appears to be the result of some selections based on local beliefs and superstitions. The incidence is relatively low in Borneo, China and Taiwan where there is a prejudice against such animals, but in Bali and parts of Thailand more than 15% of buffaloes are white as these animals are considered auspicious.

Hair whorls of swamp buffaloes are characteristic and distinctive, and could be used as an important means for identification of an animal. The colour of the skin varies from jet black to light pink. The skin may be unpigmented in localized areas, black or slate-grey being the most common in dark-colour animals. Leucoderma is known to occur in buffaloes. The aetiology of leucoderma is unknown and the condition does not affect production in any manner.

The sheath of the male swamp buffalo adheres close to the body except at the umbilical end. In the river buffalo the sheath is more pendulous and is somewhat similar to that of zebu cattle. There is no tuft of hair at the preputial opening in the buffalo. The scrotum of the male buffalo, both of the swamp and the river types, is much smaller than that of male cattle of similar size. In the swamp buffalo, there is no constriction near the attachment of the scrotum to the abdominal wall, but a distinct neck can be seen in the river buffalo.

Among the river buffaloes, a few well-defined breeds with standard qualities and with specific physical characters, that differentiate them unmistakably from other types, can be found only in India, Pakistan, Italy and Latin America. These are all milk breeds, but their number forms only a very small fraction of the total buffalo population in these countries. The vast majority of buffaloes is nondescript. These vary greatly in size, weight and general features. These animals are the product of natural breeding without any positive selection criteria for improving production.

Average milk yield of milch type buffaloes is 1750 litres/lactation of 305 days yielding a dry matter of 2.09 kg/litre which is almost double to that of indigenous or cross bred cow.

4.2 Appearance

The swamp buffalo is slate grey, droopy necked and ox-like. It has massive backswept horns (BSTID, 1981). There are no clear differences between breeds of swamp buffalo, except for body size. The riverine buffalo is usually black or dark grey, with tightly coiled or drooping straight horns (BSTID, 1981). They are generally large in body size. There are greater differences between riverine breeds than between the breeds within the swamp type. The body weight of a female buffalo of Murrah breed ranges from 430 to 500 kg, (Ganguli, 1981).

4.3 Nutrition and Life Cycle

Buffaloes are grazers (Pathak, 1992) and they graze on a wider range of plants than cattle do (BSTID, 1981). They utilize low-grade roughage more efficiently than cattle do. Buffaloes have slower ruminal movements, a smaller rate of outflow from the rumen and higher bacteria population in rumen fluid. This leads to a longer exposure of the feed and consequently a more complete digestion. The buffalo rumen also has a higher production of volatile fatty acids than the rumen of cattle. This might be one of the factors contributing to the higher fat content in buffalo milk (Ganguli, 1981).

The buffalo has an exceptionally long productive life. A normal healthy female buffalo could have as many as nine to ten lactations (Ganguli, 1981).

4.4 Heat Tolerance

Buffaloes are less tolerant of extremes of heat and cold than various breeds of cattle. The body temperature of a buffalo is lower than that of a cow in spite of the fact that its black skin absorbs much heat and its skin has only one-sixth the density of sweat glands than that of a cow skin. This explains why buffaloes like to wallow in water when the temperature and humidity are high (BSTID, 1981). Regulation of body temperature in this way influences feed intake, reproduction and milk production.

4.5 Temperament of Dairy Buffalo

A comparative study of temperament was done among Murrah buffalo, crossbred cows and Red Sindhi (Indian breed) cows. Nayak and Mishra (1984) showed that these buffalo had a higher percentage of docile animals. Almost 50% of this group of Murrah buffaloes were docile. About 7% of the group were aggressive. The rest were classed as restless or nervous animals. However, in another study Roy and Nagpaul (1984), compared Murrah buffalo with Karan Swiss and Karan Fries dairy cows (two Indian breeds of crossbred cows)and reported that the buffalo had higher temperament score (more aggressive temperament) than the dairy cows. The temperament scores for all three groups decreased with increasing parity between the 3rd and 5th lactation (Roy and Nagpal, 1984).

4.6 Growth Characteristics

The birth weight, growth rate and maturity of an animal are important parameters of the growth characteristics. The depth of scientific data on these aspects of the water buffalo except its birth weight, constitutes a serious weakness in the breeding programmes. These characteristics have been discussed here with special reference to genetic and environmental factors, which are causing variation in these characters.

4.6.1 Birth weight

The birth weight of water buffalo is higher than that of all domestic breeds of cattle except Friesian cattle. If we compare the birth weight of Egyptian, Indian and Pakistan buffaloes, the birth weight of Egyptian buffaloes is significantly higher than those in the Indian subcontinent. This clear difference in birth weight between the breeds is mainly due to the difference in the genetic factors and different climatic conditions under which they have been raised.

The birth weight of buffalo varies with respect to the sex of the calf and the calving sequence. The average birth weight of male buffalo is significantly higher than the female buffaloes. During subsequent calvings, the birth weight of calves increases up to the fourth calving when the dams attain full maturity. In a study, birth weight of the male calves increased in subsequent calvings and there were no significant increase in the weight of female calves. The average birth weight ranges between 27.3 and 33.2 kg.

4.6.2 Growth rate

The growth rate of the domestic water buffalo is lower than that of cattle. It is a slow maturing animal and its growth continues till 10th year, although rate of growth is slower after the 5th year. The delayed maturity affects the age at first calving and ultimately the total productive life of the animal. But very little work and experiments have been done on this aspect. The average longivity and useful life of Murrah buffaloes is 127 months and 83 months respectively. For increasing the useful life of buffaloes, it is very essential to reduce the age at first calving and the calving interval. Among the non-genetic factors, feeding aspect of buffaloes need attention. In India, there is a general neglect in the feeding of buffaloes during the early stages of their growth period, which adversely affects their subsequent growth rate and delays their maturity.

4.6.3 Maturity

The sexual maturity of male and female buffaloes is attained in about 2–3 years depending on the type of breed, management practice and feeding. The average age of buffalo bull at the first service is 3.5 years while the heifer comes into heat at 2–2.5 years of age. The oestrus cycle in river type of buffalo is around 21 days and the duration of heat varies from 21–24 hr. In Indian buffaloes the average heat cycle length is 37 days and oestrus could be for 24 hr. It has been observed that the onset of heat is not very pronounced and often it is difficult to detect. This is the main reason for the highest percentage of the conception failures during the first service. There is considerable scope for reducing the age of first conception by proper feeding and management practises, particularly the careful checking of heat.

4.6.4 Age at first calving

The age at first calving of buffaloes is generally higher than that reported. By proper feeding and management, the age at first calving can be reduced and at the same time the expenses in rearing heifers from birth to calving would also be reduced. Under average village conditions in India, the feeding of buffalo calves is very much neglected and as a result, their maturity is delayed and their age at first calving is 4–5 years, which is higher than that of farm bred animals under proper feeding and management.

Late maturity and delay in conception increase the age of the heifers at first calving and thereby raise their cost of feeding and management considerably and also affects their life time production adversely. The breeding of heifers at an earlier age will definitely reduce the age at first calving and increase the efficiency of life-time milk production.

In the Indian subcontinent, the average age of buffalo heifers at the first calving is 40 months with a range of 30–50 months. In the Murrah buffalo, the age at first calving is 44 months (on an average). In Indian buffalo in Military Dairy Farms it is 42 months while at IARI Delhi it is found to be 45 months. The Bhadhawari type of buffalo takes 50 months. According to a study on the optimum age at the first calving it was found that the buffalo heifers calving between 42 and 48 months and producing more than 1800 kg milk in first lactation are most economical; Table 4.1 gives observation recorded on the age at the first calving at various places in India.

Table 4.1 Age at the First Calving

Type, breed	Place	Age at 1st calving in months	
		Range	Average
Murrah	India	–	44.3
Indian buffalo	India	–	30.4
Indian buffalo (farm bred)	Terai state farms, IVRI, Agri. college Agra, military farms IARI, Delhi	41.0–51.0	45.7
Indian buffalo	Military dairy farms	–	42.0
Nili	Military farms, India	–	37.4
Murrah	Dist. demonstrtion farm Mathura	–	42.0
Murrah	Agri. Instt. Allahabad	–	43.4
Murrah	Liv. res. station Mathura	–	42.3
Murrah farm bred buffaloes in organized dairy farms	Military farms	40.4–41.98	37.1
Murrah	Military dairy farms in India	–	40.6
Bhadawari	Livestock farm Bharari, Jhansi	–	50.7

Source: Fahimuddin M. (1986).

4.7 Characteristics of the Draught Buffalo

Buffaloes provide 30 to 40% farm power. Buffalo can be utilized in small holdings as work animal since tractor requires a minimum of 4 hectares of land for economic operations. It is being used for ploughing, puddling rice fields and levelling, pumping water, thrashing grains, pressing sugar cane for extracting juice and so on. This could be because of loose fetlock joints, which can easily move in the paddy fields in mud. Generally, the draught buffalo is judged by its temperament, general appearance, health condition, body conformation and physical characters of body parts related to working efficiency including colours and markings.

4.7.1 Temperament

A buffalo with good temperament is always most desirable for effective handling. Animals have actually been selected for this character because those which are unmanageable are naturally not kept by the farmer. Being kindly and closely cared for by their owners, these animals are mostly tame and trustworthy. However, they are sensitive to a foreign environment and are suspicious of strangers and unfamiliar treatment; their reaction is sometimes violent and harmful.

4.7.2 General appearance and ideal conformation

A good buffalo must have a big body, a deep and round chest and an abdomen which slightly reduces progressively towards the pelvis, making the animal look massively trunky and heavy. The head should be long and slender with a wide forehead and muscular mandibles. The facial veins are distinct. The muzzle is wide, rectangular, and always moist. Big nostrils indicate breathing efficiency. The mouth and lips are proportional to the head. The eyes are small and brilliant, not bulging or deeply buried in the sockets. The outer canthus should form an acute angle on both sides. The ear pinnae are small with tapering, not blunt ends, and should be attached close to the horn bases. The horns of the swamp buffalo are generally of two types; long horns which extend laterally and backwards or short horns of about half the length of the long ones. Loose or pendulous horns occur frequently and are inherited as a recessive. Polled swamp buffaloes are extremely rare (Fischer and Richards, 1965). Short horned buffaloes are said to be stubborn and more difficult to handle. They are used mostly in the drier land such as in north east Thailand. The long horned ones are more favoured in the paddy land with much water. Probably, such heavier horns are useful in balancing and forward pushing while the animal is working in the water (Bodhipaksha *et al.,* 1980). A massive neck with a broad base near thick, long shoulders is desirable. The shoulders should have a steep angle of 45° or less. The withers should be prominent but not so high as to form a hump as in zebu cattle. The back should be flat, broad and straight from the withers to the tailhead. A sagging back is undesirable. The chest should have a big, expanded rib cage with narrow intercostal spaces for respiratory efficiency. The abdomen should be big and bulky but not sagging. The flank should be deep and full. The buffalo flank is relatively narrower than that in cattle. The loin should be flat and thick and the rump round and muscular (Niamsri, V., Bodhisiri, L., and Yadi, Y. personal communication, 1984). The limbs should be relatively short in comparison with those of cattle but very muscular on the upper parts. The lower legs should be lean with powerful joints and big hooves. The ideal hoof should look like half a coconut shell with perfect approximation of its thick claws. The "crab claw" type is undesirable. The front legs in the standing position should be straight and vertical while the hind legs should look, from the side, like a "cat's hind leg". The tail should be long enough to hang down just beyond the hock with a rich tuft of hair at its end. The tail base should be thick and broad, covering the perineum completely. The prepuce should not hang loosely down under the abdomen but should conform to the abdominal contour with its end hanging down only a few centimeters. A loose prepuce risks being injured when the animal works under the water or walks in bushy areas (Macgregor, 1941).

4.7.3 Colours and markings

The hair whorls are useful for identifying an animal as well as indicating if the animal is in good or bad omen. Hair whorls show a great individual variation in number, position and direction.

4.7.4 Movement and specific characteristics

A good buffalo walks briskly and is always in the leading position. Thus, it has a better opportunity to feed while grazing, taking advantage of the others.

The swamp buffalo has an interesting eating habit. If food is plentiful, it is choosy. When availability of food is limited, it can eat almost anything on land or under water. The animal is an efficient weed eradicator of Asia.

The animal wallows to cool and protect itself from being disturbed by heat and insects. It is more often done during the hot and insect infested seasons. Wallowing keeps the buffalo in good health. Those without opportunity to wallow are usually in poor condition. The act of wallowing has been well described by Tulloch (1974).

CHAPTER 5

BREEDS OF BUFFALOES

Breeds of animal have originated due to the natural processes of selection and evolution. Many of them have been specifically developed by institutions, visionary individuals, cultural and social groups. These breeds have been developed for a variety of considerations of which economic, cultural and social reasons were important. The domestic water buffalo is classified as the river and the swamp types. They belong to the same species but have different habitats. The swamp buffalo has 48 chromosomes and the river has 50. They interbreed and produce fertile hybrid progeny. The disparity in chromosome numbers has not been explained so far. The swamp buffalo is mainly a draught-cum-meat animal which has swept back horns and is found principally in eastern half of Asia. The river buffalo usually has curled horns and is found in western half of Asia, Brazil, Russia and parts of CIS countries, Italy and a few east European countries. It prefers clean water, rivers, irrigation canals and ponds to wallow. This type is docile and has specially been developed for milk production with high fat content in Indo-Pak subcontinent.

Some of the well-known milch breeds of river buffaloes are Murrah, Nili-Ravi and Surti in India, and Nili-Ravi and Kundi in Pakistan. The Egyptian, Iraqi and Italian buffaloes also have high potential for milk and are derivatives of this sub-group. Some other breeds of milk type found in India are Jaffarabadi, Bhadawari and Mehsana, which are primarily used for milk and work and are also classified as dual purpose.

Brief description of some of the important breeds of buffalo is given in the following paragraphs.

5.1 Indian Subcontinent

5.1.1 Indian buffalo breeds

5.1.1.1 *Murrah (plate 1)*

The breeding tract of this breed extends over the entire north-west of Indian subcontinent. These animals have long been selected for milk and curled horns. The resulting breed was named Murrah, meaning curled. Murrah is considered to be the best milch-cum-meat breed of buffaloes.

Distribution

Its home tract stretches around the southern part of Haryana comprising the districts of Rohtak, Jind, Hisar and Gurgaon, and the Union Territory of Delhi. However, this breed has spread to almost all parts of the country and is being bred either in pure form or is being used for grading up local buffaloes. In fact, this breed has even found an important place in the livestock industry of many developing countries like Bulgaria, the Philippines, Malaysia, Thailand, China, Indonesia, Bangladesh, Nepal, former USSR, Myanmar, Vietnam, Brazil and Sri Lanka, and is being bred there extensively.

Description

The animals are massive, stockily built with a short broad back, light neck and head. The horns are short and tightly curved. The udders are well developed. The hips are broad, and fore and hind-quarters are drooping. The tail is long, reaching to fetlocks. The colour is jet black with white markings on tail, face and extremities.

Physical characteristics

The body weight of an adult female is 350–700 kg (average 516 kg) and that of a male, 450–800 kg (average 567 kg). Averages of length, height and heart girth of males are 150, 142 and 220 cm, respectively, and of females 148, 133 and 202 cm respectively.

Husbandry

Buffaloes of this breed are traditionally managed in domestic conditions together with the calf. They are hand-milked twice a day. They are fed different kinds of roughages; barley and wheat straw, cornstalks, sugarcane residuals. In addition, they are given concentrate mixtures. If grazing is available, they graze all day long. They are naturally mated. Some villages also provide artificial insemination.

Performance and performance traits

Total lactation milk yield, 305 day milk yield, lactation length and dry period average 1678.4 kg, 1675.1 kg, 307.0 days and 187.6 days, respectively, in first lactation; and 1751.8 kg, 1660.1 kg, 298.7 days and 154.8 days, respectively, for overall lactations. On an average, milk contains about 7.3% fat. Service period averages 177.1 days in first parity and 136.3 days in overall parities; age at first calving (months) 50.6 ± 2.0; calving interval varies from 430–604 days. Products: milk, ghee, cream, meat and skin (Bhat and Taneja, 1987).

5.1.1.2 ***Nili-Ravi*** *(plate 1)*

Earlier authors (Oliver, 1938; Handa, 1938) referred to the Nili and Ravi as strains of the Murrah which differed little from it, except in geographical location, *i.e.* the Sutlej and Ravi river valleys, as distinct from Rohtak, Karnal, Gurgaon and Delhi for the true Murrah. However, in 1938 (ICAR, 1939), Nili was shown as a breed at the first All-India Cattle show and after the second and third shows, the Nili and Ravi were briefly described in ICAR bulletin No. 46 (1941) as distinct breed types. The name Nili means blue and refers to the blue water of river Sutlej. Ravi gets its name from the river Ravi. However, Nili and Ravi are no longer considered as distinct breeds both in India and Pakistan.

Distribution

In India, these buffaloes are found in Fazilka, Ferozepur, Jira and Makhu tehsils of Ferozepur district, and Patti and Khemkaran tehsils of Amristar district of Punjab. The breeding tract is spread all along the Sutlej river on the Indo-Pak border.

Description

The skin and hair of Nili-Ravi are usually black although brown is not uncommon. Wall eyes and white markings on forehead, face, muzzle, legs and tail switch are common in Nili. Pink markings on udder and brisket are sometimes found in Ravi but more rarely in Nili. The horns are short, broad at the base and closely curled back behind the base. The body is massive and barrel-shaped with a deep frame. The head is long, bulging at the top and is depressed between eyes. The muzzle is fine but with wide nostrils. The double chin is conspicuous. Nili has a more convex profile with more coarse hair on head and face, and conspicuous double chin. The horns are fine and tightly coiled and circular in cross-section. In Ravi, the horns are oval in cross-section; transverse rings or pits are more conspicuous. The legs are short and bony. The rump is broad, long and slightly sloping. Pin-bones are prominent and set well apart. The tail is long, touching the ground. The udder is well developed extending far forward and backward. The milk-veins are prominent, long and tortuous.

Physical characteristics

The average adult Nili-Ravi female buffalo weighs about 450 kg and male nearly 600 kg. Height at withers of adult male is 140 cm, body weight is 700 kg. Height at withers of adult female is 135 cm, body weight is 454 kg.

Husbandry

Buffaloes of this breed are traditionally managed under domestic conditions together with the calf. They are hand-milked twice a day. They are fed different kinds of roughages: barley and wheat straw, cornstalks, sugarcane residuals. In addition, they are given concentrate mixtures. If grazing is available, they graze all day long. They are naturally mated. Artificial insemination is available at the state farms and in some villages.

Performance and performance traits

Average age at first conception is 1125.05 ± 37.12 days. Average age at first calving is 1359 days. First lactation milk yield, pooled lactation milk yield and 305 day milk yield average about 1483, 1850 and 1820 kg respectively. Average lactation length is about 294 days. Fat content varies from 5.1 to 8%. Dry period is about 151 days and service period about 202 days. Average first calving interval is 520.50 ± 2.58 days and overall calving interval about 488 days. Average number of services per conception is 2.38. Products: Milk, ghee, cream, meat and skin.

5.1.1.3 *Surti (plate 1)*

The home tract of this breed is southwestern part of Gujarat comprising Kheda, Vadodara, Bharuch and Surat districts.

Distribution

This breed is concentrated between the Mahi and Sabarmati rivers in Gujarat.

Description

The skin is black or reddish and the hair silver-grey to rusty-brown. Below the knees and hocks, the hair colour is generally whitish grey. The tuft of tail is sometimes white. Animals with white markings on forehead, legs and tail-tip are preferred. The horns are flat, of medium length, sickle-shaped and are directed downward and backward and then turn upward at the tip to form a hook. Animals are of medium size with straight back. The head is elongated, fairly broad and rounded between the horns. The face and muzzle are clean and sharply narrowed below the eyes, with big nostrils and muzzle. The eyes are round and bulging. The ears are of medium size with reddish colour inside. The neck is long and thin in females; thick and massive in males. The legs are of medium size with black hooves. The barrel is well-built, of wedge shape and medium size. The rump is broad and slightly sloping. The pin bones are wide apart. The hocks are prominent and strong. The tail is fairly long, thin and flexible. The udder is well-developed. The teats are of medium size and squarely placed.

Physical characteristics

Mature females and males weigh around 383 and 500 kg respectively. Average body length, height and heart girth are 142, 130 and 190 cm, respectively, in males, and 119, 125 and 184 cm, respectively in females.

Husbandry

Buffaloes of this breed are traditionally managed under domestic conditions together with the calf. They are hand-milked twice a day. They are fed different kinds of roughages (barley and wheat straw, cornstalks, sugarcane residuals). In addition, they are given concentrate mixtures. If grazing is available, they graze all day long. They are naturally mated. Some villages also provide artificial insemination.

Performance characteristics

The milk produced by this breed is known to have slightly higher fat (%) 7.9.

Performance and performance traits

Total lactation milk yield and 305 day milk yield average 1396.5 and 1065.3 kg, respectively, in first lactation, and 1285.4 and 1289.5 kg, respectively, in overall lactation. Lactation length, dry period, service period and calving interval are 352.2, 242.1 and 584.6 days, respectively, in first parity, and 344.7, 185, 142.6 and 534.7 days, respectively, in overall lactations. Number of services per conception ranges from 1.5 to 3.0. Fat ranges from 7.5 to 8.3%, which is slightly higher than that of other breeds mainly because of high proportion of fodder in the feed.

5.1.1.4 *Mehsana (plate 2)*

The home tract of Mehsana buffalo lies in Mehsana district of Gujarat state of India. Mehsana breed has been derived from crosses between Murrah and Surti.

Distribution

Concentrated around the town of Mehsana. It is also available in Banaskantha and Sabarkantha districts. Typical animals are seen in the towns of Mehsana, Patan, Sidhpur, Vijapur, Kadi, Kalol and Radhanpur.

Description

It shows a wide variation in appearance and colour. It is believed to have been produced by crossing Murrah with Surti and subsequent crosses. Animals are black or brown/fawn with white markings on the face, legs or tip of the tail. The horns vary from coiled to sickle in shape. Mehsana has a longer body than Murrah, lighter limbs, and longer and heavier head. The chest is deep with broad brisket. The shoulders are broad and blend well with the body. The legs are medium to short in length. The barrel is long and deep with well sprung ribs. The tail is medium thick and long. The udder is well developed and well set. The teats are fairly thick, long and pliable. The Mehsana females have the reputation of being persistent milkers and regular breeders.

Physical characteristics

Body weight of adult male is 565 kg. Body weight of adult female is 484 kg. Average of body length, height and heart girth are 153.7 cm, 133.7 and 200.6 cm respectively, in males, and 141.7 cm, 127.5 cm and 189.3 cm respectively, in females.

Husbandry

Buffaloes of this breed are traditionally managed under domestic conditions together with the calf. They are hand-milked twice a day. They are fed different kinds of roughages: barley and wheat straw, cornstalks, sugarcane residuals. In addition, they are given concentrate mixtures. If grazing is available, they graze all day long. They are naturally mated. Some villages also provide artificial insemination.

Performance and performance traits

Age at the first service is about 830 days and at the first calving about 1266 days. Total lactation milk yield, 305 day milk yield, lactation length and dry period average 1940 kg, 1893 kg, 308 days and 179 days respectively, in first lactation, and 1988 kg, 1912 kg, 317 days and 167 days respectively, in all lactations. Milk contains about 7% fat. Average service period is about 161 days. Calving interval is about 476 days. This breed is reputed for its persistency of milk production.

5.1.1.5 *Jaffarabadi (plate 2)*

The Jaffarabadi is the heaviest of all the Indian breeds of buffaloes. This breed is found in Gir forest of Kathiawad in Gujarat. It is named after the town of Jaffarabad and is also called Bhavanagri or Gir.

Distribution

It is one of the most important breeds in Gujarat. This breed is located principally between the Mahi and Sabarmati rivers in north Gujarat. Some breeding stock has been exported to Brazil.

Description

The colour is usually black. The horns are heavy and broad, sometimes covering eyes; are inclined to droop on each side of the neck and turn up at the points into an incomplete coil. The forehead is very prominent and bulging. Animals are massive and long barreled. The body is long and not compact. The head and neck are massive. The dewlap is well developed. Average adult weight of males and females is 590 and 454 kg respectively. Butter-fat percentage is high.

Physical characteristics

Height at withers of adult male is 126 cm, body weight is 1000 kg. Height at withers of adult female is 129 cm, body weight is about 700 kg, some individuals may weigh as much as 2000 kg.

Husbandry

Buffaloes of this breed are traditionally managed in domestic conditions together with the calf. They are hand-milked twice a day. They are fed different kinds of roughages: barley and wheat straw, cornstalks, sugarcane and other crop residuals. In addition, they are given concentrate mixtures or oil cakes. If grazing is available, they graze all day long. They are naturally mated. Some villages also provide artificial insemination.

Performance and performance traits

Age at the first service is around 1000 days and at the first calving is 1361.7 days. Averages of total lactation milk yield and lactation length are 2151.3 ± 130.53 kg and 319.44 ± 17.31 days, respectively, in first lactation, and 2238.7 ± 74.87 kg and 305.1 ± 9.61 days, respectively, in overall lactations. Averages of dry period, service period and calving interval are 144.9 ± 8.4, 93.4 ± 0.69 and 440.3 ± 14.32 days respectively. Average number of services per conception is 1.5. Average fat is 7.68%. Males are good draught animals for hauling heavy loads.

5.1.1.6 *Bhadawari (plate 2)*

The home tract of this breed is Bhadawari estate, part of Bah tehsil in Agra district and the adjoining areas of Gwalior in India.

Distribution

It is raised in the Agra and Etawa districts of Uttar Pradesh and in Bhind and Morena districts of Madhya Pradesh.

Description

These animals are usually copper coloured. The hair on the body are scanty, black at the root and reddish brown at the tip. The legs from hoof to knee and hock are usually brownish white. The tail switch is white or black and white. The horns are flat, compact, growing backward and then upward, turning inward with slightly pointed tips. Animals are medium sized with wedge-shaped body. The head is comparatively small, bulging out between the horns. The forehead is slightly broad and deep in the middle. The dewlap is absent. The chest is well developed. The barrel is short but well developed with short and stout legs having black hooves. The tail is long, thick and flexible. The udder is not so well developed. The testes are of medium size though not of uniform length. It is believed that the fat percentage of milk is very high (13%) in this breed. Males are usually used for draught and can stand heat better than the other breeds.

Physical characteristics

Average body length, height and heart girth are 116.9, 122.8 and 184.5 cm, respectively, in adult males, and 115.0, 123.1 and 184.3 cm, respectively, in adult females. Average birth weight of calves is 25.3 ± 0.23 kg. Body weight of adult males is around 475 kg and that of adult females around 425 kg.

Husbandry

Buffaloes of this breed are traditionally managed under domestic conditions together with the calf. They are hand-milked twice a day. They are fed different kinds of roughages (barley and wheat straw, cornstalks, sugarcane residuals). In addition, they are given concentrate mixtures. If grazing is available, they graze all day long. They are naturally mated. Some villages also provide artificial insemination.

Performance and performance traits

Age at first calving is around 1477 days. Bhadawari buffaloes are low milk producers. Average lactation milk yields is 780 ± 25.4 kg and 300 days milk yield is 711 kg, in first lactation, and 903 kg and 812 kg, respectively, in overall lactations. On an average, these buffaloes produce milk for 272 days. Lactation period varies from 140 to 350 days. Average dry period is about 190 days. Fat varies from 6 to 12.5%. Total solids in milk are around 17%. Average service period is 179 days and interval between successive calvings is 478.7 ± 11.55 days.

5.1.1.7 *Nagpuri (Syn. Ellichpuri)* (plate 3)

The home tract of this breed is Vidarbha region of Maharashtra state and the adjoining areas of Andhra Pradesh and Madhya Pradesh of India.

Distribution

This breed is raised in the Nagpur, Wardha and Berar districts.

Description

The animals are usually black although white patches are often found on face, legs and tail. The distinguishing feature of the breed is long, flat, curved horns carried back on each side of neck, nearly to the shoulders. The face is long and thin. The neck is somewhat long. The limbs are lighter and the tail is comparatively short, reaching below the hocks

Physical characteristics

Height at withers of adult male is 140 cm, body weight is 520 kg. Height at withers of adult female is 130 cm, body weight is 364 kg.

Husbandry

Buffaloes of this breed are traditionally managed under domestic conditions together with the calf. They are hand-milked twice a day. They are fed different kinds of roughages: barley and wheat straw, cornstalks, sugarcane residuals. In addition, they are given concentrate mixtures. If grazing is available, they graze all day long. They are naturally mated.

Performance and performance traits

Females are fairly good milkers and produce about 780 to 1520 kg of milk. Lactation length average 286 days. Fat varies from 7.0 to 8.5%. Average service period is about 116 days; average dry period is 129.1 ± 4.85 days. Inter calving period ranges from 350 to 721 days. Products: Milk, ghee, cream, meat.

5.1.1.8 *Toda (plate 3)*

Toda breed of buffaloes is named after an ancient tribe (Toda) of southern India.The Toda buffaloes are exclusive to the district of Nilgiris. Their spread is limited to a small geographical area of 2545 sq km lying between latitudes 11° 15" and 11° 30" N and longitudes 76° 15" and 77 °E. Walker (1986) described district of Nilgiris as an island, a discription used to emphasize its geographical separation from the surrounding low lands. This geographical separation had helped in the evolution of the breed in genetic isolation and in the process, the breed had adapted itself to the high rainfall and high humid zone of the plateau of the Nilgiris.

Distribution

This breed is raised in the Nilgiri hills of Madras. These buffaloes are quite distinct from other breeds and are indigenous to Nilgiri hills.

Description

Animals have long body, deep and broad chest, and short, strong legs. The head is heavy with horns set wide apart, curving inwards, outwards and forwards. Thick hair coat is found all over the body. The animals are gregarious in nature. Toda buffaloes are good milkers, yielding daily 4.4 to 8.8 litres of very rich milk.

Physical characteristics

Height at withers of adult male is 160 cm; body weight is 380 kg. Height at withers of adult female is 122 cm; body weight is 300 kg.

Husbandry

The breed is semi-wild and raised under semi-nomadic conditions, with total grazing.

Performance and performance traits

Age at the first calving varies from 1200 to 1400 days. Toda buffaloes produce around 500 kg of milk in a lactation of about 200 days. The mean fat is 8.22 ± 0.08% and protein 4.45 ± 0.12%. Calving interval is around 480 days. Average carcass weight was 27.17 ± 3.94 kg in calves, 50.83 ± 6.83 kg at 1.5 to 2 years, 69.00 ± 7.50 kg at 2.5 to 3 years, and 142.13 ± 10.10 kg in adults.

5.1.1.9 *Pandharpuri* (*Syn. Dharwari*) *(plate 3)*

Distribution

The Pandharpuri buffalo is a native of Kolhapur, Solapur, Sangli and Satara districts in south Maharashtra. These buffaloes are named after the name of the geographical area, *i.e.* Pandharpur block in Solapur district.

Description

The Pandharpuri buffalo is a medium-size animal with long narrow face. Skin colour is usually black but it varies from light black to deep black. White markings are also found on the forehead, legs and tail in a few cases. Frontal bone is comparatively long and straight. Nasal bone is very prominent, long and straight. Horns are very long extending beyond shoulder blade, sometimes up to pin bones and are of three types: (i) *bharkand* (53%) curving backward and usually twisted; (ii) *toki* (36%) curving backward, upward and usually twisted outwards; and (iii) *meti* (11%) long, flat, running downward from sides of head and usually twisted. Neck is long and thin. Switch of the tail is white in majority of animals and extends just below the hock. Hooves are usually black. Hook bones are prominent and well placed. Legs are long and straight. Udder is of medium size, compact, somewhat hidden in between hind-quarters with firm attachment. Shape of udder is mostly trough (41%) followed by bowl (31%) and round (17%) shape. Udder texture is smooth and collapsible. Teats are cylindrical and squarely placed. Mostly teat tips are rounded and sometimes pointed. Milk-vein is prominent. Animals are mild in temperament and easily manageable.

Physical characteristics

Averages of body length, height and heart girth of Pandharpuri females are 132.9 ± 130 and 192.8 cm respectively. Adult body weight of females is around 416 kg.

Performance and performance traits

Pandharpuri buffaloes produce, on an average, 1502 kg of milk in a lactation period of 330 days. Total lactation and 305 day milk yield average 1197 and 1142 kg, respectively, in first lactation. Milk contains about 7% fat and 9.28% SNF. These buffaloes have high reproductive efficiency and low feed intake. Averages of age at puberty, age at first conception, age at first calving, dry period, service period and intercalving period are 795, 945, 1255, 144, 165 and 465 days respectively.

Husbandry

Animals are usually housed in the open close to human dwellings. Natural service is practised. Paddy straw, sorghum straw (*jowar kadbi),* dry mixed grasses, green grasses, sugarcane leaves and sugarcane tops are mainly used as fodder.

5.1.1.10 *Marathwada (plate 4)*

Distribution

Marathwada buffaloes in the Marathwada region of Maharashtra state are entirely different from that of western and northern types, and clearly represent a very ancient indigenous type characterized with lighter built and long flat horns. These buffaloes are mainly found in Parbhani, Nanded, Bid and Latur districts of Maharashtra.

Description

Marathwada buffaloes are light to medium built with compact stature, and have adult weight of 300 and 370 kg. Coat colour varies from grayish-black to jet black, although white markings on forehead and on lower parts of the limbs with white switch of tail are not uncommon. Horns are medium in length, parallel to neck, reaching up to shoulder but never beyond shoulder blade like those commonly seen in Pandharpuri buffaloes and usually not flat. Forehead is moderately broad, and eyes are generally red tinged. Neck is short. Legs and feet are properly sets which in males suit for draught and transportation in hilly tract. Tail is of moderate length reaching up to hock.

Performance and performance traits

Average birth weight of male and female calves is 24.5 and 23.7 kg respectively. Adult body weight ranges from 320 to 400 kg. Age at first calving is around 1670 days. Lactation yield ranges from 845 to 960 kg in a lactation of about 302 days. Gestation period, calving interval and dry period in Marathwada buffaloes are 310, 430 and 134 days respectively.

Husbandry

Farmers maintain mixed herds of cattle and buffaloes. Generally, animals are housed in open, close to farmers' residence. Sorghum and paddy straw, grasses, sugarcane leaves and tops are usually fed to animals. Concentrate is offered only to milking females.

5.1.1.11 *Sambalpuri (Syn. Kimedi and Gowdoo) (plate 4)*

Breeding tract of this buffalo is referred to be the Sambalpur area in Orissa.

Distribution

This breed is also found around Bilaspur in Madhya Pradesh.

Description

Black in colour with white switch on tail, narrow and short horns, curved in a semi-circle, running backward, then forward at the tip.

Husbandry

Buffaloes of this breed are traditionally managed under domestic conditions together with their calf. They are hand-milked twice a day. They are fed different kinds of roughages: barley and wheat straw, cornstalks, sugar cane residuals. In addition, they are given concentrate mixtures. If grazing is available, they graze all day long. They are naturally mated. It is a good healthy draught animal with a rapid pace and it is comparatively the most productive breed of the region. Some exceptional buffaloes may yield as high as 2300 to 2700 kg in about 340 days (Narayana Reddi, 1939).

Performance and performance traits

Lactation duration 350 days; milk yield <1200 kg; Products: Milk, ghee, cream and meat.

5.1.1.12 *Tarai (plate 4)*

Distribution

This breed is found in the Tarai area of Uttar Pradesh, particularly between Tankpur and Ramnagar.

Description

Black to brown colour coat; sometimes there is a white blaze on the forehead, tail switch is white. Horns are long and flat with coils bending backwards and upwards.

Physical characteristics

Height at withers of adult male is 127 cm; body weight is 375 kg. Height at withers of adult female is 120 cm; body weight is 325 kg.

Husbandry

Buffaloes of this breed are traditionally managed under domestic conditions together with the calf. They are hand-milked twice a day. They are fed different kinds of roughages: barley and wheat straw, cornstalks, sugarcane residuals. If grazing is available, they graze all day long. The breed is well adapted to the difficult climatic and feeding conditions of the Tarai region. Sometimes it is crossbred with the Murrah.

Performance and performance traits

Lactation duration 250 days; milk yield 450 kg; milk fat 6.6–8.1%; milk protein 4.2–4.6%; Products: milk, ghee, cream and meat (Cockrill, 1974; FAO, 2003).

5.1.1.13 *Godavari (plate 5)*

Godavari breed is a result of crossing native buffaloes of Andhra Pradesh with Murrah bulls.

Distribution

The home tract of this breed is Godavari deltaic areas of Tanuku, Bhimavaram, Narasapur, Ramchandrapuram, Kothapeta and Alamuru taluqs, parts of Tadepalligudem and Kovvur taluqs, and Krishna deltaic areas of Gudlavalluru, Gudivada, Avanigadda, Kankipadi and Vuyyura taluqs (Bhat and Taneja, 1987).

Discription

This breed has attained uniformity and almost reached the production level of Murrah. The animals are medium statured with compact body. The colour is predominantly black with a sparse hair coat of coarse brown hair. The head is clean cut with a clean face, convex forehead and prominent bright eyes. The horns are short, flat, curved slightly downward, backward and then forward with a loose ring at the tip. The chest is deep with well-sprung ribs. The barrel is massive and long with straight back and a broad level rump. The udder is of medium size. The tail is thin and extends below the hocks with or without a white switch.

Physical characteristics

Average body length, height and heart girth is 143, 128 and 192 cm, respectively, in female buffaloes, and 151, 143 and 204 cm, respectively, in male buffaloes.

Performance and performance traits

Godavari buffaloes are reputed for high fat with daily average milk yield of 5 to 8 litres and lactation yield of 1200 to 1500 litres. The best animals even produce about 2000 litres milk in a lactation.

The animals breed regularly and have a short calving interval compared to Murrah. They are hardy and possess good resistance against majority of the prevailing diseases.

5.1.1.14 *Parlakhemundi*

Distribution

The Parlakhemundi buffaloes are of 3 sub-types, *viz.* Manda, Deshi and Jerangi. Manda and Jerangi are distributed in the plateau region whereas Deshi is found in the plains of Ganjam and Koraput districts of Orissa.

Description

The skin colour of Parlakhemundi buffaloes is light slate grey with a reddish hue on the ventral surface of the body including udder and teats. Albinism (rose and solid white skin with white hairs) is exhibited by some Manda and Deshi strains while it is not seen in Jerangi. Manda and Deshi have white chevrons (one below jaw and another above brisket), white freckles or spots on either cheek, white or pale white eyebrows and white stockings on legs. Manda and Deshi have often grey, white or greyish white hair coat while Jerangi have only grey hair. The animals have light head, long narrow face, with prominent eyebrows, small tubular ears, long and thin neck, flat and long crescentic horns with transverse grooves. The body is long and barrel shaped supported by thin but strong limbs with heavy fore quarters and light hind quarters.

Physical characteristics

The body weight at maturity has been reported as 332, 284 and 277 kg respectively, for Manda, Deshi and Jerangi (Patro and Mishra, 1986). Average body length, height at withers, heart girth are 114.69 ± 2.167, 122.89 ± 1.470 and 184.48 ± 3.952 cm respectively.

Performance and performance traits

Females yield about 3–4 kg of milk/day.

5.1.1.15 *Jerangi (Syn. Zerangi)*

Distribution

These buffaloes are found in Jerangi hills in Ganjam district of Orissa and northern parts of Vishakhapatnam district of Andhra Pradesh.

Description

Black in colour, with small horns running backwards. It is a small animal and has a maximum height of just 114 cm.

Husbandry

Buffaloes are traditionally managed in domestic conditions together with the calf. If grazing is available, they graze all day long. They are naturally mated. It is a draught animal with a rapid pace (Cockrill, 1974; FAO, 2003).

5.1.1.16 *Manda*

This is an improved local breed, resulting from the selection of Indian breeds of buffaloes.

Distribution

Herds of Manda buffaloes are found around Kakriguma, Laxmipur, the hills of Damanjodi, the Patraput area and the Petta hills near Jeypore, Gupteswar, Jolapur, Balimela, Pedabial and the Arku valley hills of Koraput district in Orissa.

Description

Uni colour: grey, brown.

Performance and performance traits

Average body length, height at withers and heart girth are 115.42 ± 2.723, 124.46 ± 1.851, 174.50 ± 3.330 cm, respectively.

5.1.1.17 *Kalahandi (Syn. Peddakimedi)* *(plate 5)*

Distribution

It is a breed of eastern hilly part of Andhra Pradesh and adjoining areas of Orissa. These animals were brought by Peddakimedi people from Kalahandi, Sansthanam of Orissa. Little wood (1936) named this breed as Peddakimedi.

Description

The colour is grey or ash grey with white switch of medium length tail. The forehead is slightly protuberant. Due to light colour it tolerates sun heat better than dark colour buffaloes (Narayana Reddi, 1939). Milk yield is quite satisfactory and production is less expensive. The animal is of docile temperament but quite hardy in work and good for draught purpose to carry heavy loads on to the hill tops.

Physical characteristics

The average height at withers of male and female Kalahandi buffaloes is 121.61 ± 0.52 and 118.53 ± 0.3 cm respectively. The average body length of male and female Kalahandi buffaloes are 121.55 ± 0.65 and 116.51 ± 0.40 cm respectively (Dash *et al.*, 2005).

5.1.1.18 *Other Indian breeds*

Apart from the above mentioned breeds, some other breeds are found like South Kanara in Mangalore region on the west coast of India; Sikamese in Sikkim;

Assamese in Assam and Kujang in Orissa state. These breeds are generally poor producers of milk but highly adapted to the ecology of the area.

Body weights at various ages and body measurements of some Indian breeds is summarized in Table 5.1 and 5.2 respectively.

Table 5.1 Body Weights (kg) of Various Breeds of Buffaloes

Breed		Birth weight			Adult weight	
		Male	Female	Overall	Male	Female
Bhadawari	Mean	27	25	25.3 ± 0.23	475	425.7 ± 7.72
	No.	–	–	860	–	49.0
	Range	–	–	24–27		300–540
Mehsana	Mean	29.5 ± 0.51	28.5 ± 0.45	29.0	565.4	484.2
	No.	209	252	461	55	314
	Range	16–44	14–40	14–44	400–602	315–580
Murrah	Mean	31.7	30.0	30.3	567	516
	No.	2186	6043	8574	–	1761
	Range	28–34	26–33	26–34	450–800	350–700
Nagpuri	Mean	29.0 ± 0.32	28.1 ± 0.14	28.6 ± 0.27	520	363.5
	No.	–	–	1028	–	200
	Range	27–30	25–29	25–30	–	340–400
Nili-Ravi	Mean	35.1	34.5	34.8	567	454
	No.	182	222	404	–	–
	Range	27–39	27–38	27–39	–	–
Pandharpuri	Mean	28.0 ± 0.91	25.6 ± 0.74	26.8	–	416.2 ± 10.10
	No.	17	26	43	–	11
Surti	Mean	26.3	24.5	25.2	500	382.6 ± 12.14
	Range	24–40	23–29	23–30	–	318–414
Toda	Mean	27.9 ± 0.60	28.05 ± 0.59	27.9 ± 0.43	380	380
	No.	57	43	100	–	–

Table 5.2 Body Measurements (cm) of Various Breeds of Buffaloes

Breed		Male			Female		
		Length	Height	Heart girth	Length	Height	Heart girth
Bhadawari	Mean	116.9	122.8	184.5	115.0	123.1	184.3
	No.	90	90	90	169	169	169
Jaffarabadi	Mean	127.7	126.1	207.7	132.6	129.1	200.9
	No.	58	58	58	1419	1419	1422
Mehsana	Mean	153.7	133.7	200.6	141.7	127.5	189.3
	No.	55	55	55	314	314	314
Murrah	Mean	150.0	142.0	220	148.0	132.7	202.4
	No.	–	–	–	1372	1698	1746
Nagpuri	Mean	180.0	140.0	210.0	128.6	122.8	181.8
	No.	–	–	–	413	413	413
Nili-Ravi	Mean	160.0	140.0	230.0	165.4	134.2	207.7
	No.	–	–	–	150	140	120
Pandharpuri	Mean	–	–	–	132.9	130	192.8
	No.	–	–	–	201	201	201
Surti	Mean	142.0	130.0	190.0	118.8	124.9	184.0
	Range	–	–	–	25	25	25
Toda	Mean	–	–	–	132.7	121.8	180.4
	No.	–	–	–	131	131	129

5.1.2 Pakistan

5.1.2.1 *Kundi* *(plate 5)*

The word *kundi* means fish-hook in Sindhi language. Ware (1942) first described Kundi as a distinct breed while admitting that some breeders still consider it a geographical type of Murrah. The Kundi breed is distributed in the forest tract along the river Indus, in the rice-growing region of north Sindh and in the swampy and rice-growing tracts of Karachi and Hyderabad districts of Pakistan.

Distribution

Widespread in Sindh region of south Pakistan

Description

Kundi animals are generally jet black (85%), although light brown are not uncommon (15%). Horns are thick at the base, inclined backward and upward, and end in a moderately tight curl. The forehead is slightly prominent. The face is hollow and eyes are small. Hind-quarters are massive. Mammary glands are capacious with prominent milk-veins; teats are squarely placed.

Physical characteristics

Kundi buffaloes are smaller than Nili-Ravi with females weighing 320–450 kg. Height at withers of adult male is 135 cm, body weight is 700 kg. Height at withers of adult female is 125 cm, body weight is 600 kg.

Husbandry

Buffaloes are traditionally managed under domestic conditions together with the calf. They are hand milked twice a day. They are fed different kinds of roughages: barley and wheat straw, cornstalk, sugarcane residuals. In addition, they are given concentrate mixtures. If grazing is available, they graze all day long. They are mated mainly through natural mating. Some villages also provide artificial insemination.

Performance

Lactation duration 320 days; milk yield 2000 kg; milk fat 7.0%; milk protein 6.0% (Alexiev, 1998; Tunio, 1999).

5.1.2.2 *Nili-Ravi (plate 6)*

Distribution

All over Pakistan

Description

Has already been dealt in 4.1.1.2

5.1.3 Bangladesh

5.1.3.1 *Bangladeshi*

The buffaloes of Bangladesh are of importance primarily as work animals. Their potential as milk animals and as a source of good meat is questionable.

Description

Black in colour, white spot on the forehead and tail-switch in some cases. Curled and short horns. Indigenous Bangladeshi buffaloes of the river type are found in the South-west. In the remaining parts of the country they are either swamp or crosses of Nili-Ravi and Murrah (Faruque, 2000).

5.1.4 Nepal

5.1.4.1 *Lime*

The pure Lime breed is believed to have originated from the wild area and has been domesticated throughout the known history of Nepal. The Lime buffalo is estimated to constitute 35% of the total indigenous buffalo population in the hills and mountains of the country.

Description

Animals are of light brown colour, small body size, characteristic chevrons of grey or white hair below the jaws and around the brisket, small sickle-shaped horns, curved towards the neck. Height at withers of adult female is 115 cm, body weight is 399 kg.

Distribution

The breed is found in the mountains, high hills and hill river valleys in Nepal. It is not found in the Terai plane.

Husbandry

Mainly raised under migratory conditions or semi-stall systems. The breed is a voracious eater and is fed only low quality feedstuff such as rice, wheat and millet straw. Small farmers exchange breeding animals within and between villages. Among the migratory herds, male and females are grazed together and mate freely during the breeding season from June to November. Females are legally banned from slaughter; only culled animals are slaughtered for meat.

Performance and performance traits

Lactation duration 351 days; milk yield 875 kg; milk fat 7.0%; products: milk, ghee, meat, swiss-cheese, yoghurt, leather (Rasali, 1997; Rasali, 1998a,b).

5.1.4.2 *Parkote*

The hill buffalo of Nepal, named Parkote buffaloes, are the typical buffalo of the mid-hills and river valleys. However, due to the traditional practise of crossbreeding with Lime buffalo as well as recent crossbreeding with Indian Murrah, their population in pure form is now declining. At present the pure bred population is estimated at only 25% of the indigenous population in the hills and mountains of Nepal.

Description

The Parkote are dark in coat colour and of medium-built body size, with sword-shaped horns directed laterally or towards the back. Black skin, black muzzle, black eyebrows. Usually they have no markings on the legs.

Distribution

The breed is raised in the mountains, high hills and hill river valleys of Nepal. Height at withers of adult female is 114 cm, body weight is 410 kg.

Husbandry

Mainly raised under migratory conditions or semi-stall system. The breed is a voracious eater and is fed only low quality feedstuff such as rice, wheat and millet straw. Small farmers exchange breeding animals within and between villages. Among the migratory herds, male and females are grazed together and mated freely during the breeding season from June to November. Females are legally banned from slaughter; only culled animals are slaughtered for meat.

Performance and performance traits

Lactation duration 351 days; milk yield 875 kg; milk fat 7.0%; Products: milk, ghee, meat, swiss-cheese, yoghurt, leather (Rasali, 1997; Rasali, 1998a,b).

5.2 Iran

5.2.1 Iranian *(plate 6)*

Buffaloes of this breed found in southern Iran are of a uniform type, dark grey in colour, weighing about 400 kg and 140 cm in height (Rastagir, 1950). The females gives up to 2500 liters of milk in 300 days. Males are generally used for draught and meat. The buffaloes generally show great diversity in adult size, weight and colour. Most animals show white patches on head and hocks.

5.2.2 Azari

Distribution

These buffaloes are spread in 5 states of Iran, Khuzestan, Ardebil, east Azerbaijan west Azerbaijan and Gilan (FAO, 1997).

Description

Black in colour, short horns growing backwards. Height at withers of adult male is 137 cm, body weight is 650–700 kg. Height at withers of adult female is 133 cm, body weight is 500 kg.

Husbandry

Housing differs from region to region. They are generally untethered in summer and tied up in winter. In some areas, milking females are tethered all year round. Average slaughter weight is 300 kg, at the age of 15 months. Carcass yield is 50%. Overall growth rate is 420 g/day.

Performance and performance traits

Lactation duration 200–220 days; milk yield 1200–1300 kg; milk fat 6.6%.

Products: milk, yoghurt, fresh cream, fresh cheese, butter, ice-cream, rice pudding, churned yoghurt, dried whey, ghee. In Iran, the price of buffalo milk is twice that of cows' milk. Buffalo skin is used in the leather industry. Buffalo manure is used for fuel in rural areas (Latifova, 2000; Turabov, 1991; Turabov, 1997a,b; Naderfard and Qanemi, 1997; Marmarian, 2000; Qanemi 1998; Borghese, 2005).

5.3 Iraq

5.3.1 Iraqi or Kuhzestani *(plate 6)*

Iraqi buffaloes are widely distributed, with foci of concentration in Baghdad, Basra, Amara, Diwaniya, Hilla and other towns, where they are used for milk production. The largest concentration is still found in the Amara area of Shatul Arab.

Description

There is great diversity in colour, which varies from slate black to almost white. White patches on the head, legs and tail are common. The horns are generally sickle shaped. The face is slightly dished. The withers are high and the croup and haunch bones are prominent. The body is generally elongated. The udder is big in good specimens, is carried well back and has good-sized and well-placed teats.

Physical characteristics

Average measurements are: height at withers, 143 cm; length from point of shoulder to pin bone, 153 cm; heart girth, 209 cm (Williamson, 1949; Bhat, 1975). Milk yield varies from 1600 to 2500 liters in a lactation period of 200–300 days.

Husbandry

Buffaloes of this breed are raised out doors all through the year. They are housed in paddocks made of local plants (reeds, brushes, palm leaves) with a wall on one side, and three open sides. They are hand fed at the time of milking, morning and evening, with available green forage. They are also fed any type of by-products waste of sugarcane, reeds from marshy land, home baked wastes. Those that swim in ponds and rivers are also fed aquatic plants. Milking is done by hand in 95% of cases and in a few cases with movable milking machines, there are no milking establishments. Male buffaloes are very hazardous, strong and difficult to handle and always aggressive to humans. In a few cases, for tilling operations, they are castrated. Females are very sensitive to non-familiar persons and reduce milk yield with non-familiar milkers. Generally females are also not docile. Average slaughter weight is 400 kg, at the age of 12 months. Carcass yield is 50%. Overall growth rate is 580 g/day.

Performance and performance traits

Lactation duration 200–270 days; milk yield 1300–1400 kg; milk fat 6.6%; Products: milk, yoghurt, fresh cream, fresh cheese, butter (National Buffalo Project, 1988; Magid, 1996; Saadat, 1997; Borghese, 2005).

5.4 CIS Countries

5.4.1 Caucasian

Description

The conformation of the Caucasian buffalo breed is as follows: large head with coarse features and narrow forehead; large well-developed thick horns, occasionally polled; long broad ears; long neck; back short or of medium length, broad, slightly roof-like or straight; the small of the back is long, broad and straight; the rump is raised and the belly voluminous. Buffalo cows of the Caucasian breed have a rounded and bowl-like udder with well-developed glandular tissue; teats are long and correctly set. The hair coat is not dense. The prevailing colour is dark brown to black; some individuals have a white spot on the forehead. The tail tip is white. Legs are long and thick. Among the faults of conformation are: front legs too close together, sloping and roof-like rump.

Distribution

In Azerbaijan, they are found everywhere. It is also found in Georgia, Armenia, Ukraine and Russia

Physical characteristics

The live weight of adult cows recorded in the state herdbook is 500 kg; that of bulls is 650–700 kg. The measurements of cows are as follows (cm): height at withers 132–136.1, width of chest 40.7–43.7, oblique body length 144.1–144.5, depth of chest 70.0–75.3, heart girth 194.5–196.6 and shank girth 20.3–20.8 cm. The animals have acquired greater massiveness and length of body, while cases of acute hindquarters have lessened.

Husbandry

The buffalo breeding farms in Azerbaijan at present use the tied stall system in winter and camp and pasture management in summer. Calves are reared en masse by group suckling. Machine milking is practised. To carry out effective breeding work, directions for buffalo breeding have been worked out and implemented, including judging guides, standards for animal appraisal and recording in herdbooks, instructions for artificial insemination and for rational feeding and stall barn housing, measures to prevent barrenness and to promote productive buffalo breeding.

Performance and performance characteristics

Lactation length of Caucasian buffalo cows is 270–300 days, calving interval 350–400 days and service period 60–100 days; milk yield of buffalo cows in test farms in the first lactation is 1150–1250 kg, in the second 1250–1500 and in the third and later lactations 1500–2500 kg, with a fat content of 8.0–8.1% and protein 4.7–5.0%; products: gatyg (yogurt, sour milk), kaimag (cream), shor, kyasmig (curds, cottage cheese) and airan (buttermilk). In recent years buffalo milk standardized to 3.2% butterfat is being used in the dairy industry, as well as for producing high-quality cheese and kaimag, which is packed in small portions and is in great demand. Buffalo meat plays an essential role in the meat balance of Azerbaijan. Slaughter yield is 43–53%. At the age of 18 months animals weigh 300–320 kg and 350–360 kg after intensive fattening. Hides are valuable raw material for leather production. The average hide weight is 25–30 kg and the maximum 50 kg; the thickness is 4–9 mm.

5.5 European/Mediterranean Countries

5.5.1 Turkey

5.5.1.1 *Anatolian* *(plate 6)*

The Anatolian buffalo has been raised in Turkey for centuries, originating from Indian migration (7th Century), together with the expansion of Islam.

Description

Black in colour, long hair, with variation in tail length and frequent white switch. Height at withers of adult male is 138 cm, body weight is 200–500 kg. Height at withers of adult female is 138 cm, body weight is 200–500 kg.

Distribution

Concentrated in the Black Sea region, North of Middle Anatolia, Thrace, Hatay, Mus, Kars, Dyarbakir, Afyon, Sivas.

Husbandry

In dairy farms, housing differs from region to region. If grazing is available, the three to five buffaloes owned by the family are taken to graze together with the other buffaloes from the village. Mating and calving occur at the pasture. Generally on the ground floor of each house there are barns to keep the buffaloes in winter. The barns have no windows and the doors are tightly closed. Young animals are never taken outdoors in winter in the cold climate. Buffaloes are slaughtered together with cattle. Milking is done by hand except at the two existing research stations. Average slaughter weight is 300–350 kg, at the age of 18–20 months. Carcass yield is 53–55%. Overall growth rate is 400 g/day.

Performance

Lactation duration 220–270 days; milk yield 700–1000 kg; milk fat 6.6–8.1%; milk protein 4.2–4.6%; products a semi-hard cheese called "peyaz peyneri" is made from buffalo milk. Ayran is a drink with water and buffalo yoghurt. Buffaloes are raised for milk production only as source of income that does not require any expenditure, *i.e.* in the areas that have natural feeding conditions. The price of buffalo milk is only slightly higher than the price for cows' milk. Meat production is all converted into sausages. The price of buffalo meat is 10% less than the price for beef (Sekerden *et al*., 1996a, b; Sekerden *et al.,* 2000, Borghese, 2005).

5.5.2 Italy

5.5.2.1 *Italian*

The major concentrations of Italian buffaloes are in the region of Campania, particularly in Caserta and Salerno provinces. Apart from this, Italian buffaloes are found in the provinces of Frosinone and Latina in the region of Lazio and the province of Foggia in the region of Puglia. The animals are short-limbed and broad in proportion to their length. They are suitable for both meat and milk production. The average milk yield ranges from 1400 to 3000 kg. The adult body weight varies from 450–600 kg.

5.5.3 Bulgaria

5.5.3.1 *Bulgarian Murrah*

From 1962 to 1990, Murrah buffaloes from India were imported into Bulgaria and a new population of buffalo was created by upgrading the local buffalo. The buffalo population in Bulgaria has dramatically declined since the 2nd World War, with the advent of Holstein and mechanization. Furthermore, after 1989, privatization forced the cooperative buffalo farms to close down. The private sector is composed of small units which has made selection and recording more difficult.

Description

Black or black and brown or dark grey in colour. Body weight of adult male is 700 kg. Body weight of adult female is 600 kg.

Distribution

All over Bulgaria, also found in Romania and South America.

Husbandry

Buffaloes are traditionally managed under domestic conditions together with the calf. They are hand-milked twice a day. Some milking machines are now available. During winter, they are kept in sheds and are fed different kinds of roughages: barley and wheat straw, cornstalks. In addition, they are given concentrate mixtures, sometimes mixed with beet pulp. During the summer, they graze all day long in the marshy areas and in the evening they return to their sheds. They are mated mainly through natural mating. Some villages provide artificial insemination. In State Buffalo farms (200–400 buffaloes). They are managed according to their condition: heifers, lactating, pregnant, dry. Milking buffaloes are kept in closed sheds and tied up. During winter, they are allowed outside in paddocks for part of the day, in summer they are allowed to graze. They are always given concentrate mixture in addition to roughage. AI is used on all buffaloes. Average slaughter weight is 400 kg at the age of 16 months. Carcass yield is 50.4%. Overall growth rate is 750 g/day.

Performance and performance traits

Lactation duration 270–305 days; milk yield 1800 kg; milk fat 7.5%; protein (%) 4.5; products yoghurt and milk by-products. Processed meat products are very important: all kinds of salami and sausages, Pastarma, lukanska and flat sausages. (Peeva *et al* 1991; Peeva, 1996; Alexiev, 1998).

5.5.4 Mediterranean

The Mediterranean buffalo originate from the Indian buffalo. It was introduced into Europe with the advent of Islam and the Arab occupation as well as through other central European conquerors in the 6th and 7th centuries. The buffalo population in Europe has been dramatically declining since the second world war, with the advent of Holstein and mechanization.

Description

The animal has black, black and brown, dark grey coat. Horns are flat at the bottom, backwards and slightly outwards pointed, and backwards straightened; the top is pointed inwards. They have a compact conformation with a deep and wide chest as well as a developed pectoral. The back is short. The rump is short. The udder is of medium size with squarely placed quarters and halves; the teats are cylindrical. Where machine milking is popular (only in Italy) udders are more regular and better shaped. Size, weight and productivity vary a lot according to the environment and management. Average herd size is below five breedable buffaloes in most countries, except in Italy where it is 90. The proportion of breedable females to total buffaloes is about 45% except in Italy where it is 62%, since males have little market potential. The body weight of the adult female is 450–650 kg.

Distribution

Italy, Romania, Greece, Albania and Republic of Macedonia, Bulgaria, Switzerland and Hungary.

Husbandry

The most common housing system is the one referred to as traditional, consisting of keeping buffaloes indoors at night and confined in fenced areas during the day. In the favourable season they are allowed to graze during the day. In Italy, they are housed loose in paddocks all year long, with the same modern systems used for dairy cows. One-third of Italian buffaloes are also put out on pasture in the favourable seasons, or green forage "cut-and-carry" such as alfa-alfa can also be used. Maize silage, concentrates and by-products are the basic foodstuffs in Italy. Performance varies very much depending on the area. There is no common practise to wean buffalo calves. When milking is done by hand, both male and female calves suckle from the dam. In some cases they suckle from a dairy cow. This results in a wide difference in daily gain up to weaning, as well as weaning weight and age. Males are now in greater demand as meat producers, therefore increased attention is being paid to their feeding and health. Average daily milk yield reveals a huge variability, mainly depending on the feeding system. It can range from 3 to 4 kg milk/day for poorly fed animals to 15 kg/day in intensive management systems. In Bulgaria, Romania, Republic of Macedonia, Greece and Albania, extensive management systems are employed. Average slaughter weight is 250–400 kg, at the age of 12–15 months.

Performance

Lactation duration 270 days; milk yield 900–4000 kg; milk fat 8.0%; milk protein 4.2–4.6%.

5.6 Egypt

5.6.1 Egyptian *(plate 7)*

Buffaloes were introduced into Egypt from India, Iran and Iraq approximately during the middle of the 7th Century. The distinction between the different types of Egyptian buffaloes is only environmental. It is the most important and popular livestock for milk production in Egypt.

Description

The animal is blackish grey, horn form varies from lyre to sword-shaped. The head is long and narrow, the jaws are long and strong. Ears are long and drooping. The neck is rather long, thin and straight. The forelegs are rather short and heavy boned. Ribs are wide, deep and well sprung. The rump is sloping and the tail setting is low. Height at withers of adult male is 178 cm, body weight is 600 kg. Height at withers of adult female is 144 cm, body weight is 500 kg.

Distribution

All over the country, mainly in peri-urban areas and the Nile delta.

Husbandry

The farmer keeps manure in a solid state inside the animal enclosure. The solid manure is taken twice a year and spread in the fields before planting. The animals are slaughtered only in slaughterhouses, following the Islamic practise of cutting the jugular vein. Milking is done by hand, twice a day, mainly by women. Average slaughter weight is 500 kg, at the age of 18–24 months. Carcass yield is 51%. Overall growth rate is 700 g/day.

Performance

Lactation duration 210–280 days; milk yield 1200–2100 kg; milk fat 6.5–7.0%; products: the following cheeses are produced with the addition of cow milk: Domiati, Karish, Mish, Rahss (El Kirabi, 1995; Nigm, 1996; Ragab and Abdel Salam, 1963; Mokhtar, 1971; Askar *et al.,* 1973; Borghese, 2005).

5.7 Latin American Breeds

5.7.1 Brazil

Brazil is emerging as a major buffalo nation in Latin America with the largest concentration (1.1 million) of buffaloes. There are other ten countries with sizable buffalo populations among Latin American countries. Most of the animals are progeny of the stock imported from Indo-China in 1890; of Murrah, Jaffarabadi, Nagpuri, Surti, Nili-Ravi from Indian sub-continent during 1907–1962; and of Italian buffaloes from Italy. Most of these are used for milk and meat, while carabaos are used for work and meat. These breeds have undergone a period of mixing. At the present time, they are classified as:

5.7.1.1 *Brazil Murrah* (*plate 7*)

Brazil Murrah derived from imports of Murrah breed of India.

5.7.1.2 *Latin Jaffarabadi*

These animals are essentially of Jaffarabadi breed imported from India and are presently non distinguishable from original imported animals. In addition there are crosses of a large number of breeds imported from Indian subcontinent and from Italy which have been interbred and are used for meat and milk.

5.7.1.3 *Mediterrano* (*plate 7*)

Mediterrano derived from imported Italian buffaloes and some Murrah admixture, currently bred within the stock as purebred. These three breeds are bred for milk and its products, and for meat.

5.7.1.4 *Swamp buffaloes* (*plate 7*)

Swamp buffaloes imported from Indo-China are now raised as a distinct breed group and has been given the name Latin carabao.

5.7.1.5 *Other breeds*

Besides this, imports of Nagpuri, Surti, Nili-Ravi have also been made and these are raised both as pure breds or as crossbreds. Table 4.3 gives country-wise distribution of various breeds and their utilities.

Table 5.3 Distribution of Various Breeds of Buffaloes in Latin America, their Utilities and Countries of Origin

Country	Breeds	Date of introduction	Type of management and production	Country of introduction and breed
Brazil	Murrah	1890/1895	Extensive for meat	Carabao *via* French Guyana
	Jaffarabadi	1907	with some feedlot	from Indonesia; Murrah,
	Mediterraneo	1962	units. For dairy	Jaffarabadi from India; and
	Carabao		production intensive management is practised.	buffaloes from Italy (1920)
Venezuela	Mediterraneo	1920	Milk and meat Bulgaria, and India	Trinidad, Italy,
		1940/67/72	Extensive, stall feeding for milk/ cheese cincho	
Trinidad	Murrah	1900	Work, meat	India
	Jaffarabadi	1924		
	Nagpuri	1938		
	Surti	1949		
	Bhadawari			
	Nili-Ravi			
	Mixture of these			
Peru	Mediterraneo		Dairy/meat	Brazil
	crossbreds	1966		
Paraguay	Mediterraneo	1974	Stall fed for milk	Brazil
	crossbreds			
	Murrah			
Argentina	Brazilian		Milk production	Rumania, Italy, Brazil and India
	Mediterraneo	1910		
	Murrah	1980		
	Jaffarabadi	1983		
Ecuador	–	1986		–
Colombia	–	1967	Meat, milk and work	Trinidad
Surinam	–	1890/95	Work and meat	French, Guyana, India
Cuba	–	1983, 1986	Meat/milk	Trinidad

Source: Bhat (1997).

5.8 Swamp Buffalo

Cockrill (1967) prefers to group all the Swamp buffalo types together as one breed, as no distinct breed has so far been evolved. Like the river-type nondescript buffaloes there are also many variants to be found in swamp buffaloes. The large swamp buffalo of Thailand may weigh well over 900 kg while the carabao of the Philippines or the small water buffalo of Borneo may weigh only 370 kg or even less.

Distribution

These are distributed from the state of Assam of India to Yangtze valley in China. The major concentrations are mainly in south-east China, Myanmar, Malaysia, Laos, Cambodia, Thailand, Indonesia, the Philippines and Vietnam.

Description

The skin is grey to slate blue, generally dark grey; white animals occur frequently. Black spots may be present on the skin of white animals. Eyes are not pink. The animals are not true albinos. They are largely having swept back horns and are similar in general appearance across the countries except the size. The horns grow laterally and horizontally in young animals and curve round in a semi-circle as the animal gets older. They are massively built, heavy bodied and stockily built; the body is short and the belly large. The forehead is flat, orbits are prominent with a short face and wide muzzle.

They weigh 300 to 500 kg when fully grown. Attempts to classify the swamp buffalo into breeds has not been very rewarding because of large diversity in conformation and colour. However there are some types that are recognized as being distinct. The differences are mostly in horn size, body size and colour. Local strains have been described in Thailand and Laos. White animals occur with a frequency of 1% in the Philippines and 5% in Malaysia. It is primarily used as a work animal in paddy cultivation. It is also used for pulling carts and for hauling timber in jungles. Milk yield of swamp buffaloes is very low (150–200 kg) in a lactation period of 150–200 days.

A characteristic of the Swamp buffalo, which it shares with only one other breed, the Surti, is the presence of two light grey chevrons, one just below the jaw and other on the chest above the brisket.

Plate - I, Breeds of Buffaloes

♂ = Male, ♀ = Female

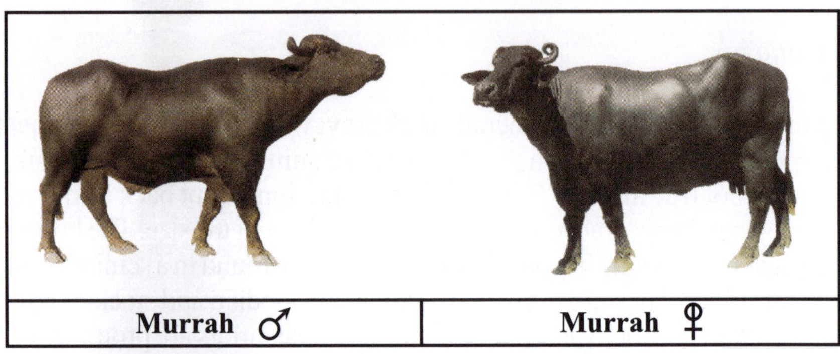

Murrah ♂	Murrah ♀

Nili-Ravi ♂	Nili-Ravi ♀

Surti ♂	Surti ♀

Plate - 2, Breeds of Buffaloes

♂ = Male, ♀ = Female

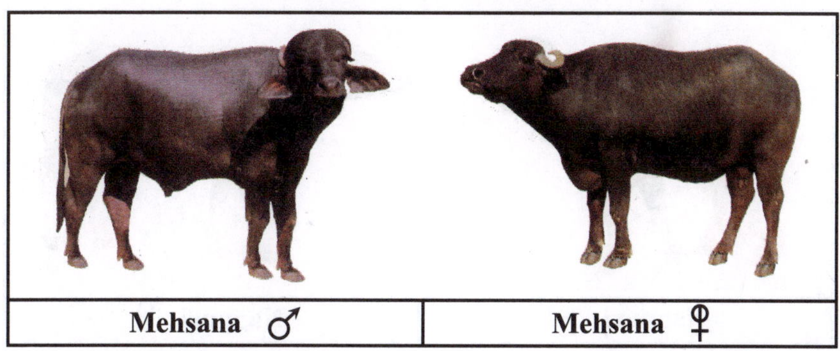

Mehsana ♂	Mehsana ♀

Jaffarabadi ♂	Jaffarabadi ♀

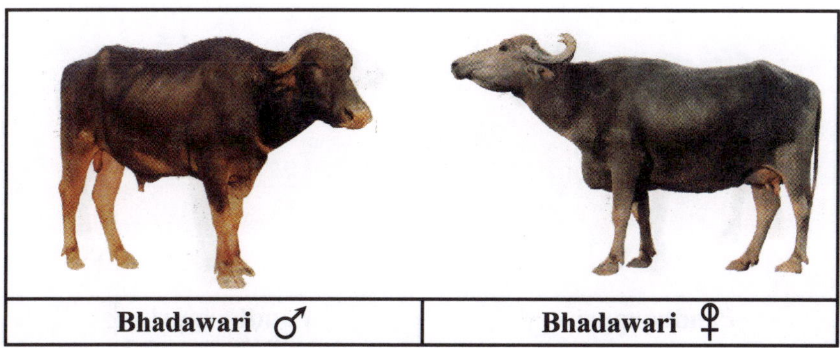

Bhadawari ♂	Bhadawari ♀

Plate - 3, Breeds of Buffaloes

♂ = Male, ♀ = Female

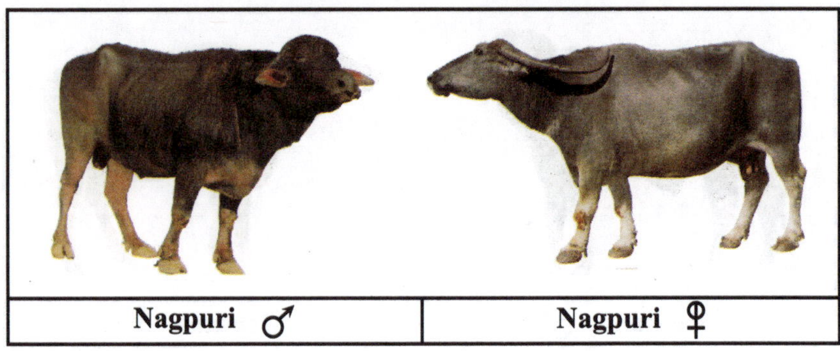

Nagpuri ♂	Nagpuri ♀

Toda ♂	Toda ♀

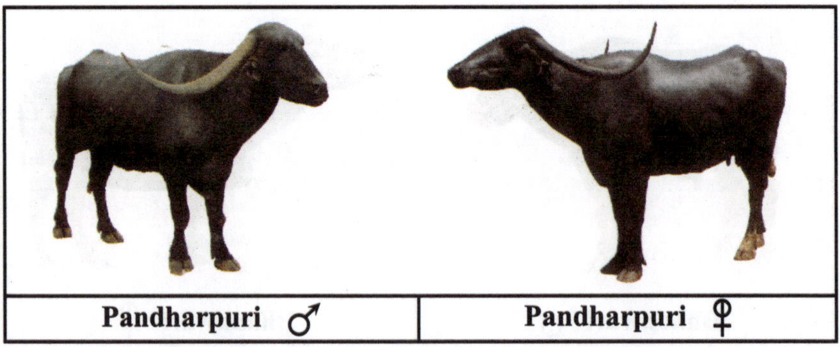

Pandharpuri ♂	Pandharpuri ♀

Plate - 4, Breeds of Buffaloes

♂ = Male, ♀ = Female

Marathwada ♂	Marathwada ♀

Sambalpuri ♂	Sambalpuri ♀

Tarai ♂	Tarai ♀

Plate - 5, Breeds of Buffaloes

♂ = Male, ♀ = Female

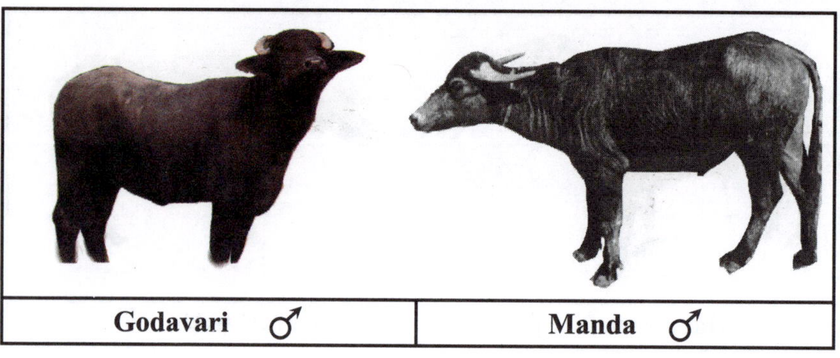

Godavari ♂	Manda ♂

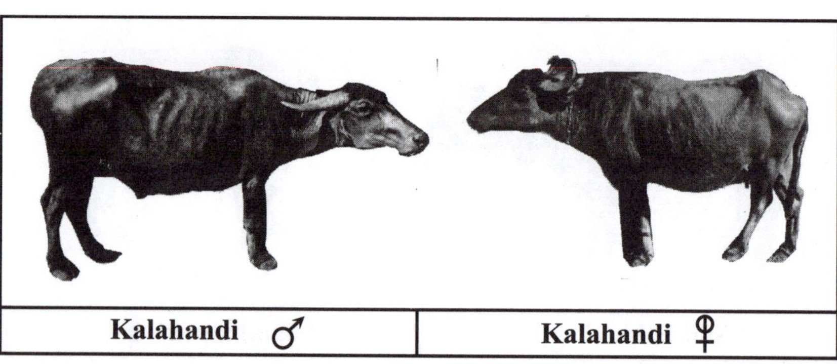

Kalahandi ♂	Kalahandi ♀

Kundi (Pakistan) ♂	Kundi (Pakistan) ♀

Plate - 6, Breeds of Buffaloes

♂ = Male, ♀ = Female

Nili-Ravi (Pakistan) ♂	Nili-Ravi (Pakistan) 

Iranian	Iraqi

Turkish Anatolian

Plate - 7, Breeds of Buffaloes

♂ = Male, ♀ = Female

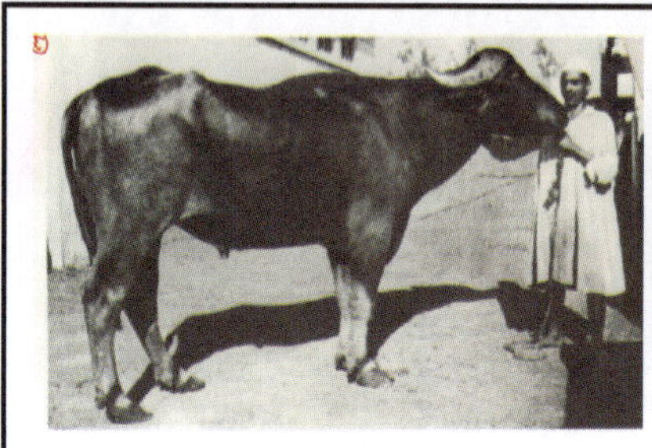

Egyptian ♂	Egyptian ♀

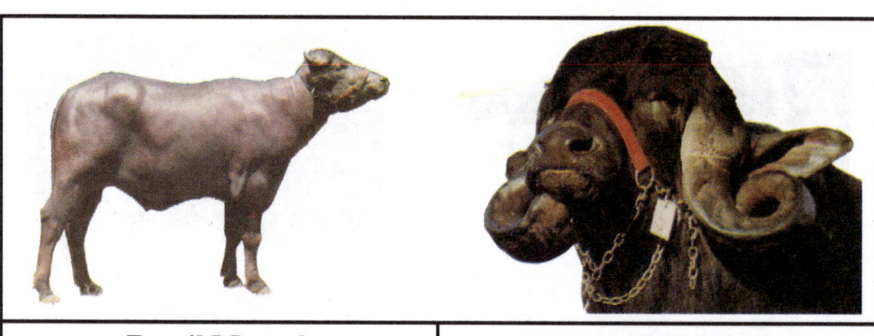

Brazil Murrah	Latin Jaffarwadi

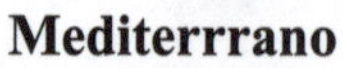

Mediterrrano	Swamp Buffalo

Plate - 8, Breeds of Buffaloes

Seal from Mohanjo-Daro, showing Male Swamp Buffalo.

Buffaloes Wallowing in Water

CHAPTER 6

GENETICS

The buffalo occupies an important place amongst domesticated animals of the tropical countries as provider of dairy produce, meat and draught power. Since this animal plays a substantial role in the economy of the country, large scale development programmes have been initiated to improve the buffalo.

6.0 Basic Genetics

6.1 Selection Characteristics

6.1.1. Qualitative characteristics

6.1.1.1 *Blood groups*

The research on blood groups and biochemical polymorphism has been used for improving their productivity through proper identification in progeny tests. The blood group and protein variants have their use in the study and development of breeds of livestock, their origin and evolutionary history.

Work on blood groups in buffaloes was initiated by Singh (1942) at the Indian Veterinary Research Institute (IVRI) in India. He demonstrated four blood groups using naturally occurring antibodies. The first among these was tentatively designated as J-factor. Chet Ram *et al.* (1964) showed that this factor had overlapping specification with J-factor of cattle. It was present in red cells and serum, serum alone or was completely absent. Its inheritance was controlled by three alleles.

Khanna (1973) studied blood groups in Bhadawari, Marathwara, Murrah, Nagpuri, Nili, Pandharpuri and Surti breeds by using 22 blood typing reagents.

Breed differences in frequencies of these factors. 14 factors were allocated to 7 blood group systems; the other factors were designated as belonging to independent systems.

The presence of a factor was dominant over its absence except in J system. In this system 3 alleles in the descending order of dominance were involved. The results obtained by Khanna *et al.* (1968) on blood groups are summarized in Table 6.1.

Table 6.1 Blood Group Systems and Factors

Blood group system	Antigenic factor	No. of alleles recognized
A	AC	5
B	BDY	8
F	FG	4
N	NV	3
O	OW	4
Q	OX	4
J	J	3
L, P, R, S, T, Z, A' and BL	P, R, S, T, Z, A' and B'2 each	

Source: Khanna *et al.* (1968).

Iannelli *et al.* (1973) identified, two paternal half-sibs in water buffalo which, when used as recipients, produced antibodies of the same specificity as those stimulated when used as donors.They further concluded that the apparent serological paradox was presumably due to an antigenic determinant located beneath the surface of erythrocytes.

Amano (1978) found six blood antigenic factors in swamp buffaloes in Indonesia. These are: Wh_1, Wh_2, Wh_3, Wh_4, Wh_5 and WbJ at frequencies of 1.5,13.0, 23.2, 21.7, 0.0 and 15.9% respectively. He also detected some of these blood groups in crossbred (swamp × river) buffaloes, but none were detected in the river or Indian buffaloes. This evidence suggests that swamp and river buffaloes are immunogenetically different.

6.1.1.2 *Biochemical polymorphism*

(a) *Haemoglobin*

In buffaloes, two haemoglobin components were reported by Giri and Pillai (1956). However, Khanna and Braend (1968) recorded two rare phenotypic variants. One variant exhibited three bands.

The structural difference of haemoglobin of adult buffaloes was studied by Balani and Barnabas (1965). The two haemoglobin components designated as Hb1 and Hb2 were found to differ from each other in the alpha chains. These were called α_2I $_2$ and $_2$II $_2$. Researchers identified both qualitative and quantitative variations of the Hb phenotype. For this, 398 river buffalo samples, were randomly collected in distinct breeding areas of the Campania region. High resolution analytical systems were used on polyacrylamide gel isoelectric focusing and HPLC were used to determine the ratio between HBA1 and HBA2 globin chains; restriction endonuclease analysis was performed to assess whether quantitative variations in Hb bands were related to an unusual number of -globin genes. In the two buffalo subpopulations, allele frequencies of the and -globin systems were calculated, and F statistics (FIS, FIT and FST) were estimated as parameters of genetic diversity. The results suggested that: (i) as shown by RFLP analysis, only a couple of associated -globin genes account for the quantitative variations recorded at the phenotypic level; (ii) as expected, in the -globin gene system (HBA), the frequency of haplotype B (HBA-B) largely exceeded that of haplotype A (HBA-A) (95.1% vs 4.9%); (iii) the frequency of the usual allele at the -locus is 0.6, as opposed to 0.4 of the slow variant; the most significant component of variation of the genetic system of hemoglobin is between individuals within the same locus.

(b) *Transferrin*

Three transferrin phenotypes controlled by 2 co-dominant alleles were reported in buffaloes by Khanna (1969), Naik *et al.* (1969) and Basavaiah (1970). Khanna (1973) reported four transferrin types, *viz.* TfDD, TfKK, TfDN and TfKN, controlled by the three co-dominant alleles, *viz.* TfD, TfK and TfN, in order of decreasing anodic mobilities.

Each allele controlled expression of three bands at 7.6 pH. TfK was the most frequent allele (gene frequency ranged from 0.74 to 0.95). TfD and TfN were present in low frequencies. TfN was confined to Nili and Surti breeds only.

(c) *Potassium types*

George and Balakrishnan (1980) observed 2 distinct potassium types, *viz.* low (LK) and high (HK), in buffaloes. In LK type animals the potassium value was low (25 to 52 meq/litre of red cells) and the sodium was high (55 to 90 meq/litre of red cells). In HK type, the potassium value was high (65 to 95 meq/litre) and sodium low (10 to 45 meq/litre). In both the types, the total concentration of electrolytes (Na+K) remained almost constant (100 to 115 meq/litre of red cells).

Preliminary analysis of data revealed that the levels of potassium and sodium in RBC were controlled by a single Mendelian gene with the LK allele being dominant over HK. George (1981) reported that criss cross immunization of red cells from LK and HK buffaloes produced sera with titre ranging between $^1/_4$ and $^1/_{32}$ in LK animals, whereas HK buffaloes did not produce sera with any antibody activity. To produce monovalent reagent specific to antigens of HK types, antisera raised in LK animals were repeatedly absorbed with 4 different panel LK red cells of 20 animals each. All possible combinations of absorptions did not give a unitary reagent specific to HK cells. It was suspected that some of the panel animals might be heterozygous for LK type, which might be responsible for removing the HK specific antibody as well. The absorption studies, however, indicated that there might be multiple antigens for the HK type.

(d) *Albumin*

Khanna and Braend (1968) and Juneja and Chaudhary (1971) reported three phenotypes controlled by 2 co-dominant alleles Alb^F and Alb^S. The frequency of AlbF was quite low (0.05 to 0.29) than that of Alb^S.

(e) *Amylase*

Khanna (1973) described the existence of amylase isozymes, *viz.* Am^C, Am^A and Am^B, in order of decreasing anodic mobility. Each isozyme was characterized by one band. In the presence of calcium chloride ions, an additional slower zone of amylase activity exhibiting one band was also observed.

(f) *Carbonic anhydrase*

Khanna and Tandon (1978) described Ca^C, Ca^F and Ca^S isozymes in the order of decreasing mobility in Murrah breed. The gene frequency of Ca^F was highest followed by that of Ca^S and Ca^C.

(g) *Ceruloplasmin*

This protein is monomorphic (Khanna, 1973). Amano (1978) examined 22 electrophoretic loci and found that only 6 loci showed polymorphism Alb (albumin), Tf (transferrin), AmI (serum amylase I), Hb (haemoglobin), CA (carbonic anhydrase) and PepB (peptidase B). From the analysis of genetic distances, they found distinct biochemical genetic differences between swamp and river buffaloes. For example, in swamp buffaloes at the Alb locus they reported Alb^A, Alb^B and Alb^X alleles, but only the Alb^A and Alb^X alleles were detected (with frequencies of

0.57 and 0.43 respectively), Alb^B allele was not detected. In river buffaloes the respective frequencies were 0.55 for AlbB, 0.05 for Alb^X and 0.40 for Alb^A.

6.1.1.3 *Physical traits*

(a) *Colour*

As far as colour markings are concerned, a white chevron (sometimes two) below the neck of the grey buffaloes is commonly found. White stockings on the feet of buffaloes can be found, but this trait is not common. Traditionally, Thai farmers regard the animal with white stockings on four feet as good luck. No study has been conducted on the inheritance of these traits.

(b) *Horns*

In general, swamp buffaloes have much bigger and longer horns than riverine. However, a few polled buffaloes have been reported in Thailand. Variation in horn size and setting is often observed. Most of the swamp buffaloes have horns, which extend outward and then curl backward into semi-circles, but remain on the same plane as the forehead. A few buffaloes have loose horns or drooping horns. Buffalo horns are usually long, flat and thick; however, in some cases they may be short and thick. The nature of the inheritance of buffalo horns is not known at present.

(c) *Hair whorls*

Hair whorls are commonly found on various parts of the buffalo's body. They usually appear on both sides of the shoulder, both sides of the hip, the face and the forehead. The number of hair whorls and the combination of their positions seem to be unique to each animal. In Thailand, hair whorls have been used as a means of identifying animals for registration. Exact knowledge on the inheritance of this trait in buffaloes is not available.

6.1.1.4 *Studies on cytogenetics*

Cytogenetic reports on buffalo are very few. Dutt and Bhattacharya (1952) were the first to report diploid number of chromosomes as 48 from squash preparations. Fischer (1974) reported the diploid number to be 50, using cell-culture technique. Subsequent studies at IVRI (Chakrabarty, 1977; Gupta and Raychaudhuri, 1978) confirmed the findings of Fischer (1974). Chakrabarty (1977) established that there were 24 pairs of autosomes comprising five pairs of submetacentric and 19 pairs of acrocentric chromosomes. X chromosome was the longest acrocentric chromosome. Y was acrocentric but not morphologically distinct from the small pairs of autosomes. The longest autosome in their study was 6.92 ± 0.35% and the smallest 2.10 ± 0.11% of the genome. The X and Y chromosomes contributed about 6 and 2% of the total haploid set respectively. Balakrishnan *et al.* (1980) reported that Y chromosome was not the smallest acrocentric chromosome as is usually believed, and that its length is between 19th and 20th autosomes.

Goswami *et al.* (1980) reported that constitutive heterochromatin was located around centromere in all the 38 acrocentric autosomes. Very faint to no heterochromatin was noted in 5 pairs of metacentric chromosomes. Of the sex chromosome, Y exhibited a very distinct and large band at the centromeric region. An interesting feature of the X chromosome was that a band of intercalary heterochromatin was located near the centromeric region. The chromosome was, however, heterochromatic negative and was little darker than other small acrocentric autosomes.

Yadav and Balakrishnan (1982) showed that submetacentric chromosomes were C-band positive. X chromosome had a very large C band, besides an intercalary band. These scientists reported XX/XY chimeric X chromosome trisomy in a Murrah buffalo with hypoplastic ovary. Pericentric inversion of the fourth metacentric chromosome of a buffalo and secondary constriction of submetacentric autosome has also been reported. Yadav and Balakrishnan (1984) used two sets of heterosexual triplets to study sex chromosome chimerism in the buffalo. In set 1, the female and one of the two males were studied while in set 2 only the female was available. All the animals showed sex chromosome chimerism. There are no previous reports of sex chromosome chimerism in buffaloes. Yadav and Balkrishnan (1985) using one set of heterosexual twins and two co-sibs of a buffalo triplet showed that the degree of chimerism expressed as proportion of XY cells ranged from 42.1–59.3% in females (freemartins) and 38.7–59.0% in males. The mean frequency of XY cells was 50.7% in females and 48.9% in males.

Fischer (1974) reported the chromosome number in swamp buffalo to be 48 (5 metacentric and 18 acrocentric pairs and the acrocentric X and Y), while Murrah buffalo had one additional acrocentric pair of chromosome, making the diploid number to be 50. Investigations on Swamp × Murrah buffaloes with 50, 75, 87.5 and 93.75% Swamp buffalo blood, respectively, revealed that the diploid chromosome number of the first two groups was 49, whereas the second two groups had the same karyotype as that of purebred Swamp buffaloes. All the hybrids showed normal fertility.

Bongso *et al.* (1983) studied the meiotic chromosome behaviour in Murrah, Swamp and Murrah × Swamp hybrids aged between two and two and a half years. Chromosome sets ranging from 22 to 26 (most frequent, 24 and 25) with many cells carrying univalent, bivalent and multivalent configurations were observed in hybrids, whereas the meiotic cells in Murrah and Swamp showed chromosome sets exclusively of 25 and 24 (bivalents) respectively. Histological examination of the hybrid testis revealed a large proportion of degenerating spermatocytes and abnormal spermatid in the process of spermiogenesis suggesting that the various synaptic associations leading to unbalanced gametes may be responsible for the degenerating germ cells in the hybrids. The unbalanced meiotic products will probably lead to selection against such spermatozoa or early embryos after fertilization. Due to large percentage of germinal epithelial cells in F_1 hybrids being wasted, the fertility of backcross and F_2 generations would be subnormal. This, however, needs to be confirmed.

Chavananikul *et al.* (1994) investigated the karyotype by lymphocytic culture and Giemsa staining technique of 50% Murrah–50% swamp, 25% Murrah–75% swamp, 75% Murrah–25% Swamp and 87.5% Murrah–12.5% Swamp and reported that 50% Murrah crosses possessed 49 chromosomes, while 25% Murrah crosses had both 48 and 49 in the ratio of 1:1. The 75% and 87.5 Murrah had 2 types but different karyotypes of 49 and 50 chromosomes in the ratio of 1:1.

There is dearth of knowledge about the incidence of various chromosomal abnormalities in buffaloes. Yadav and Balakrishnan (1982) reported a trisomy-X in River buffaloes. The extra chromosome was a large acrocentric chromosome approximately the same length as the X-chromosome. It was assumed that animal must have a 51, XXX chromosome complement. Both the above animals had problems associated with fertility.

Yadav *et al.* (1990) reported a 49, XO chromosomal complement in buffaloes. Monosomy-X syndrome is unfavourable to viability and may also be a major cause of foetal losses in buffaloes.

In various mammalian species specific chromosomes are involved in nucleolus formation. The unique DNA templates are localized to specific sites on these chromosomes and are referred to as nucleolus organizer regions (NORs). These DNA regions code for ribsomal RNA production and the transcriptionally active regions can be revealed using silver staining. Parkash and Balakrishnan (1992) studied the localisation and frequency NORs in metaphase chromosomes of Murrah, Surti and Mehsana buffaloes. Six pairs autosomes (numbers 3, 4, 6, 21, 23 and

24) carried the sites of NORs at their termini in all the three breeds. The frequency of NOR(P) for chromosomes 3, 4, and 6 were different, lowest for chromosome 4 and highest for chromosome 6. The distribution of genotypes (heterozygous and homozygous) with respect to these three pairs of chromosomes was variable in three breeds. The observed genotypic frequencies for chromosomes 3 and 4 differed significantly from that of chromosome in Murrah and the pooled population. The possibility of use of NORs in the detection of origin of chromosome rearrangements and in studying genetic diversity was discussed. On another occassion it was studied that the locations of genes coding the synthesis of ribosomal RNA at different stages of mitotic cell division in Murrah breed. Activity of rDNA was highest at interphase which decreased at prophase, metaphase and anaphase. At metaphase silver staining was localised at the nucleolus organising regions of specific chromosomes only. Positive silver staining indicated the functioning and not merely the presence of rRNA cistrons.

Parkash and Balakrishnan (1993) traced the inheritance of NORs for chromosome pairs 3, 4 and 6. The patterns of NORs expected from different types of parental matings agreed with the observed patterns in the progress. The authors concluded that NORs followed mendelian inheritance and suggested that these heritable chromosome loci could be used as genetic cellular markers.

Iannuzzi *et al.*, (2005) investigated forty phenotypic female river buffalos with reproductive problems and two phenotypic male co-twins of females, raised in the provinces of Caserta and Salerno (southern Italy) and subjected for cytogenetic investigation. Of the 42 animals studied, 10 freemartins (8 females and 2 males) were found with variable percentages of male and female blood cells, the majority however showing similar percentages of both. Of the eight females, six showed normal body conformation, vagina and clitoris, while two showed some male traits (tight pelvis). The two males were apparently normal with only a reduced sizeof one testicle in one animal. Clinical observations performed in the internal reproductive organs of the female carriers by both rectal palpation (5 females) and direct observation after mating (3 females) revealed serious damages varying from complete lack of internal sex ducts (closed vagina) to hypoplasia of Müllerian ducts and absence (or atrophy) of ovaries. All freemartin females were sterile. This is the first detailed description of freemartinism in river buffalo by combining clinical and cytogenetic analyses.

6.1.1.5 *Restriction fragment length polymorphism (RFLP)*

RFLP is a promising recombinant DNA technique for the study of cryptic variation at the nucleotide level. This technique allows study of genetic variation in terms of nucleotide sequences and permits quantification of variation in terms of nucleotide diversity. These variations, once established, can be correlated with different metric traits and used as selection criteria. This approach can also lead to identification of quantitative trait loci (QTL) which can be manipulated in breeding programmes.

(a) *Repetitive DNA sequences*

All eukaryotic genomes contain many DNA sequences that are repeated a few to millions of times, forming a class of highly repetitive DNAs called the satellite DNAs. These satellite DNAs are supposed to have a role in the determination of chromosomal architecture of cell nuclei (John and Miklos, 1979). They have some role in gene expression and regulation (Britten and Davidson, 1969). Eight types of satellite DNAs have been reported in cattle (Makaya *et al.*, 1978). These satellites share some of the sequences and hence indicate a common origin (Plucienniczak *et al.,* 1982). It has been shown that some of the satellite DNA sequences are conserved in different families which is suggestive of their evolutionary relationship. There are no reports available in buffalo regarding satellite DNAs, their organization and sequence homology. Bhat *et al.,* (1990) began a study of buffalo satellite DNAs to generate baseline data on this species. These studies will also throw some light on the evolutionary relationship between cattle and buffalo; they include the investigation of variation and organization in a buffalo population, with reference to highly repetitive DNA sequences, using restriction enzyme and Southern blot analyses. Restriction endonucleases Pst I, Bam HI, Hind III, Eco RI, Bgl II, Pvu II, Msp I and Taq I were included in the study (Table 6.2). A homologous satellite DNA Fragment (Buffalo Pst 1–0.7 kb fragment) was used to explore the organization of different satellites. The Southern hybridization was carried out at higher stringency as the probe used was a homologous sequence. These findings are suggestive of the presence of 3 or more satellites in the buffalo genome. Pst I, Hind III, Eco RI, Pvu II and Msp I restriction enzymes revealed 9, 10, 12, 7 and 8 DNA fragments, respectively. Some of the Southern blots revealed variable periodicities. Pst I showed 0.7 kb periodicity, Hind III digest revealed some doublets, EcoRI showed many sites in the buffalo genome. Pvu II revealed a multimeric series of 1.43 kb in most of the fragments. Msp I restriction enzyme revealed 2 series, one of 100 bp and another of 200 bp. Interrelationship between various fragments of different satellites could also be observed.

There was no variation in satellite DNA among individuals as well as between sexes with reference to Pst I, Bam HI, Hind III, Eco RI, Bgl II, Pvu II, Msp I and Taq I, though they included both hexacutters as well as tetracutters.

No site variation exists in buffalo satellite DNA with reference to these enzymes (Kumar *et al.*, 1991).

Table 6.2 Molecular Sizes (kb) of the Repetitive DNA Bands by Different Restriction Enzyme Digestions as Revealed by Ethidium Bromide Staining

Pst I	Bam HI	Hind III	Eco RI	Bgl II	Pvu II	MspI	Taq I
–	–	–	–	–	–	–	3.94
–	–	–	–	3.55	–	–	–
–	–	3.35	–	3.30	–	–	3.40
–	–	–	–	–	–	–	3.16
–	2.98	2.98	–	3.00	–	–	2.90
–	2.63	–	2.37	–	2.48	–	2.50
2.09	–	–	–	–	2.19	–	–
–	–	–	–	1.90	1.97	–	–
–	–	–	1.58	1.70	1.68	–	–
1.32	–	–	1.42	–	–	–	1.35
1.20	1.27	1.27	–	–	–	–	–
–	–	1.19	–	–	–	–	–
–	–	–	–	–	–	1.01	–
–	–	0.88	–	–	–	–	–
0.70	–	0.84	0.79	–	–	0.78	–
–	–	–	–	–	–	0.69	–
0.50	0.56	–	–	–	–	0.57	0.58
–	–	–	–	–	–	–	0.52
–	–	–	–	–	–	0.45	0.45
–	–	–	–	–	–	0.42	–
(5)	(4)	(6)	(4)	(5)	(4)	(6)	(9)

Figures in parentheses are total numbers of bands.
Source: Bhat, P. N. Personal Communications (1991).

Buffalo DNA reveals the same periodicities with Pst I restriction enzyme using homologous Pst I-0.7 kb fragments as was shown with cattle Eco RI-1.4 kb fragment. It can be concluded that buffalo Pst I-0.7 kb fragment has high sequence homology with cattle Eco RI-1.4 kb fragment. The differences in intensities of different bands suggest loss of repetitive DNA in the process of formation of metacentric chromosomes by centric fusion. With this background, further studies will certainly help to unravel the molecular nature of the evolutionary process at the divergence level of cattle and buffalo.

(b) *Unique DNA sequences*

The use of cloned genes coding for mRNA and proteins as probes increases the likelihood of detecting an RFLP having a direct effect on an economic trait. Once a direct effect on a trait of economic importance has been assigned to a particular RFLP allele or haplotype, a few generations of selection can bring this allele to a higher frequency in a population. Although it is difficult to identify the direct effects, once identified, they offer potential for selection for improved economic traits.

The genetic variability in unique DNA sequences (structural genes) can be studied with the help of molecular biology techniques. Tools are now available to identify these genetic markers for their future manipulation in breeding plans.

Structural genes code for mRNA and proteins and are of two different types consensus sequences or variable sequences across different species as well as within species. Hence, representatives of each of these types the thyroglobulin gene (a sequence supposed to be more conserved) and the major histocompatibility complex (a less conserved gene complex having both conserved and highly variable sequences) were studied.

(c) *Thyroglobulin locus (Tg locus)*

Thyroglobulin is a large dimeric protein which is the biosynthetic precursor of thyroid hormone (Vassart *et al.,* 1983). Thyroid hormone is essential for normal growth and neurological development in mammals (Mercken *et al.,* 1985). In cattle, the thyroglobulin locus covers an area of approximately 200 kb, which is split into 35 exons (Georges *et al.,* 1987) with an exonintron ratio of $^1/_{20}:^1/_{25}$. The mRNA of thyroglobulin gene is of 8400 bp size. Because of the high exon-intron ratio, this is an ideal locus for finding RFLP markers as one cDNA probe can screen a large area of the genome. Another reason for exploring this locus is the synteny already established between Tg locus and oncogene c-myc (Crews *et al.,* 1982). The conservation of this linkage in diverse species like mouse and man can allow one to speculate about this linkage group in buffalo.

Thyroglobulin cDNA clones of cattle origin as well as human cDNA clones of the major histocompatibility complex (class II region) were used for probing the buffalo genomic DNA.

An analysis of buffalo thyroglobulin gene (Tg) showed that all the 4 cDNA clones (of cattle Tg gene) successfully hybridized with buffalo genomic DNA; hence, it can be concluded that there exists a high sequence homology between cattle and buffalo genomes with reference to Tg locus. This gene was further studied region wise; the 5' region and the 3' end. Comparison of nucleotide diversities in the two species revealed that buffaloes are 13 times more variable than cattle with reference to the Tg locus. This conclusion was drawn on the basis of the hybridisation patterns and nucleotide diversity. The right region with a nucleotide diversity of 0.0096 was more variable than the left region, which has a diversitv of 0.0047. Msp I restriction enzyme did not reveal any polymorphism in either region. Probably Msp I sites are conserved in the buffalo Tg gene and hence not useful for studying the buffalo Tg locus.

Eco RI revealed a polymorphic pattern with the right region only, while Hind III resolved polymorphic patterns with both the probes.

(d) *Histocompatibility locus (MHC locus)*

The major histocompatibility locus has been associated with different diseases. The cattle MHC system known as BoLA (bovine lymphocyte antigen) locus is reported to be associated with many diseases such as mastitis (Solbu *et al.*, 1982), tick infestation (Stear *et al.*, 1984) and occular squamous cell carcinoma (Caldwell and Cumberland, 1978). This association has also been reported in many other species. The goat MHC segregates with viral arthritis (Ruff and Lazary, 1985), whereas the horse MHC system has association with sarcoid tumours (Meredith *et al.*, 1986) and sweet itch (Lazary *et al.*, 1985). The MHC system also affects reproduction and fertility (MacCluer *et al.*, 1988). The 'ped' gene detected in mice and pigs controls the rate of development (Warner, 1986) and is either within MHC or linked to it.

Using the human MHC system the nucleotide variation in the different classes of genomic DNA in buffaloes was studied. It has been established that the MHC homologous region is present in buffaloes and has been named BuLA (for buffalo lymphocyte antigen). The subregions which hybridized with DQb and DRa were named the DQb and DRa subregions of BuLA. The DQb subregion of buffaloes was analysed with Pst I, Bam HI, Hind III, Eco RI, Pvu II and Msp I (Table 6.3). All the enzymes tested gave complicated hybridization patterns and all individuals differed from each other with regard to the band pattern, whereas the DRa subregion revealed mostly a nonpolymorphic pattern except with Msp I. Nucleotide diversity studied at the DQb subregion revealed that the nucleotide site varied once in 60 bp. MHC of buffalo showed a diversity of 0.0148, indicating that variation at nucleotide site exists every 68 bp. In BuLA, DQb is more polymorphic than DRa (Kumar *et al.*, 1993).

Buffaloes have a highly polymorphic DQ subregion in the class II region of the MHC. The DR subregion was mostly monomorphic with the enzymes tested except Msp I. Beta chain genes are much more polymorphic than alpha chain genes in buffaloes.

Table 6.3 Bands Detected by Hybridization with DQ β after Digestion of Buffalo Genomic DNA with Various Restriction Endonucleases

Restriction endonucleases	No. of polymorphic bands	Molecular sizes (kb)	No. of non-polymorphic bands	Molecular sizes (kb)
Pst I	18	35.1, 8.5, 6.5, 5.7, 4.5, 4.3, 4.1, 3.5, 3.4, 3.2, 2.9, 2.5, 2.1, 1.8, 1.6, 1.4, 1.3	3	1.24, 0.85, 0.61
Bam HI	2	2.82, 2.70.	5	17.78, 7.33, 6.38, 5.19, 2.10
Hind III	21	25.1, 21.4, 13.8, 9.1, 7.8, 7.1, 6.2, 6.0, 5.6, 5.5, 5.4, 5.1, 5.0, 4.3, 2.7 2.6, 2.4, 2.2, 1.8, 1.6, 1.5	1	3.47
Eco RI	16	15.3, 11.7, 10.0, 8.7, 8.4, 8.1, 6.8, 6.4, 6.2, 6.1, 5.6, 5.2, 4.7, 4.5, 4.1, 3.8		
Pvu II	10	7.4, 6.1, 4.4, 3.9, 3.8, 3.5, 2.9, 2.7, 2.6, 1.8	1	5.37
Msp I	15	12.3, 8.9, 7.8, 5.1, 3.9, 3.6, 3.1, 2.8, 2.5, 2.1, 1.9, 1.7, 1.5, 0.8, 0.6	3	16.4, 6.3, 1.1

Buffaloes have a highly polymorphic DQ subregion in the class II region of the MHC. The DR subregion was mostly monomorphic with the enzymes tested except Msp I. Beta chain genes are much more polymorphic than alpha chain genes in buffaloes.

6.1.1.6 *Studies on polymorphism of Mitochondrial DNA–light on homology between species*

Mitochondrial DNA (mt DNA) is maternally inherited and changes in the nucleotide sequence occur faster than in nuclear DNA (Hutchison *et al.*, 1974; Brown, 1980). The analysis of mt DNA has yielded another source of information concerning genetic variability in organisms. Reports are available about mtDNA polymorphism based on restriction endonuclease analysis in domestic animals, *e.g.* cattle, pigs, domestic fowl (Laipis *et al.*, 1982; Watanabe *et al.*, 1985; Wakana *et al.*, 1986).

Bhat *et al.* (1990) reported the polymorphism of mitochondrial DNA (mtDNA) in buffalo.

Table 6.4 gives the number of cleavage sites for 13 restriction endonucleases and the molecular length of fragments. There was no site variation with respect to the 12 enzymes described except in Bgl I which clearly differentiated the mtDNA between individuals.

Table 6.4 Number of Cleavage Sites of Buffalo mtDNA for 13 Restriction Endonucleases and the Molecular Lengths of Fragments

Restriction	No. of endonuclease	Molecular lengths cleavage sites of fragments (kb)
Ava I	3	10.96, 3.47, 1.32
Ava II	3	11.22, 3.8, 1.3
Bgl I		
A	2	9.7, 6.6
B	3	8.6, 6.6, 1.1
Bgl II	3	11.2, 4.22, 0.8
Dra I	7	6.2, 3.7, 1.9, 1.9, 1.45, 0.55, 0.55
EcoRI	4	4.67, 4.30, 3.99, 1.95
Hind III	3	8.71, 5.01, 2.66
Hpa I	3	9.4, 5.13, 1.67
Kpn I	0	–
Pst I	1	16.4
Sal I	0	–
Xho I	2	13.1, 3.3
Bam HI	4	8.4, 3.9, 3.6, 0.5

Source: Bhat *et al.*, (1990).

Scarcely any of the restriction fragments of buffalo mtDNA matched those of cattle mtDNA, indicating minor homology between cattle and buffaloes at mitochondrial mtDNA level.

Pansin *et al.*, (2005) devised for detecting polymorphisms in the D-loop regions using the PCR followed by restriction enzyme digestion to reveal restriction fragment length polymorphism. Total DNA were extracted from the white blood cell of swamp buffaloes. Restriction endonuclease cleavage patterns of mitochondrial DNA (mt DNA) of local swamp buffalo from north east of Thailand were analyzed using three enzymes (Ava II, Bam HI and Hae III) which recognize 4 and 6 nucleotides. Among the 141 animals were analyzed, Ava II, BamH I and Hae III revealed different polymorphisms. Ava II showed two types of cleavage pattern. BamH I showed two types of cleavage pattern, and Hae III showed four types of cleavage pattern. The polymorphic fragments patterns for these three restriction endonuclease (Ava II-BamH I-Hae III) revealed eleven types of D-loop mtDNA in all the animals examined which were from three groups of three different locations. These preliminary results indicated that polymorphism in the D-loop region of swamp buffalo mtDNA can be revealed by PCR and restriction enzyme analysis. And these enzymes would be able to show the genetic difference among the swamp buffaloes.

6.1.1.7 *Milk protein polymorphism*

Milk protein polymorphism in farm animals has been reported in lactoglobulin, α-lactalbumin and casein. It was observed that polymorphism in κ casein protein, controlled by two alleles, the slower one with a very low frequency. There are two phenotypes of κ casein (Aschaffenberg, *et al.,* 1968).

κ–*Casein*

Pipalia (2001) conducted a study to determine polymorphism at κ-casein locus of buffalo. Caseins are milk proteins existing in several molecular forms (αs1, αs2, β and κ) with variant alleles of each. κ-casein variant B is reported to be favourable for milk quality and considered to be included in breeding strategies of dairy animals. PCR-RFLP is a fast and efficient method for discrimination between different κ-casein variants and is also useful for genotyping bulls. DNA samples from 4 different buffalo breeds (18 Jafarabadi, 29 Surti, 44 Mehsana and 5 Pandharpuri) were subjected to PCR amplification using bovine κ-casein primers and the PCR product of 379 bp was digested with Hind III and Hinf I. The PCR products from all the DNA samples showed only 2 bands 225 and 154 bp on Hind III digestion and only 288 and 91 bp fragments on Hinf I digestion. Thus, all the animals were genotyped as BB.

Whey protein gene polymorphism

The quality of milk is largely dependent on milk proteins. These proteins exhibit very high degree of polymorphism in cattle, and correlations have been described between allelic variants of these fragments with milk protein genes, milk yield, and quality of milk proteins. All major milk proteins have genetic variants transmitted through simple Mendelian inheritance without dominance. Generally, these variants have been detected by the techniques of gel electrophoresis, as the amino acid substitution or deletion has resulted in a change in the charge of the protein molecule. Of the six major milk proteins two whey proteins, *viz.* α-lactoglobulin (α-la) and β-lactoglobulin (β-lg) also have considerable influence on the quality and quantity of milk production.

β -*Lactoglobulin locus*

To date, no biological function for β -lg has been discovered. The amino acid composition is such that the protein is of high nutritional value, but the molecular properties, particularly the acid stability, lead to the supposition that some other more specific function for this protein exists.

Eight variants of the β -lg have so far been reported in cattle. The possibility of using milk protein genetic variant as biochemical markers to predict future performance of the animal has resulted in much work on their influence on various products variables. The effect of genetic variants on manufacturing properties of the milk are more consistent than those on milk yield and gross composition.

A significant effect of β -lg genotype on β -lg concentration in the milk (genotypes AA>AB>BB) has been consistently found. Data reported so far suggest that synthesis of β -lg decreases in the order β -lg A, B, C genotypes and hence, there is higher whey protein in milk of animals with β -lg A genotype. Studies have also shown that β -lg B milk genotypes have a higher casein concentration than β -lg A in milk. In general, it appears that as the β -lg B gene is substituted for the A gene, the concentration and proportion of β -lg and also total whey protein decreases while casein content increases, although total protein content remains constant. Several studies have demonstrated that β -lg B milk has a higher fat content than β -lg A milk. Cows with β -lg A genotype have higher protein yield in their milk than β -lg B genotype. Milk from animals with β -lg A coagulates faster than β -lg B. However, more firmer curd is formed with milk having β -lg B protein. Profitability of cheese making was increased by breeding for increased proportio of the B genetic variants of α -casein, α -casein and β -lg in the dairy cow population. An additional advantage is that κ casein B and β -lg B milk has greater heat s tability when concentrated and is, therefore, more suitable for manufacture of evaporated milk.

DNA samples of buffalo yielded 265 bp fragment encompassing the site for *Hae* III exhibited monomorphic pattern in all the buffaloes. The pattern was neither of A allele nor of B allele specific for cattle, thereby indicating that buffalo β -lg gene with respect to *Hae* III sites is different from cattle gene. Earlier reports on the basis of electrophoretic mobility, molecular weight, indistinguishable crystal form, sedimentation data and ultraviolet absorption properties, the amino acid composition and peptide pattern analysis had shown it to be identical to β -lg B allele of cattle. Restriction site of *Hae* III which differentiated A and B β -lg alleles of cattle is situated in exon-4. It is 92 bp down stream to the first nucleotide of 26 primer situated in the same exon. The second *Hae* III site in the 265 bp β -lg amplicon corresponds to 74 bp down stream to first *Hae* III site in the intron-4 as exon-4 is only of 111 bp, thereby yielding 166 and 99 bp fragments of A allele and 92, 77 and 99 bp fragments of B allele sequentially. It can be concluded that the cattle β -lg B protein is similar to β -lg of buffalo, and the restriction pattern of 265 bp amplified products of buffalo β -lg is probably similar to β -lg B of cattle as for exon-4 is considered but they differ with respect to intron-4 sequence. For cattle β -lg B the second *Hae* III site is 72 bp down stream from first *Hae* III restriction site, whereas in buffalo the location of this site is probably ~150 bp or ~30 bp away from the first *Hae* III restriction site situated in the intron-4. In order to confirm this conclusively, detailed sequence studies are warranted.

Amplification of α la gene fragments

Two primers corresponding to 5 region (intron 1) of the gene of cattle α -la were selected for amplification of 306 bp fragment. This fragment contains the only known mutation in the protein gene wherein codon CGG of arginine at position 10 of protein had mutated to CAG of glutamine in case of B allele, thereby generating a site for enzyme *MspI/HpaI* in B allele. On analysing the fragment size of the different amplified samples of α -la of cattle and buffalo no variation in fragment size was observed among the cattle and buffaloes.

MspI digest of the α -la 306 bp amplicon exhibited monomorphic pattern in all the animals screened. A characteristic cattle α -la B allele pattern with 220 and 86 bp fragments was observed in all the samples. The present observations of monomorphism of α -la gene of Indian buffalo (*Bubalus bubalis*) is in agreement with findings of Bhattacharya *et al.* (1963). This study clearly indicated that the buffalo α -la has restriction site for *Msp*I at the same site where it is for B allele of cattle, thereby yielding fragments pattern similar to α -la B of cattle. Although in electrophoretic mobility α -la of buffalo is similar to α -la A of cattle, at genic level it seems more probable that it is like α -la B variant of cattle.

The frequency criterion of α -la of cattle unequivocally designates α -la B as the original type of the allele in genus *Bos* and probably original to *Bubalus bubalis* too, in view of the fact that when amino acid residue at position 10 of α -la in cattle and buffalo are compared, arginine is at this position of α -la B of cattle and α -la of buffalo is evident from restriction pattern with Mspl (Fig. 6.1 and 6.2). Similar observation of arginine at position 10 of α laprotein was also made in Italian water buffalo. Further, if B allele is the original one then a single mutation in a protein coding region of gene will explain the evolution of buffalo allele from B variant, whereas for a variant two which are less probable, one corresponding to tenth position of protein and other replacing glycine for aspartic acid at a different site.The second mutation is not in exon-5 of the gene, because 100% sequence homology between exon-5 of α -la cattle and buffalo has been shown.

These data indicate that buffalo α -la is showing more sequence homology with α -la B allele of cattle rather than with A and probably it is the variant of α -la B allele which has descended from the common ancestor in antiquity.

Dayal *et al.,* (2005) reported that α -lactalbumin (α -la) is a major whey protein found in milk. Polymorphs of α -la gene are reported to be significantly associated with milk production and constituent traits. Therefore, the present study was undertaken to detect polymorphism in α -la at the genic level and to explore allelic variability at this locus. A total of 196 animals, belonging to four breeds of riverine buffalo *viz.* Bhadawari, Mehsana, Surti and Murrah were included in this investigation. Two fragments *i.e.* 133 bp (Exon 1) and 159 bp (Exon 2) of α -la gene were amplified by polymerase chain reaction and subsequently, single strand confirmation polymorphism (SSCP) study was carried out to identify different allelic pattern and genotypes of the animal included in the study. Both fragments of α -la gene were polymorphic in all the four breeds of riverine buffalo. Number of genotypes and allele varied breed to breed for both the fragments. In case of 133 bp fragment, four alleles A, B, C and D were found among different breeds of buffalo whereas in 159 bp fragment, five alleles namely A, B, C, D and E were found in different breeds. Nucleotide sequence data of different alleles showed the presence of both silent as well as functional mutation leading to variability in polypeptide chain.

6.1.1.8 *Growth hormone (GH)*

The exogenous bovine growth hormone (bGH) performs versatile functions in body when injected in animals. Since the GH is galactopoietic, it improves the biological efficiency of milk production without jeopardizing milk quality and processing characters The galactopoietic action of growth hormone is homeostatic regulation of metabolism by redirecting the partitioning and use of absorbed nutrient to milk production. There is 6–35% increase in milk production by exogenous supplementation of GH in lactating dairy animals. GH/rbGH orchestrates homeototic control over intermediary metabolism of glucose, fat and protein in combination with its direct and indirect effects on mammary secretory tissues. This encourages increased synthesis and secretion of milk by the mammary gland.

GH gene is a single copy gene and consists of exons separated by four intervening sequences and spans on 2 kb length of the genome. Cattle GH gene has been mapped to chromosome 19. Since bGH is involved in overall growth

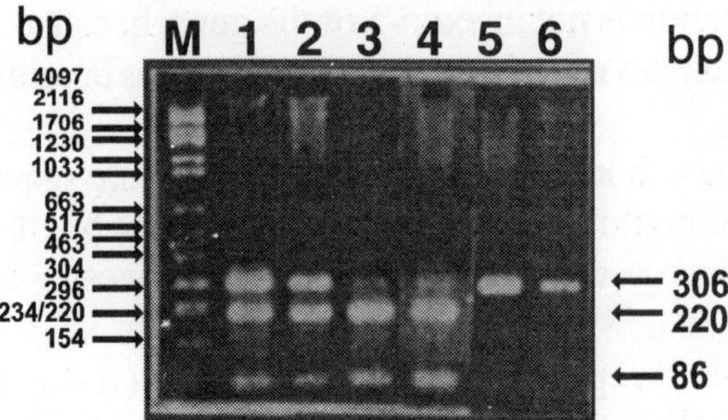

Fig. 6.1 Analyis of 306 hp amplificates of α latabumin gene by PCR-RFLP with Mspl Lanes 1–6 (A) and 1–6 (B) Hariana cattle samples with genotypes. AB in lanes 1, 2, 6 (A) and 1–2 (B), BB in 3–5 (A) and 3– 4 (B) and AA in 5–6 (B) Lane 8–14 (A) Murrah buffalo samples with genotype similar to BB of cattle. Lane M (A and B) pBR328 DNA digested separately with Bam HI Bg II and Hinf 1 and mixed in a ratio of 2:3:3 respectively, and used as maker. On left side of photograph fragment size of maker and on right side fragment size of amplificates are marked respectively.

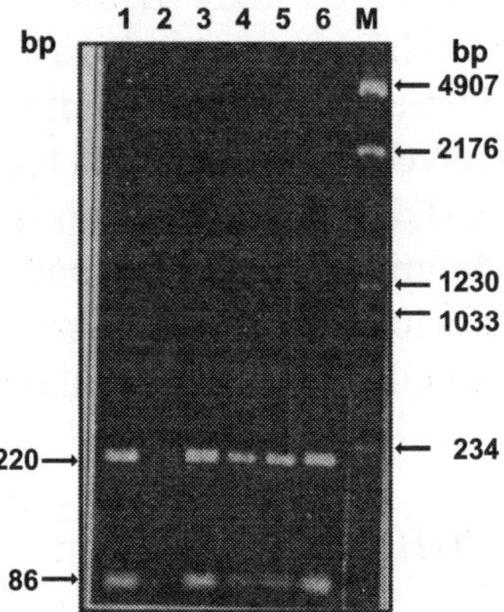

Fig. 6.2. Analysis of the α lactabumin PCR product of buffalo DNA with Msp I enzyme. Lanes, M-pBR 328/bg I maker. I to 6 PCR product digested with Msp 1. On the left and right sides of the photograph fragment sizes of digested amplicon and maker, respectively are indicated.

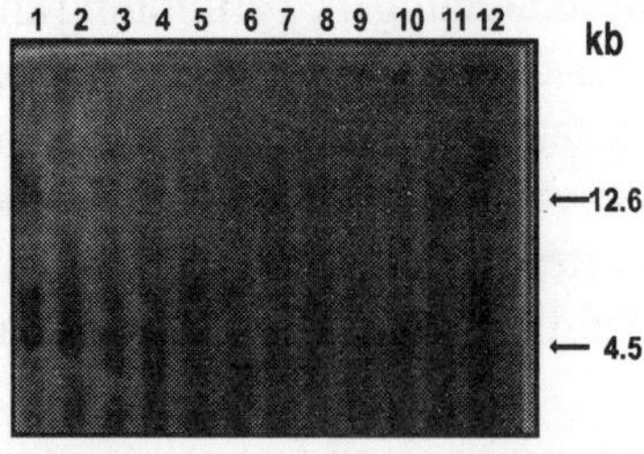

Fig. 6.3 Southern blot of buffalo genomic DNA's hybridized with ovine growth hormone cDNA probe.

and lactation of animals, an RFLP linkage with quantifiable traits such as milk and meat production has been postulated. No information is available on the RFLPs associated with GH of buffalo.

The information generated will be useful for searching markers of higher production. RFLP with Eco RI was observed at the locus with 4.5 kb fragment being most prevalent (95.5%). The other and less prevalent fragment was 12.6 kb (4.5%). The 4.5 kb fragment was observed in all the animals, none of the animals was homozygous for 12.6 kb fragment (Fig. 6.3).

Biswas *et al.* (2003) studied the polymorphism at growth hormone locus to observe its effect on birth weight of Sahiwal, Holstein-Friesian, Jersey and crossbred cattle and Murrah, Bhadawari, Jaffarabadi, Nagpuri and Surti buffaloes maintained at different organized herds in Uttar Pradesh. A 223 bp fragment of the gene was amplified and digested with Alu I restriction enzyme. Two alleles, L and V with three genotypes LL, LV and VV were observed in Jersey, Holstein and cross bred cattle. Sahiwal cattle and buffalo were monomorphic for this locus producing only one genotype LL and one allele L. The effect of genotype on birth weight was significant. LV genotype had higher birth weight than other genotypes in Holstein Friesian.

6.1.1.9 *Microsatellite analysis*

Ryan *et al.*, (1988) measured genetic variation in 105 African buffalo from four populations in South Africa to investigate the effects of habitat fragmentation. Levels of heterozygosity, allelic diversity, genetic differentiation among populations were quantified using seven polymorphic microsatellite markers. There was a significant correlation between the amount of genetic variation and population size, and differentiation was detected among all populations measured by F_{ST} and R_{ST}. They used likelihood analysis to infer the effective population sizes of each population and to determine whether the fragmented populations were historically differentiated from one another. The genetic estimates of census size were consistent with historical records, and no historical genetic differentiation could be inferred in the original population before fragmentation. The results indicated that in the light of conservation management of fragmented buffalo populations, particularly where natural gene flow is no longer possible, a radically different stratagy is needed.

Tanaka *et al.*, (1999) isolated satellite DNA sequences from the water buffalo (*Bubalus bubalis*) after digestion with two restriction endonucleases, Bam Hl and Stu l. These satellite DNAs of the water buffalo were classified into two types by sequence analysis: one had an approximately 1,400 bp tandem repeat unit with 79% similarity to the bovine satellite I DNA; the other had an approximately 700 bp tandem repeat unit with 81% similarity to the bovine satellite II DNA. The chromosomal distribution of the satellite DNAs was examined in the river type and the swamp type buffaloes with direct R-banding fluorescence *in-situ* hybridization. Both the buffalo satellite DNAs were localized to the centromeric regions of all chromosomes in the two types of buffaloes. The hybridization signals with the buffalo satellite I DNA on the acrocentric autosomes and X chromosome were much stronger than that on the biarmed autosomes and Y chromosome, which corresponded to the distribution of C-band-positive centromeric heterochromatin. This centromere specific satellite DNA also existed in the interstitial region of the long arm of chromosome 1 of the swamp-type buffalo, which was the junction of the telomere-centromere tandem fusion that divided the karyotype in the two types of buffaloes. The intensity of the hybridization signals with buffalo satellite II DNA was almost the same over all the chromosomes, including the Y chromosome, and no additional hybridization signal was found in noncentromeric sites.

Kumar *et al.*, (2006) studied the genetic variation and relationships among eight Indian riverine buffalo breeds. Twenty-seven microsatellite loci were used to define genetic variation and relationships among eight Indian riverine buffalo breeds. The total number of alleles ranged from 166 in the Toda breed to 194 each in the Mehsana and the Murrah. Significant departures from the Hardy–Weinberg equilibrium were observed for 26 locus-breed combinations due to heterozygote deficiency. Breed differentiation was analysed by estimation of FST index (values ranging from 0.75% to 6.00%) for various breed combinations. The neighbour joining tree constructed from chord distances, multidimensional scaling (MDS) display of FST values and Bayesian clustering approach consistently identified the Toda, Jaffarabadi, and Pandharpuri breeds as one lineage each, and the Bhadawari, Nagpuri, Surati, Mehsana and Murrah breeds as admixture. Analysis of molecular variance refuted the earlier classification of these breeds proposed on the basis of morphological and geographical parameters. The Toda buffaloes, reared by a tribe of the same name, represent an endangered breed from the Nilgiri hills in South India. Divergence time of the Toda buffaloes from the other main breeds, calculated from Nei's standard genetic distances based on genotyping data on seven breeds and 20 microsatellite loci, suggested separation of this breed approximately 1800–2700 years ago.

CHAPTER 7

BREEDING

7.1 Systems of Breeding

The art of breeding lies in the proper application of principles of heredity to animal improvement. The problem of animal improvement may be approached in two ways:

i. Modification of environment as better feeding, management and disease control.

ii. Genetic improvement, which is permanent, e.g. selection and mating systems.

Systems of breeding do not create any new gene. They sort out old genes into new patterns. Success, therefore, depends upon the proportion of favourable genes present in the foundation stock. Genes that are not present in the foundation animals can sometimes be found in other strains or populations and can be introduced through crosses.

Breeding	
Inbreeding	**Outbreeding**
i. Close breeding	i. Crossbreeding
ii. Line breeding	ii. Outcrossing
	iii. Top crossing
	iv. Back crossing
	v. Grading
	vi. Species hybridization

7.2 Inbreeding

Inbreeding is the mating of males and females that are related. We consider animals to be related only when they have one or more ancestors in common in the first 4 to 6 generations of their pedigree. The intensity of inbreeding depends on the degree of relationship, *e.g.* mating of son to dam or brother sister mating are called close breeding in contrast to cousin matings or those which are not closely related.

Measurement of relationship between individuals help us to understand the intensity of inbreeding. Two animals no nearer related than the average of their breed have a relationship of zero. Two animals with exactly similar genotype have a relationship of 100. They are alike in 100% of the genes, *e.g.* identical twins. The degree of relationship therefore, ranges from 0 to 100.

Relationship may be of two kinds, direct and collateral. You are directly related to your father. That is you and your father have more genes in common (50%) than do unrelated members of the human population. Similarly, one half of your genes are identical with those of your mother. You and your cousins are collateral relatives, because you both have some ancestors in common. Your cousin probably has some identical genes that came to each of you from your common grand parents.

7.2.1 Coefficient of inbreeding

When animals which are related, in other words those having genes in common, are mated, more homozygosity results in the offspring in relation to average animals of the same breed in the foundation stock. Inbreeding, therefore, increases homozygosity or decreases the heterozygosity in individuals.

The average percentage increases in homozygosity or decreases in heterozygosity in an inbreed animal in relation to an average animal of the same breed of the foundation stock is known as the coefficient of inbreeding. It is obtained by multiplying the relationship among parents by ½; since the new generation produced is once further removed from the common ancestors and a further halving of the genetic material occurs. The formula for the coefficient of inbreeding of individuals (Fx) is

$$Fx = S\,[(^{1}/_{2})\ n'1 + n+1]$$

n: number of generations or halvings from the sire to the common ancestor.

n' : Number of generations from the dam to the common ancestor.

The coefficients are not absolute but relative measures. It measures the probable similarity of germ cells. It is useful for study of breeds and lines within a breed and analysis and comparison of individuals, groups and breeds for the part inbreeding plays in their respective developments.

Effect of inbreeding

Inbreeding is the mating of animals, which are related or have more number of similar genes. Hence, it increases the likelihood of similar genes becoming paired. In other words it increases the percentage of homozygotes and reduces the proportion of heterozygotes. Inbreeding makes the genes, favourable or unfavourable, homozygous. When the animals are homozygous for a number of traits, the regularity of inheritance is assured, i.e. it fixes the characteristics. A high degree of homozygosity increases the prepotency of the inbred individuals i.e. the ability of a parent to impress its characteristics uniformly on its offspring.

7.2.2 Reasons for inbreeding

1. To promote genetic purity and thereby increase prepotency.
2. To bring undesirable recessives to light and give the breeder an opportunity of culling them from the stock. When a sire is mated to 20 of its daughters, if it does not throw out any recessive character, it may be reasonably stated that the sire is not heterozygous for character under question.
3. To develop inbred lines for nicking ability.
4. For regrouping the genetic material.

It is generally believed that inbreeding reduces vigour. The reasons for this are:

i. The recessive genes become homozygous during inbreeding.
ii. If overdominance exists, where a2 a2 is superior, inbreeding diminishes the quality of trait and a2 a2 becomes a1 a1 and a2 a2 during inbreeding. To offset the bad affects of inbreeding, it is desirable that it is practiced only in herds that are better than average i.e. where frequency of desirable genes is more. It may be practical in herds where an outstanding sire has been used. It is also necessary that the breeder should know the merits and demerits of this system before he practices it. Inbreeding should not be practiced in grades or in commercial herds below average for the sake of economy in a single sire herd.

7.2.3 Line breeding

Line breeding is a form of inbreeding, but so directed as to keep the relationship of the individuals very close to as in admired ancestor. The admired ancestor is usually a male since it can give more offsprings during its life time than a female. When we say that an animal is a line breed, the question immediately arises, line bred to what?

When and why line breeding? When a sire, used on good dams produce offsprings better than their dams, the breeder should line breed at once strongly to this sire, while the animal is yet alive. It can be used on its daughters and grand daughters generation after generation. Often an animal is old or dead before its superiority is recognized. If its sons and daughters are mated to unrelated individuals, within three to four generations, the influence of the outstanding animals is so scattered that no one descendent is like the original individual. So, line breeding holds the expected amount of inheritance from the admired ancestor as a constant level instead of letting it to be halved every generation. If at the time of death, there are no relatives more closely related than 50%, we cannot produce animals more close to it than that, but it may be possible by inbreeding to keep and maintain that level.

Line breeding builds up homozygosity and prepotency. It tends to hold the gain made by selection while attempt is made to make further gain.

Line breeding is specially useful where there is much epistasis, where a desired characteristic depends on a combination of genes and where the combination tends to get scattered at each generation. These genes can be made homozygous in different lines and lines can be crossed for their combining or nicking ability.

When progress by inbreeding comes to a standstill, line breeding makes additional progress possible.

7.2.4 Dangers of line breeding

Line breeding tends to make genes, good or bad, homozygous rapidly. Hence choosing of the ancestor (sire) to line breed is very important. Those that are definitely superior should alone be selected. Besides rigid selection, culling of the undesirable recessives is highly essential. Line breeding should be practiced only in herds distinctly superior to the general average of the breed.

7.2.5 Prepotency

Prepotency is the ability of an individual to stamp its characteristics on its offsprings to such an extent that they resemble their parents more closely than is usual. It is the property of the characteristic and not the individual breed or sex. When two individuals are mated, one may have more influence than the other on the offspring. Similarly, some lines and breeds are more prepotent than others. However, prepotency cannot be passed on from one generation to another unless it is possessed by both sires and dams.

A high degree of homozygosity and the possession of a high percentage of dominant genes are the inherent qualities that will enable an animal to stamp its own characteristics on majority of its offsprings. A perfectly homozygous animal produces only one kind of gametes and all its offsprings will receive exactly the same gene from it. Any genetic difference between the offsprings would depend entirely on their having received different genes from the other parents. If the parent is homozygous for several dominant genes, all the offsprings will resemble it irrespective of what they received from the other parent. Here, prepotency is the maximum.

Measure of prepotency

Inbreeding and increase of homozygosity is the only means of making animals prepotent for characteristic. The more the animals are inbred the more they become homozygous for a number of genes. The inbreeding coefficient then is the best estimate of an animal's prepotency. Prepotency, however is not transmissible from parent to offspring.

7.3 Outbreeding

Outbreeding is the mating of animals distinctly less closely related to each other than the average of the population, i.e. those that have no common ancestors in the preceding 4 to 6 generations of their pedigree. It is just the opposite of inbreeding. It promotes the pairing of unlike genes by mating animals that belong to different families, breeds or species. Thus it increases heterozygosity and variability. The chief reasons for outbreeding are:

i) To bring about an increase in vigour. Vigour includes almost anything that pertains to desirability, *e.g.* rate of gain, efficiency of gain, fertility, general strength etc.

ii) To make full use of dominance of characteristic.

iii) To introduce new genes in a closed population. If a certain breed or family is deficient in a certain trait, the quickest and most certain method of improving that trait is to introduce genes through cross breeding to some stock known to be superior in that trait.

iv) To start new breeds with a broad genetic background.

v) To produce market animals making use of heterosis.

Outbreeding includes:

(a) cross breeding,
(b) outcrossing,
(c) back crossing,
(d) top crossing,
(e) grading and
(f) species hybridization.

7.4 Crossbreeding

Cross breeding is the mating of two animals, which are pure bred but belonging to different breeds. It is widely practiced in swine, sheep, and poultry and less so in cattle and horses. The main purpose in this is to produce commercial stock where the individual merit for economic traits is promoted. The breeding value of the individual, however, is lowered.

When the crosses are used for breeding purposes, their offsprings are more variable than the crossbreds and generally average, somewhat, lower in individual merit, below their purebred grand parents.

Judicious crossing of breeds that complement each other might result in increased vigour. The economy of crossbreeding, therefore, depends upon whether the increase in production is more than enough to balance the possible confusion regarding the breeding value of the crossbred individuals and also increase in cost of replacement of pure bred stock under a cross breeding system. It is more profitable where fertility is highest and females can be kept for long and the cost of their replacement is lowest. Mainly for these reasons, it is mostly practiced in swine, poultry and sheep.

In a crossbreeding system, the males are to be discarded because of their lowered breeding value. The heterosis in females can, however, be utilized by crossing it with a third different breed and rotating the same in a systematic manner. This is known as triple crossing, or rotational crossing.

Three breeds are used in this system. The females or crosses are used on sire of pure breeds in rotation. The crossbreds will soon come to have $^4/_7$ of inheritance of the breed of immediate sire, $^2/_7$ from the breed of maternal grand sire and $^1/_7$ of the hereditary material of the other pure breed. Heterosis is thus continuously maintained.

Crisscrossing is another method proposed for utilizing heterosis in dams, without incurring the full decline in average individual merit which usually occurred when crossbreds are mated.

7.5 New Breeds From Crossbreeds

Crossbreeding has been utilized for developing several new breeds of livestock. It offers a broad genetic basis from which by a process of selection and inbreeding, new gene combinations can be made for specific purposes.

Mehsana is a new breed of buffalo developed in Gujarat by crossing Murrah male and Surti female.

The question of development of new breeds arises only when we are not satisfied with the existing breeds as regards their utility value and when we feel that their value can be enhanced by making new gene combinations from two breeds which is likely to complement each other in different traits.

7.6 Outcrossing

Outcrossing is the mating of the animals, that are members of the same breed, but show difference in the herd. Intense inbreeding makes the genes homozygous and at the same time closes the door for further improvement. It fixes the deleterious genes also. An outcrossing brings into the herd new genes and gives an opportunity for selection and further improvement of the herd.

Outcrossing is a useful procedure where it is desired to change the type of the herd rather drastically, when necessitated by market demands.

7.7 Top Crossing

Top crossing is the mating of a male of a certain family to females of another family of the same breed. It is the same in principle as grading up, except that top crossing is usually applied to different families within a pure breed, whereas grading up is applied to continued use of sires of one pure breed starting with foundation females which are of another breed or mongrel stock.

7.8 Back Crossing

Back crossing is the mating of a crossbred animal back to one of the pure parent races which were used to produce it. It is commonly used in genetic studies, but not widely used by breeders. When one of the parents possesses all or most of the recessive traits, the back cross permits surer analysis of the genetic situation than an F_2 does.

A heterozygous individual of the F_1, when crossed with a member of the homozygous recessive parent race, the offsprings group themselves into a phenotypic ratio of 1:1; on the other hand if the individual of the parent race were to be homozygous dominant, all the offsprings will be phenotypically alike.

7.9 Grading Up

Grading up is the continuous use of purebred sires on females of another breed or mongrel stock, to raise them quickly to the level of the purebred sires. When the purebreds are relatively scarce, this is the only quickest way available for improving the mongrel stock. In grading up, generally, the first cross shows a marked improvement over the original stock. Further improvement by each successive cross is progressively less. In a purebred, which is stationary in level, the mongrel stock by the seventh generation almost reaches that of the purebred.

7.10 Species Hybridization

By crossing two different species sometimes we get good fertile individuals. The mule is a good example of a commercially important species hybrid. *e.g.,* Mare × Jack = Mule; She Ass × Stallion = Hinny. Male mules are always sterile as far as is yet known. A few cases of fertile mare mules have, however been reported. But these are very rare. Hinny is generally inferior to mule as work animal. It is also sterile.

Horse having 32 pairs of chromosomes and ass 31 pairs, the mule comes to possess 63 single chromosomes in all. The mare mules have given birth to mule foals and horse foals when bred to jack and stallion respectively. The inference is that the mare rules essential function as mares as far as the genetics of their eggs is concerned.

If all the horse chromosomes were extruded in the polar body, these mules will function genetically as asses. But no case of this sort has been reported. True breeding of mules as such seems also theoretically impossible.

European cattle and American bison when crossed, produce sterile males and fertile females. By back crossing the females to bison and cattle, attempts are being made to form a new breed of cattle, the Cattalo.

7.11 Heterosis or Hybrid Vigour

Heterosis or Hybrid vigour is a phenomenon in which the crosses of unrelated individuals often result in progeny with increased vigour much above their parents. The progeny may be from the crossing of strains, varieties, or species. One of the explanations for this increased vigour is that genes favourable to production are usually dominant over their opposites. As a species or breed develops, it becomes homozygous for some dominant genes. They also have few unfavourable recessive ones. When one breed is crossed with the other one parent supplies a favourable dominant gene to offset the recessive one supplied by the other and vice versa.

The offspring, therefore, has a larger number of dominant genes than does either parent and is likely to be more vigorous.

Another explanation for hybrid vigour is overdominance, where a heterozygous condition is much more superior to any of the homozygous conditions.

Heterosis is much employed to produce commercial stock where the individual merit is promoted, but the breeding value is lowered. The successful exploitation of heterosis depends upon how superior the crosses are over the purebreds and whether it is worth the confusion caused in lowering the breeding value of the individual and the cost of replacement of purebred stock. For these reasons it is more commonly practiced in poultry, swine and sheep where the fertility is high and the cost of replacement of purebred stock is likely to be low.

CHAPTER 8

SELECTION

The process in which certain individuals in a population are preferred to others for the production of the next generation is known as selection. Selection in general is of two types: natural, due to natural forces, and artificial, due to the efforts of man.

No new genes are created by selection. Under selection pressure there is a tendency for the frequency of the undesirable genes to be reduced whereas the frequency of the more desirable ones is increased. Thus, the main genetic effect of selection is to change gene frequencies, although there may be a tendency also for an increase in homozygosity of the desirable genes in the population as progress is made in selection.

8.1 Natural Selection

The main force responsible in nature for selection is the survival of the fittest in a particular environment. Natural selection is of interest because of its apparent effectiveness and because of the principles involved.

Natural selection can be illustrated by considering the ecology of some of our wild animal species.

Some of the most interesting cases of natural selection are those involving man himself. All races of man that now exist belong to the same species, because they are interfertile, or have been in all instances where matings have been made between them. All races of man now in existence had a common origin, and at one time probably all men had the same kind of skin pigmentation. As the number of generations of man increased, mutations occurred in the genes affecting pigmentation of the skin, causing genetic variations in this trait over a range from light to dark or black. Man began to migrate into the various parts of the world and lived under a wide variety of climatic conditions of temperature and sunshine. In Africa, it is supposed, the dark skinned individuals survived in larger numbers and reproduced their kind, because they were better able to cope with environmental conditions in that particular region than were individuals with a lighter skin. Likewise, in the northern regions of Europe, men with white skins survived in a greater proportion, because they were better adapted to that environment of less intense sunlight and lower temperatures. But Eskimos who lived in the polar regions of the North are also dark skinned. This is because, Eskimos are more recent migrants from Asia to the polar region and compared to the Negro in Africa and the whites in Europe, they have not lived so long in that region.

Further, evidence is available that there is a differential selection for survival among humans for the A, B and O blood groups. It has been found that members of blood group A have more gastric carcinoma (cancer) than other types and that members of type O have more peptic ulcers. This would suggest that natural selection is going on at the present time among these different blood groups, and the frequency of the A and O genes might be gradually decreasing unless, of course, there are other factors that have opposite effects and have brought the gene frequencies into equilibrium.

Natural selection is a very complicated process and many factors determine the proportion of individuals that will reproduce. Among these factors are differences in mortality of the individuals in the population, especially early in life; differences in the duration of the period of sexual activity; the degree of sexual activity itself; and differences in degrees of fertility of individuals in the population.

It is interesting to note that in the wild state, and even in domesticated animals to a certain extent, there is a tendency toward an elimination of the defective or detrimental genes that have arisen through mutations, through the survival of the fittest.

8.2 Artificial selection

Artificial selection is that which is practised by man. Under this, man determines to a great extent which animals to be used to produce the next generation of offspring. Even in this, selection seems to have a part. Some research workers have divided selection in farm animals into two types one known as automatic and the other as deliberate selection.

Litter size in swine can be used as an illustration to define these two terms. Here, automatic selection would result from differences in litter size even if parents were chosen entirely at random from all individuals available at sexual maturity. Under these conditions, there would be twice as much chance of saving offspring for breeding purposes from a litter of eight than from a litter of four. Automatic selection here differs from natural selection only to the extent that the size of the litter in which an individual is reared influences the natural selective advantage of the individual for other traits. In deliberate selection, this term is applied to selection in swine for litter size above and beyond that which was automatic. In one study by Dickerson (1973) involving selection in swine most of the selection for litter size at birth was automatic and very little was deliberate; the opportunity for deliberate selection however among pigs was utilized more fully for growth rate.

Definite differences between breeds and types of farm animals within a species proves that artificial selection has been effective in many instances. This is true, not only from the standpoint of colour patterns which exist in the various breeds, but also from the standpoint of differences in performance that involve certain quantitative traits. For instance, in dairy cattle there are definite breed differences in the amount of milk produced and in butterfat percentage of the milk.

8.2.1 Basis of selection

(a) Individuality

Selection on the basis of individuality means that animals are kept for breeding purposes on the basis of their own phenotype. Selection may be made for several traits, such as coat colour, conformation, performance, or carcass quality. In the past, the emphasis in selection probably was based on coat colour and conformation, although performance and carcass quality have received more attention in recent years.

Most of the breeds of livestock are characterized by a particular coat colour or colour pattern, and this is one of the requirements for entry into the registry associations. Selection for coat colour has been practised because of its aesthetic value rather than its possible correlation with other important economic traits.

Attempts to relate variations of coat colour to performance within a breed have not met with success although many livestock men feel that there is a relationship. There is a strong belief of horse breeders that there is a strong relationship between colour and temperament which has no basis as per the evidence. There is however evidence that animals of some colours are better able to cope with certain environmental conditions, such as high temperatures and intense sunlight in some regions of the tropics or in the south and the south-western portions of the United States. Coat colour in some instances is closely related to lethal and undesirable genes in farm animals. Further, other species such as the mouse, dog, cat, mink, and fox, also show such relationships. Certain coat colours are the trademark of the some breeds of livestock. This is probably because this can be easily recognized. It is thus important that the breeder must conform to the breed requirements for this trait otherwise he will not be in the purebred business for long.

Type and conformation have been used as the basis of selection for many years throughout the world. Type may be defined as the ideal of body construction that makes an individual best suited for a particular purpose. This basis of selection has merit in some instances. The conformation of a draught horse is such that he is better suited to pulling heavy loads than he is to racing. On the other hand, the reverse is true of the thoroughbred.

The performance of individuals has also been given some attention in the development of some of our breeds of livestock. For many years thoroughbred horses have been selected for breeding purposes for their speed. Dairy cows have been selected for their ability to give large amounts of milk and butter fat. In beef cattle and swine, however, less attention has been paid to selection for performance and carcass quality until recently.

Increased emphasis is now being placed on selection for performance and carcass quality, because breeders realize that the type or conformation of an individual is not the best indicator of its potential performance or its carcass quality. Appropriate measures of these traits must be applied before progress can be made in selection for them.

The correlation between type and carcass quality is greater in some instances than is the correlation between type and performance. The meatiness of hogs by a visual inspection, can be assessed but this is not reliable. Better methods are backfat probes on live animals, actual weighings and measuring of lean meat in the carcass.

The fact that type and performance are not usually closely related, indicates the importance of selecting separately for the important traits in livestock production. If the correlation between type and other traits is low, it means that they are inherited independently and that they can be improved only if selection is practised for each of them.

Individuality for certain traits should always be given some consideration in a selection program. However, it is more important in some instances than in others. It is most important as the basis of selection when the heritability of a trait is high, showing that the trait is greatly affected by additive gene action. High heritability estimates also suggest that the phenotype strongly reflects the genotype and that the individuals that are superior for a particular trait should also possess the desirable genes for that trait and should transmit them to their offspring.

The greatest disadvantage of selection on the basis of individuality is that environmental and genetic effects are sometimes difficult to distinguish. Much of the confusion may be avoided by growing or fattening of the offspring being compared for possible selection purposes under a standard environment. Even then, it is still possible to mistake some genetic effects for environmental effects. This is less likely to happen, however, in the outstanding individuals than in those that have a mediocre record. For instance, a bull calf placed on a performance test may make a poor record because of an injury or because of sickness while on test. But if he makes an outstanding record, it is certain that he possessed the proper genes and in the right combination as well as the proper environment to make the good record. It cannot always be certain, however, whether an individual with a mediocre record would have done better even if adverse environmental factors had not interfered. We can be certain that his record is poor and by culling on this basis, elimination of the genetically poor individuals is possible. This chance is worth taking, even though we may discard some genetically superior individuals occasionally.

Studies of selection on the basis of individuality within inbred lines of swine have shown that selection favoured the less inbred litters. This is another way of saying that selection probably favoured the more heterozygous individuals, and this may be true also in many cases where inbreeding is not involved to a great extent. Chance combinations of genes may make an individual outstanding, but his offspring may be inferior, because he cannot transmit his heterozygosity to his offspring. The breeder should avoid keeping superior individuals from very mediocre parents and ancestors. For breeding purposes, it would be much more desirable to keep superior individuals from parents and ancestors that themselves were outstanding.

(b) Pedigrees

A pedigree is a record of an individual and his ancestors that are related to him through his parents. Earlier, the information included in a pedigree has been simply the names and registration numbers of the ancestors, and little has been indicated as to the type and performance of the ancestors. Pedigrees now include information on the size of the litter at birth and weaning.

If full information is available on the ancestors as well as the collateral relatives, it may be of importance in detecting carriers of a recessive gene. Such information has been used to a great extent in combating dwarfism in beef cattle.

A disadvantage of the use of the pedigree in selection against a recessive gene is that there are often unintentional and unknown mistakes in pedigrees that may result in the condemnation of an entire line of breeding when actually the family may be free of such a defect. On the other hand, the frequency of a recessive gene in a family may be very low, and records may be incomplete. Then later, it will be found that the gene is present.

Another disadvantage of pedigree selection is that the individuals in the pedigree, especially the males, may have been selected from a very large group, and the pedigree tells us nothing about the merit of their relatives.

Still another disadvantage of pedigree selection is that a pedigree may often become popular because of fashion or fad and not because of the merit of the individuals it contains. The popularity of the pedigree may change in a year or two, and the value of such a pedigree may decrease considerably or may even be discriminated against. If popularity is actually based on merit, there is less danger of a diminution of value in a short period of time.

In using pedigrees for selection purposes, weight should be given to the most recent ancestors. This is because the percentage of genes contributed by an individual's ancestors is halved in each new generation. Some breeders place much emphasis on some outstanding ancestor for which three or four generations has been removed in the pedigree, but such an ancestor contributes a very small percentage of the genes the individual possesses and has very little influence on type and performance, unless linebreeding to that ancestor has been practised.

An individual's own performance is usually of more value in selection than its pedigree, but the pedigree may be used as an accessory to sway the balance when two animals are very similar in individuality but one has a more desirable pedigree than the other. Pedigree information is also quite useful when the animals are selected at a young age and their own type and conformation is not known. Pedigree is useful in identifying superior families if good records are kept and are available.

(c) Collateral relatives

Collateral relatives are those that are not related directly to an individual, either as ancestors or as their progeny. Thus, they are the individuals brothers, sisters, cousins, uncles and aunts. The more closely they are related to the individual in question, the more valuable is the information for selection purposes.

Complete information on collateral relatives, gives an idea of the kinds of genes and combinations of genes that the individual is likely to possess. Information of this kind has been used in meat hog certification programs, where a barrow and a gilt from each litter may be slaughtered to obtain carcass data. This is done, because otherwise the animal himself has to be slaughtered if information on his own carcass quality is to be obtained, information on collateral relatives is also used in selecting dairy bulls, since milk production can be measured only in the cows even though the bull transmits genes to his offspring for this trait. Information on collaterals has been used in the All India Coordinated Research on Breeding wherein information on slaughter traits has been used from full brothers for selecting bulls for future breeding.

d) Progeny test

Selection on this basis means that we estimate the breeding value of an individual through a study of the traits or characteristics of its offspring. In other words, the progeny of different individuals are studied to determine which group is superior, and on this basis the superior breedings are given preference for future breeding purposes. If information is complete, this is an excellent way of identifying superior breeding animals.

Progeny tests are very useful for determining characteristics that are expressed only in one sex, such as milk production in buffalo or egg production in hens. Even though the bull does not produce milk nor does the rooster lay eggs, they carry genes for these traits and supply one-half of the inheritance to each of their daughters for that particular trait.

Progeny tests are also useful in measuring traits which cannot be measured in the living individual. A good example of this is carcass quality in cattle, sheep, and hogs.

Progeny tests are also being used at the present time by experiment stations in studies of reciprocal recurrent selection. This type of selection is used to test for the "nicking ability" of individuals and lines and is based on the performance of the line cross progeny. Selection of this type is for traits that are lowly heritable and in which non-additive gene action seems to be important.

In comparing individuals on the basis of their progeny, certain precautions should be taken to make the comparisons fair and accurate. In conducting a progeny test, it is very important to test a random sample of the progeny. It would be more desirable if all progeny could be tested, but where this cannot be done, as in litters of swine, those nearer the average of the litter should be tested. It is also important that the females to which a male is mated should be from a non-selected group. One would expect the offspring of a sire to be superior if he is mated to the outstanding females in the herd. Such a practise would be misleading in comparing males by a progeny test, since much of the superiority of the offspring of one male could come from the dams and not from the sire. Some breeders prefer using a rotation of different dams when testing males, but this is practical only in swine, where two litters may be produced each year.

Using a large number of offspring in testing a sire increases the accuracy of the test rule. Where the number of females in a herd is limited, the number of males that may be progeny tested will be less as the number of matings per sire is increased. The point is, then that the breeder must make some decision as to how many sires to test and how many progeny must be produced to give a good test. The number of offspring required for an accurate progeny test will depend upon the heritability of a trait, with fewer offspring being required, when the trait is highly heritable, and more being required when it is lowly heritable.

To make accurate progeny tests, it is also important to keep the environment, as nearly as possible, the same for the offspring of the different sires. In progeny testing in swine, for instance, confusion would result when the progeny of one sire were fed in dry lot during the summer and the progeny of another were fed on pasture. This would be particularly true in progeny testing for rate of gain, where pigs fed with modern rations often grow considerably faster in dry lot than on pasture. When this environmental condition are not controlled, the inferior sire might actually be thought to be superior.

Progeny tests in most of our farm animals have certain definite limitations. In cattle especially, it takes so long to prove an animal on a progeny test that he may be dead before the test is completed and his merit actually known. Progeny tests may be now easily done in swine than in other farm animals, but even in this case the males are usually disposed of by the time they are thoroughly progeny tested.

The process of progeny testing may be speeded up by testing males at an earlier age than they would ordinarily be used for breeding purposes. By hand-mating them to a few females, or by using them on a larger number of females by artificial insemination, harmful effects that might occur from overuse at too early an age may be prevented.

Too often, farmers send their old sires to market just as soon as their daughters are old enough to breed, in order to prevent inbreeding. This practise has resulted in much loss of good genetic material for livestock improvement. Actually a sire is not proved until his daughters come into production. Rather than being slaughtered, a sire that has proved himself to be of high genetic merit should be used more extensively. It is true that his usefulness in a particular herd may be finished when his daughters are of breeding age, but he should be sent to another herd to be used for additional breeding purposes. To be proved, a sire must have completed a satisfactory progeny test record of some kind. He may be considered proved if he has offspring who have completed one year's record, but this varies with the traits involved. This may be a lactation record, or one of litter size, egg production, or birth and weaning weights, fleece yield and quality. A sire so tested may be said to be proved whether his offspring are good or poor. Before buying a proved sire to use in a herd, a breeder should not neglect to find out if he has been proved good or a poor producer. Newer methods of progeny testing may be developed that are superior to those already available. For instance, the semen of a buffalo bull that has been proved highly superior could be collected at regular intervals, frozen, and stored for later use, even after his death. In swine, it might be possible to get quicker progeny tests on females by weaning their pigs at two or three weeks of age and breeding them again as soon as possible to produce three litters per year. Superovulation, by the injection of certain hormones, a female can be made to produce hundreds of eggs instead of the usual one or few. Embryo transfer technique has made possible using extra ova to other females, where the fertilized ova may develop to birth and possess the characteristics of the mother which ovulated the egg. The success of the embryo transfer transplantation of ova has been limited, but future studies may make it more practical. If this could be done, it would be possible for an outstanding female to have many offspring in one year, rather than one or just a few.

8.2.2 Methods of selection

The amount of progress made, regardless of the method used, depends upon the size of the selection differential (selection intensity), the heritability of the trait, the length of the generation interval and some other factors. The net value of an animal is dependent upon several traits that may not be of equal economic value or that may be independent of each other. For this reason, it is usually necessary to select for more than one trait at a time. The desired traits will depend upon their economic value, but only those of real importance need to considered. When too many traits are selected for at one time, less improvement, in any particular one is expected. Assuming that the traits are independent and their economic value and heritability are about the same, the progress in selection for any one trait is only about 1/n times as effective as it would be if selection were applied for that trait alone. When four traits were selected for at one time in an index, the progress for one of these traits would be on the order of $^1/_2$ (not $^1/_4$) as effective as if it were selected for alone. For the selection of superior breeding stock, several methods can be used for determining which animal should be saved and which should be rejected for breeding purposes. Three of these methods which are generally used are given as below.

8.2.2.1 *Tandem method*

In this method, selection is practised for only one trait at a time until satisfactory improvement has been made in this trait. Selection efforts for this trait are then relaxed and efforts are directed toward the improvement of a second, then a third, and so on. This is the least efficient of the three methods practised in respect of the amount of genetic progress made for the time and effort spent by the breeder.

The efficiency of this method depends a great deal on the genetic association between the traits selected for. When there is a desirable genetic association between the traits, improvement in one by selection results in improvement in the other trait not selected for, the method could be quite efficient. If there is little or no genetic association between the traits, the efficiency would be less. Since a very long period of time would be involved in the selection practised, the breeder might change his goals too often or become discouraged and not practise selection that was intensive and prolonged enough to improve any desirable trait effectively. A negative genetic association between two traits, in which selection for an increase in desirability in one trait results in a decrease in the desirability of another, would actually nullify or neutralize the progress made in selection for any one trait indicating a low efficiency of the method.

8.2.2.2 *Independent culling method*

In this method, selection may be practised for two or more traits at a time, but for each trait a minimum standard is set that an animal must meet in order to be selected for breeding purposes. The failure to meet the minimum standard for any one trait causes that animal to be rejected for breeding purposes. Let us assume that Pig A was from a litter of 9 pigs weaned, weighed 77 kg at 5 months, and had 1.3 inches of backfat. For Pig B, let us assume that it was from a litter of 5 Pigs weaned, weighed 94 kg at 5 months, and had 0.95 inches of backfat at 84 kg. If the independent culling method of selection were used, Pig B would be rejected, because it was from a litter of only five pigs. However, it was much superior to Pig A in its weight at five months and in backfat thickness, and much of this superiority could have been of a genetic nature. Thus in practise, there is likelihood to cull some genetically very superior individuals when this method is used.

The independent culling method of selection has been widely used in the past, especially in the selection of cattle and sheep for show purposes, where each animal must meet a standard of excellence for type and conformation regardless of its status for other economic traits. It is also used when a particular colour or colour pattern is required. It is still being used to a certain extent in the production of show buffalo/cattle and sheep. It does have an advantage over the tandem method, when selection is practised for more than one trait at a time. Sometimes, it is also advantageous, because an animal may be culled at a young age for its failure to meet minimum standards for one particular trait, when sufficient time to complete the test might reveal superiority in other traits.

8.2.2.3 *Selection index*

This method is based on the separate determination of the value for each of the traits selected for and the addition of these values to give a total score for all the traits. The animals with the highest total scores are kept for breeding purposes. The influence of each trait on the final index is determination by how much weight that trait is given in relation to the other traits. The amount of weight given to each trait depends upon its relative economic value, since all traits are not equally important in this respect, and upon the heritability of each trait and the genetic associations among the traits.

The selection indices is more efficient than the independent culling method, as it allows the individuals which are superior in some traits to be saved for breeding purposes even though they may be slightly deficient in one or more of the other traits. If an index is properly constructed, taking all factors into consideration, it is a more efficient method of selection than either of the other two described earlier, because it should result in more genetic improvement for the time and effort made its use.

Selection indices seem to be gaining in popularity in livestock breeding. The kind of index used and the weight given to each of the traits is determined to a certain extent by the circumstances under which the animals are produced. Some indices are used for selection between individuals, others for selection between the progeny of parents from different kinds of matings, such as line-crossing and crossbreeding, and still others for the selection between individuals based on the merit of their relatives, as in the case of dairy bulls, where the trait cannot be measured in that particular individual.

CHAPTER 9

PRODUCTION CHARACTERISTICS AND GENETIC IMPROVEMENT

9.1 Growth

The study of body weights at various ages provide a measure of growth and size, and reflect the suitability of breed or strain to a particular ecosystem and is important from the point of view of early maturity and general adaptability.

Birth weight

The birth weight of buffalo calf is between 27 and 41 kg in large breeds (Ragab and Singh *et al.*, 1971; Nautiyal and Bhat, 1977; Basu and Rao, 1979) and 24 to 30 kg in medium size breeds (Venkatachar and Sampath, 1978; Chowdhury and Barhat, 1979).

Bhalla *et. al.* (1967) did not observe any effect of month/season and year/period on birth weight while others reported these to have significant effect (Sharma and Sarker, 1968; Misra *et al.* 1970; Singh *et al.* 1971; Rathi *et al.*, 1973; Basu, *et al.* 1978; Alim and Taher, 1979; Johari and Bhat, 1979; Nautiyal and Bhat, 1979). Summer born calves were heavier than those born in autumn. On the contrary, Fahmy (1972) reported that calves born during summer (33-34 kg) were lighter than those born in other months (34–36 kg).

Significant effect of sire on birth weight has been reported by Tomar and Desai (1965), Arya and Desai (1969), Rathi *et al.* (1973) and Basu and Rao (1979).

Most workers (Sharma and Sarker, 1968; Amble *et al*., 1970; Misra *et al.*, 1970; Basu *et al.* 1978; Alim and Taher, 1979; Nautiyal and Bhat, 1979) observed the differences between sexes to be significant; males having higher birth weight than females.

Galal and Fahmy (1969) and Fahmy (1972) reported significant effect of parity. The birth weight increased from first to six/seventh calving. Weight and age of dam at calving also influenced the birth weight; it being higher in females with higher age and weight at calving (Tomar and Desai, 1967; Misra *et al.*, 1970; Fahmy, 1972; Basu *et al.*, 1978; Alim and Taher, 1979). The above discussed reports suggest that month/season, year/period, sire, sex, parity and weight of dam significantly influence birth weight of the calf.

The heritability of birth weight in Indian buffaloes ranged between 0.16 and 0.74 (Ragab and Abd-El-Salam, 1963; Tomar and Desai, 1965; Arya and Desai, 1969; Basu and Rao, 1979). However, most of the heritability estimates based on data adjusted for significant non-genetic effects were low, 0.16 ± 32 (Tomar and Desai, 1967; Tomar, 1969; Sreedharan and Nagarcenkar, 1978; Basu and Rao, 1979). Nautiyal and Bhat (1979), however, reported a slightly higher estimate of 0.56 ± 0.09 (using adjusted data) for this trait.

The phenotypic correlations of birth weight with body weights up to first calving though positive, ranged between low to moderate (0.02-0.34) and a gradual decline in association was noted with an increase in age (Arya and Desai, 1969; Johari and Bhat, 1979; Nautiyal and Bhat, 1979). Similar trend was noted for genetic correlations. The phenotypic correlations of birth weight with age at first calving and first lactation yield were also low and nonsignificant (Arya and Desai, 1969; Marwaha, 1974). The low genetic and phenotypic correlations of birth weight with body weights at later ages and of birth weight with age at calving or milk yield suggest that birth weight can not be used to predict either the mature body weight, age at first calving or milk yield

Weight at various ages

Body weights at different ages across the breeds are given in Table 9.1. Examination of body weight at 24 months across the breeds revealed that among Indian breeds, the Nili-Ravi had the highest weight followed by Murrah, while Surti had the lowest weight. Egyptian heifers had, however, lower weights at 24 months than Nili-Ravi. Reports on body weights in male calves are generally not available because males except those to be retained for breeding are culled as early as possible.

Rathi *et al.* (1973) reported that weights at 9,12 and 18 months were signiflcantly influenced by seasons of birth while those up to 6 months and at two years were not affected. Marwaha (1974) did not observe any effect of month of calving on body weights at 3, 6 and 12 months of age. However, Basu and Rao (1979) reported significant effect of season of birth on body weights at 3, 6 and 12 months. The calves born in winter weighed more at 3 and 6 months than those born in other seasons. The weight at 12 months was highest in autumn-born calves (189 kg) followed by winter (188 kg), summer (186 kg) and monsoon (177 kg). These results suggest that calves born in cooler months (October to March) had higher rate of gain and thus had higher weights at 3, 6 and 12 months of age than those born in hotter months. This could also be due to availablity of better quality greens during the cooler months.

Table 9.1 Body Weights (kg) at 3, 6, 12 and 24 Months in Females

Breed	Body weight at				
	3 months	6 months	12 months	24 months	First calving
Murrah (Basu and Rao, 1979)	70.56	112.14	185.09	283.21	446.42
Murrah Grade (Johari and Bhat, 1979)	77.79	127.10	212.59	326.05	461.91
Nili-Ravi (Sharma and Basu, 1984)	86.33	144.50	243.25	395.22	531.06
Surti (Basavaiah, 1978)	58.09	79.64	109.37	205.13	–

The heritability estimates of body weights at various ages ranged between 0.27 ± 0.09 and 0.72 ± 0.11 (Rathi *et al.*, 1973; Marwaha, 1974; Basu and Rao, 1979). Nautiyal and Bhat (1979) reported heritability estimates of 0.49 ± 0.09, 0.42 ± 0.08, 0.39 ± 0.09 respectively for body weights at 3, 6 and 12 months (Table 9.2). These results suggest that selection would be effective in improving body weights/growth.

The phenotypic and genetic correlations among the adjacent body weights, in general, were high and with increase in age, a gradual decline in the correlation values was noted. The genetic correlation, in general, were higher than the phenotypic correlations (Nautiyal and Bhat, 1979; Johari and Bhat, 1979). The correlations of body weights at later ages with weight at flrst calving were positive and ranged between moderate and high (Nautiyal and Bhat, 1979; Sharma and Basu, 1984) thus suggesting that higher weights at 1 and 2 years would lead to higher weight at first calving.

The genetic and phenotypic correlations of body weights at various ages with age at first calving were, however, negative (Tomar and Desai, 1965; Arya and Desai, 1969; Marwaha, 1974), suggesting that increase in weight would lower the age at first calving.

Table 9.2 Heritability, Genetic and Phenotypic Correlations among Various Growth Traits in Indian Buffaloes

	Body weight at								
	Birth	**12 week**	**26 week**	**52 week**	**18 month**	**24 month**	**30 month**	**36 month**	**1st calving**
Birth	0.56 ± 0.09	0.21 ± 0.02**	0.11 ± 0.02**	0.10 ± 0.02**	0.09 ± 0.02	0.06 ± 0.02	0.09 ± 0.02*	0.14 ± 0.03*	0.08 ± 0.03**
12 week	0.19 ± 0.11	0.49 ± 0.09	0.74 ± 0.02**	0.41 ± 0.03**	0.33 ± 0.02**	0.23 ± 0.02**	0.22 ± 0.02**	0.24 ± 0.03**	0.17 ± 0.03**
26 week	0.18 ± 0.12	0.85 ± 0.04*	0.42 ± 0.08	0.62 ± 0.02**	0.50 ± 0.02**	0.40 ± 0.02**	0.37 ± 0.02**	0.33 ± 0.03**	0.22 ± 0.03**
52 week	0.26 ± 0.13*	0.45 ± 0.11**	0.89 ± 0.31**	0.39 ± 0.09	0.55 ± 0.02**	0.57 ± 0.02**	0.51 ± 0.02**	0.46 ± 0.04**	0.21 ± 0.03**
18 month	0.10 ± 0.13	0.30 ± 0.12*	0.63 ± 0.08**	0.88 ± 0.04**	0.38 ± 0.08	0.73 ± 0.02**	0.63 ± 0.02**	0.47 ± 0.03**	0.18 ± 0.03**
24 month	0.10 ± 0.13	0.25 ± 0.12*	0.65 ± 0.07**	0.68 ± 0.08**	0.98 ± 0.00**	0.40 ± 0.08	0.75 ± 0.16**	0.60 ± 0.03**	0.19 ± 0.03**
30 month	0.11 ± 0.13	0.24 ± 0.13	0.58 ± 0.09**	0.53 ± 0.11**	0.89 ± 0.07**	0.92 ± 0.02**	0.37 ± 0.08	0.64 ± 0.03**	0.25 ± 0.03**
36 month	0.43 ± 0.16*	0.24 ± 0.19	0.32 ± 0.19	0.67 ± 0.13**	0.68 ± 0.09**	0.49 ± 0.16**	0.67 ± 0.12**	0.20 ± 0.09	0.40 ± 0.03**
First calving.	0.19 ± 0.17	0.18 ± 0.18	0.31 ± 0.19	0.34 ± 0.19	0.29 ± 0.19**	0.50 ± 0.15**	0.68 ± 0.12**	0.64 ± 0.12**	0.15 ± 0.16

*P<0.05, **P<0.01, Above the diagnoal are phenotypic, below the diagonal are genetic correlations; along the diagonal are heritability estimates.

Source: Nautiyal and Bhat (1979).

The moderate to high heritability estimates for body weights at various ages and positive correlations among them, suggest that these two facts should be utilised in formulating the breeding plan by giving appropriate weightage to body weights in the selection criterion so that higher body weights/growth rates leading to early maturity could be achieved.

Weight at first heat

Basu *et al.* (1984) reported average weight at first heat in Murrah buffalo heifers as 336.6 ± 2.8 kg with coefficient of variation of 12.3%. The frequency distribution of weights at first heat showed that 41.7% of the heifers weighted between 321 and 360 kg. Positive phenotypic correlations of weight at first heat with age at first calving (0.25) and weight at first calving (0.42) were observed. The heritability estimate for the trait was 0.30 ± 0.19.

Weight at first calving

The weight at first calving in Murrah ranged between 449 and 525 kg (Singh, 1967; Gokhale, 1974; Johari and Bhat, 1979; Basu *et al.*, 1984).

Differences between farms, periods and sires for weight at flrst calving were significant (Tomar and Desai, 1969; Rathi *et al.*, 1971; Johari and Bhat, 1979). Age at first calving also significantly lnfluenced the weight at first calving (Basu *et al.*, 1984).

Tomar and Desai (1969), Johari and Bhat (1979) and Reddy (1980) reported heritability estimates of 0.34 ± 0.14, 0.23 ± 0.05 and 0.19 ± 0.06 respectively for this trait.

The phenotypic correlations between weight and age at first calving ranged between 0.15 and 0.39 (Tomar and Desai, 1965; Arya and Desai, 1969; Marwaha, 1974; Reddy, 1980). The genetic correlations were positive and high, 0.68 to 0.74 (Reddy, 1980) thus suggesting that selection for either of the trait would lead to correlated positive change in the other.

The phenotypic correlations between weight at first calving and first lactation yield were between 0.15–0.45 (Venkayya and Anantkrishnan, 1957; Gokhale, 1974; Marwaha, 1974). The genetic correlation between weight at first calving and first lactation milk yield (300 days) was 0.74 ± 0.09 (Gokhale, 1974). The high genetic correlation suggested that animals with higher weight at calving would have higher first lactation yield.

Growth rate

Growth rate, in general, was linear from birth to 36 months (Amble *et. al.*,1970; Rathi *et al.*, 1973; Nautiyal and Bhat, 1977). Average daily gains for different periods of growth have been reported by a number of workers. Nautiyal and Bhat (1977) reported average absolute monthly gains of 14.9, 16.4 and 15.1 kg for 0–3, 3–6 and 6–12 months age interval respectively. The average monthly gains, thereafter, up to 36 months for different period varied between 9.8 kg and 11.3 kg. The overall average monthly gain from 0–36 months was 12.1 kg.

Sharma and Basu (1984) reported the mean daily body weight gains for different age intervals as 647 g (birth to 12 weeks), 657 g (12 to 24 weeks), 552 g (24 to 36 weeks), 512 g (36 weeks to one year), 416 g (1 to 2 years) and 271 g (2 years to age at first calving) in Nili-Ravi buffaloes.

The results of body weight gains suggest that period up to one year of age could be economically utilised in feed lot for obtaining maximum growth rate. It seems that present potential of the species is around 500 g per day up to one year, and this can be improved further through selection of fast growing males.

9.2 Reproduction

The reproductive traits generally studied are age at oestrus, number of services per conception, number of calves produced per unit of time, age at first calving, service period, dry period and calving interval. Some of these parameters have been used to measure the breeding efficiency.

Breeding behaviour

Bhat *et al.* (1983) observed that buffalo cows continue to come in heat regularly in all months, highest being in October and lowest in April (Table 9.3). There were no differences between months and between seasons with regard to percent animals exhibiting oestrus and subsequent conception. The view earlier held that there was seasonality in conception (Goswami and Nair, 1964; Rao and Rao, 1968; Roy *et al.*, 1968) due to seasonal variation in oestrus cycle pattern, therefore, has to be rejected. The lower conception rate during summer (April to September) was attributed to poor semen quality of buffalo bulls. It was observed that hot-humid months affect the process of spermatogenesis of male buffalo adversely. The use of frozen semen during these months, is recommended to overcome the lower conception.

Table 9.3 Month and Season-wise Occurence of Heat and Conception Rate in Murrah Buffaloes

Month of insemination	Heat (%)		Conception rate (%)	
January	8.8		26.0	
February	8.3		26.2	
March	7.2	25.2b	23.6	23.6b
April	5.9		19.6	
May	7.4		19.8	
June	7.4	24.8c	16.2	19.6c
July	7.1		17.3	
August	7.2		18.2	
September	8.1	28.6d	21.2	22.0d
October	13.1		24.3	
November	10.3		28.3	
December	9.2	28.8a	24.9	26.5a

a: winter (November to January), b: spring (February to April), c: summer (May to July), d: autumn (August to October).
Source: Bhat *et al.* (1983).

Age at first oestrus

Very few reports are available on age at first oestrus in buffaloes. Basu *et al.* (1984) reported the age at first heat to be 30.6 months in Murrah buffaloes. Nagpuri buffaloes come in heat late at 42.48 months of age (Kaikini and Paragaonkar, 1969).

Age at first calving

A large variation in age at first calving across the breeds was noted; it being highest in village buffaloes (Amble *et al.*, 1958; Chhikara *et al.*, 1978). The averages based on large numbers in Murrah and Nili-Ravi were between 40 and 44 months (Gokhale, 1974; Sreedharan and Nagarcenkar, 1978; Johari and Bhat, 1979; Reddy, 1980), while that in Egyptian buffaloes was around 40 months (El-Seikh, 1967; Alim, 1978). In Surti, Bhadawari and Nagpuri buffaloes, the age at first calving was slightly higher, 46–54 months (Table 9.4).

Nonsignificant effects of month/season of calving on age at first calving have been reported by Rathi *et al.*, (1971); Gokhale (1974), Johari and Bhat (1979), whereas Kanaujia *et al.* (1974) and Reddy (1980) reported it to be significant. However, differences between farms and between years/periods for age at first calving were significant (Sharma and Singh 1978 a,b; Johari and Bhat, 1979; Reddy, 1980; Dutt and Taneja, 1994). The significant farm differences were due to differential availability of inputs at these farms in addition to varying climatic conditions. The farm differences could also be genetic in nature due to differences in genetic merit of sires used at these farms. In view of natural service practised at these farms, the chances of unequal distribution of superior germplasm were more.

Table 9.4 Age at First Calving (months)

Breeds	Location	No. of observation	Mean ± S.E.	References
1	2	3	4	5
Murrah				
- do -	Punjab	1340	51.4 ± 0.3	Amble *et al* (1958)
- do -	Deccan	279	39.9 ± 0.3	Bhadula and Desai (1973)
- do -	Lucknow	400	43.0 ± 0.2	Lal (1975)
- do -	Mumbai	529	43.6 ± 6.4	Gudi and Khedkar (1977)
- do -	Karnal	162	42.1 ± 0.5	Reddy and Mishra (1980)
- do -	Hisar	323	54.5 ± 1.4	Jain and Taneja (1982)
- do -	Military farms	1192	41.3 ± 0.1	Goswami and Nair (1965)
- do -	Military farms	502	40.3 ± 0.3	Krishnamacharyalu and Prabhu (1973)
- do -	Military farms	3250	42.4 ± 0.1	Johari and Bhat (1979)
- do -	Military arms	2465	42.5 ± 0.3	Reddy (1980)
Nili-Ravi				
- do -	Punjab villages	196	53.2 ± 0.6	Amble *et al.* (1958)
- do -	Military farms	59	40.5 ± 0.5	Ambe *et al.* (1970)
- do -	Military farms	425	41.4 ± 0.4	Reddy and Taneja (1984)
Bhadawari				
- do -	Bharari	80	50.7 ± 0.8	Singh and Desai (1962)
- do -	Nagpur	70	54.2 ± 0.8	Deb and Kadu (1977)
- do -	Nagpur	70	44.3 ± 0.2	Kadu *et al* (1978)
Pakistan (Nili-Ravi)				
- do -	Pakistan	338	47.0	Ashfaq and Mason (1954)
Egyptian				
- do -	Egypt	713	38.7 ± 0.2	Sidky (1953)
- do -	Egypt	240	39.9 ± 0.5	Alim (1978)

The heritability estimates for this trait varied between 0.12 and 0.34 (Reddy, 1980; Mangrurkar and Desai, 1981; Kalsi and Dhillon, 1982; Gurung and Johar, 1982). Jain (1982) Dutt and Taneja (1994) reported the heritability for age at first calving to be 0.53 ± 0.21 and 0.43 ± 0.08 respectively. Large differences in heritability estimates were due to differences in number of sires used, the number of progeny per sire and the variation between individuals within halfsib groups apart from differences in genetic merit of sires used in these herds (Table 9.5). It can however be concluded that this trait has low heritability.

Table 9.5 Heritability Estimates of Various Economic Traits of First Lactation in some Breeds of Indian Buffaloes

Breed	No. of obs.	Method of estimation	H2 ± SE	References (s)
		Age of first calving		
Murrah	2089	PHS	0.120 ± 0.060	Gokhale (1974)
- do -	804	- do -	0.200 ± 0.100	Sharma (1982)
- do -	1144	- do -	0.154 ± 0.068	Kumar (1984)
- do -	3250	- do -	0.374 ± 0.060	Johari and Bhat (1979a)
- do -	2465	- do -	0.290 ± 0.150	Reddy (1980)
Surti	95	- do -	0.475 ± 0.378	Basavaiah *et al.* (1983)
Bhadawari	167	- do -	0.452 ± 0.299	Sharma and Singh (1978a)

Low phenotypic correlations between age at first calving and first lactation milk yield have been reported by various workers (Dutt *et al.*, 1965; Tomar and Desai, 1968; Gokhale, 1974; El-Arain 1986). However, Dutt *et al.* (1965) reported negative phenotypic association between age at first calving and milk production up to 6 (-0.74), 8 (-0.55) and 10 years (-0.46) of age. Low and negative phenotypic correlations between these two traits were also reported by Rathi *et al.* (1971) and Gogoi *et al.* (1985) the genetic correlations between these two traits were reported to range between -0.98 and 0.13 (Dhinsa, 1963; Sane *et al.* 1972a; Gokhale, 1974; Gogoi *et al.* 1985; El-Arain, 1986; Dutt, 1991). Most of the estimates, however, had high standard errors. Gokhale (1974) reported genetic correlation between age at first calving and 300 days and total milk yield as -0.66 ± 0.11 and -0.76 ± 0.09. The negative genetic association between these two traits suggested that decrease in age at first calving would significantly increase the milk yield.

Gogoi *et al.*, (2002) studied the effects of non-genetic factors on age at first calving (AFC) of Murrah and Surti buffaloes and their heritability estimates on farms of Assam. The least square mean of AFC of Murrah buffaloes was significantly higher than Surti buffaloes (53.88 ± 0.48 vs. 51.51 ± 1.18 months, $P<0.01$). Farm had a significant effect on the AFC of Murrah buffaloes. The effect of season and period of birth on AFC was not significant for any of the breeds. It was concluded that the AFC of Murrah and Surti buffaloes under the agroclimatic conditions of Assam was satisfactory, although considerable farm variation exists due to the differences in the availability of feeds, health management practices and possible variation in climatic conditions among farms.

Rakshe (2003) conducted a study to determine the effect of age at first calving and subsequent period of breeding on the performance of buffaloes. 81 Surti, 62 Murrah and 37 crossbred heifers were grouped according to age at first calving, and the effect of this age was studied on subsequent performance by critically investigating the records of 489, 222 and 188 calvings, respectively. Murrah buffaloes had greater average age at first calving compared to Surti and crossbred heifers. More than 48% of the Surti buffaloes calved between the ages of 33 and 47 months. In Murrah and crossbred buffaloes, the percentages were 29 and 43%, respectively. Most of the Surti buffaloes did not give maximum milk yield in the first lactation irrespective of the age at first calving. The milk yield gradually increased from 2nd to 8th lactations. Murrah buffaloes calving at an early age (4 years) produced more milk than those that calved after 57 months of age. Similar observations were found in crossbred buffaloes. No effect of age at first calving was observed on milk days, dry days and calving interval. Serving the buffaloes in 3rd or 4th oestrus periods was better in Surti and Murrah breeds, whereas in crossbreds, no specific statement could be made. Serving buffaloes in 2 oestrus periods after calving resulted in good performance.

Meena and Jat (2003) studied calving records of 510 Surti buffaloes spread over a period of 25 years (1973–1997) collected from LRS Vallabhnagar, RAU, Bikaner, India. Least squares means and their standard error of total calves born and female calves reaching milking herd were estimated in relation to period, season, age at first calving (AFC) and first lactation milk yield (FLMY). The overall means for total calves born to each buffalo and total female calves reaching milking herd from each buffalo were 3.55 ± 0.16 and 0.68 ± 0.07, respectively. The least squares analysis of variance indicated that the effect of period, AFC and FLMY was significant on total calves born and total female calves reaching milking herd whereas season had no effect on either trait.

Service period

The average first service period in Murrah buffaloes ranged from 115 to 230 days with an overall average of 132 days (Kohli and Malik, 1960; Goswami and Kumar, 1968; Kanaujia *et al.*, 1974; Johari and Bhat, 1979; Reddy, 1980; Jain and Taneja, 1982). The average first service period in Nili-Ravi, Bhadawari and Egyptian buffaloes was 201, 193 and 198 days respectively (Ragab *et al.*, 1956; Sharma and Singh, 1978; Reddy and Taneja, 1984). In general, the service period in Murrah buffaloes at well organised farms was much lower than other breeds.

The significant farm and year/period differences for service period were reported by Kanaujia *et al.* (1975), Johari and Bhat (1979) and Reddy (1980). Similarly, significant differences between months/seasons were also observed

(Kanaujia *et al.*, 1974; Basu *et al.*, 1978; Sharma and Singh, 1978; Reddy and Taneja, 1984). Luktuke (1957), observed that buffaloes calving during autumn and winter showed post-partum oestrus 50–60 days.The period between calving and first fertile heat was 80–90 days in animals calving in autumn and winter and 100 days in those calving in spring. Reddy (1980) observed that average first service period was lower (132–142 days) in July to November calvers than December to April calvers (203–239 days). This suggested that the buffaloes calving during December to April months should be carefully watched for heat in order to reduce the service period.

The heritability estimates for service period were, in general, low and ranged between 0.08 ± 0.06 (Reddy, 1980) and 0.22 ± 0.28 (Jain, 1982). The phenotypic correlations between service period and milk yield were between 0.16 and 0.55 (Kohli and Malik, 1960; Gill and Dev, 1970; Basu and Ghai, 1978; Reddy, 1980). The positive and significant correlations suggested that higher milk yield was associated with longer service period.It was concluded that this trait needs to be improved by better management.

Dry period

The averages of first dry period in Murrah buffaloes (Table 9.6) ranged from 153 to 217 days (Jawarkar and Johar, 1974; Gurnani *et al.*, 1976; Kanaujia, 1978; Reddy, 1980; Jain and Taneja, 1982) with an overall average of 160 days. Chaudhary and Shaw (1965) reported much shorter first dry period of 98 days in Pakistan's Nili-Ravi buffaloes whereas longer dry period (202 days) for this breed was reported by Reddy and Taneja (1984). Average dry period pooled over the lactations in Bhadawari, Marathwada and Nagpuri buffaloes was 156.4 (Singh and Desai, 1962), 134.2 (Hadi, 1965) and 129.7 days (Kadu *et al.*, 1978). In Egyptian buffaloes, it ranged from 200 (Alim, 1978) to 296 days (Alim and Ahmad, 1954).

The variation between farms, between years and between months influenced dry period significantly. The average dry period for July to November calvers (145–180 days) was lower than December to April (191–237 days) calvers (Reddy 1980). Both significant (Kumar, 1984; El-Arain, 1986, Gupta, 1988) and nonsignificant (Dutt *et al.* 1983) effects of season of calving on first dry period have been reported.

Jawarkar and Johar (1974) estimated the heritability of first four dry periods as 0.20, 0.01, 0.20 and 0.02 respectively in Murrah. In Egyptian buffaloes, Asker *et al.* (1953) estimated the heritability of dry period as 0.18. Kumar (1984) and

Dutt (1991) reported the heritability estimate to be around 0.0 and 0.8 to 0.7. Most of the estimates reported had high standard errors which suggest that the trait can be improved through better breeding management, the heat detection and breeding control.

Table 9.6 Averages of First Dry Period (days) in Some Breeds of Indian Buffaloes

Breed	No. of obs.	Mean±SE	Reference (s)
Murrah	301	138.10 ± 14.56	Arora *et al.* (1962)
- do -	210	151.52 ± 8.92	El-Arian (1986)
(MDF)	2539	183.40 ± 2.00	Johari and Bhat (1979)
- do -	450	204.51 ± 4.20	Dutt *et al.* (1983)
Surti	69	211.81 ± 14.84	Basavaiah *et al.* (1983)

Gestation length

The gestation length in buffaloes is somewhat longer than cows with a range between 299 and 346 days (El-Sheikh, 1967; Joshi *et al.*, 1969). Perera and de Silva (1985) observed that the mean gestation length for Lanka buffaloes (316.3 ± 6.5 days) was significantly longer than that for Murrah (309.9 ± 6.5 days).

Ghanem *et al.* (1955) observed significant difference in length of gestation between paternal and maternal half-sibs and full-sibs. The heritability estimates for gestation length were reported as 0.32 ± 0.02 (Joshi *et al.*, 1969) in Murrah and 0.32 ± 0.13 (Ghanem *et al.*, 1955) in Egyptian buffaloes. Most of the other estimates reported were low with high standard error.

Calving interval

The first calving interval in Murrah, Nili-Ravi and Egyptian buffaloes (Table 9.7) varied between 479 and 508 days (Alim, 1978; Johari and Bhat, 1979; Reddy and Taneja, 1984; Dutt, 1991). The overall calving interval, however, was lower (430–457 days) than the first calving interval (Kanaujia *et al.* 1975; Lall, 1975). Patro and Bhat (1979) reported a decline of about 95.2 days in calving interval from first to the seventh lactation. Basavaiah *et al.* (1983) reported the average first calving interval to be 583.3 days in Surti buffaloes (Based on 69 records only).

Significant effects of farms, years/periods on calving interval have been reported (Kanaujia *et al.*, 1975; Basu and Ghai, 1978; Johari and Bhat, 1979; Reddy, 1980; Kumar, 1984; Singh and Yadav, 1986). Month/season of calving also influenced this trait significantly (Gurnani *et al.*, 1976; Basu *et al.*, 1978; Reddy, 1980; Kumar, 1984; El-Arain, 1986; Singh and Yadav, 1986).

Most of the heritability estimates reported for this trait were low and non-significant. The phenotypic correlations between first calving interval and first lactation milk yield (300 days) were between 0.20 and 0.38 (Dhinsa, 1963; Gill and Dev, 1970; Basu and Ghai, 1978; Basu *et al*., 1978; Reddy, 1980; Dutt, 1991). These significant correlations suggested that buffaloes with higher lactation yield would have longer calving interval. Such a phenotypic association was expected since prolonged calving interval which was directly related to feeding, management and heat detection practises in the farm.

Table 9.7 Average of First Calving Interval (days) in some Breeds of Indian Buffalo

Breed	No. of obs	Mean ± SE	Reference (s)
Murrah	1144	494.48 ± 05.51	Kumar (1984)
- do -	628	495.37 ± 81.63	Singh and Basu (1988)
(MDF)	2579	479.50 ± 02.41	Johari and Bhat (1979a)
- do -	2027	483.15 ± 07.62	Reddy (1980)
Nili-Ravi	368	508.39 ± 10.10	Reddy (1980)
Surti	69	583.33 ± 20.84	Basavaiah *et al*. (1983)

9.3 Production

Part lactation

Part lactation records are used to predict the lactation milk yield and are also useful in early evaluation of sires. Detailed systematic studies on part lactation records in buffaloes are lacking and more work needs to be done.

Four methods of extending part lactation records have been used by various workers. These are (i) Ratio, (ii) Multiple regression, (iii) Modified regression and (iv) Regression of remainder of the lactation on the last test. Errors of multiple regression and regression estimates on the last test are minimum, modified regression inter-mediate while error of the ratio estimate was the largest.

Kumar *et al.* (1979) extended part lactation records using ratio, regression and modified regression methods and compared their efficiency. Prediction of total yield using various cumulative monthly yields by three methods showed that predicted values using modified regression methods were close to the average yield than those obtained by regression or ratio methods. The regression method was however, better than the ratio method. The precisions were better when early segments of the lactation curve were used.

Kumar *et al.* (1978, 1979) observed that heritability of monthly milk production increased with the stage of lactation up to sixth month and declined thereafter Table 9.8. The heritability estimate for 6th month yield was highest, 0.26 ± 0.06. The differences between the heritability estimates for 3rd, 4th, 5th and 6th month were, however, small; therefore, the period between 3rd and 6th month could be classed as one which reflected the totality of production. The heritability estimates for cumulative monthly yield showed an increase from 1st to 7th month and declined steadily thereafter. The heritability estimates for first 5 months to first 9 months yield were higher while that of first 4 month cumulative yield was almost the same as that for total yield. In view of higher heritability estimates of early segments of lactation curve, selection on the basis of first 4 months to first 7 months cumulative production will be as effective as that based on total yield.

Table 9.8 Heritability Estimates of Monthly and Cumulative Monthly Milk Yield and Genetic and Phenotypic Correlation between Cumulative Part and 300 day Milk Yield.

Month of lactation	$h^2 \pm$ S.E.	Cumulative months	$h^2 \pm$ S.E.	Genetic correlation	Phenotypic correlation
1	0.13 ± 0.05	1	0.13 ± 0.05	0.94 ± 0.03	0.63 ± 0.02
2	0.18 ± 0.05	1–2	0.18 ± 0.06	0.90 ± 0.04	0.71 ± 0.02
3	0.23 ± 0.06	1–3	0.22 ± 0.06	0.91 ± 0.03	0.77 ± 0.01
4	0.23 ± 0.06	1–4	0.26 ± 0.07	0.92 ± 0.03	0.80 ± 0.01
5	0.25 ± 0.07	1–5	0.27 ± 0.07	0.93 ± 0.02	0.84 ± 0.01
6	0.26 ± 0.07	1–6	0.28 ± 0.07	0.94 ± 0.02	0.87 ± 0.01
7	0.20 ± 0.06	1–7	0.29 ± 0.07	0.95 ± 0.02	0.89 ± 0.01
8	0.13 ± 0.05	1–8	0.27 ± 0.07	0.96 ± 0.02	0.93 ± 0.01
9	0.22 ± 0.07	1–9	0.26 ± 0.09	0.98 ± 0.01	0.97 ± 0.01
10	0.10 ± 0.08	1–10	0.26 ± 0.07	0.99 ± 0.00	0.99 ± 0.00

Source: Kumar *et al.* (1978, 1979).

The genetic correlations among early months of lactation were higher than those among the later months. The correlations among the first seven months segments were near unity. The genetic correlations between different cumulative part yields and total yield showed steady increase with each added month of lactation (0.90 to 0.99). Similar was the trend for phenotypic correlations which ranged between 0.63 and 0.99. In view of high heritability estimates of middle segments of lactation curve and their high genetic and phenotypic correlations with total yield, these cumulative part records could be used in sire evaluation and thus would give higher genetic gains than expected from selection based on total yield.

Persistency

The individual variation in lactation curve in buffaloes in a herd, due to inherent variability in their milk yield, may be ascribed to differences in either of the following or a combination thereof; initial maximum yield, persistency of production and the lactation length. The initial maximum yield and the length of lactation can directly be determined from milk records but the persistency of lactation needs to be quantified. Persistency refers to the degree with which the level of milk production after the attainment of peak is maintained as lactation advances.

Bhat *et al.* (1982) estimated persistency of milk yield using ratio of production in different segments of the lactation, to find out whether the methods used in cattle could be satisfactorily used for measuring persistency in buffaloes. The method P_2 (ratio between four segments of 11 weekly intervals) gave relatively more precise estimate of persistency (0.98 ± 0.002). Order of lactation had significant effect on persistency of milk yield estimated by four out of eight methods. In all these four methods, the persistency values for first lactation were always higher than in succeeding lactations.

Farms had significant effect on estimates of persistency which could be due to (i) differences in segments of lactation period utilised, (ii) variation in number of animals completing 300 days of lactation within each farm, (iii) differences in genetic constitution of the herds, (iv) prevailing geo-climatic conditions, and (v) levels of management available at different farms. The effect of year of calving on persistency of milk yield was significant in 4 out of 8 methods, which could be due to differences in the level of management, changes in herd size and changes in level of production due to variation in climate.

The heritability estimates of persistency index from all the eight methods were nearly zero indicating that persistency was primarily influenced by managemental factors rather than the heredity.

Lactation milk yield

The average first lactational milk yield in Murrah buffaloes varied between 1540 and 1868 kg (Gokhale, 1974; Bhalaru and Dhillon, 1978; Sreedharan and Nagarcenkar, 1978; Patro and Bhat, 1979; Reddy, 1980) while in Nili-Ravi, it was 1707 kg (Reddy and Taneja, 1984). Alim (1967) reported the lactation milk yield to be 1771 kg in Egyptian buffaloes,whileFahmy *et al.*(1975) reported a range of 1360 to 2267 kg for this breed. The first lactation averages for Bhadawari (1165 kg), Nagpuri (926 kg) and Surti (1364 kg) were much lower (Table 9.9).

Average milk yield in Nili-Ravi buffaloes in Pakistan varied between 1969 and 2731 kg (Ashfaq and Mason, 1954; Cady *et al.* 1983; Khan, 1986) while the average milk yield in Kundi breed was reported to be 1781 kg (Wahid, 1976). Cady *et al.* (1983) observed that these were 1.2% of all buffaloes at the government farms which had more than 4000 kg milk and this proportion of high yielding buffaloes suggested reasonable opportunity for selection of dams to produce bulls.

Analysis of data on 5098 lactation records of 1200 animals from CIRB from 1965 to 1996 were utilized for estimation of repeatability of various economic traits. The repeatability estimates were: 0.338 ± 0.015 for total lactation milk; 0.309 ± 0.015 for 301 days or less lactation milk yield; 0.311 ± 0.015 for peak yield; 0.266 ± 0.014 for milk yield per day of lactation length, and 0.216 ± 0.016 for milk yield per day of calving interval. The repeatability for lactation length, dry period and calving interval was 0.160 ± 0.015, 0.251 ± 0.015 and 0.160 ± 0.015, respectively.

Table 9.9 First Lactation Yield

Breed	No. of Obs.	Mean ± S.E.	Reference (s)
Murrah	2045	1540.7 ± 9.8	Gokhale (1974)
- do -	576	1709.5 ± 16.4	Kanaujia and Singh (1975)
- do -	817	1867.4 ± 20.9	Kumar and Bhat (1978)
- do -	1352	2041.5 ± 5.2	Singh *et al.* (1990)
Surti	62	2203.0 ± 51.7**	Singh and Singh (1977)
- do -	95	1237.7 ± 56.3	Basavaiah *et al.* (1983)
- do -	477	1421.2 ± 15.1	Nagar Cenkar *et al.* (1985)
Bhadawari	37	1165.0 ± 27.7	Singh and Singh (1977)
- do -	167	925.9 ± 23.9	Sharma and Singh (1987b)
Nagpuri	73	926.3 ± 17.1	Belorkar *et al.* (1977)
Marathwada	23	845.7 ± 46.2	Singh and Singh (1977)
- do -	40	1512.0 ± 79.0	Alim (1957)
- do -	170	1360.0 ± 2267.0	Fahmy *et al.* (1975)

** The lactation length in Surti under reference wasvery high, being 426 days.

Gopal Dass and Sadana (2000) analysed 1176 first three lactation records of Murrah buffaloes maintained at the National Dairy Research Institute, Karnal and the Punjab Agricultural University, Ludhiana, spread over a period of 11 years to assess the influence of farm, season and period of calving and parity on peak yield, 305 days yield, complete lactation yield, lactation duration, dry period and calving interval. The least squares means were 12.04 ± 0.14 kg, 1934.77 ± 19.81 kg, 2067.74 ± 23.86 kg, 313.19 ± 2.77 days, 172.84 ± 5.52 days and 492.38 ± 5.82 days, respectively for the traits under study. Farm, period of calving, season of calving and parity had influence ($P<0.01$) on all the traits except non-significant effect of farm on peak yield and complete lactation yield and period of calving on dry period. Significant ($P<0.05$) effect of period of calving on complete lactation yield and dry period and season of calving on peak yield and dry period and parity on dry period was also observed.

Gopal Dass and Sharma (2000) observed daily milk yields of 404 Murrah buffaloes in their first lactation over a period of 15 years (1975–89) at the National Dairy Research Institute, India, were utilized for predicting 305-day lactation yield using 2 sampling (systematic and stratified random) methods at 3 (fortnightly, monthly and bimonthly) sampling intervals. All sampling schemes and intervals overestimated the actual 305 day yield. The effects of season (5% level of probability) and period (1% level of probability) on predicted yields in systematic and stratified random schemes of 3 samplings were significant. The regression of milk yield on age at first calving had no significant effect.

Arora *et al.* (1962) reported that milk yield increased from first (1733.6 kg) to sixth (2563.6 kg) lactation in Murrah. Almost similar observations were made by Sane *et al.* (1972) except that lactation milk yields reported by them in first (1564.7 kg) and sixth (1849.9 kg) lacatations were lower. Patro and Bhat (1979) observed a steady increase from first lactation onwards with a peak in fourth lactation (Table 9.10).

Table 9.10 Milk yield (kg) with Lactation Length (days) up to First 6 Lactations

Lactation number	Milk yield Mean ± S.E.	300 day yield Mean ± S.E.	lactation length Mean ± S.E.
1	1618.5 ± 22.78	1573.4 ± 22.21	297.8 ± 2.91
2	1880.0 ± 32.64	1790.4 ± 27.74	300.0 ± 3.82
3	1964.0 ± 37.02	1878.0 ± 36.25	298.3 ± 4.32
4	2039.5 ± 50.18	1963.9 ± 45.56	291.3 ± 5.48
5	2024.3 ± 73.94	1959.7 ± 67.51	290.8 ± 7.86
6	1823.7 ± 97.38	1767.5 ± 89.00	270.0 ± 10.08

Source : Patro and Bhat (1979)

The milk yield up to ninth lactation was maintained close to the peak lactation suggesting that buffaloes could be maintained in the herd up to ninth lactation (16 years of age) with reasonable economic returns. These authors suggested that stage of lactational maturity corresponded with the development and increased functioning of active secretary tissue of the udder, which was attained around fourth lactation. As a species, the lactational maturity in buffaloes was attained between fourth and sixth lactation.

Significant effects of farms and years/periods on milk yield have been reported (Khanaujia *et al.* 1975; Bhalaru and Dhillon, 1978; Kumar and Bhat, 1978; Patro and Bhat, 1979; Reddy, 1980). Patro and Bhat (1979) observed the farm differences to be significant for all the six lactations studied. The lower/higher yield at the farms were attributed to corresponding lactation length. The higher yields at some farms were not only due to longer lactation length but could also be due to better management and to the presence of genetically high yielding stock.

Singh and Singh (1967), Kumar and Bhat (1978) and Jain and Taneja (1982) reported the effect of month/season of calving to be nonsignificant for Murrah and Polihronov *et al.* (1967) for Bulgarian buffaloes. These authors observed higher milk yield for buffaloes that calved during least calving season (January-June). Reddy (1980) reported that first lactation yield was higher for animals calving from January to April (1966 to 1733 kg) and lower for August to December calvers (1548 to 1611 kg). The higher milk yield for January to April calvers was attributed to the availability of sufficient amount of green fodder during these months.

Bhat and Patro (1978) examined the effects of age and weight at calving, preceding dry period and preceding service period on milk yield adjusted for effects of farm, period and season of calving. Age at calving had positive and significant effect on milk yield up to third lactation, while weight at calving and preceding service period had an effect up to fifth and fourth lactations respectively. Longer preceding dry periods were not favourable for higher milk yield. The age and weight at calving preceding service and dry period and lactation length together explained 34.1% of the total variability in milk yield. These factors, without lactation length, explained only 9.7% of variability. Jain (1982) studied the effects of lactation length and first service period individually and jointly in addition to periods, sires and seasons on yield. The effect of lactation length was highly significant when considered alone and with service period. The models with lactation length explained 55% of the total variation in milk yield. Addition of service period in the model did not contribute anything to the total variability. These results suggest that lactation length was a major source of variation in milk yield.

Most of heritability estimates (Table 9.11) for first lactation milk yield in Murrah buffaloes barring a few which were moderate were between 0.08 ± 0.04 and 0.19 ± 0.08. The heritability estimate for Nili-Ravi was 0.28 ± 0.20 (Bhullar, 1974). Jain (1982) demonstrated the impact of adjustment for variation in lactation length on the magnitude of heritability of milk yield by considering models with and without lactation length. The heritability estimate for milk yield when only periods and seasons were considered, was 0.08 ± 0.07.

Table 9.11 Heritability Estimates for Milk Yield

Breed	No. of obs./pairs	$h^2 \pm$ SE	(Method)	Reference (s)
Murrah/ Murrah grades	413	0.16 ± 0.11	(ISRD)	Mahadevan (1960)
- do -	618	0.48 ± 0.15	(PHC)	Bhadula (1966)
- do -	503	0.28	- do -	Amble *et al.* (1970)
- do -	886	0.42 ± 0.05	- do -	Gurnani and Nagarcenkar (1972)
- do -	1433	0.12 ± 0.06	- do -	Basavaiah (1978)
- do -	2735*	0.08 ± 0.04	- do -	Johari and Bhat (1979)
- do -	1144	0.18 ± 0.07	- do -	Kumar (1984)
Nili Ravi	204	0.28 ± 0.20	- do -	Bhullar (1974)
Nili Ravi grade	356	0.32 ± 0.16	- do -	Bhullar (1974)
Nili Ravi	357	0.08 ± 0.10	- do -	Reddy (1980)
Egyptian	64	0.18 ± 0.16	- do -	Asker *et al.* (1953)
Egyptian	2281	0.48	- do -	Asker *et al.* (1967)

PHC = Paternal half sib correlation method
ISRD = Intrasire regression of daughter on dam method.
* = Murrah, Murrah grades, Nili-Ravi and Nili-Ravi Grades.

Significant phenotypic correlations between first lactation milk yield and first lactation length (0.42 to 0.65) were reported by Dutt *et al.* (1965), Johari and Bhat (1979), Patro and Bhat (1979) and Reddy (1980). Higher correlation values (0.64 to 1.0) between these two traits have also been reported (Gill and Dev, 1970; Rao *et al.*, 1976; Basu andGhai, 1978).The phenotypic correlations between lactation yield and 300 day yield over the first six lactations were high, 0.85–0.98. The genetic correlations were also high 0.66–1.0 (Patro and Bhat, 1979) which was expected. The genetic correlation between first lactation yield and first lactation length was 0.46 ± 0.24 (Johari and Bhat, 1979) while that between milk yield and lactation length over the first six lactation were not significantly different from zero (Patro and Bhat, 1979). This was because most of the heritability estimates for these traits were low and had high standard errors.

Singh and Nivasarkar (2000) studied the production and reproduction potential of Bhadawari animals and the factors (farm, parity, calving problem, sex of calf, season and period) influencing various traits in Bhadawari buffalo. The least square means of birth weight (BW), age at first calving (AFC), first lactation total milk yield (FLTMY), first lactation milk yield 300 days (FLMY 300 days), pool lactation milk yield 300 days (PLMY 300 days), pool lactation total milk yield (PLTMY), lactation length (LL), milk yield per day of lactation length (MY/LL), milk yield per day of calving interval (MY/CI), dry period (DP), gestation period (GP), service period (SP), and calving interval (CI) were 25.5 ± 0.4 kg, 1540.7 ± 46.6 days, 693.2 ± 63.1 kg, 678.2± 53.9 kg, 650.4 ± 449 kg, 657.9 ± 48.6 kg, 284.8 ± 10.7 days, 2.45 ± 0.12 kg, 1.52 ± 0.10 kg, 213.1 ± 22.7 days, 308.9 ± 1.8 days, 213.3 ± 26.7 days and 524.7 ± 25.9 days, respectively

Karthikeyan *et al.* (2002) assessed the performance of the Toda buffalo in the Nilgiris, Tamil Nadu, India. A total of 578 milk recordings were carried out from 143 animals twice daily at monthly intervals and 235 milk samples were collected for the estimation of milk constituents. The overall mean daily milk yield was 2.53 ± 0.06 kg. The mean lactation length and estimated lactation yield were 198.25 ± 5.25 (n=68) days and 603 kg, respectively. The fat, protein and total solids in milk were 8.28 ± 0.12, 4.29 ± 0.05 and 16.06 ± 0.14% respectively. The average ages at first calving and calving interval were 46.92 ± 0.36 (n=169) and 14.26 ± 0.36 (n=103) months respectively. There was a marked seasonality in calving with a peak in August. The distinguishing features of the Toda buffaloes are high milk fat content, regularity in reproduction and longevity of buffalo cows.

Lactation length

The first lactation length (Table 9.12) in Murrah and Nili-Ravi was around 300 days (Gokhale, 1974; Basu and Ghai, 1978; Reddy, 1980); El-Arain, 1986; Dutt, 1991). However, higher averages of 321, 323 and 349 days in Murrah buffaloes were reported by Singh and Basu (1988), Jain (1982) and Gupta (1988). In Bhadawari, Nagpuri and Surti, it was between 276 and 295 days (Singh and Desai, 1962; Belorkar *et al.*, 1977). Higher lactation length (359–426 days) in Surti buffaloeswasreported (Singh and Singh, 1977; Basavaiah *et al.* 1983; Rao *et al.* 1985). Alim (1967) reported the first lactation length as 325 days in Egyptian buffaloes.

Table 9.12 Averages of First Lactation Length (days) in some Breeds of Indian Buffaloes

Breed	No. of obs.	Mean ± SE	Reference (s)
Murrah	716	348.52 ± 4.50	Gupta (1988)
(MDF)	2842	295.40 ± 1.00	Johari and Bhat (1979b)
- do -	2138	297.80 ± 2.91	Patro and Bhat (1979b)
- do -	2226	296.39 ± 3.17	Reddy (1980)
Surti	1307	359.41 ± 2.23	Rao *et al.* (1985)

Significant effects of farms on lactation length have been reported by Kanaujia *et al.* (1975) and Patro and Bhat (1979) while Reddy (1980) and Dutt (1991) reported it to be nonsignificant. Significant differences between months/seasons for lactation length have been reported by Tomar and Tomar (1960), Patro and Bhat (1979) and Jain and Taneja (1982) for Murrah; Polihronov *et al.* (1967) for Bulgarian and Roy Choudhury *et al.* (1971) for Italian buffaloes. These authors observed that off-season calvers (December to June) had longer lactation length. However, Desai and Kumar (1964), Singh (1966), Reddy (1980) Dutt (1991) observed the effect of month/season year of calving on lactation length to be nonsignificant.

Patro and Bhat (1979) reported heritability estimate of 0.11 ± 0.05 for first lactation length. Higher heritability estimates of 0.28 ± 0.11 (Sharma, 1982) and 0,.31 ± 0.06 (Singh and Tiwana, 1983) for lactation of length have also been reported. The estimates for subsequent lactation lengths were lower than the first. Most of the estimates reported by other workers were also low and had high standard errors (Table 9.13).

Table 9.13 Heritability Estimates of First Lactation Length

Breed	No. of obs.	Method of estimation	h^2 ± SE	Reference (s)
Murrah	2735	PHS	0.111 ± 0.040	Johari and Bhat (1979b)
-do -	910	- do -	0.31 ± 0.06	Singh and Tiwana (1983)
Surti	95	- do -	0.076 ± 0.29	Basavaiah *et al.* (1983)

Negative phenotypic correlations between lactation length and dry period have been reported by Gill and Dev (1970) and Basu and Ghai (1978) while Reddy (1980) and Jain (1982) reported it to be low. The phenotypic correlations between lactation length and calving interval and between lactation length and service period were, in general, high, 0.51 to 0.76 (Gill and Dev, 1970; Basu and Ghai, 1978; Reddy, 1980; Jain, 1982) suggesting that longer lactation period would increase calving interval. The high correlation between first lactation length and calving interval are expected because lactation length is part of the calving interval and a major cause of variation in it. Similarly, service period is a cause of variation in lactation length. These correlation values suggest that increase in service period shall lead to an increase in lactation length and increase in lactation length shall increase calving interval, which is not financially viable and needs to be avoided in the managment of the herd.

Measures of efficiency of milk production

The first lactation milk yield is the most commonly used selection criterion in dairy animals. The economic merit of dairy animals, however, is also influenced by many other characters like age and weight at first calving, lactation length and calving interval. Although, selection index procedure is the most efficient known method of combining many traits into a single measure of net merit, it needs facilities of a computer and can not be easily used under field conditions. It is therefore, essential to develop some simple measure of milk production efficiency which takes into account variation caused by factors like lactation length, calving interval, age at first calving and has high genetic correlation with milk yield reflected by higher heritability than that of milk yield. Some measures of milk production efficiency developed and studied are: milk yield per day of age at calving, milk yield per kg of weight at first calving, milk yield per day of lactation length and milk yield per day of calving interval.

The average milk yield per day of lactation length ranged from 4.06 kg to 6.62 kg (Ram *et al.*, 1976; Singh and Singh; 1977. Dutt, 1991, in Murrah buffaloes. Amble *et al.* (1970) reported it to be 5.50 ± 0.34 and 6.10 ± 0.34 and 6.10 ± 0.22 kg in Nili-Ravi buffaloes at two farms. Singh and Singh (1977) reported it as 6.04, 3.31, 3.48 and 5.17 kg, respectively, in Nili, Bhadawari, Marathwada and Surti buffaloes. Elsawaj *et al.* (1964) reported an average yield per day of first lactation length as 4.84 kg in Egyptian buffaloes.

The milk yield per day of calving interval ranged between 3.02 and 4.00 kg (Amble *et al.*, 1970; Singh, 1976; Rao *et al.*, 1976; Bhalaru and Dhillon, 1978; Bhat *et al.*, 1982; Dutt, 1991) in Murrah buffaloes, while it was 3.70 ± 0.24 and 4.00 ± 0.20 kg in Nili-Ravi buffaloes at two farms (Amble *et al.*, 1970).

Bhalaru and Dhillon (1978), Reddy (1980), Bhat *et al.* (1982), Dutt (1991) estimated genetic parameters for some measure of milk production efficiency and compared these with heritability estimates for milk yield. Most of the estimates reported by Reddy (1980) and Bhat *et al.* (1982) for milk production efficiency parameters were low and many of these were not different from zero. The trend however, was that the heritability estimates for milk yield per day of lactation length and calving interval were higher than those for first lactation yield. Bhalaru and Dhillon (1978) observed that heritability of milk yield per day of first lactation length (0.29 ± 0.10) was higher than that of first lactation yield (0.19 ± 0.09).

The genetic correlation between these two traits was 0.90 ± 0.05. The coefficient of variation, however, for milk yield per day of lactation length was low (22%) as compared to first lactation yield (27%). Dutt (1991) estimated the heritability of milk yield per day of lactation length to be 0.33 ± 0.08 as against 0.23 ± 0.07 for first lactation milk yield. In view of the higher heritability of milk yield per day of first lactation length and higher genetic relationship with milk yield, selection on the basis of this trait would result in higher genetic improvement than expected from selection for milk yield.

The heritability of milk yield per day of first calving interval was (0.19) equal to that of first lactation milk yield. The heritability estimate for milk yield per day of age at second calving was 0.31 ± 0.10, and it had high genetic correlation (0.83 ± 0.08) with first lactation yield. The coefficient of variation was also higher (32.1%) when compared to first lactation yield. However, selection for milk yield per day of age at second calving would increase the generation interval. It would, therefore, be desirable to select on the basis of milk yield per day of lactation length, which also adjusts for variation in milk yield due to lactation length.

Bhalaru and Dhillon (1981) evaluated the usefulness of some measures of efficiency of milk production in buffaloes as selection criteria as compared to selection based on either first lactation yield or an index incorporating age at first calving, first lactation yield and first calving interval. The results of accuracy of different criteria of selection revealed that selection index was the most accurate criterion of selection (R=0.603). Out of different measures of efficiency of milk production, the milk yield per day of age at second calving had the highest correlation (R=0.462) with net merit, followed by milk yield per day of first lactation length (R=0.415) and first lactation milk yield (R=0.294).

The results of comparative efficiency of different selection criterion revealed that selection based on milk yield per day of age at second calving and per day of first lactation length were expected to be 23 and 31% less efficient than the selection index. The selection based on first lactation yield, would, however, be 51% less efficient than the index selection. Milk yield per 100 kg body weight at first calving and yield per day of first calving interval were the least efficient criteria of selection. Perusal of expected genetic gains in net merit (per generation) showed that selection based on an index was expected to result in maximum gain in the net merit as compared to any other selection criteria. The index selection was expected to bring about maximum decrease in age at first calving days (23.7) and maximum increase in first calving interval (3.2 days). The genetic progress in first lactation yield would be 30.5 kg. As compared to this, selection based on first lactation milk yield will result in much lesser decrease in age at first calving (6.6 days) and a much smaller increase (0.8 days) in first calving interval. The genetic improvement in first lactation milk yield will almost be of the same magnitude as that from use of selection index. The selection based on yield per day of lactation length would be expected to bring about maximum increase in first lactation yield (33.4 kg) and decrease of 1.6 days in first calving interval. The decrease in age at first calving (8.5 days) will, however, be relatively small.

Herdlife

The average productive life (sum of first three or four calving intervals) in Murrah buffaloes ranged between 1378 and 1884 days (Sharma and Singh, 1975; Tomar and Basu, 1981; Kalsi and Dhillon, 1982). Sharma and Basu (1986) reported the herdlife in Murrah buffaloes to be 1947.04 ± 72.57 days; herdlife was the difference between the date of first and the last calving. Dutt *et al.* (1965a) estimated the productive life (date of birth to date of disposal) as 83.29 ± 5.22 months and longevity (date of first calving to date of culling) as 126.87 ± 5.33 months for Murrah buffaloes. Kalsi and Dhillon (1982) reported the productive llfe (sum of first four claving intervals) for large size dairy buffaloes of unspecified breed to be 1359 ± 83.40 days.Dutt (1991) reported the number of lactations completed and herdlife in military farm buffaloes to be 5.78 ± 0.15 and 2401.30 ± 63.65 days respectively.

Lifetime milk production and yield per day of life

Lifetime milk production (based on 3 lactation) has been reported to range between 4498.04 ± 26.20 and 5251.7 ± 29.64 kg (Johari and Bhat, 1979b; Patro and Bhat, 1979b; Gokhale and Nagarcenkar, 1980) for Murrah buffaloes. Dutt *et al.* (1965) reported the average milk yield up to age of 6, 8, 10 years and up to age at culling to be 3289.34 ± 115.14, 6167.39 ± 182.04, 9067.39 ± 224.84 and 9671.90 ± 692.06 kg, respectively. Sharma and Basu (1986) defined lifetime milk yield as the amount of total milk up to death or disposal and reported the average to be 8913.50 ± 381.30 kg. Dutt (1991) reported the lifetime milk yield to be 9993.95 ± 338.30 kg.

Milk yield per day of productive life (first to fourth lactation) was reported to be 4.70 ± 0.06 kg (Umrikar and Deshpande (1985), and herd life to be 4.30 ± 0.90 kg (Sharma and Basu, 1986) and 4.0 ± 0.07 kg (Dutt, 1991) in buffaloes. It has been reported the average daily lifetime milk yield for Murrah and Nili-Ravi breeds to be 5.36 ± 0.46 and 4.72 ± 0.56 kg, respectively.

Umrikar and Deshpande (1985) observed that farm, period and season of calving had highly significant effect on milk yield per day of productive life in Murrah buffaloes. Johari and Bhat (1979b) and Patro and Bhat (1979a) also reported significant differences between farms in respect of lifetime milk production (based on 3 lactations).

It was observed that effects due to farms were highly significant for lifetime milk yield, lactation period and dry period in Murrah and Nili-Ravi buffaloes. The effect of season of calving was non-significant for lifetime milk yield and lifetime average daily milk yield. Farm × season interaction did not influence lifetime milk yield and lifetime average daily milk yield, but significantly affected lifetime lactation period and dry period. They also reported that period × season interaction had no significant effect on lifetime traits. Dutt (1991) reported the farm and year of calving effects for number of lactations completed herdlife, lifetime milk yield and milk yield per day of herdlife to be significant.

Genetic and Phenotypic Parameters of Lifetime Performance Traits

Sundaresan *et al.* (1954) reported correlation of age at first calving with lifetime production up to 5, 7 and 10 years of age to be 0.39, -0.39, -0.79 and -0.35, respectively. The correlations of first lactation with lifetime milk production up to 5, 7 and 10 years of age in Murrah buffaloes were 0.51, 0.42 and 0.30, respectively. Alim (1957) studied records of 294 buffaloes maintained at commercial dairy herd at Alexandria, and reported that early first calving of buffalo heifers did not affect lifetime production. The very early bred heifers tended to have lower yields and shorter calving intervals.

Dutt *et al.* (1965a) using the records of 53 foundation and 96 farmbred Murrah buffaloes kept at Mathura dairy farm reported negative and significant correlations of age at first calving with milk yield up to 6, 8 and 10 years to be -0.69, -0.45 and -0.21, respectively. They also observed that age at first calving had non-significant positive correlations with milk yield up to age at actual culling (0.12) and productive life (0.06) but positive and significant correlation with longevity (0.23). The correlations of 305 days milk yield with milk yield up to 6, 8 and 10 years and up to age at culling were positive and significant (0.33, 0.38, 0.50, 0.24) while that with longevity was positive and nonsignificant (0.19).

Kalsi and Dhillon (1984) using the lactation records of 650 buffaloes sired by 90 bulls from military farms, reported the heritability estimates for total life, total lactation period, productive life, lifetime milk production, milk yield per day of total life, milk yield per day of productive life and milk yield per day of total lactation period to be 0.14 ± 0.11, 0.17 ± 0.11, 0.17 ± 0.11, 0.13 ± 0.11, 0.30 ± 0.13, 0.27 ± 0.12 and 0.26 ± 0.12, respectively. The lifetime milk production had positive and high (rg>0.80) genetic correlations with various measures of efficiency of lifetime production and negative genetic correlation (-0.45, -0.75) with total and unproductive life. The genetic correlations among various measures of efficiency of lifetime production were positive and high (rg>0.81). Phenotypic correlations of the lifetime milk production with total life, productive life, total lactation period and various measures of efficiency of lifetime production ranged between 0.39 and 0.85. The phenotypic correlations among various measures of efficiency of lifetime milk production were positive and high (rp>0.75).

Umrikar and Deshpande (1985) using records of Murrah buffaloes from military farms reported that heritability estimate for milk yield per day of productive life to be 0.12 ± 0.02. Milk yield per day of productive life had highly significant and positive phenotypic (rp>0.41) correlations with all early lactational traits except dry period, which had highly significant and negative correlation (-0.26 ± 0.02). The genetic correlations of milk yield per day of productive life with early lactational traits ranged between 0.55 ± 0.16 and 1.00 ± 0.00, while that with dry period, the estimate was negative (-0.86 ± 0.07).

Dutt (1991) reported the heritability estimates for lifetime traits in Murrah buffaloes to be 0.17 ± 0.07 for herdlife, 0.26 ± 0.08 for lifetime milk yield and 0.26 ± 0.08 for milk yield per day of herdlife. The phenotypic and genetic correlations varied from 0.65–0.78 to 0.89 ± 0.05–0.98 ± 0.05) among them.

Examination of genetic and phenotypic parameters of the lifetime performance traits reported by various workers indicated that the heritability estimates for lifetime milk yield and average lifetime milk yield were higher than various measures of longevity traits viz. number of lactations completed, herdlife, lifetime lactation period etc. The genetic and phenotypic correlations of lifetime milk yield with longevity traits were higher than those of average lifetime milk yield with longevity traits. The higher heritability and correlations suggested that lifetime milk yield was a better representative trait among the lifetime performance traits.

Lactation curve

It is well known that the milk yield after parturition in an animal rises to a maximum in first few weeks, tends to maintain around this level for a short period and then gradually declines till secretion finally stops. This pattern of milk secretion can be represented mathematically with respect to time and is commonly known as lactation curve.

Various mathematical models have been developed to explain both the ascending and desending phase of the lactation curve. The exponential function (Brody *et al.* 1923) Yt=Ae-kt, which described only the rate of decline of milk secretion was the first to be used. Subsequently, the parabolic exponential function (Sikka, 1950) Yt=Aebt+ct2 , gamma type (Wood, 1967) Yt=At=Atbe-ct and inverse polynomial (Nelder, 1966) Yt=t(A+bt+ct2)-1 were developed to forecast the milk yield at any stage of lactation.

Dave (1971) studied lactation curve in two herds of buffaloes. The animals were divided into three groups based on the lactational duration, (i) below 250 days, (ii) above 250 days and less than 360 days and (iii) above 360 days. A polynomial of second degree was fitted to the lactation curve. The milk yield generally showed linear trend and that 75% to 96% of the total variation was accounted by linear regression coefficient of yield on time. The linear trend showed an increase in tendency from short lactation to long lactation period. In the short lactation period, yield increased with the passage of time but showed downward trend earlier. In the longer lactation period, the milk yield showed continuous increasing tendency and remained constant thereafter.

Kumar and Bhat (1979) fitted exponential, parabolic exponential, gamma type and inverse polynomial functions on average lactation records of six lactations. The R^2 values for these functions showed that gamma type function gave the best possible fit followed by inverse polynomial, parabolic exponential and exponential. These functions explained variability up to 99.0–99.3, 98.0–98.5, 94.5–96.5 and 75.3–79.6%, respectively, by interactive procedure. It was suggested that the gamma type function followed by inverse polynomial should be preferred over others for describing the lactation curve in Indian buffaloes.

Life time production

Chaudhary and Shaw (1965) in Nili-Ravi buffaloes of Pakistan and Asker *et al.* (1971) in Egyptian buffaloes reported average production life to be 72.0 and 76.8 months respectively.

Dutt *et al.* (1965) in Murrah buffaloes recorded yield up to first three lactation as 3308 ± 116 kg while Iqbaluddin *et al.* (1970) recorded it to be 4848 ± 112 kg and 4475 ± 120 kg, respectively, at two farms in Murrah buffaloes. Patro and

Bhat (1979) reported the milk yield in first three lactations to be 5131 kg in Indian buffaloes. The analysis of data in their study revealed that by the end of 4th lactation more than 50% of the buffaloes left the herd and only less than 1% completed 10 lactations.

Prediction of life time milk yield

Dutt *et al.* (1965a) estimated the accuracy of predicting the lifetime milk yield using early traits (age at first calving and 305 day first lactation milk yield) in Murrah buffaloes . Lifetime milk yield in their study was defined as total milk yield up to 6, 8, 10 years and up to age at culling. The accuracies of prediction involving age at first calving and 305-day first lactation milk yield for various lifetime milk yield were 71.9, 51.0, 57.6 and 20.0% respectively.

Jain and Taneja (1984) predicted the lifetime milk yield using age at first calving and milk yield in different first three lactations in Murrah buffaloes. Lifetime production was defined as the cummulative total yield in first three (LTP-1) and first five lactations (LTP-2). They observed that second lactation yield was better predictor of LTP-1 (R^2 40.24% vs 60.56%) and LTP-2 (R^2 26.79% vs 43.49%) than the first lactation production. Both taken together, further increased the accuracy of prediction (82.40% for LTP-1 and 61.15% for LTP-2) . When age at first calving was combined with first lactation milk yield, it explained 46.50% of the variation in LTP-1, indicating that addition of age at first calving did not increase the accuracy of prediction.

Sharma and Basu (1986) fitted polynomial regression up to third degree to explain the variation in various lifetime performance traits due to the age at first calving in Murrah buffaloes. Herdlife was defined as the difference between the date of first and last calving up to disposal or death. Age at first calving explained 0.26, 5.32, 3.51 and 6.53% variation in lifetime milk yield, milk yield per day of herdlife, lifetime profit and profit per day of life, respectively.

It can be inferred that from the reports discussed that higher accuracies of prediction as reported by Dutt *et al.* (1965a) and Jain and Taneja (1984) in Murrah buffaloes were mainly because of the fact that lifetime milk yield in their studies was restricted to age (6, 8 and 10 years) and number of lactations (3 or 4) and thus resulted in higher phenotypic correlations of early economic traits with lifetime milk yield as defined in their study. With the increase in age/number of lactations, the reduction in accuracy of prediction for lifetime production was observed. It was apparent that polynomial regression analysis did not explain significantly more variation in the lifetime performance traits.

Optimum levels of early economic traits for maximum lifetime yield

The importance and impact of optimum levels of various early economic traits like age at first calving, first lactation milk yield and first lactation length on the performance of a dairy herd is well established.

Bhadula and Desai (1972) studied the effects of age at first calving (with a class interval of 3 months) on total milk yield up to first three lactations. They observed that the total milk yield during first three lactations was practically similar for animals dropping their first calf at 33 to 45 months of age, after which there was a continuous decline in yield. The total milk yield of first three lactations for buffaloes dropping their first calf at 33 to 45 months of age ranged between 4862.70 and 5135.10 kg. The age for the late calvers to complete 3 lactations was about 86 months, whereas for those dropping their first calf between 33 and 36 months of age it was only about 74 months. The production period up to 3 lactations was higher in the early calvers than those in the late calvers.

It was recommended that the replacement stocks for breeding purpose should necessarily be selected from the stock that dropped the first calf between 33 and 39 months of age. The selection of male stock for breeding purpose must be made from the stock whose dams and/or half sisters are early calvers, calving within about 36 months of age.

Singh (1976) determined the economic levels of milk production in first lactation, yield per day of first lactation length and yield per day of first calving interval in relation to lifetime economics of buffaloes classified into four groups on the basis of age at first calving. Lifetime economics was calculated on the basis of net expenditure incurred on maintaining each of the buffaloes since birth to its completion of fifth lactation and returns thereoff. They reported that buffaloes having first lactation yield over 1808, 1859, 1924 and 2155 litres, respectively in the groups up to 42 months, over 42 to 48 months, over 48 to 54 months and over 54 months of age at first calving were predicted as the economic ones. The economic levels for per day of first lactation in these four groups were 4.99, 5.76, 5.72 and 6.15 kg, respectively, and these levels for per day of first calving interval were 3.39, 4.08, 4.21 and 4.27 kg, respectively. It was concluded that the predictions based on complete first lactation production could be used for predicting the buffalo's economy accurately.

Singh and Tomar (1981) reported economic levels of various early economic traits on the basis of lifetime milk production and herdlife (age at disposal). It was found that age at first calving (below 42 months), higher first lactation production (over 2000 kg) and lower first calving interval (not more than 15 months) would result in significant increase of herdlife and lifetime milk production.

Sharma and Basu (1986) fitted polynomial regression up to third degree using lifetime records of 804 Murrah type buffaloes to determine optimum levels of various economic traits. Regression curves of lifetime traits on age at first calving showed a linear progression and continuously decreased with increase in age at first calving indicating higher lifetime milk production and longer stability for early calvers. They observed no lower limit to age at first calving, perhaps because the optimum age of calving was outside the range observed in their study.

From the reports reviewed, it can be concluded that lower age at first calving, preferably below 42 months and higher first lactation milk yield, preferably over 1800 kg were desirable for lifetime production.

9.4 Draught Power (work)

The buffalo has been used as a work animal for thousands of years. Its body build, enormous strength, docile temperament, amenability for easy training and capacity for long sustained work makes it an excellent animal for haulage and many kinds of agricultural work. The water buffalo is the classic work animal of Asia, an integral part of this continent's traditional village farming structure. Probably the most adaptable and versatile of all work animals, it is widely used to plough, level land, plant crops, puddle rice fields, cultivate field crops, pump water, haul carts, sleds and shallow draft boats, carry people, thresh grain, press sugarcane, haul logs, and much more. Even today, water buffaloes provide 20–30% of the farm power in South China, Thailand, Indonesia, Malaysia, the Philippines, and Indo-China. Millions of peasants in the far east, middle east and near east maintain draught buffaloes. For them it is often the only method of farming food crops. As fuel becomes scarce and expensive in these countries, the buffalo is being used more frequently as a draught animal. In 1979 water buffalo prices soared in rural Thailand because of the increased demand.

Although Asian farms have been increasingly mechanized in the last 20 years, it has often proved difficult to persuade the farmer to replace his buffalo with a tractor since the buffalo produces free fertilizer and does not require diesel fuel. Now there is renewed official interest in draught power. Because of demand for work animals, buffalo shortages have become a national development problem. Indonesia's transmigration schemes are also handicapped by shortages of animal power. For many small farmers the buffalo represents capital. It is often the major investment they have. Buffalo energy increases their productivity and allows them to diversify. Even small farms have work animals that, like the farmer himself, subsist off the farm. Tractors usually require at least four hectares for economical operation, which precludes their use on most small holder farms. Further, the infrastructure to maintain machinery is often not readily available.

The working buffaloes are slower in movement in comparison to oxen; they usually cover 3.2 km per hour against 4.8 to 6.4 km per hour by draught bullocks. However, the buffaloes can draw heavier loads than oxen (Fahimuddin, 1975). Liu (1978) and Nagpaul *et al.* (1984) studied the draught efficiency of crossbred vis-a-vis zebu and buffalo bullocks for various farm operations like carting and ploughing. The pulse rate, temperature and respiration rate increased steadily after the start of work in all the categories of bullocks. The buffalo bullocks started showing signs of fatigue after two hours of work, whereas zebu and crossbred bullocks did not exhibit these symptoms even after 4 hours of work. Season had a significant effect on rate of increase in temperature, pulse rate and respiration rate in these categories of bullocks. It was observed that buffalo bullocks during summer covered a distance of 2193 m in one hour of carting as against 2682 m per hour by crossbred bullocks. In winter, buffalo bullocks covered 2768 m per hour as against 2645 m by crossbreds. It is essential that draught capacity of various breeds of river and swamp buffaloes used for milk should be investigated. The improvement or otherwise in draught capacity as a consequence of selection for milk also needs to be investigated.

A survey by Cockrill (1968) revealed that in more than a score of Asian countries buffaloes provide the most efficient and the cheapest source of power. Both swamp and river buffaloes make good work animals, but for paddy cultivation in the rice growing areas in eastern Asia the swamp buffalo is unmatched for efficiency. They are used for ploughing the soil, for puddling the earth and harrowing the fields after flooding and for making the bed ready for planting the rice seedlings. The large hooves and great flexibility of the pastern and fetlock joints enable the swamp buffalo to work with ease in thick, muddy, water logged paddy fields. Their productivity also makes them very well adapted for this kind of work in slush and mud. It is a common sight to find the buffalo working contentedly and with ease in knee-deep, muddy water in rice fields. Although possessing a massive body frame and large hooves, buffaloes are quite nimble-footed and work with ease in very small plots without damaging the low mud walls, called 'bunds', forming partitions between the plots. Apart from roughing and harrowing, buffaloes are used for many other types of work such as threshing, drawing water for irrigation, expressing oil from oilseeds or juice from sugarcane in sugar mills, puddling clay for preparing bricks, logging and haulage of loads on sledges or carts. In Pakistan and Sabah, buffaloes are widely used as pack animals and for riding. Buffaloes play a role even in races and games in Indonesia and Sabah.

In northern India buffalo has replaced bullocks for haulage of sugarcane from the fields to the factory. Its introduction into Latin American countries was motivated by their relative ease of haulage of sugarcane to the factories.

Although the majority of working animals are males, female buffaloes, dry or with a poor milk yield, are also sometimes put to work. This practise is not uncommon in Egypt, parts of Indonesia and the Philippines. For drawing heavy loads buffaloes are hitched singly or in pairs to crudely built wooden carts. They are used extensively for this purpose in India, Pakistan, Indonesia and other countries, particularly in and around urban areas. On good roads, it is not unusual to find a pair of buffaloes drawing a load of about 2 mt (2.02 tons) at a slow pace of 3.2 km an hour. As buffaloes are incapable of working for long on hot days when they are exposed to direct sunlight during the hot summer they need to be put to work at night or in the early morning. As work animals male buffaloes are used both castrated or uncastrated. The practise varies from country to country, following local beliefs and customs. In India and Pakistan castration is not the usual practise and uncastrated animals seldom, if ever, become intractable. Castration, if carried out, is done at a late age as early castration is believed to produce a poor working animal. Docility and mild manner make the buffalo so easy to manage at work that in many instances no restrictive controls are necessary. Whips or chains are seldom used to drive the animal.

For working on hard ground, hooves of buffaloes need protection against excessive wear by shoeing. Thin iron plates are used in India and Pakistan. Shoes made of straw and 'trompas' shoes made from old motor tyres are used for this purpose in Taiwan and East Java respectively.

1. Selection indices

In various farm animal species, selection indices are used to estimate the breeding value of future breeding stock. A selection index estimates the value of an individual for an aggregate genotype, whereas the latter in most applications is expressed in financial terms and is a linear combination of additive genetic merit for component traits weighted by their respective economic weights. Selection indices have been constructed for buffaloes by Tomar and Desai (1969), Kanaujia *et al.* (1974), Gokhale and Nagarcenkar (1980), Bhalaru and Dhillon (1978) and Johari and Bhat (1978).

Johari and Bhat (1978) constructed indices involving traits of growth, reproduction and production. Comparative study of relative efficiency of these indices indicated that selection index incorporating birth weight, weight at six months, weight at one year, weight at first calving, first lactation milk yield and lactation length had the highest relative efficiency and was slightly superior than the index based on all the nine traits. It was recommended that sequential selection would be more useful for genetic improvement. At one year of age, the heifers should be selected using an index which incorporated body weight at birth, six months and one year. When body weight at first calving and age at first calving become available, the index incorporating body weights at six months, one year and first calving, and age at first calving could be used. At the completion of first lactation record, the index which incorporated body weights at six months, one year and first calving, age at first calving, first lactation milk yield and lactation length could be used; it was as efficient as the best index. The authors suggested that reproductive traits should be improved through efficient feeding, management and improved sexual and health control operations.

Several authors have suggested several indexes which could be useful, but so far none of indexes have used for obtaining genetic improvement in buffaloes for milk yield or any other quantitative traits.

2. Response to selection

Alim (1953) estimated the genetic gain for milk yield through dams of buffalo cows and bulls as 5.7 and 13.4 kg, respectively, in a herd of Egyptian buffaloes. The total generation length through four paths was 27 years and annual gain in milk yield was 0.71 kg (0.04% of the herd average). Asker *et al.* (1955) did not observe any effect of selection for age at first calving or calving interval in Egyptian buffaloes over a period of 20 years. Genetic improvement in milk yield from selection of dams of heifers was 60.3 kg per generation. The annual genetic gain in milk yield was 8.1 kg which was 0.8% of the herd average.

Agarwala (1956) estimated the annual genetic gain in butterfat yield in a herd at Allahabad as 0.10 kg which was about 0.1% of the average first lactation butterfat yield of 95.9 kg. Reddy and Taneja (1982) estimated genetic gain in milk yield in Murrah herd at Military Farm, Jabalpur. The estimates of genetic superiority through four paths, *viz.* sire to daughter, sire to son, dam to daughter and dam to son were 7.92, 4.91, 80.45 and 438.94 kg, respectively, which were equal to 1.49, 0.92, 15.12 and 82.47% of the total. In this study, the genetic contribution from dam to daughter path was much higher than sire to daughter path. In selection of bulls also, maximum emphasis on dams' performance was indicated by the highest genetic superiority through dam to son path (82.47%). Thus in selecting both cows and bulls, the emphasis was mostly on their dams' performance. This was expected as the selection of bulls was done on the basis of dam's performance. To maximize the rate of genetic gain, a combination of selection of outstanding bull mothers and progeny testing of bulls should be practised. The annual genetic gain in 300 day first lactation milk yield was 15.88 kg, equal to 0.99% of the herd average of 1611.26 kg. If this procedure was adopted Singh (1983) estimated the annual genetic gains for milk yield using progeny testing, half-sib testing, pedigree and individual selection in a theoretical study for population sizes of 400, 1600 and 10000 buffaloes. An increase in genetic gain with increase in population size was noted in progeny testing and half-sib methods. The annual genetic gains for herd size of 10000 were 1.04, 1.00, 0.07 and 0.06 % of the herd average respectively. Gain from half-sib testing was almost equal to progeny testing because of smaller generation interval in half-sib testing. Expected gains from pedigree and individual selections were negligible.

Most of the studies reviewed above indicate that selection of both bulls and heifers should be done on the basis of dam's performance. For making improvement in milk yield, it is essential that outstanding bull mothers are selected and progeny testing of young bulls is taken up to maximize the rate of genetic gain. Considering a heritability estimate of 0.15 for milk yield in 300 days and different selection intensities (culling at 10, 20 and 30% of herd), the improvement (ΔG) per year will be between 3.76 and 9.60 kg (Table 9.14).

Table 9.14 Rate of Genetic Improvement in Milk Yield

% culled	i	h2	p	ΔG per generation (kg)	ΔG per year (kg)
10	0.195	0.15	900	26.33	3.76
20	0.340	0.15	900	45.90	6.56
30	0.498	0.15	900	67.23	9.60

Ashfaq (1961) reported the results of a study of progeny testing of the buffalo bulls maintained at the Government Buffalo Breeding Farm, Bahadarnagar in Pakistan during 1947–51. The analysis showed that progeny testing was feasible. A comprehensive study was conducted to explore the extent of genetic and environmental sources of variation in the productive and reproductive traits of a purebred herd of Nili-Ravi buffalo (Khan 1986). The effectiveness of selection and inbreeding in altering the milk producing ability of the animals over a period of 44 years was also studied (Khan and Ahmed, 1991).

The data showed that the average estimated genetic superiority of the dams of she buffalo (I_{CC}) was 11.34 kg or 0.598% of the herd average per generation; the average estimated genetic superiority of dams of bulls (I_{CB}) was 84.20 kg of milk or 4.44% of the herd average per generation. Since progeny testing was not practised in this herd, the other two paths of genetic superiority, *viz.* I_{BB} and I_{BC}, amounted to be zero. The pooled genetic superiority was 95.55 kg of milk or 5.04% of the herd average for a generation interval of 7.38 years. The genetic improvement per year, thus amounted to 3.278 kg of milk or 0.171% of the average first lactation milk yield (1,869 ± 38.8 kg and heritability, 0.137 ± 0.083). Out of the genetic improvement in milk yield, 88% was due to the selection of the dams of the sire (I_{CB}).

The estimate of annual genetic improvement in milk yield through selection was not only markedly low but was also much less than 1% of the mean suggested by Rendel and Robertson (1950) as the maximum rate of genetic improvement without progeny testing. Thus it may be concluded that

(a) although selection remained ineffective to bring about the desired changes over a long period of 44 years, yet there was scope for six fold improvement in production through direct selection;

(b) selection of buffaloes to produce heifers (I_{CC}) was ineffective because the selection of dams to produce heifers and bulls was primarily done on the basis of type and conformation with little attention to milk yield since 1936. It was obvious that these characters had little or no correlation with traits of economic importance; and

(c) the logical consequence of the above finding is that progeny testing programme for buffalo needs to be initiated. This was obvious from the fact that 12% genetic gain was from the selection of dams of cows (I_{CC}) and 88% was obtained through the selection of dams of bulls (I_{CB}). This conclusion is further substantiated by the findings of Cady *et al.* (1983) who reported that at government livestock farms, 1% of all buffaloes produced more than 4000 kg milk and this proportion of high-yielding buffaloes gave good opportunity for the selection of dams to produce bulls (I_{CB}). Analysing data of 1457 Nili-Ravi buffaloes to estimate age specific probabilities of cows culled for low milk yield, reproductive failures, poor health and old age. Reproductive failure was identified as the principal reason for culling of buffalo cows. Reproduction and health accounted for $^{2}/_{3}$ of all culling, leaving a relatively narrow margin for culling based on low milk yield.

3. Co-ordinated research project on buffaloes

An All India Co-ordinated Research Project on Buffalo Improvement was initiated by the Indian Council of Agricultural Research in 1970–71. with the main focus on progeny testing of bulls. Two centres each for Murrah and Surti were established at different locations in the country. The detailed objectives and the breeding programme of the project have been described by Nagarcenkar (1978). The project envisaged evaluating buffalo bulls both under farm and field conditions. The number of bulls of the two breeds with 20% superiority above their contemporary average have been identified and these are being used for improvement of the herd and in elite matings for production of young bulls for testing. Elite dams with more than 2000 kg milk for Murrah and 1200 kg for Surti have been identified and used in mating with the tested bulls. So far 40 bulls of Murrah breed have been progeny tested with daughters' performance of 2000 kg/305 days to 2700/305 days for various proven bulls. The frozen semen of these bulls is being used for improvement of buffalos in the breeding tract.

4. Progeny testing of bulls in Pakistan

It was with this background that the progeny testing programme in Pakistan was started at the Livestock Production Research Institute, Bahadarnagar, Okara in 1979. It was a station testing programme; 2 batches of bulls with 4 bulls in each batch were put to test mating during 1979–80 and 1981–82, respectively, allowing random herd mates to each bull. In batch I, four bulls were put on test mating herd in 1979 and 40–50 conceptions from each bull were randomly obtained. The daughters of these bulls who attained sexual maturity were bred accordingly. The bulls were ranked according to the method described by Jain and Malhotra (1971). These bulls were recommended as sires for the production of future candidate bulls after preserving sufficient doses of their semen. By 1982, four bulls which were the sons or grandsons of proven sires out of top-class bull mothers were put on test mating in the test herd, *i.e.* buffaloes other than elite cows. Out of 94 daughters, 28 completed their first lactation while the others were culled for various reasons. Since the progeny testing programme was confined only to the Livestock Production Research Institute, Bahadarnagar, Okara, where small number of buffaloes were present, accuracy was expected to be low and selection was not large enough to allow significant genetic progress. Hence, field recording of farmers' herds, as an adjunct to institutional programme was initiated. A co-ordinated field progeny testing project was started in 1984 and 1995. The new programme was extended to 3 buffalo populations, *viz.* buffaloes at the government farms, buffaloes with progressive farmers and buffaloes from the selected village population.

Asghar (1987) highlighted the important features of this programme. About 800 breeding buffaloes at different livestock experiment stations and military farms and 4,000 buffaloes kept by the farmers in the home tract of Nili-Ravi buffalo were registered. The criterion for registration was more than 10 litres of milk daily during their peak period. The sons from elite buffaloes and sires assessed for breeding value by Cady *et al.* (1983) were taken as base for progeny testing programme. The buffaloes from these populations were inseminated with the semen from the proven sires or the best available sires to produce candidate bull calves. The topmost 35–40 bulls were put on test mating every year in all buffalo populations till 80–100 conceptions were obtained from each bull. Such bulls were ranked by contemporary comparison. Mean lactation yield of 1770 buffaloes at the government and military farms was 1873 ± 14. The top 12.8% of these buffaloes were selected as bull mothers and the average lactation yield of these elite buffaloes was 2760 ± 13 kg. The average milk yield of 3177 registered buffaloes from the field was 2074 ± 10 kg, and 9.9% of these buffaloes having average milk yield of 3215 ± 8 kg were selected as bull mothers. In 1983–84, 16 bulls were put on test mating on all the buffalo populations registered under the project. These bulls attained body weight of 420–570 kg at the age of 24 months. The fertility in these bulls was 52.7%. At the end of this phase of the project, a total of 185 contemporary daughters had completed at least one lactation each. The analysis of the data revealed non-significant difference in 305-day milk yield of buffaloes kept at the government farms and at the private farms. Hence, the two sets of data were pooled and the bulls were ranked accordingly.

The differences among the bulls were significant ($P<0.05$) but the top 3 bulls of the 4 tested (B-62, PD 210, B-63 PD 284 and B-65 PD 175) did not differ from each other. Until now, 7 batches of 124 young candidate bulls have completed test mating and the data are being analysed.

5. Milk Composition

Buffalo milk has comparatively more fat and solid-not-fat (SNF) than cow milk Table 9.15. The mean fat and SNF percentages in composite Murrah buffalo milk have been reported as 7.0±0.04 and 10.1±0.03 with a variation of 10% and 4.9% respectively (Singh *et al.*, 1979). Pal *et al.* (1971) reported that the protein, lactose, mineral together constituted 9.60%. In Egyptian buffaloes, Alim (1978) reported the fat content as 6.6 while protein, lactose, mineral content together was 9.4%.

Table 9.15 Composition of Buffalo and Cow Milk

Species	Fat (%)	Protein (%)	Lactose (%)	Total solids (%)
Buffalo	7.64	4.36	4.83	17.96
European cow	3.90	3.47	4.75	12.82
Zebu cow	4.97	3.18	4.59	13.45

Source : The Water Buffalo: FAO (1977)

Singh *et al.* (1979) observed that lactational fat percentage was affected significantly by year and season while parity of calving did not influence the fat percentage. However, year, season and parity of calving affected the lactational SNF percentage significantly. The monthly fat percentage was significantly influenced by stage of lactation and parity while the monthly SNF percentage was not affected either by stage of lactation or parity of calving. The fat percentage showed a continuous increase from first to tenth month. The influence of season of calving on fat percentage is well recognized. Both the fat and SNF percentage were maximum among autumn calvers and minimum among rainy season calvers (Singh *et al.*, 1979). However, Ragab *et al.* (1958) reported higher fat percentage in summer calvers than in winter calvers in Egyptian buffaloes. In contrast, SNF percentage was higher in winter than the summer calvers. No explanation for this has been given.

Pal *et al.* (1971) reported that per cent protein was highly heritable (0.74) while per cent ash (0.29), percentage total solids (0.35) and percentage solids-not-fat (0.39) were moderately heritable. The repeatability estimates for fat (0.71), ash (0.68), total solids (0.72) and SNF percentage (0.68) were high, while those for protein (0.42), lactose (0.36) and casein percentage (0.57) were moderate. However Singh *et al.* (1979) estimated heritability for percentage fat and SNF as 0.41 ± 0.28 and 0.26 ± 0.23 respectively. The repeatability estimates for percentage monthly fat in their study ranged from 0.20 to 0.49 and percentage SNF from 0.19 to 0.37.

Table 9.16 Mean Fat and SNF percentage in Different Stages of Lactation and Parity in Murrah Buffaloes

Month after	Fat (%)	SNF (%)	Lact No.	Fat (%)	SNF (%)
1	6.40 ± 0.04	10.20 ± 0.03	1	6.93 ± 0.14	10.14 ± 0.09
2	6.51 ± 0.03	10.10 ± 0.03	2	6.83 ± 0.15	10.16 ± 0.08
3	6.69 ± 0.03	10.05 ± 0.03	3	7.02 ± 0.08	10.10 ± 0.04
4	6.87 ± 0.03	10.10 ± 0.03	4	7.02 ± 0.06	10.09 ± 0.04
5	6.95 ± 0.03	9.96 ± 0.03	5	6.94 ± 0.08	10.05 ± 0.66
6	7.06 ± 0.03	10.00 ± 0.03	6	6.73 ± 0.10	10.09 ± 0.07
7	7.16 ± 0.07	10.03 ± 0.02	7	6.87 ± 0.20	9.95 ± 0.15
8	7.17 ± 0.04	10.00 ± 0.10	8	6.72 ± 0.32	10.30 ± 0.29
9	7.43 ± 0.05	10.05 ± 0.03	–	–	–
10	7.43 ± 0.05	10.07 ± 0.03	–	–	–

Source: Singh *et al.* (1979).

Pal *et al.* (1971) estimated the genetic correlations among milk constituents to be positive (0.28 to 0.97%). The highest correlation values were observed between total solids and other constituents. Singh *et al.* (1979) reported positive correlation between fat and SNF percentage (r=0.57). However, the correlations between milk yield and constituent percentage were negative, and ranged between 0.26 and 0.77. These results suggest that selection for milk yield would lower milk constituent percentage and vice-versa. Although the increase in milk yield will decrease the milk constituent percentage, total fat, protein and SNF yield would increase as all yields are positively correlated with milk yield (Pal *et al.*, 1971).

6. Meat production

Studies were conducted on the meat-production characteristics of weaned male Murrah buffalo calves at the IVRI in India. Groups of 9-month-old calves weighing

75 kg were divided into two groups, one lot being on *ad lib.*, high-grain ration and the other on *ad lib.* roughage type, leguminous hay and wheat straw ration. On attainment of 300 kg body weight the animals were slaughtered for evaluation of carcass quality. Animals fed the high grain ration demonstrated growth rates of 0.555–0.628 kg per day, and attained a slaughter weight of 300 kg at 16 to 18 months of age, whereas those on the roughage type ration gained at the rate of 0.410 to 0.450 kg per day and reached the same slaughter weight in 20 to 22 months. The dressing percentage of the high-grain-fed animals ranged from 56.8 to 59.6 (mean 57.4), while that of animals fed roughage-type rations varied from 50.8 to 53.5 (mean 50.9). The lean fat bone ratios in the grain-fed and roughage-fed groups were 55:26:28 and 59:19:21 respectively. If the buffalo is properly managed and fed as a meat producing animal and slaughtered at 16 to 20 months of age it can produce meat that would be comparable with beef obtained from cattle, both in quality and in quantity.

Krishnan and Nagarcenkar (1979) observed average daily gain in body weight in buffalo calves as 474 g from birth to slaughter and 513 g from 6 months to slaughter under conventional concentrate feeding. The corresponding averages when calves were fed concentrate mixture containing urea and molasses were 421 and 386 g respectively. The differences in average daily gain under two feeding regimes were significant. For meat studies, the calves were slaughtered in weight groups of 140, 180 and 200 kg. The dressing percentage was highest (58.7%) under conventional concentrate feeding in 180 kg weight group, while it was lowest (53.5%) under experimental concentrate feeding in 140 kg weight group.

In China, Liu (1978) made a comparison of fattening efficiency between triple crossbred ($^1/_2$ Nili-Ravi $^1/_4$ Murrah $^1/_4$ local) buffaloes and $^1/_2$ Santa Gertrudis grades under same feeding conditions, both slaughtered at 18 months of age. The dressing percentages were 53.0 and 59.9, muscle percentages 43.2 and 42.1 and bone ratios 1:4.5 and 1:4.4 respectively. The buffalo meat was as palatable as beef. In another feeding experiment, triple-cross buffalo steers maintained on elephant grass for 100 days gave a daily weight gain of 0.8 kg and the animals consumed 67.3 kg of elephant grass for each kilogram increase in body weight. The animals when slaughtered at 24 months of age gave dressing percentage of 53.6 with muscle percentage of 43.0.

A report on comparative palatability of cooked meat of water buffalo with that of local crossbred Jamaica Red × Sahiwal steer and an imported carcass of European beef steer, revealed that meat from water buffalo scored an average of 76.6 points vs 70.7 and 69.3 points, respectively, for the local and imported beef. The buffalo meat is considered to be of attractive colour, fresh and tasty. Buffalo veal is considered a delicacy. Calves are usually slaughtered for veal between 3 and 4 weeks of age and the dressed percentage is 59.66.

9.5 Performance of Buffalo Crosses

9.5.1 River × River buffalo crosses

In Gujarat (India) farmers have been breeding Surti buffalo cows with Murrah or Jaffarabadi bulls. Mehsana, a buffalo breed, is said to have been evolved from crossing of Surti buffalo cows with Murrah bulls. A pilot study on crossing Surti females with Murrah bulls was undertaken at the National Dairy Research Institute, Karnal, India (Table 9.17).

Table 9.17 Performance of Murrah, Surti and their Crosses

Traits	Murrah	Murrah × Surti	Surti
Age at calving (months)	42.00	41.00	48.20
Milk yield, 305 days (kg)	1471.50	1415.60	1109.80
Lactation length (days)	324.10	321.20	294.90
Dry period (days)	157.90	106.70	165.10
Calving interval (days)	468.90	411.30	487.80
Efficiency of milk production (kg/day)	3.20	3.40	2.30

Source: Basu and Sarma (1982).

Murrah and Surti buffaloes were raised contemporarily. In crossbreds, a significant increase in milk yield and improvement in reproduction traits was observed (Basu and Sarma, 1982). In terms of efficiency of milk production (milk yield per day of calving interval), the crossbreds (3.4 kg) were even better than the Murrah (3.2 kg). In Bulgaria, crossing of native Bulgarian females with Murrah buffalo males imported from India was initiated in 1962. Murrahs were superior to Bulgarian buffaloes both for body weight and milk yield. The Murrah × Bulgarian crosses were close to Murrah for body weights at different ages (Table 9.18).

Table 9.18 Performance of Murrah, Bulgarian and their Crosses

Traits	Murrah		Bulgarian		F_1 Murrah × Bulgarian	
	X	SE	X	SE	X	SE
Weight at birth (kg)	34.10	0.4	31.90	0.2	34.30	0.2
Weight at 6 months (kg)	170.10	1.7	132.30	0.8	160.20	2.7
Weight at 24 months (kg)	508.40	6.8	432.80	6.2	507.60	5.7
300 day milk yield (kg)	1822.80	75.7	1197.60	13.2	1491.50	18.7
300 day fat yield (kg)	145.20	7.2	91.40	1.1	113.90	1.1
Lactation length (days)	306.20	14.7	296.80	4.9	288.00	5.1
Dry period (days)	156.30	19.7	151.10	3.8	156.40	5.2
Fat (percentage)	7.95		7.59		7.64	

Source: Nagarcenkar (1978).

The milk yield and total fat in the crossbreds were close to mid-parent value (Nagarcenkar, 1978).

9.5.2 Crossbreeding of swamp buffaloes

There are no identified breeds of swamp buffaloes, although variations in coat colour, horn type and body weight are substantial between them. It is possible to crossbreed the swamp buffaloes with the river buffalo breeds. Research on crossbreeding of the swamp buffaloes with the riverine breeds, especially the Murrah, has been conducted in different countries, *e.g.* China, Malaysia, the Philippines, Thailand and Vietnam. The number of chromosomes in the swamp buffalo is 48 (2n=24 pairs), while that in the Murrah breed is 50. The number of chromosomes in the F_1 crossbred is 49 and, in spite of this change in chromosome numbers, both males and females are fertile. From a limited number of observations, the F_2, F_3 and F_4 resulting from backcrosses to the Murrah also appear to be fertile.

9.5.2.1 *Growth performance*

Buffalo crossbreds Philippine Carabao × Murrah or Phil-Murrah and Philippine Carabao × Nili-Ravi or Phil-Ravi were heavier than the Philippine Carabaos. The data from 322 Carabaos and crossbreds (102 Philippine carabaos, 182 Phil-Murrah crossbreds and 38 Phil-Ravi crossbreds) were analyzed. The variations in weights among the test animals were primarily due to breed of sire and locations. Regardless of sex, the crossbreds were heavier than the Carabaos at birth, and at 12, 24 and 36 months of age. Harisha and Azmi (1989) also reported that the least-square mean weights of swamp × Murrah buffaloes were consistently heavier than those of swamp × swamp buffaloes at birth, and 6,12,18, 24,36 and 48 months of age ($P<0.05$).

9.5.2.2 *Milk production*

Chantalakhana (1978) reported the performance of imported Murrah, Nili-Ravi and their crosses with swamp buffaloes in South-East Asian countries. The Murrah adapted well to the local environmental conditions, levels of feeding and management, and produced about two to three times more milk than the swamp buffalo cows. The crossbred buffaloes are superior in working ability to the swamp type and are nearly equal in milking ability to the Murrah. Age at first calving in Murrah breed in these countries varies between 4 and 4.5 years. However, in crossbred buffaloes it is around 3.2 years. Murrah generally produced 3.5 to 5 kg of milk per day during their 200–300 days of lactation.

Pei-Chien (1978) reported the performance of Murrah (imported from India), Nill-Ravi (from Pakistan) and their crossbreds with local swamp buffalos of China. Milk yield in Murrah born and reared in China for an average lactation period of 213 days was 1,382.9 kg. In general, the performance of Murrah grades was satisfactory. They were heavier and stronger than the local animals, especially in the hindquarters, good for working and produced more milk. Triple crosses produced using Nili-Ravi bulls on Murrah × local F_1 cow, had an average body weight of over 200 kg at 6 months of age, the body weight being 300 and 450 kg at 1 and 2 years of age. An average daily weight gain of 900 g was achieved in these triple crosses.

Liu *et al.* (1985) reported the performance characteristics of Chinese buffaloes, imported Murrah and Nili-Ravi buffaloes and their crosses. The Chinese buffaloes can be classified as large (560–784 kg), medium (519–544 kg) and small (402–452 kg). They have the ability to thrive on coarse feed and live under rough management conditions. However, their milk and meat production is not high although the milk is rich in fat (7.8–11.3%). The average milk yield in Chinese buffaloes is 520 (210 days) to 751 kg (300 days).

The performance of exotic breeds, Murrah and Nili-Ravi, with respect to growth, production and reproduction compare well with those in their native homes and are higher than the averages reported for Chinese buffaloes. The average lactation yields for Murrah and Nili-Ravi are 1,976 (272 days) and 2,076 kg (275 days) with fat percentage of 6.7 and 7.2 respectively.

The average lactation yield of Murrah × local F_1 progeny cows is 118.2% (271 days); and of the F_2 is 1540.3 (291 days) with a fat percentage of 7.5. Use of Nili-Ravi bulls on $^1/_2$ Murrah grade resulted in reduction in age at puberty (605 days) compared to $^1/_2$ Murrah grade (669 days). Similarly, the post-partum oestrus period (71 days) and calving interval (382 days) in the triple crosses were lower than in Murrah (94 and 455 days), Nili-Ravi (128 and 466 days) and $^1/_2$ Murrah grade (171 and 540 days). The average milk yield in these triple crosses was higher (2662 kg, 311 days) than those in Murrah grades (1540 kg in F_2 progeny), and approached the level of Murrah and Nili-Ravi. In general, growth, milk yield and confirmation of triple crosses appeared to be reasonably good; hence, the triple crosses are proposed to be used as the foundation stock for development of new breed of milk-meat buffaloes in China.

In the Philippines the milk production of Phil-Ravi F_1 and Phil-Murrah F_1 is about 2.4 and 2.7 times more than the production of the Philippine carabao (Table 9.19). The lactation length of the Philippine carabao averages 200.8 days compared to the lactation of 348.5, 259.5 and 233.7 days for Murrah, Phil-Murrah and

Phil-Ravi respectively. The average daily milk yield of swamp buffaloes was 1.94 kg based on 10 lactation records; of the F_1 of swamp × Murrah, 3.73 kg on 12 lactation records; of the F_1 of swamp × Nili-Ravi, 4.3 kg on 60 lactation records; of the purebred Murrah, 6.60 kg on 81 lactation records; and of the purebred Nili-Ravi, 7.2 kg on 25 lactation records. The average lactation period was 235.9 days for swamp buffaloes; 276.9 days for swamp-Murrah F_1; 270.8 days for swamp-Ravi F_1; 237.1 days for purebred Murrah; and 261.0 days for purebred Nili-Ravi. The F_1 crossbreds of swamp and riverine buffaloes on an average produced about twice as much milk daily as that produced by swamp buffaloes.

Table 9.19 Average Milk yield (kg) and Lactation Length (days) of the Philippine Carabao, Murrah, Phil-Murrah F_1 and Phil-Ravi F_1

Parameters	Philippine carabao n = 21	Murrah n = 21	Phil-Murrah F1 n= 29	Phil-Ravi F1 n = 12
Milk yield	259.40c	1804.40a	705.60b	623.10b
Lactation length	200.81b	348.48a	259.45b	233.67b

Source: Shrestha and Parker (1992).

In Sri Lanka, upgrading of local buffalo cows with Murrah and Surti from India resulted in a significant increase, of approximately 290%, in milk yield over the local buffaloes, which yielded 355 kg per lactation. Murrah were superior to Murrah grade only by 241 kg (16%) in milk production with no apparent difference in lactation length. Assuming a generation interval of 6 years and a turnover of 5 generations, one could expect an improvement of approximately 34 kg milk per year through the upgrading programme.

The crossbreds between swamp and Murrah in Taiwan generally produce 3.5–5 kg of milk per day during 200–300 days of lactation, with a peak yield of around 7 kg per day. The F_1 crossbreds on an average yield 3.4 kg per day (Table 9.20). However, in the Philippines only 1.6 kg of milk per day during 300 days is obtained from the cross-breds.

Table 9.20 Performance (means) of Murrah and Crossbred Buffaloes in Taiwan (Liu, 1978)

Traits	Murrah	F_1 cross	G_1 backcross to Murrah
Birth weight (kg)	33	32	32
Weight at 10 months (kg)	301	265	270
Total milk yield (kg)	1518	824	1015
Lactation period (days)	240	240	240
Milk yield per day (kg)	6.3	3.4	4.2
Peak milk yield (kg)	8.0	4.6	6.0

Source: Liu, 1978

While these data indicate promising results from crossbreeding of swamp buffaloes with the riverine breeds, one of the most serious questions is how to proceed in the breeding programme without the risk of eliminating the local indigenous swamp buffaloes. It is also important to recognize that, under small farm conditions in most Asian countries, there remain various socio-economic aspects which need to be carefully examined concerning the suitability and farmer acceptance of crossbred buffaloes.

9.5.2.3 *Reproductive performance*

Prepubertal and pubertal performance

The Phil-Ravi F_1 and Phil-Murrah F_1 crossbred heifers reach puberty at 593 (19.8 months) and 460 (15 months) days earlier, respectively, than the Philippine carabao heifer. The weight advantage of the crossbreds over that of the Philippine carabao is about 58 kg for Phil-Ravi F_1 and 80 kg for Phil-Murrah F_1 (Table 9.21). In both the genotypes, follicular development occurred much earlier than the commencement of first behavioral oestrus as determined by a vasectomized bull together with other physiological manifestations of heat.

Table 9.21 Pubertal Performance (means + SD) of the Philippine Carabao, Phil-Murrah F_1, and Phil-Ravi F_1 Crossbreds

Parameters	Breed		
	Philippine carabao N = 21	Phil-Murrah F_1 n = 21	Phil-Ravi F_1 n = 7
Age at development of 1st follicle (days)	966 ± 193a	684 ± 84b	676 ± 89b
Age at the commencement of 1st behavioral estrus (day)	1348 ± 406a	881 ± 177b	761 ± 120b
Age at first ovulation1 (days)	1367 ± 424a	907 ± 172b	774 ± 114b
Weight at puberty (kg)	282 ± 32b	362 ± 48a	340 ± 60a

1 Age at puberty, based on palpation of corpus luteum to assess ovarian development.
Means with different superscripts within row are statistically significant (P<0.05).
Source: Momongan *et al*., 1994

In terms of the development of the reproductive tract, the Phil-Murrah F_1 and Phil-Ravi F_1 crossbreds exhibited no significant difference in anatomical measurements compared with those of the Philippine carabao, indicating that size and weight of the animal at puberty had no bearing on the development of the reproductive tract in buffalo (Table 9.22). No significant difference was observed among the Philippine carabao, and Phil-Murrah and Phil-Ravi F_1 crossbreds in diameter of the cervix, size of the body and horns of the uterus, as well as the size of the ovaries.

Liu (1978) also claimed that Murrah and Murrah-swamp crossbreds reached age of puberty earlier than the local Chinese swamp buffaloes. He reported the age at puberty for Murrah to be 431 days (range: 314–643); for Murrah × Chinese swamp 674 days (range: 384–1203); and for Chinese swamp buffalo 1045 days (range: 659–1387).

Table 9.22 Anatomical Measurements (means ± SD, cm) of the Reproductive Tract of the Philippine Carabao, Phil-Murrah F_1 and Phil-Ravi F_1 Crossbreds at Puberty (Momongan *et al.*, 1994)

Parameters	Breed		
	Philippine carabao N = 21	Phil-Murrah F_1 n = 21	Phil-Ravi F_1 n = 7
Diameter of the cervix	2.26 ± 0.54	2.12 ± 0.53	1.80 ± 0.29
Diameter of the body the uterus	2.22 ± 0.44	2.19 ± 0.52	1.83 ± 0.36
Diameter of the mid-portion of the horns of the uterus			
Left horn	1.74 ± 0.33	1.83 ± 0.48	1.61 ± 0.40
Right horn	1.75 ± 0.33	1.84 ± 0.49	1.61 ± 0.40
Dimensions of the ovaries (L × W)			
Left ovary	1.95 × 1.59	1.90 × 1.52	1.84 × 1.26
Right ovary	1.90 × 1.60	1.90 × 1.47	1.83 × 1.24

L = length, W = width.

Postpubertal and postpartum reproductive performance

The Phil-Murrah and Phil-Ravi F_1 crossbreds conceive and give birth to calves one year earlier than the Philippine carabaos. The ages at first calving were greater than those reported by Momongan (1985) 44.8 months (range: 32.4–61.7) for river buffaloes and 54.6 months (range: 45.7–63.6) for swamp buffaloes. Momongan *et al.* (1984) reported the age at first calving for carabaos under village conditions as 63.6 months and Momongan (1985) as 66.0 months. There is no significant difference in the number of services per conception and in the length of gestation period among Philippine carabaos, Phil-Murrah and Phil-Ravi F_1 crossbreds.

After calving, there is no significant difference between the Philippine carabao, Phil-Murrah F_1, and Phil-Ravi F_1 crossbreds in terms of length of time for the uterus to involute, the recurrence of follicular development, manifestation of first behavioural estrus and development of corpus luteum in the ovary. However, it seems that more Phil-Murrah F_1 crossbreds exhibited silent heat because there were instances when CL could be palpated but the animals were not observed to show signs of heat. The Phil-Ravi F_1 crossbreds have shorter service period and calving intervals compared to Phil-Murrah F_1 crossbred and Philippine carabao, although these differences are not statistically significant (P>0.05). The calving interval reported in this study was longer than that reported by Momongan (1985) who recorded it as 482.4 days (range: 430–551) for river buffalo and 567 days (range: 448–726) for swamp buffalo. In a study conducted under the village conditions, Momongan (1985) reported that the calculated calving interval for 82 multiparous cows was 26.3 ± 12.2 months (range: 11.5–57.5).

Draught animal power

According to Liu (1978), about 45,000 crossbreds, between swamp buffalo and Murrah, had been produced in southern China with the intention of producing more milk and meat, in addition to draught power. These crossbreds were equal in heat tolerance to the local swamp buffaloes and were superior in draught power, both in terms of speed and area per unit of time at ploughing.

De Los Santos and Momongan (1987) conducted two experiments to evaluate the draughtability of the Philippine carabao as compared with Phil-Murrah and Phil-Ravi crossbreds. In one experiment, each buffalo was simultaneously made to carry a pack load equivalent to 20% of its body weight and made to walk a premeasured distance of 6 km. This was repeated 12 times. In second experiment, each buffalo was simultaneously made to pull a sledge with a load equivalent to 50% of its body weight over a fixed distance of 1.6 km. This was repeated 20 times. Results indicated that, in general, the crossbreds were comparable to the Philippine carabaos as draught animals in terms of resistance to work stress and docility. There was no significant difference among the three genotypes in terms of initial and final pulse rates, respiration rates and body temperatures (PRT). As expected, the load stress increased the PRT significantly ($P<0.01$) after work within breeds. However, among genotypes, the differences were not significant indicating that all the three genotypes react similarly to work stress.

Garillo *et al.* (1987) compared the ploughing ability of the Philippine carabao and Phil-Murrah crossbreds steers. Four mature Philippine carabaos and four Phil-Murrah crossbred steers (weighing about 520 kg and 4–7 years old) were evaluated for draught power to plough under wet and dry land conditions. PRT changes before, during and after the ploughing were recorded. There was no significant difference ($P > 0.05$) between the ploughing ability of the Philippine carabao and Phil-Murrah crossbred in terms of depth, width and velocity of ploughing, as well as draught force and drawbar horsepower. Moreover, no difference was observed on the PRT measurements of the Philippine carabao and Phil-Murrah crossbreds as influenced by ploughing either under wet or dryland conditions.

Sarabia and Momongan (1993) evaluated the draughtability of the Philippine carabao and Phil-Murrah F_1 crossbred under lowland rainfed condition. A total of eight untrained intact buffalo bulls, four Philippine carabaos and four Phil-Murrah F_1 crossbreds, with an age range of 4–8 years, were used. The animals underwent a training period. They were subjected to back-riding, pulling sledge and ploughing activities. Their responses to back-riding and ploughing activities were not significantly different ($P > 0.05$). The Philippine carabao pulled the sledge continuously for 42.75 minutes, before it stopped and refused to move forward. The Phil-Murrah F_1 lasted for only 38.35 minutes. The difference in time was significant. The animals were subjected to draught activities such as pulling sledge and ploughing. Each buffalo was simultaneously allowed to pull a sledge with a load equivalent to 50% of its body weight until it refused to pull its load. Results showed that no significant difference was observed among the behavioural (first manifestation of refusal to walk) and physiological (onset of salivation and panting) responses of the two genotypes to stress. Furthermore, no significant difference in the ploughing ability between the two genotypes was observed in terms of pull (kN), draught force (kN), speed (m/sec), draught power (kW or hp), depth and width of ploughing and area ploughed/hr. This indicates that the crossbreds are adaptable to work as the Philippine carabaos when exposed to similar conditions.

9.5.3 Crossbreeding to combine draughtability and milk production

During the past two decades there have been various attempts in different countries to cross swamp buffaloes with Murrahs to produce crossbreds for draught and milk production. Although buffalo crossbreeding and selection among crossbreds for milk production appear to be promising from the genetic point of view, it is worth-while considering the comparative economic advantage of milk production from buffaloes versus dairy cattle crossbreds or purebreds in many countries, especially in the south east Asian countries. Buffaloes, both crossbreds and purebreds, on an average, produce less milk than, say, the Holstein crossbreds. *e.g.*, Murrah in Thailand averaged 806 kg per lactation, while Holstein crossbreds in the same environment (F_1 and F_2) produced at least two to three times as much during the same year.

It is questionable whether a high priority should be given to the selection of swamp buffaloes for milk production. However, if the development goal is to improve the nutrition and health of rural people through upgrading of swamp buffaloes for household milk consumption, the breeding goal becomes a dual-purpose animal both for draught and milk production. Many reports indicate the superiority of crossbreds in milk production but information on their draughtability and acceptance by farmers is still lacking. In one region of the Philippines 39% of the farmers rated the crossbreds poorer than, 24% better than and 14% equal to swamp buffaloes.

It should also be noted that some part of the superiority of milk yield of the crossbreds over swamp buffaloes is due to differences in feeding and management. The performance of crossbreds under village farm conditions needs to be evaluated. What will be the nutritional requirements of working buffaloes if they are also used for milking? Can the village farmers provide the additional feeds required without negative socio-economic impacts on other aspects of farm production? These are some of the questions, which need to be researched upon before embarking on large scale massive crossbreeding programmes with swamp buffaloes.

CHAPTER 10

ADAPTABILITY

10.1 Adaptation–in General

Animals of diverse origin vary widely in their capacity to withstand severe climatic shifts. Environmental adaptation, involves complementary homeostasis of internal body's biophysical norms of: temperature, osmolarity and acid-base balance against the ever fluctuating analogous environmental (exogenous) conditions. These three norms are controlled by several interactions of physiological functions, *i.e.* body temperature, skin temperature, respiration rate, pulse rate etc., these phenomena are related with each other and are product of environmental factors climate and genetic constitution of the animal.

Buffaloes are more sensitive than cattle to direct solar radiation and high ambient temperatures. This is due to several factors: (i) their dark body colour is conducive to heat absorption when the animals are exposed to the direct rays of the sun during the hotter part of the day; (ii) the relatively small number of sweat glands per unit area of skin is inimical to high heat loss by sweating, buffalo skin has one-sixth the density of sweat glands than that of cattle, hence buffaloes dissipate heat poorly by sweating, and (iii) the thick epidermal layer of the skin reduces heat loss by conduction and radiation.

Buffaloes have a lower level of respiration rate, body temperature and pulse rate than cattle. Under shade, they have a better cooling mechanism than cows. The black pigmented skin assists the absorption of infra-red rays with a subsequent rise in body temperature, while in thin skinned animals heat penetrates easily, resulting in over heating of the body and interference with efficient elimination of heat.

Despite these shortcomings, the buffalo is well adapted to the hot and humid climates of the tropics and sub-tropics. This adaptability is attributable principally to the semi-aquatic nature of the buffalo which makes it seek water and immerse its body as a means of reducing the heat load. In situations where there is no ready access to water, shade serves equally well, because evaporation along the respiratory tract is an efficient mechanism of heat loss in buffaloes (Sastry, 1983). The limited genetic studies on heat tolerance in the species suggest that the trait is heritable and that it influences production performance. Sweat gland density and sweating volume coefficient are known to be important traits for efficient heat dissipation and good thermoregulation, but their correlations with heat tolerance indices in buffaloes have not been found to be statistically significant (Nagarcenkar and Sethi, 1981).

The fact that buffaloes are found in widely differing geographical conditions suggests that this species is adaptable to a wide range of environmental conditions. This is particularly true of the river buffalo. While the buffalo is remarkably versatile, it has less physiological adaptation to extremes of heat and cold. Body temperatures of buffaloes are actually lower than those of cattle, but buffalo skin is usually black and heat absorbent and only sparsely protected by hair. If worked or driven excessively in the hot sun, a buffalo's body temperature, pulse rate, respiration rate, and general discomfort increase more quickly than those of cattle. Buffaloes prefer to cool off in a wallow rather than seek shade. They may wallow for up to 5 hours a day when temperature and humidity are high. Immersed in water or mud, chewing with half-closed eyes, buffaloes are a picture of bliss.

In shade or in a wallow buffaloes cool off quickly, perhaps because of black skin, rich in blood vessels conducts and radiates heat efficiently (*plate–8*). Nonetheless, wallowing is not essential. Experience in Australia, Trinidad, Florida, Malaysia and elsewhere has shown that buffaloes grow normally without wallowing as long as adequate shade is available.

High humidity affects buffaloes less than cattle. In fact, if shade or wallows are available, buffaloes may be superior to cattle in humid areas. In southern Brazil, trials comparing buffalo and cattle on subtropical riverine plains have favoured the buffalo.

The swamp buffalo is not as adaptable as the river buffalo. They are found only in swampy and marshy areas, and in hot climates. Swamp buffaloes must have almost unlimited access to water to keep them cool.

Water buffaloes are well adapted to swamps and to areas subject to flooding. They are at home in the marshes of southern Iraq and of the Amazon, the tidal plains near Darwin, Australia, the Pontine Marshes in south-central Italy, the Orinoco Basin of Venezuela, and other areas. In the Amazon, buffaloes (Mediterraneo and swamp breeds) are demonstrating their exceptional adaptability to flood areas. Buffalo productivity outstrips that of cattle, with males reaching 400 kg in 30 months on a diet of native grasses.

The buffalo is considered essentially as animal of the plains, but the native buffaloes of Nepal are found distributed from the plains of the Terai region up to an elevation of 2700 m. Murrah buffaloes from the plains of India have been introduced into the mid-altitude region of Nepal. These animals have easily adapted to this hilly environment and are thriving. In Bulgaria, buffaloes are used for drawing snow-ploughs. A herd of swamp buffaloes is thriving at Kandep in Papua New Guinea, 2500 m above sea-level.

The best milk breeds of buffaloes in India and Pakistan are mostly confined to areas where the summer temperature rises above 46 °C (115 °F) and the winter temperature falls below 4 °C (39.5 °F). Nondescript buffaloes are scattered all over India and Pakistan. The buffalo is capable of maintaining itself in good condition, where there is barely enough sustenance to maintain life such as stubble fields or marshy lands with sedge, reeds, water weeds and grasses. It can also work and supply milk when fed the poorest of diets.

10.2 Heat Tolerance

10.2.1 Thermoregulatory mechanism

When exposed to direct solar radiation or if made to work in the sun during hot weather, buffaloes exhibit signs of great distress. On account of the dark skin and the sparse coat of hair on the body, there is great absorption of heat from direct solar radiation. Besides this, there is also less efficient evaporative cooling from the body surface of the animal due to a rather poor sweating ability. If kept in shade and rested or put to work at a slow pace unexposed to the sun during the hot weather, their tolerance to heat is of no mean order. Their thermoregulatory mechanisms function efficiently and are more effective in the shade than they are in cattle, when the speed of recovery from the effect of stress is taken as a measure of efficiency (Badreldin and Ghany, 1952; Mullick, 1960). Results of investigations by Pandey and Roy (1969) on seasonal changes in body temperature, cardiorespiratory and hematological attributes and body water and electrolytic status of the buffalo under conventional farm management lend further support to their view. Pandey and Roy (1969) emphasized that the plethora of changes observed in the buffalo are orderly manifestations of various physiological adjustments necessary for adaptation to higher environmental temperature and that these changes are within physiological limits.

Under cool, comfortable conditions the average rectal temperature, pulse rate and respiration rate of the buffalo are lower than those of the zebu, but exceed on exposure to hot-humid conditions. Buffaloes become more hyperthermic on exposure to direct sun due to black skin and scanty hair on the body. Body temperature rises sharply from 0.3 °C to 2.4 °C depending on the environmental temperature and duration of exposure (Badreldin and Ghany, 1952; Mullick, 1960). However, temperature falls sharply to normal on shifting the exposed animals to sheds, splashing of cold water or wallowing. The effect of stress on exposure to a hot-arid environment is more severe in buffalo heifers than in zebu heifers.

The rise in climatic temperature reduces heat dissipation by convection and radiation due to the drop in the temperature gradient (skin temperature-air temperature). The first response of the animal is to elevate skin temperature by vasodilation of skin to bring more heat from the core by fast circulating blood to the skin. This mechanism is effective in buffaloes than in cattle due to the absence of intervening hair coat in buffaloes. However, this mechanism is not effective with high rise in air temperature. With failure of heat dissipation by physical (non evaporative) channels, the animal increases water vaporization from its surfaces, skin and respiratory tract. Since buffaloes have apocrine sweat glands they rely less on active sweating for skin vaporization and depend more on respiratory vaporization by forcing more ventilation rate.

10.2.2 Physiological adaptation

This fast output of water from the surface of the respiratory system enforces mobilization of water from the interstitial fluid (ISF) and blood plasma (BP), to replenish this with water, continuous flow of water from the rumen (the enormous water store in the ruminants), the gut in general, to the blood stream is generally accelerated to maintain the vital normal plasma volume (PV). With this water mobilization, Na^+ and K^+ ion concentrations are interrupted which call for the rearrangement of these concentrations to maintain normal osmosis in the body fluids and cells. The kidney and the relevant hormones (ADH and corticoids) are activated by the neurohormonal system to control the water and ionic concentrations through urine adjustments. This tedious train of events and physiologically elaborate functions as response to heat stress is augmented if the ruminant is exposed to lack of drinking water (dehydration) Ashour and Shafie, (1992).

Buffaloes have higher water turnover rates than *Bos taurus* and *Bos indicus* cattle (Siebert and MacFarlane, 1969). They are also less efficient users of water, as evidenced by their higher intakes of water per unit dry matter intake, higher urine outputs and lower percentage kidney re-absorption of filtered water (Moran, 1978; Moran *et al.,* 1979). These must be taken into account especially during the hot season.

With rapid increase in respiration rate (panting) there was faster depletion of CO_2 by increased ventilation, which caused a disturbance of "acid-base (pH)" balance towards alkalosis. In order to repair this respiratory alkalosis and maintain the vital norm of pH, the thermo-cardio respiratory responses get activated.

10.2.3 Effect on behaviour, health and production

Heat stress and/or dehydration evokes behavioral responses, mainly seeking shade, swimming or wallowing and eagerness for water intake when provided. A stressful status is reflected by low appetite and drop or seizing of rumination.

Milk yield of buffaloes is also adversely affected during summer. If those buffaloes in the optimal period of their lactation are splashed on the body with water during the summer season there is an increase in milk production, but in buffaloes not given this treatment the milk yield becomes irregular and declines.

Animals that are put out to graze in direct, hot sunshine eat less during the day, but may eat more at night. There is also a marked reduction in fertility. The animals that come into heat tend to do so mostly during the cooler hours of the night, and the heat periods themselves tend to be shorter and feebler or silent, with little or no outward indications.

10.2.4 Effect on Reproduction

Though the thermoregulatory mechanism of the buffalo is reasonably efficient, there are reports of the reproductive system being adversely affected by climatic changes. The quality of the buffalo semen deteriorates like that of cattle, sheep and goat semen during summer, but unlike other species the Murrah buffalo fails to retain the quality of its semen at a high level during winter also (Mukherjee and Bhattacharya, 1953; Kushwaha *et al*., 1955; Sengupta *et al*., 1963).

Poor reproductive performance in buffaloes is clearly a functional adaptation to the environment. In swamp buffaloes, this is further aggravated by difficulties in oestrus detection and long calving intervals in the female and delays in attaining sexual maturity in the male. Thus, swamp buffaloes in Malaysia can be used for effective service only at 3.5 to 4 years of age (Bongso *et al*., 1984). The exact time of completion of spermatogenesis and age at puberty are not definitely known in the different swamp buffalo populations. They also respond equally well to good management. Thus, if buffaloes are fed and managed as meat animals from birth, it in possible to attain live weights of not less than 300 kg at 2 years of age, with a quality of meat that is comparable to that of cattle. Likewise, machine milking of buffaloes, when practised with normally efficient routines from the commencement of milk production, results in milk of superior hygienic quality than hand-drawn milk; it also prevents the abnormal elongation of teats which occurs when force is exerted to open the teat sphincter under crude hand milking methods.

Some degree of seasonal variation in breeding efficiency is usual with most domestic livestock, but the variation is more marked in the buffalo. A seasonal trend of reproduction among the Philippine buffaloes was first reported by Villegas (1928). Subsequently, this peculiarity of breeding of the buffalo was reported by many workers from India, Pakistan, Egypt, Bulgaria and Italy. Analysing the data from various farms in India and Pakistan. It was observed that there was practically no incidence of oestrus during the hot months and that the oestrous symptoms were not pronounced during the adjacent periods. Bhat (1983) noted that this observation could not be supported from the data of military dairy farms. They showed that female buffaloes came in heat regularly during summer and that there was no particular season of breeding for Murrah buffaloes, although there was a tendency for better performance during the cooler months. They noted that buffaloes in the marshes of southern Iraq breed throughout the year, but more so in the spring and a little less so in the autumn.

The depression of the breeding function of buffaloes during hot months led many to suspect that it was in some way related to high environmental temperatures or to the temperature-humidity complex of the tropical summer. Others considered that in addition, there might be a photoperiodic effect. Roy *et al.* (1968) conducted a series of systematic investigations. From the results of the earlier experiments it appeared that air temperature and humidity had a direct effect on breeding efficiency as improvements in breeding performance were obtained by alterations in housing conditions to provide relief from heat stress. Subsequent investigations have suggested that the photoperiod, rather than the heat stress factor, is primarily involved. Further studies and more critical evaluation of the experimental results led this team to believe that manifestation of oestrous symptoms is poor during summer and that with proper husbandry practices it is possible to make the buffalo breed fairly uniformly all through the year. There are many questions relating to the reproduction of the buffalo that still remain to be investigated, and it is very likely that a number of factors singly or in combination are operational in causing the marked seasonal breeding behaviour.

Generally speaking, competent management to secure good adaptation to the environment includes the provision of adequate and balanced feed for milk, meat and/or draught production, the adoption of appropriate measures for disease prevention and parasite control, protection against extremes of climate, and the use of suitable techniques to maintain high reproductive efficiency.

10.2.5 Managerial improvements

Suitable protection against thermal stress is thus an essential requirement for good buffalo husbandry. Most studies of the environmental physiology of the buffalo have demonstrated that shade is the most important protective measure (Karam Shah *et al.,* 1979; Sastry,1982) This should be accompanied by appropriate housing with protection from solar radiation through suitably planted trees, shrubs or creepers. Experience in the Indian subcontinent suggests that loose housing is to be preferred to the more conventional cattle sheds/barns, provided that the walls and other roof supporting structure are minimal. Otherwise their surfaces tend to retain and reflect radiation into the house thus causing a considerable increase in temperature. This, in turn, gives rise to a greater evaporation of moisture from the floor wash-water urine and dung, leading to higher levels of humidity.

Loose houses may take the form of simple shelters with grass thatch, bamboo or asbestos; these are most suitable from the point of view of both comfort and cost. The animals retire to them when they receive inadequate protection from the sun during the hotter part of the day. Feeding and watering arrangements are facilitated and more effective management can be achieved. During the nights, the drop in ambient temperature in a loose house is sufficiently large to influence feed consumption. It has been found (Singh, 1983) that in contrast to the well known trends of lowered feed consumption/utilization by animals exposed to high temperatures in climatic chambers, the relief experienced by animals at night under field conditions is conducive to high dry matter intakes.

Thus, the most important consideration in the alleviation of climatic stress in buffaloes is protection from exposure to direct sun. Other supplementary measures that are usually practised include wallowing, showering or splashing water on the animals. The wallow has the disadvantage that it may become fouled with excreta unless there is a continuous flow or regular change of water. Moreover, the village practice of taking buffaloes to ponds for wallowing could be counterproductive if the animals are made to walk to and from the pond in the hot sun. Where good management obtains, buffaloes are not allowed out in the sun from mid-morning to late afternoon when solar radiation is most intense.

Showering or splashing water three or four times during the hottest part of the day for about 3 min at a time is as effective as wallowing in alleviating thermal stress in buffaloes. The provision of cooled drinking water in earthen pitchers of the type commonly used in villages in India also helps. The work of Radadia *et al.* (1980a,b) has served to confirm these observations.

During night, buffaloes may be left in the open, provided they are protected against predators. The drop in ambient temperature from day to night enables the animals to dissipate by radiation the heat they would have absorbed during the day. Even where housing is provided in the middle of an open paddock or pasture, it is commonly observed that the animals spend most of the nighttime in the open area. Only when it is very cold or during rains do they move into the house.

The adoption of such managerial improvements on a substantial scale at the small farm level poses a number of practical problems. The more important of these are the difficulties encountered by small farmers and landless labourers in producing and/or obtaining the minimal amounts of concentrates and forages of the quality, that would be required for calf feeding. Both economic and infrastructural considerations loom large in the resolution of these difficulties. Several solutions are available for this which need government support.

10.2.6 Effect of temperature on buffalo calf

Buffalo calves under one year of age are more prone to heat stress than adults; the principal symptoms are loss of appetite, lassitude, high body temperature, swollen fetlocks and a watery discharge from the eyes. Well ventilated housing offering suitable protection against extremes of heat and cold is essential. Feeding and management that would stress the animals should be avoided. High mortality in buffalo calves is a universal problem, often justifying the oft-repeated statement that "the buffalo calf has little desire to live". It is all the more reason why particular care should be taken during the short critical period after birth when the calf may not be consuming enough colostrum, either because it is not inclined to suckle or because the cow is not inclined to nurse, or the farmer is not keen to feed the required quantity of milk. Suckling may have to be assisted where necessary or the colostrum should bc hand fed to the calves if natural means fail. In older calves under-feeding is the principal stress factor. Thus where milk production from the dam is important, as is usually the case with river buffaloes, the calf is used principally as a means of securing milk let down and it is therefore often subject to under feeding. Deliberate starvation is also known to be practised with male calves. The replacement rate in buffaloes is generally low because of their long reproductive life, and this contributes to large numbers of calves being allowed to die. However, on large and well managed enterprises, the calves are weaned at about the fourth day, bucket-fed thereafter on limited amounts of whole milk and supplemented with good quality concentrates and forages. In these circumstances, mortality is minimal so long as good hygienic practices and proper feeding and watering schedules are observed.

10.3 Cold Tolerance

Although generally associated with the humid tropics, buffaloes have been reared for centuries in temperate countries such as Italy, Greece, Yugoslavia, Bulgaria, Hungary, Romania, and in the Azerbaijan. Buffaloes are also maintained on the high, snowy plateaus of Turkey as well as in Afghanistan and the northern mountains of Pakistan.

The buffalo has greater tolerance to cold weather than is commonly believed. The current range of the buffalo extends as far north as 45° latitude in Romania, and the sizable herds in Italy and the Soviet Union range over 40 ° N latitude. Cold winds and rapid drops in temperatures cause illness, pneumonia and sometimes death. Most of the animals in Europe are of the Mediterranean breed, but other river-type buffaloes (mainly Murrah from India) have been introduced to Bulgaria and the CIS countries which indicates that river breeds at least have some cold tolerance.

The low temperatures recorded during winter in semi-arid tropical and sub-tropical areas are, generally speaking, not severe enough to warrant elaborate houses or other measures to protect buffaloes against the cold (except, of course, in the case of very young stock). The common village practise, however, is to confine the animals inside closed houses at night, while on large farms loose housing is often provided with extra protection in the form of hessian drapes over doors and windows. Studies at the Haryana Agricultural University, Hisar, suggest that such practises are unnecessary and only cause extra expenditure without any special benefit. The common observation of a drop in milk yield during a cold spell is usually a function of management, failure caused by the discomfort experienced by attendants, milkers and persons responsible for harvesting fodder for the animals. Inadequate care of the animals, incomplete milking and reduced feeding, rather than cold stress per se, are the principal contributory factors. Failure during the winter to produce semen of high quality as is produced in spring may be the result of the sensitivity of the buffalo to a cold environment.

Since buffaloes have sparse hairs they have no way to engage an efficient insulative coat to check heat loss from their body. Even the neonatal and young calves with their denser hair cover gain less value as the pili muscles are weak in performing hair erection function to increase the thickness of the coat in accordance with the degree of coldness. The animal has to increase the metabolic rate to compensate for faster heat loss. The subcutaneous fat layer, however, is a good substitute of the outer coat for insulative heat preservation, if it is developed to effective thickness, a character waiting anatomical and physiological studies. The subcutaneous fat has abundant blood supply, rejecting the erroneous impression of feeble circulation. The plenty of blood permits effective deposition and utilization of fat in this depot. Vasomotion in skin is also a primary physiological response to fluctuations in environmental temperature. Since the skin of the buffaloes is very thick(0.6–10.0 mm) vasoconstriction plays a great part in heat conservation in cold weather.

Very few reports are available about the heritability of environmental adaptation phenomena, more particularly of thermo-cardio-respiratory responses in buffaloes. Some workers studied the inheritance of heat tolerance in buffaloes by measuring change in the body temperature and pulse rate before and after exposure to direct solar radiation for 6 hours. They reported heritability estimates of 0.24 ± 0.19, -0.05 ± 0.12, 0.39 ± 0.24 for initial body temperature, initial pulse rate and heat tolerance index (Iberia heat tolerance test).

CHAPTER 11

REPRODUCTION

Reproduction response in a species is result of management processes which eco-system provides for the survival and growth of a species. The buffalo has been at disadvantage as the cow was the backbone of agricultural motive power, hence the attention on buffalo could never grow till agricultural power shifted to electro-mechanical means. Therefore the parameters of reproduction and production are natural to the species and need not be compared to cattle.

In general, reproductivity has been better studied in the river buffalo as compared to the swamp buffalo affected by many factors such as nutrition, management and environment. Comprehensive studies on reproduction of buffalo are necessary to improve and manage buffalo programme

Buffaloes have a reputation for being a sluggish breeder, but the average animal is so poorly fed that its reproductive performance is unrepresentative of its capabilities. Without reasonable nutrition the animals cannot reach puberty as early in life or reproduce as regularly as their physiology or genetic capability would normally allow.

The reproductive organs of the buffalo conform to the general mammalian type and there is considerable similarity in the anatomy of the general organs of the buffalo (*Bubalus bubalis*) and the ox (*Bos taurus* and *Bos indicus*).

11.1 Male Reproductive System

11.1.1 Anatomy

The genital organs of buffalo bull are smaller as compared to those of a cow bull. The testis of a male buffalo measure 10 to 12 cm in height. The circumference at the maximum breadth is 30 to 34 cm. The hair on the scrotum is absent there are no preputial hair. The cauda epididymis is smaller and not very distinct.

The length of the penis of an adult male buffalo in front of sigmoid flexure is 80.15 ± 2.83 and the diameter is 1.85 ± 0.15 cm. The glans penis gets tapering and smaller while glaea glandis is poorly developed in buffalo bulls.

11.1.2 Physiology of reproduction in buffalo bulls

Spermatogenesis in the buffalo starts quite early in life. Meiotic division of the spermatogonial cells has been observed as early as, at one year of age in the Indian buffalo (Dutt and Battacharya, 1952). It is likely that spermatogenesis may start even earlier. There are, however, reports from Italy and Russia of the buffalo bulls being put to service at around two years of age and this is not much different from the age at which male cattle are used for breeding. The age at first service of the buffalo bull is around three years in Iraq although some well-fed and well-developed bulls may be put to stud at two years. In India, Pakistan and Egypt the buffalo bull is generally not used as stud before 3 to 3½ years of age. The duration over which the buffalo bull may be retained for service is as yet a subject of individual opinion as no systematic study appears to have been made on this aspect of husbandry. Macgregor (1941) stated that in the buffalo, by six or seven years, there is frequently a loss in potency with a rising proportion of unsuccessful services but desire continues until he is 12 years or more, during which time his muscular strength seems to increase. Complete senility, *i.e.*, loss of muscular strength as well as desire, does not manifest itself until he is over 15 years of age. Others have expressed the view that the buffalo bull can be retained for service up to 10 to 15 years of age. Buffalo bulls are used for breeding in Iraq up to ten years of age. In Italy, bulls are replaced after only four or five years of service.

Divergent views have been expressed about the number of female buffaloes to be allowed per bull. According to Macgregor (1941) a good buffalo bull of the river type can serve 100 cows a year, but it is unusual to allow more than 12 cows to each bull as each of them will be served several times during her heat. Hafez (1952) stated that a buffalo bull can serve 50 females in a year. Lazarus (1946) opined that the buffalo bull should be used sparingly so that he does not serve more than 75 times a year. The number of females allotted to one buffalo bull in Iraq villages varies widely from 12:1 to 600:1, but in the majority of villages it is less than 100:1.

In Egypt, 75% of the services in buffaloes occur during the four months of the year that constitute the breeding season. A bull is used at least three times a week during this period. With such heavy service there is deterioration in semen quality and consequent lowering of fertility (Asker and El-Itriby, 1958). In Egypt with three semen collections per week from the buffalo bull there was a comparative deterioration in the semen quality. With semen collection twice a week for a period extending over three years, there was no deterioration in semen quality or sex vigour in Murrah buffalo bulls. Sex vigour of the buffalo bulls declines during the hot summer in north Indian subcontinent and improves with the onset of the cold season. With the introduction of artificial insemination (AI) and wider use of frozen semen technique, this aspect has lost much of its significance in most countries where milk production is in high focus and husbandry has improved.

Behaviour of buffalo bulls

Buffalo bulls can be easily trained for donation of ejaculate in the artificial vagina. The thrust given by buffalo bulls is not as forceful as in case of cow bulls. Bhosrekar (1993) observed that if young buffalo bulls are trained just after the onset of monsoon, they learn mounting and donation of ejaculate better as compared to the summer months. In summer, they tend to be lethargic and unwilling to learn, but those which have been trained for the semen collection programme do not pose any problems except lengthening of reaction time.

11.1.3 Factors affecting libido and semen quality

The size and weight of testes are significantly less in buffaloes than in cattle and this is reflected in relatively poor libido and in semen production and quality. The scrotal circumference and testicular volume have large and positive correlations with body weight and age. The age at first ejaculation in buffalo males is very late. The time lag between testosterone secretion (12 months) to support spermatogenesis and onset of spermatogenesis (16 months) and the achievement of puberty (2.5–4 years) is long. The possibility of reducing this age through better feeding and possible administration of exogenous gonadotrophic hormones needs to be examined. It may be possible to obtain usable semen at 2.5 years of age.

Administration of exogenous gonadotrophins has produced variable results. Intramuscular injections of prostaglandisn (PGF_2) 1 hr before semen collection reduced the reaction time and resulted in increased ejaculate volume without affecting the total and live sperm concentration.

The major problems of male fertility are poor libido, quantitative and qualitative reduction in buffalo semen during summer, and a wide range of variability in libido, semen production, and quality freezability among bulls. The spermatogenic efficiency of the buffalo is only 43%. Although these problems are mostly related to nutrition and the physical environment, especially heat stress, there are large genetic differences which could be exploited in improving the performance of males.

11.1.4 Artificial insemination

Artificial insemination (AI) has been used so far as a breeding tool on a field scale only in a few buffalo-rearing countries. It has been practised on the largest scale in India where it is claimed that the first buffalo calf in the world bred by means of AI was born at the Allahabad Agricultural Institute on 21 August 1943.

The male buffalo can be trained more easily than male cattle for donating semen by artificial service. The buffalo is less choosy concerning the teaser and readily mounts an anoestrous female or even a male buffalo in the service crate. Coat colour or the physiological status of the female buffalo in the service crate does not alter the reaction time of the buffalo bull (Prabhu and Bhattacharya, 1954; Prabhu, 1956). The thrust given by male buffaloes during service is less vigorous than male cattle. As in the case of male cattle, collection in the artificial vagina is also the most convenient method of obtaining semen from the buffalo. However, temperature of the artificial vagina needs to be modified suitably. With the temperature of the artificial vagina reaching 41 °C (105.8 °F) there is bursting and breaking of buffalo spermatozoa, but a temperature of 39 °C (102.2 °F) causes no damage to the cells (Mahmoud, 1952). Although it is possible to collect semen from the buffalo by the massage technique, the animal is less responsive to this method of ejaculation than male cattle, and also requires a longer period for training. The semen collected by this method is relatively poorer in quality. The capillaries of the rectal wall are more fragile in the buffalo than in cattle and hence manipulation through the rectum while massaging must be done with great care as otherwise bleeding may result as a consequence of the breaking of the capillary vessels. Electro-ejaculation from the buffalo although successful, is not likely to come into general use in the near future.

11.1.5 Semen studies

(a) Morphology of buffalo spermatozoa

Morphological characteristics of the spermatozoa of the buffalo and cattle differ. Macgregor (1941) was possibly the first to report that the head of the buffalo spermatozoa is more rectangular than that of cattle and that the stained portion of the head is also slightly narrower in the former species. According to Rao (1958), spermatozoa of the buffalo possess distinct morphological characteristics by which they can be readily differentiated from spermatozoa of other domestic animals. Comparative morphology of the spermatozoa of Egyptian buffaloes and indigenous cattle can be seen in Table 11.1. The morphology of spermatozoa changes with the age of the buffalo (Venkataswami and Vedanayagam, 1962) and with ageing in different diluents.

Table 11.1 Comparative Morphology of the Spermatozoa of Buffaloes and Cattle

Spermatozoa measurements	Buffaloes			Cattle		
	Mean (m)	SD (m)	CV (%)	Mean (m)	SD (m)	CV (%)
Head length	7.436	0.442	5.9	9.126	1.326	14.5
Head breadth (anterior)	4.264	0.520	12.3	4.732	0.494	10.4
Head breadth (posterior)	3.172	0.442	14.0	2.730	0.520	19.4
Ratio head breadth (anterior) to head breadth (posterior)	1.340	0.330	24.8	1.790	0.320	1.7
Neck	0.442	0.208	47.1	0.650	0.338	51.5
Length of middle-piece	11.648	0.936	7.9	12.558	0.624	4.9
Breadth of middle-piece	1.092	0.286	26.2	1.006	0.268	26.8
Length of tail	42.82	3.042	7.1	46.280	6.084	13.2

Source: Mahmoud (1952).

Compared to cattle spermatozoa, the spermatozoa of the buffalo have an inherently poorer metabolic activity as judged by O_2 uptake and fructolysis. Variation in semen quality due to changes in the climate is also more pronounced in buffaloes than in cattle, as evidenced by changes in the respirometric activity of spermatozoa.

The depressing effect of the phosphate ion, and the stimulating effect of the chloride ion on the O_2 uptake pattern, as can be found in cattle spermatozoa, are not observable in buffalo sperm cells. This would suggest that either the influence of certain ions like phosphate and chloride on aerobic metabolism of buffalo sperm cells is intrinsically different from that of their effect on cattle sperm, or that the seminal plasma of the buffalo has properties, which in some way obliterate the specific effects of the chloride or phosphate ions on O_2 utilizations *in vitro* (Sinha *et al*., 1966). The work of Roy *et al*. (1960) lends experimental support to the second hypothesis.

Buffalo spermatozoa are richer in all lipid components than cattle spermatozoa, while seminal plasma is poor in lipid contents as compared to cattle. Triglycerides are the major lipids in buffalo semen and phosphotidyl choline is the major phospholipid in buffalo sperm and seminal plasma. The buffalo sperm's oxidative metabolism is depressed by high level of alkaline phosphatase esters and hence reduces its survival *in-vitro*. The glycosidase activity is lower in buffalo semen. The hyaluronidase as well as glucuronidase activity is lower in buffalo semen, which might be the explanation for lower fertility of buffalo semen. The oxygen uptake and aerobic fructolytic index is inherently low in buffalo spermatozoa and hence sperm motility falls more rapidly under aerobic conditions.

(b) Semen characteristics

Unlike cattle semen which has a yellowish tinge, the semen from healthy buffalo bulls is milky white or milky white with a light tinge of blue (Shukla and Bhattacharya, 1949; Mahmoud, 1952). Cattle semen gives a colour reaction of various shades with DOPA (3–4 dihydroxyphenylalanine) (Mukherjee, 1964), but such a reaction is not detected with buffalo semen.

Volume, sperm concentration, initial motility and speed of travel of spermatozoa are generally lower in buffalo than in cattle semen. Pawan Singh and Singh (2000) found a beneficial effect on buffalo semen production with regards to ejaculate volume, mass activity and total sperm per ejaculate when the experimental animals were given multiple showers and artificial acration. Prajapati *et al.* (2000) observed that semen quality improved significantly in the those group of buffaloes which were subjected to forced exercise, due to a reduction in sperm abnormalities in all seasons, specially during summer and monsoon seasons.

The epithelial cells cycle in buffaloes is shorter as compared to cow bulls and spermatogenesis takes only 43 days as compared to cow bulls (56 days).

Buffalo bull semen is rich in calcium, chloride, acid soluble phosphorus, and alkaline phosphatase. The alkaline phosphatase might adversely affect the viability of buffalo spermatozoa during storage. The total electrolyte content in buffalo semen is less as compared to bull semen. The altered citrate chloride ratio may be the cause of the intrinsically poor buffering capacity of buffalo semen. The buffalo semen has low level of ascorbic acid and hence lower effective life of spermatozoa. The buffalo semen also has lower protein and total cholesterol as compared to cattle semen.

Roy *et al.* (1960), based on the comparative study on the semen biochemistry of buffalo and cattle (Table 11.2), observed that significantly higher concentration of calcium, esterified phosphate and the phosphate-splitting enzymes in buffalo semen create conditions that adversely influence the viability of spermatozoa in stored semen.

Table 11.2 Differences in some Chemical Attributes of Buffalo and Cattle Semen

Constituent	Buffalo	Cattle	Statistical
	(mg/100 ml semen or seminal plasma)		
Total reducing substances	700 ± 52	769 ± 42	
Fructose	355 ± 17	611 ± 39	**
Calcium	40 ± 2	25 ± 5	**
Chloride	373 ± 55	249 ± 26	*
Inorganic phosphate	6.4 ± 0.6	5.9 ± 0.5	
	(6.3 ± 0.4)	(5.6 ± 0.4)	
Acid-soluble phosphate	72 ± 3.9	29 ± 3.2	**
	(64 ± 2.2)	(27 ± 2.9)	**
Total phosphorus	103 ± 8.9	74 ± 2.5	**
	(95 ± 7.2)	(42 ± 4.8)	**
Acid phosphatase activity	308 ± 44	145 ± 11	**
(Bodansky unit)	(307 ± 41)	(167 ± 11)	**
Alkaline phosphatase activity	252 ± 37	134 ± 14	*
(Bodansky unit)	(266 ± 42)	(152 ± 18)	*

Note: Figures in parentheses indicate values in seminal plasma. *Significant at 5 % level. **Significant at 1% level.

Source: Roy *et al.* (1960).

(c) Semen dilution and preservation

Several of the diluters, which proved very satisfactory for extension and storage of cattle semen, did not give as satisfactory results when used for buffalo semen. Egg-yolk phosphate buffer (EYP) has proved unsuitable for use with buffalo ejaculate. Egg-yolk citrate (EYC) is fairly satisfactory for dilution and storage of buffalo semen.

Kampschmidt's glucose sodium bicarbonate egg-yolk extender (with or without addition of sulphamezathene), appeared to be satisfactory for AI of the buffalo. Higher fertility results have also been reported with the use of Kampschmidt's diluent (Gokhale, 1958). Citric acid whey (CAW) is effective as diluent for storage of buffalo semen at refrigeration or deep freezing temperatures. The spermatozoa in CAW are extremely sensitive to the pH of the diluent and sperm motility is maximal at pH 6.8.

Very satisfactory results have been reported from Pakistan with the use of homogenized milk as a diluent for buffalo semen.

The overall assessment of a large number of trials on different extender formulations, level of cryoprotectives, methods of glycerolization, cooling and equilibration time, freezing and thawing rates vis-à-vis freezability of buffalo semen and subsequent fertility, indicated Tris-yolk-glycerol to be the best. Optimum level of the glycerol in the extender is around 6.5–7%. It is possible to do glycerolization of buffalo semen in Tris-based extender at room temperature (20 °C) without any apparent adverse effect on sperm motility. The cooling and equilibration time most suitable is around 6 hr. Cryopreservation on liquid nitrogen vapour is the simplest and most effective method.

11.1.6 Frozen semen technology

There are two main reasons why deep frozen semen is considered appropriate and efficient for improving reproduction of buffalo. They are:

1. The animals are raised by small farmers who are scattered over vast rural areas. The use of breeding bulls or fresh semen for artificial insemination is slow. In contrast to this, deep frozen semen can be used efficiently in selection programmes and for the control of sexually transmitted diseases.
2. The swamp buffalo ejaculates a very small volume of semen. Deep freezing permits doses to be dispensed accurately and with appropriate judgment; valuable semen can be made available for extended usage.

Deep freezing of buffalo semen is now the routine method of choice in many countries. Although efforts to deep-freeze buffalo spermatozoa by the conventional dry ice-alcohol method started as early as 1955, yet the real breakthrough came in 1973 when vapour freezing technique was used with good freezability and fertility of frozen semen (Roy and Bhat, 1973).

(i) Extenders

The relative merit of different extender formulations vis-à-vis freezability of buffalo spermatozoa was also studied by several workers. Citrate-yolk-glycerol was found superior to citric acid whey (CAW) glycerol as well as reconstituted skim milk-glycerol in terms of post-thaw sperm motility (Elimwadi *et al.,* 1974; Bandyopadhayay *et al.,* 1974; Singh *et al.,* 1980; Bhandari *et al.*, 1982).

More recently, a new twitter ion buffer based on extender Tes-Tris-yolk-glycerol was found significantly superior to Tris-yolk-glycerol in terms of post-thaw motility, quantitative aerosomal damage and enzyme leakage following freezing (Sukhija, 1984). 10% egg yolk was a better in a tris based buffer for freezing buffalo semen compared to lower levels (Pawan Singh *et al*., 2000). Singh *et al.* (2000) investigated and confirmed that tris-based extender was better for cryopreservation of buffalo semen than the laiciphos and biociphos extenders.

Of the wide assortment of extenders tried later, in terms of post-thaw motility and fertility, Tris-yolk-glycerol was found to be most effective. This extender is prepared by first preparing buffer solution of the following chemical ingredients:

Tris 30.48 g, citric acid 17.00 g, fructose 12.52 g, distilled water 840.00 ml.
This buffer is mixed with yolk and glycerol in the following proportions:
Tris buffer (as above) 740 ml, egg yolk 200 ml, glycerol 60 ml, total 1000 ml.

Further, Gupta and Saxena (2000) found that the deep freezing of buffalo spermatozoa was significantly better (P<0.01) when tris and citrate diluents supplemented with fetal calf serum were used.

(ii) Cryoprotective agent

Glycerol, at variable concentration, has been used as a universal cryoprotective agent in most studies on freeze preservation of buffalo spermatozoa. Sengupta *et al.*, (1982) reported that partial substitution of glycerol in a Tris-based extender by alternative cryoprotective agent did result in improvement of freezability with 6% raffinise but not with either 3% lactose or 1% dimethylsulfoxide. In a later study, the advantage of reffinose substitution was not evident in a Tes-Tris-based extender. There is obviously an interaction between the additive and the extender formulation (Sukjija, 1984). Rohilla *et al.* (2000) studied the comparative effects of glycerol, ethylene glycol and propylene glycole as cryoprotective agents added in tris citric acid egg yolk extender on pre and post-freezing quality of Murrah buffalo semen and concluded that values of post-thaw motility, live spermatozoa and for intact acrosomes were satisfactory with ethylene glycol only.

The optimum level of glycerol in the extender media appears to be 6.5 to 7.0% when used as the sole cryoprotective agent, but a lower level of glycerol is compatible with other cryoprotective agents like lactose or raffinose. The minimum cooling and equilibration time consistent with good freezability was about 6 hr. Because of sheers simplicity and convenience, the LN vapour freezing method has become most popular.

(iii) Glycerolization

Although it is universally recognized that glycerol is the key cryoprotective agent in freezing of bovine spermatozoa, yet there are conflicting reports regarding the rate and temperature at which glycerol should be added. Successful glycerolization of buffalo semen at room temperature (20°–22 °C) in Tris or modified Tris-based extenders have been achieved with good post-thaw motility and least morphological and biochemical damage during freezing (Chauhan *et al.*, 1981; Sengupta *et al.*, 1982).

It was further observed that by incorporation of an anti-dilution factor-cysteine hydrochloride in the extender media, a faster rate of extension (8–10 min) of neat semen with a pre-glycerolated extender at room temperature was possible without any additional damage to the spermatozoa than was achieved with the conventional slow rate of glycerolization (25–30 min). In fact, freezability was better with a faster rate in a cysteine supplemented extender than with the slower rate in an un-supplemented control extender (Sukhija, 1984).

(iv) Equilibration time

A very crucial step in the pre-freezing semen processing technology following glycerolization is the time interval between addition of glycerol and initiation of freezing called the "Glycerol equilibration" period. Bandyopadhayay (1974) using either EYC or skim milk-based extenders found 3 hr equilibration time too short for optimum sperm survival following freezing. The 9 hr equilibration time was found to give marginally better survival rate than a 6 hr period in both the extenders. More recently, Tuli *et al.,* ((1981) found that in Tris buffer based extender, 4 hr equilibration time gave consistently better post-thaw motility than either 2 or 6 hr equilibration.

(v) Packaging system

Bovine frozen semen packaging system have undergone, over the last few decades, considerable geometric change from the conventional glass ampoules through neutral straws of different materials, diameter, length and capacity and finally to circular concentrated pellets. The majority of the subsequent workers used the 0.5 ml PVC median straws and derived satisfactory post-thaw survival and fertility.

Now-a-days 0.25 ml straws are used for freezing the semen. The sperm count is maintained at an optimum pre-freezing level of 40 to 45 million live sperms. The mini-straw freezing has a slight advantage over the medium straws in survival rates obtained after freezing.

(vi) Freezing and thawing

In early studies on buffalo semen, extended semen packaged in glass ampoules were frozen in a dry ice alcohol bath. The variable cooling rate from 5° to 15 °C and from 15° to 79 °C was manually regulated by adjusting the rate at which dry ice was added to the bath. The post-thaw survival rates in these studies were not consistently good (Sahani and Roy, 1972). Chou (1986) compared three different methods of pellet freezing *viz.,* 'Direct dripping method' 'Step freezing method' and 'Floating case method' for freeze preservation of buffalo semen at -80 °C and found the first two methods superior to the last method in terms of post-thaw motility. Because of simplification and convenience vapour freezing technique has so far been most popular with buffalo semen. Effect of different thawing rates or post-thaw motility of frozen buffalo spermatozoa have been studied by some workers (Tuli *et al.*, 1981; Sukhija, 1984; Rao *et al.,* 1986). In general, a faster thawing rate (50 °C for 30 sec) when found superior to thawing at 37 °C for 30 sec. was compared with thawing temperature of 75 °C for 9 sec; the former was found significantly superior.

In using EYL diluent, conventional pattern of freezing consists of dilution of semen in glycerol-free fraction first, cooling to 10 to 15 °C, adding glycerol fraction in stages, then loading in straws, equilibrating for 4 to 5 hr and freezing. When frozen by this method, buffalo semen in many instances cannot stand the addition of second fraction after cooling. To overcome this, citrate diluent containing lactose with 6.5 to 7% glycerol was used to dilute the semen just as in the case of tris diluent. Among the diluents described, tris diluent gave the best survival rate as compared with others. As such, this diluent is being used both for chilled and frozen semen.

Raina *et al.,* (2002) observed that the incorporation of antioxidants, especially vitamin E, had beneficial effects on the preservability of buffalo semen. Pawan Singh *et al.,* (2000) investigated that feeding of lysine and methionine produced a beneficial effect on semen quality and freezability.

(vii) Sperm numbers

In buffalo frozen semen technology, one shortcoming is the absence of any critical study to determine the minimum number of sperms required per insemination dose for optimum fertility. Most of the workers conducting fertility trials with frozen buffalo semen have rather arbitrarily kept the total sperm numbers per insemination at around 50×10^6 cells and claimed acceptable fertility rates (Elimwadi *et al.*, 1974; Reddy *et al.*, 1982; Hultnaes, 1982; Jani *et al.*, 1986; Bhavsar *et al.*, 1986). There are recent reports which indicate that with buffalo bull semen of good freezability and potential fertility the minimum sperm number can be reduced to 30×10^6 motile cells and yet satisfactory first service and overall conception rate can be obtained.

The current dose of 50×10^6 total sperms or 20–25 $\times 10^6$ motile sperms/straw/ampoule is more than double the minimum number of motile sperms required for successful conception in cattle. The inherent level of fertility of the bull also needs to be considered. Another important lacuna is the absence of any systematic information on storability of frozen semen and maintenance of fertilizing ability on long-term basis.

(viii) Evaluation of frozen buffalo semen

The spermatozoa are subjected to variable degree of stresses at different stages of cryopreservation. Of the various parameters of frozen semen evaluation, the pre-freezing and post-thaw motility had been the most widely used evaluation technique in lieu of actual fertility trial (Linford *et al.*, 1976).

It was observed that sephadex filtered Tris-Yolk extended semen exhibited significant improvement in sperm motility, live sperm count and spermatozoal penetrability through PAG column as compared to unfiltered semen. Rana *et al.* (2003) observed a significant and progressive improvement in the percentage of motile sperm when finer grades of sephadex column *i.e.* G-25 and G-200, were used for filtering semen ejaculates. These columns were efficient than others with regard to overall improvement of initialquality,freezability,post-thaw survival and keeping quality of semen.

The current world wide interest in buffalo development has opened up the possibility of commercialization of sale of frozen buffalo semen of outstanding sires for national and international market. Such a situation makes it mandatory for organizations/institutions engaged in large scale production of frozen semen for wider dissemination of superior germplasm to ensure strict quality control of the frozen material at various stages of production, processing, storage, and final use. This can be only achieved by rigid periodic andrological examination of breeding bulls and observance of utmost hygienic precautions during collection and handling of semen.

11.2 Female Reproductive System

11.2.1 Anatomy

Ovaries

As compared to the ovaries of a cow, buffalo ovaries are smaller and more tightly secured. The right ovary is found to be more active as compared to the left. It has been found that an ovary with functional corpus luteum (CL) during mid-cycle in non-pregnant animals weighed 3.96 g. During pregnancy the ovary is further activated and the weights of ovaries with or without CL were recorded as 4.23 g and 2.7 g respectively. The weights of ovaries in buffaloes with an oestrum are less than half of that in normal cycling buffaloes, *i.e.* the right and left ovary weighing 1.68 g and 1.66 g respectively. The CL in buffaloes is pinkish grey which may be due to quick conversion of carotene to vitamin A.

Fallopian tubes

The fallopian tubes of the buffalo are deep seated in the broad ligament than those of cow. Among buffaloes, Surti buffaloes have shorter fallopian tubes as compared to other breeds.

Uterine horns

Uterine horns of Indian buffaloes are smaller and are more turgid than those of cows. Among buffalo breeds, Surti buffaloes have the smallest uterine horns, while Jaffarabadi buffaloes have the largest.

Cervix

The cervix of a buffalo is smaller and thinner as compared to cows. In heifers the cervix is very small making it difficult to insert the insemination pipette or AI gun.

11.2.2 Physiology of reproduction

The age at puberty of buffalo heifers shows a wide variation, but generally much later than cattle. Basu *et al.* (1984) reported the age at first heat to be 30.6 months in Murrah buffaloes. Nagpuri buffaloes come in heat late at 42–48 months of age (Kaikini and Paragaonkar, 1969). In Egypt, buffalo heifers exhibit first oestrus after the age of 13 months (Hafez, 1955). Age at puberty of buffalo heifers in the Philippines varies between 26 and 29 months (Villegas, 1930). Female buffaloes in Cambodia attain puberty at three years of age, whereas cattle heifers in that country achieve this physiological status some six months earlier (Baradat, 1949). Buffaloes in Azerbaijan exhibit first oestrus at two to three years of age, but with good feeding, puberty is attained earlier (Gorbelik, 1935). With the wide variation in reported age for puberty in buffalo heifers from different countries, it is obvious that wide variations will be found in age of conception and the age at first calving. From available reports it appears that, in general, buffalo heifers are not bred before they are 2½ to 3 years of age. More commonly buffalo heifers are put to the bull at 3½ years or later.

11.2.2.1 *Puberty*

Age at puberty ranges from 16–22 to 36–40 months in various countries. Under field conditions, oestrus first occurs at the age of 24–36 months. In well-fed animals, puberty may be reached before 20 months.

Puberty is significantly affected by:

breed, season, climate, feeding systems, growth rate.

The average age at first calving lies therefore between 3 and 4 years, but many buffalo cows do not calve until much later in the villages.

11.2.2.2 *Seasonality*

Many believe that buffalo may be considered to be seasonally polyoestrus and a short day breeder. But experiments show that this affects more to makes being in active during hot summer season such that it appeared that.

River buffalo

The female is active from July until the end of February. The peak of first matings occurs during autumn and winter.

Swamp buffalo

Continuous cyclicity throughout the year, but a crop-associated seasonal pattern is observed.

11.2.2.3 *Oestrus, oestrous cycle and ovulation*

Oestrus in buffaloes shows marked variations. This could be due to differences in breed, environmental conditions and management. The signs of oestrus in river buffaloes from India, Pakistan and Egypt are less intense than in female cattle (Hafez, 1954; Ishaq, 1956; Luktuke and Ahuja, 1961). On the other hand, the symptoms of oestrus in the Carabao of the Philippines are more obvious than in female cattle. In 84% of cases oestrus in the Egyptian buffalo commences between 18.00 and 06.00 hr (Hafez, 1954), and in Indian buffaloes usually during the morning. According to some reports, mating desire in the buffalo ceases during the day and female buffaloes breed only at night (Macgregor, 1941; Cockrill, 1967). This may be true for wild or swamp buffaloes, but not for the river-type buffaloes of India and Pakistan. Incidence of silent heat is high in buffaloes and many oestruses may go unnoticed unless great care is exercised in detection. Srivastava *et al.* (1999) observed no significant difference in progesterone concentration during oestrus cycle due to seasonal variation they concluded that heat stress and humidity might be responsible for either on ovulation, failure of fertilization or early embryonic mortality in buffaloes, which had caused them to appear as seasonal breeders.

The length of the oestrous cycle of the buffalo is on an average around 21 days (Shalash, 1958), but considerable variations in the length of the cycle is commonly observed.

Fern-pattern formation by cervical mucus on drying is maximum during oestrus (Raizada *et al*., 1968; Sen Gupta and Sukhija, (1988). A high percentage of pregnancies results from insemination at oestrus, natural or induced, associated with typical fern-pattern formation in the cervical mucus.

Ovulation in buffaloes occurs subsequent to the cessation of oestrus as in the case of cattle and the time interval in the two species also appears to be similar (Shalash, 1958; Rao *et al*., 1960; Basirov, 1964). Shashi *et al.* (2002) observed sensualistic action of selenium to vitamin E in the reproductive performance of the heifers *i.e* early onset of oestrus and pregnancy in vitamin E+ selenium supplemented heifers as compared to vitamin E alone. The 'lutein' tissue that forms in the ovary following ovulation is white in buffaloes in contrast to the yellow colour observed in cattle. Singh (2003) observed that when progesterone releasing device in form of implant (1.9 g progesterone) was implanted intravaginal for 14 days in 10 non-cycling buffalo cows and 5 buffalo heifers, these exhibited onset sign of oestrus on days 3 to 5 after implant removal and conceived following natural service.

Symptoms to be observed for determining heat are:

- The buffalo's response to a teaser bull
- Observing the typical arborization pattern, ELISA test or electrical conductivity of vaginal mucus membrane
- Frequent micturition, raising of head typically and red nictating membrane
- A white mark on perineum below the vulva (55) this white mark appears due to frequent micturition and deposition of salt on the skin
- External changes like swelling of vulva and mucus discharge are the most reliable visible symptoms in Surti buffaloes (60). Careful observation of the non-oestrus vulva is necessary in order to detect the typical swelling of the vulva occuring in heat. It is especially important to observe wrinkles
- Mucus membrane of the vulva becomes red, moist and glossy

It has been observed that oestrus detection thrice a day could improve oestrus detection efficiency and conception in buffaloes as compared to detection twice daily.

11.2.2.4 *Breeding behaviour*

Bhat *et al.* (1983) observed that buffalo cows continue to come in heat regularly in all months, the highest being in October and the lowest in April (Table 11.3).

Table 11.3 Month and Season-wise Occurrence of Heat and Conception rate in Murrah Buffaloes

Month of insemination	Heat (%)		Conception rate (%)	
January	8.8		26.0	
February	8.3		26.2	
March	7.2	25.2b	23.6	23.6b
April	5.9		19.6	
May	7.4		19.8	
June	7.4	24.8c	16.2	19.6c
July	7.1		17.3	
August	7.2		18.2	
September	8.1	28.6d	21.2	22.0d
October	13.1		24.3	
November	10.3		28.3	
December	9.2	28.8a	24.9	26.5a

a, Winter (November to January); b, spring (February to April); c, summer (May to July); d, autumn (August to October).
Source: Bhat *et al.* (1983).

There are no differences between months and between seasons with regard to per cent animals exhibiting oestrus and subsequent conception. The view earlier held that there was seasonality in conception (Goswami and Nair, 1964; Rao and Rao, 1968; Roy *et al.*, 1968) due to seasonal variation in oestrous cycle pattern, therefore, has to be rejected. The lower conception rate during summer (April to September) is due to poor semen quality of buffalo bulls. Hot-humid months affect spermatogenesis of male buffaloes adversely. The use of frozen semen during these months is recommended to overcome lower conception rates. Markandeya and Bharkad (2002) observed that treatment with progesterone prior to GnRH therapy was more effective for the induction of oestrus and conception.

In a study on Murrah buffaloes, oestrus during gestation was observed in 6.1% of the animals and on an average the oestrus symptoms were exhibited 108.4 ± 11.8 days post-conception (Luktuke and Roy, 1964).

The breeding and calving season of various breeds of water buffalo is given in (Table 11.4).

Table 11.4 Breeding and Calving Seasons of Water Buffalo

Type of breed	Country, place	Breeding and calving season and other features of seasonal sex periodicity in buffalo
River type	–	Buffaloes show a distinct seasonal sex periodicity
Indian buffalo	Govt. farms, India	Breeding period with the maximum number of buffalo cows coming into heat; Oct to Feb the peak month being October. Approx. 61.7% of the calvings take place between July and November.
Indian village buffalo	Insemination flgs. of 13620 villages in Coimbatore district India.	Peak period of heat October to February
Indian (Murrah)	Animal genetics section, livestock res. station, Mathura north India	Buffalo cows were observed to indicate seasonal sex periodicity in respect of oestrus and conception the period with highest oestrus frequency (62.8%) was Oct to Feb with peak in Dec and this period coincided with higher conception rate, during the period from April to Sep with comparatively longer days and higher intensity of the sun, ovarian activity appears to be adversely affected. Bulls were least active sexually during humid season in respect of semen ejaculation and output.
Indian (Murrah)	Hisar, Haryana N. India	Definite seasonal periodicity in calving was observed 79.5% of the buffalo cows calved during July to Nov the minimum (1.8%) during April.
India buffalo	India	Most calving occurred during the period from June to December.
India buffalo	IVRI, Izatnagar N. India	Marked seasonal variation in the quality of the buffalo semen was observed; the quality was best during the period Sep to Oct which corresponds with the maximum ovarian activity in the female buffalo

11.2.2.5 *Gestation length*

The gestation length in buffaloes is somewhat longer than in cows, with a range between 299 and 346 days (El-Sheikh, 1967; Joshi *et al.*, 1969). Mean gestation length (± S.D.) for Lanka buffaloes (316.3 ± 6.5 days) is significantly longer than that for Murrah (309.9 ± 6.5 days) (Perera and De Silva, 1985). Within each breed, the mean gestation length for males and female calves is very similar.

The average gestation period for river buffaloes in India and Pakistan is around 307 days. Buffaloes in Egypt carry their young a week or 10 days longer. The gestation period of the swamp buffalo is considerably longer than that of the river buffalo and lasts for 330 to 340 days (Macgregor, 1941; Hua, 1957).

Ghanem *et al.* (1955) observed significant difference in length of gestation between paternal and maternal halfsibs and fullsibs. The heritability estimate for gestation length is 0.32 ± 0.02 (Joshi *et al.*, 1969) in Murrah and 0.32 ± 0.13 (Ghanem *et al.*, 1955) in Egyptian buffaloes.

11.2.2.6 *Age at first calving*

A large variation exists in age at first calving across the breeds, it being highest in village buffaloes (Amble *et al.*, 1958; Chhikara *et al.*, 1978). The averages in Murrah and Nili-Ravi is 40 to 44 months (Sreedharan and Nagarcenkar, 1978; Johari and Bhat, 1979a; Reddy, 1980), while that in Egyptian buffaloes was around 40 months (El-Sheikh, 1967; Alim, 1978). In Surti, Bhadawari and Nagpuri buffaloes, the age at first calving is slightly higher, 46–54 months.

Month/season of calving has nonsignificant effect on age at first calving (Rathi *et al.*, 1971; Johari and Bhat, 1979a; Dutt and Taneja, 1994). Differences between farms and years/periods for age at first calving are significant (Sharma and Singh 1978; Johari and Bhat, 1979a; Reddy, 1980; Dutt and Taneja, 1994). The farm differences could also be genetic in nature due to differences in genetic merit of sires used at these farms.

The heritability estimates for this trait vary between 0.12 and 0.34 (Reddy, 1980; Mangrurkar and Desai, 1981; Kalsi and Dhillon, 1982; Gurung and Johar, 1982). Dutt and Taneja (1994) reported it to be 0.53 ± 0.21 and 0.43 ± 0.08 respectively. Large differences in heritability estimates are due to differences in number of sires used, number of progeny per sire and variation between individuals within halfsib groups apart from differences in genetic merit of sires used in these herds.

Low phenotypic correlations between age at first calving and first lactation milk yield have been reported by various workers (Dutt *et al.*, 1965; Tomar and Desai, 1968; El-Arian 1986). However, Dutt *et al.* (1965) reported negative phenotypic association between age at first calving and milk production up to 6 (-0.74), 8 (-0.55) and 10 years (-0.46) of age. Low and negative phenotypic correlations between these two traits were also reported by Rathi *et al.* (1971) and Gogoi *et al.* (1985). The genetic correlations between these two traits were reported to range between -0.98 and 0.13 by Gogoi *et al.* (1985) and El-Arian (1986). Most of the estimates, however, had high standard errors. The negative genetic association between these two traits suggested that decrease in age at first calving would significantly increase the milk yield.

11.2.2.7 *Service period*

The average first service period in Murrah buffaloes is 115 to 230 days with an overall average of 132 days (Kohli and Malik, 1960; Goswami and Kumar, 1968; Kanaujia *et al.*, 1974; Johari and Bhat, 1979a; Reddy, 1980; Jain and Taneja, 1984). The average first service period is 201 days in Nili-Ravi, 193 days in Bhadawari and 198 days in Egyptian buffaloes (Ragab *et al.*, 1956; Sharma and Singh, 1978; Reddy and Taneja, 1984). In general, the service period in Murrah buffaloes at well-organized farms is much lower than in other breeds.

Significant farm and year/period differences for service period were reported by Kanaujia *et al.* (1974), Johari and Bhat (1979a), Reddy (1980) and Reddy and Taneja (1984). Reddy (1980) observed that average first service period was lower (132–142 days) in July to November calvers than in December to April calvers (203–239 days). This suggested that buffaloes calving during December to April be carefully watched for heat in order to reduce the service period.

The service period in Mehsana buffaloes is largely controlled by environmental factors and thus improvement in this trait is possible largely through improved management.

The heritability estimates for service period are, in general, low and range between 0.08 ± 0.06 and 0.22 ± 0.28. The phenotypic correlations between service period and milk yield are 0.16 and 0.55 (Singh *et al.*, 1958; Kohli and Malik, 1960; Basu and Ghai, 1978; Reddy, 1980). The positive and significant correlations suggest that higher milk yield is associated with longer service period.

It has been observed that the buffaloes that calved during the rainy season had significantly shorter service period and calving interval, since these animals showed earlier first oestrus after calving and required the least number of services per conception. On the other hand, the buffaloes that calved in spring showed post-partum oestrus after a long time due to adverse environmental stress like high temperature and showed fertile oestrus only at cooler temperatures (rainy and winter season).

11.2.2.8 *Dry period*

The averages of first dry period in Murrah buffaloes ranges from 153 to 217 (Jawarkar and Johar, 1974; Gurnani *et al.*,1976; Kanaujia, 1978; Reddy, 1980; Jain and Taneja, 1984) with an overall average of 160 days. Chaudhary and Shaw (1965) reported much shorter first dry period of 98 days in Pakistan's Nili-

Ravi buffaloes whereas longer dry period (202 days) for this breed was reported by Reddy and Taneja (1984). Average dry period pooled over the lactations is 156.4 (Singh and Desai, 1962) in Bhadawari, 134.2 (Hadi, 1965) in Marathwada and 129.7 days (Kadu *et al.*, 1978) in Nagpuri buffaloes. In Egyptian buffaloes, it ranges from 200 (Alim, 1978) to 296 days (Alim and Ahmed, 1954).

The variation between farms, between years and between months influenced dry period significantly. The average dry period for July to November calvers (145–180 days) is lower than for December to April (191–237 days) calvers (Reddy, 1980).

Jawarkar and Johar (1974) estimated the heritability of first four dry periods as 0.20, 0.01, 0.20 and 0.02, respectively, in Murrah. In Egyptian buffaloes, Asker *et al.* (1953) estimated the heritability of dry period as 0.18. Kumar (1984) reported the heritability estimate to be around 0.0 to 0.7. Most of the estimates reported had high standard errors and suggest that the trait can be improved through improvement management.

11.2.2.9 *Post partum oestrus*

The effect of month and season on prevalence of anoestrus in postpartum buffaloes was not significant. Post-partum heat in the river buffalo, when it appears early may manifest around 40 days following parturition, but in most instances appears after a period exceeding 100 days following the birth of the young (Macgregor, 1941; Shalash, 1958; Luktuke and Roy, 1964). The swamp buffalo in Malaysia exhibits first heat after parturition at about 65 days (Macgregor, 1941). Post partum reproductive cyclicity based on ovarian steroids in suckled and weaned buffaloes was studied by Arya and Madan (2001). They concluded that the service period was longer in suckled than weaned buffaloes.

Various hormonal interventions (treatment with oxytocin and PGF $_{22}$) following parturition helped in improving the reproductive performance of post parturient buffaloes.

11.2.2.10 *Calving interval*

Reports on calving interval in the buffalo present very widely differing data. Except for a report from Italy in which an interval of less than 400 days is indicated, most of the other reports give considerably longer intervals. Short calving intervals, as reported from Italy, possibly indicate a better plane of nutrition and management of the herds. There are some reports which indicate that those buffaloes which calve at an early age, have a relatively shorter calving interval than those which are late calvers (Alim and Ahmed, 1954; Salerno, 1960). Seasonal difference in calving interval has been reported in Murrah buffaloes and it has been observed that in the animals which calve between June and November the average interval is 428.7 days, but in buffaloes calving between December and May the interval is 507.1 days (Singh *et al.*, 1958).

The first calving interval in Murrah, Nili-Ravi and Egyptian buffaloes varies between 479 and 508 days (Alim, 1978; Johari and Bhat, 1979a; Reddy and Taneja, 1984). The overall calving interval, however, is lower (430–457 days) than the first calving interval (Lall, 1975). Patro and Bhat (1979b) reported a decline of about 95.2 days in calving interval from first to seventh lactation. Basavaiah *et al.* (1983) reported the average first calving interval to be 583.3 days in Surti buffaloes.

Farms/years/periods have a significant effect on calving interval (Basu and Ghai, 1978; Johari and Bhat, 1979a; Reddy, 1980; Kumar, 1984; Singh and Yadav, 1986). Month/season of calving also influences this trait significantly (Gurnani *et al.*, 1976; Basu *et al.*, 1978; Reddy, 1980; Kumar, 1984; El-Arian, 1986; Singh and Yadav, 1986). The time period and season effects were highly significant ($P<0.01$) for both calving interval and dry period.

Most of the heritability estimates reported for this trait are low and non-significant. The phenotypic correlations between first calving interval and first lactation milk yield (300 days) range from 0.20 to 0.38 (Basu and Ghai, 1978; Basu *et al.*, 1978; Reddy, 1980; Gupta, 1988). These significant correlations suggested that buffaloes with higher lactation yield would have longer calving interval. Such a phenotypic association was expected since prolonged calving was directly related to feeding, management and heat detection practises. With good management, it should be possible to produce a calf a year from a buffalo although two in three years is the current norm.

Estimates by several workers of repeatability and heritability of calving interval in the buffalo suggest that the genetic component of this attribute is very small and hence selection for a shorter calving interval is not likely to be rewarding.

11.2.2.11 *Twinning*

Twin births in the buffalo have been reported from Egypt, Italy and India. The frequency of twin births in Indian buffaloes is considerably lower (0.06%) than those in Egypt (0.2 to 0.6%) and Italy (0.3%). Cockrill (1970), however, reported a high incidence of about 1% of twinning in Surti buffaloes at Anand in India. The relatively higher incidence of twinning at Anand is attributed by Cockrill to the practice followed in that area of giving two inseminations within the same oestral period.

11.2.2.12 *Artificial insemination*

Handling of frozen semen

Deep freezing of semen is a technique for proper and long term preservation of semen. Frozen semen straws are kept in liquid nitrogen at -196 °C.

Handling of liquid nitrogen (LN) containers

The LN container should be properly protected as it is a vacuum jar and is likely to get damaged by jolts and jerks. While placing it on the ground, it should be rested on a rubber padding or cushioning. It should not be dragged, dropped or tilted.

While transporting the containers from one place to another, they should be moved on a trolley or by lifting them manually with care. The containers should be given rubber cushioning while transporting.

- Check the level of liquid nitrogen with the help of a dip stick; if the level drops, add more liquid nitrogen.
- Ensure that the canisters holding the frozen semen goblets are dipped into the liquid nitrogen.
- See that the canister handle is insulated, this will help avoid higher evaporation rate.
- Replace the cork immediately into the neck tube, or the ice formed will get into the container on pushing of the cork causing unnecessary sludge and damage to the container.
- Do not keep the cork outside for long time as it might also get wet due to condensation of water. If not wiped before replacing, ice will be formed because of continuous vaporization of liquid nitrogen which might cause difficulty in removing the cork later and damage the neck tube.
- Empty the liquid nitrogen container once every twelve months and leave to warm up for over a week.
- A protocol for SOP should be available with each worker.

Handling of semen straws during A1

Frozen semen should be properly handled once it is supplied to field staff. A few basic guidelines need to be followed while handling frozen semen.

- Check the buffalo for heat by rectal palpation before taking out the frozen semen straw. The straw once taken out should never be put back in the liquid nitrogen.

Semen straws during artificial insemination:

The points to note are listed here

- Reconfirm the source of the frozen semen. It should be from genetically highly merited bulls and free from diseases as per OIE standards.
- Confirm that the frozen semen issued has completed the quarantine period of one month.
- Do not expose the frozen semen straws to sun.
- Quickly take out the semen straws out of the container with the help of forceps and not by fingers.
- Do not lift the canister above the neck level. It should be lowered immediately back into the liquid nitrogen after taking out a straw.
- As soon as the frozen semen straw is taken out dip it into warm water at 40 °C for 15–30 sec. The straw should be totally immersed and placed horizontally.
- Wipe the straw and dry with tissue paper or clean towel before loading in the AI gun.
- Cut the laboratory plug end of the straw using sharp scissors; the cut should be at right angles so as to prevent the semen from leaking out.
- Ensure that the frozen semen is utilized within five minutes of thawing. The semen should be thawed in clean warm water preferably in a water bath and as near to the animals crush as possible.

Artificial insemination

Rectovaginal technique is preferable for inseminating the female buffalo when suitably trained technicians are available; if not, speculum method may be used, with adequate sanitary precautions. It is generally accepted that the site of deposition of semen should be the same as in female cattle. The speed of travel of spermatozoa in the reproductive tract of Indian female buffaloes is about the same as that for female cattle. Spermatozoa reach the anterior third of the fallopian tubes in 3 minutes and 20 sec following AI (Pant *et al.*, 2001).

Deposition of semen

(a) The female buffalo should be handled carefully for AI. Its rectal capillaries are very brittle and hence might bleed on rectal palpation. Therefore, back raking should be done smoothly. The technician should properly trim the nails. He should smear sufficient antiseptic cream on his arm before putting it into the buffalo's rectum.
(b) Clean the vulva thoroughly.
(c) Place the gun in the vagina at a 45 degree angle. The organs should be pushed interiorly in order to straighten the vaginal passage and remove folds. The technician should use his thumb to direct the gun to the cervix.
(d) Place the interior end of the gun half a centimeter ahead of the internal opening of the cervix in the body of the uterus and deposit the semen there.
(e) At the time of semen deposition the cervix should be kept horizontal and parallel to floor to avoid backflow of semen.

The reduced sperm concentration up to 11×10^6 spermatozoa or even lower concentration (8.5 million) per insemination did not affect the overall fertility of buffalo bull semen (Prabhakar *et al.*, 2002). But it is recommended that average dose of 30×10^6 should be used. A conception rate of 72% was not uncommon in buffaloes bred by AI in Pakistan. Lohan *et al.*, (2000) observed a poor conception rate in animals which were inseminated less than 40 days post partum.

11.2.2.13 *Conception rate*

Information available on conception rate in buffaloes either by natural service or by AI is very limited. Furthermore, different workers have assessed conception rate following different methods under widely differing husbandry practices. Consequently, it is very difficult, if not impossible, to draw any definite general conclusions. Conception rates in Indian buffaloes vary from 50 to over 90%. The average number of inseminations required per conception is 1.56. Fertility is ascertained by rectal palpation and reconfirmed by calving results.

11.2.2.14 *Number of services per conception*

The average number of services per conception in Murrah buffaloes is 1.6 to 3.1 (Luktuke and Roy, 1964; Singh and Dutt, 1964; Basu *et al.*, 1978). In Surti, the average number of services per conception is 1.67 (Shukla *et al.*, 1970), and in

Egyptian buffaloes 1.51 to 2.10 (Ragab *et al.*, 1956; Ahmed and Tantawy, 1959). In general, heifers take more services than pluriparous buffaloes (Ragab *et al.*, 1956; Rao *et al.*, 1973; Basu *et al.*, 1978).

Hafez (1953) found the average number of services required per conception in the buffalo to be 1.46. He also noticed that 55.6% of the animals conceived following the first service.

It is recommended that a reasonable number of services for conception rate should not exceed 1.5 S/C in a well managed herds.

11.2.3 Embryo transfer technology

Embryo transfer technology (ETT) has been identified as a critical biotechnological tool for increasing productivity of animals. The implementation of this technology under Indian conditions has become of paramount importance since the development plan outlays and growth predictions envisage an increased production level from our livestock, which can only be achieved, when faster multiplication of genetically superior animals is carried out. Through methods of super ovulation, embryo collection, and embryo transfer, a superior bovine female can be made to produce 5 to 10 calves in a year's time and upto 50 calves or more in her life time. This technology has a direct linkage with National Artificial Insemination Programme as it will enable us to produce superior progeny tested bulls for field artificial work in shorter time interval and at much lower costs.

Successful superovulation in cattle was reported in 1940 and the first calf produced as a result of egg transfer was born in Wisconsin in 1951. Research on embryo transplantation in buffaloes is a fairly recent development. The initial success of Drost *et al.* (1983) in the USA, who pioneered application of the technology to buffaloes, was soon followed with successful birth of buffalo calves in Bulgaria (Vlakhov *et al.*, 1985) and India (Misra *et al.*, 1988; Madan *et al.*, 1989). Studies on superovulation among buffaloes have also been carried out in Thailand (Thungtanawat *et al.*, 1981; Chantaraprateep *et al.*, 1989a; Techakumphu *et al.*, 1989; Venitkul 1989), Bulgaria (Karaivonov *et al.*, 1987; Alexiev, 1989), Malaysia (Jainudeen, 1989), Pakistan (Mehmood *et al.*, 1989) and India (Taneja *et al.*, 1990; Misra *et al.*, 1990, 1991; Singla *et al.*, 1992). The first buffalo calf by this method was born on 22nd December in 1987.

Knowledge about reproductive physiology of buffaloes is scanty. Embryo transfer can be instrumental in opening ways for more accurate understanding of numerous unknown aspects of reproduction. Information on superovulation response, recovery rate of embryos and their quality vis-a-vis the endocrine picture of both super ovulated and normal cyclic animals needs to be gained for better understanding of buffalo production.

The intricate factors responsible for biological peculiarities of buffaloes are manifold. The lower reproductive potentiality of buffaloes can be confirmed from the very stock of primordial follicles in the ovaries. The average number of primordial follicles is 19000 in Nili-Ravi buffaloes and 12000 in Surti buffaloes. These values are much lower than those for cattle (60000 to 100000) and reflect the lower reproductive potentiality of buffaloes.

Buffalo embryo transfer is difficult because of different physiology and anatomy of this species, *viz.* the total number of follicles in the ovary, superovulatory response, behavioural oestrus, development of the embryo and coiled uterine horns. This technology has to be made effective in buffalo, by undertaking more intensive studies on freezing, microsurgery of embryos, oocyte culture and *in-vitro* fertilization.

Embryo transfer is often viewed as counterpart of artificial insemination, where a superior animal, *i.e.* a donor, is superovulated, more than one embryo at a time can be recovered and transplanted into recipient females to increase the progeny number per female. The technique of AI has made it possible to increase the utility of males. It has also become possible to increase the influence of the genetically superior females by the use of embryo transfer. Introduction of a multiple ovulation and embryo transfer (MOET) scheme will increase the selection intensity through increased progeny per buffalo per annum in the herd. Kasiraj *et al.*, (2001) ascertained the success of buffalo ET under field conditions.

MOET scheme allows progeny testing of both males and females, and can increase the genetic gain per annum, over what is possible by artificial insemination alone. Under the Indo-US project the fundamental and applied research on embryo transfer technology has also been taken up in collaboration with IVRI, National Dairy Research Institute, Punjab Agricultural University etc. One buffalo embryo transfer centre has also been established by the National Dairy Development Board at the Sabarmati Ashram Goshala, Bidaj, Gujarat. More sires can be tested within a short period and at low cost by using the semen of the male to be tested on a few donors and transferring the embryos recovered to the required number of recipients.

(a) Superovulation

Two types of gonadotrophins have commonly been used to super ovulate donor buffaloes. These are pregnant mare serum gonadotrophin (PMSG) and follicle stimulating hormone (FSH), usually administered during the mid-luteal phase (9–

14 days) of the oestrous cycle. The most common method of superovulation is the injection of a single dose of PMSG (3000 IU) during the mid-luteal phase of the oestrous cycle followed 48 hr later by a leutolytic dose of PG F_2 (500 µg). Follicle-stimulating hormone and prostaglandins are used to superovulate buffaloes according to treatment regimens developed for cattle. To obtain greater superovulatory response, doses of FSH ranging from 40 to 70 mg have been used as two injections in a day. The percentage of donors exhibiting oestrus was high (about 80%) in both PMSG and FSH groups. But the prostaglandin standing heat intervals established in the PMSG-treated group was significantly longer than in the FSH-treated group. The percentage of non-ovulated follicles following PMSG treatment was considerably high. The average number of good-quality embryos per buffalo donor was not high on either PMSG or FSH treatment. A good response always depends upon a good CL and an optimum level of progesterone at the beginning of the treatment. In buffaloes the picture is different because of the absence of prominent luteal tissue. This problem can be mitigated by implantation of artificial CL like Synchromate B before initiation of treatment. Priming of ovarian tissue at an early stage of the cycle by FSH may resolve the rate of response by recruitment of a greater proportion of follicles that become atretic at the end of the first follicular wave.

It is difficult to determine the exact number of corpora lutea on the small ovaries of buffaloes, particularly when it exceeds 5 or 6, or when they are accompanied by several large tense follicles, leading to overestimation of response. Moreover, in buffaloes the CL is small, more deeply embedded and fused; it generally has a less pronounced papilla. So far the results of embryo recovery from buffaloes has been frustrating. This may be attributed to several factors. Over stimulation of ovaries can lead to a failure of the fimbriae to envelop the ovary at the time of ovulation. The problem of fluid recovery (20 to 60%) may be due to more tortuous nature of the horns.

There has been a steady progress in terms of superovulatory responses, embryo recovery and pregnancy rate following embryo transfer among buffaloes. From the early recovery rate of 0.6 transferable embryos, currently 2.0–2.4 transferable embryos are recoverable in different farms with altered superovulation protocols. Folltropin and super-ov among many compounds have comparatively given better results both in terms of follicles developed and ovulated, and embryos recovered. The lower number of follicular population and poor follicular development, as revealed through real time ultrasound scanning, explains the lower ovulatory response. The growth rate of the largest follicle and time of ovulation post-PG $á_2$ have also shown great variability in animals. Endocrine studies suggested that unovulated follicles contribute massive quantities of estrogen along with progesterone. High prolactin levels in buffaloes during summer months are also associated with poor follicular maturation. In spite of some problems related to buffalo embryo transfer, the basic embryo transfer technique has been adopted with success resulting in over 1000 pregnancies/calves.

(b) Monitoring ovarian dynamics

Ultrasound evaluation of follicular changes in superovulated buffaloes

A transrectal real time B-mode ultrasound scanner was used to study the follicular changes in 8 sexually mature multiparous, non-lactating Murrah buffaloes. The experimental animals were injected with 112 units of Super-Ov (Ausa international Ltd., USA) though 8 divided injections spread over 4 days, beginning on day 10 of oestrous cycle. Fifty percent of the total ovaries were devoid of follicles (>3 mm) 15 hr after the first Super-Ov injection, on the day of oestrus and on the day of flushing both in right and left ovaries were statistically significant ($P<0.01$). The same was found true in ovulated and non-ovulated groups. Even though the total number of follicles (< 3 mm) in the left ovary was comparatively lesser than that of the right ovary throughout the study, a statistically significant ($P<0.05$) difference was noticed only at the first observation (15 hr prior to start of super ovulation treatment). Interestingly, the left ovary had fewer ovulations than right ovary. The total number of follicles (>10 mm) size of ovulated and non-ovulated groups was 20 and 10 on the day of oestrus and 6 and 11 on the day of flushing respectively. In all, there were 16 ovulations with an average of 2 per animal. This indicates that no follicle in ovulated group and 1 in non-ovulated group had joined the pool of large (>10 mm) follicles after the day the non-ovulated group had disappeared between the day of estrus and the day of embryo flushing and perhaps had undergone atresia. The study found ultrasound imaging useful to know follicular changes among super ovulated buffaloes.

Follicular population and ovulation in buffaloes

Eight super ovulated and 3 non-super ovulated buffaloes were scanned for follicular population and size. Another 4 animals were injected with 25 mg PGF (Litalyse) and ultrasonography was performed. Follicular population at super oestrus in buffaloes is of much lower order than in cattle. Ovulation time in response to PGF treatment varies among animals.

(c) Embryo collection

An interval of 3 days is more suitable than that of 5 days between successive aspirations for recovering oocytes through TVOR in a twice weekly schedule and that repeated removal of follices though TVFA does not adversely affect the number of total follicles 3 or 5 days after IVF. Misra *et al.,* (2001) observed that superovulation and embryo collection had no effect on milk production in the buffalo. The cleavage rate and embryo development were significantly higher in oocytes from ovaries of adult buffaloes than ovaries from prepubertal and aged buffaloes (Raghu *et al.,* 2002).

(d) Embryo cryopreservation

Cryopreservation of *in-vivo* produced embryos

The freezing of embryo is not only an alternative to limited number of recipients but is practised for many other reasons also, in embryo transfer programme. Two types of protocol/programmes have been followed, *i.e.* culling the straw to -7 °C @ 1.5 °C/min and holding them at this temperature for 10 min. Straws are further cooled from -7 ° to -37 °C either @ 0.4 or 0.3 °C/min. After holding straws at -37 °C for 15 min, they are then plunged into liquid nitrogen. However, washing of embryos through 10 washes of sterile holding media and cryoprotectant (glycerol) treatment is a prerequisite to this protocol.

More than 100 cattle, buffalo and goat embryos have been frozen and kept stored in liquid nitrogen at the NDRI. The centre has been engaged in conducting pilot trials on cryopreservation and *in-vitro* produced embryos. For this specific purpose sheep and buffalo IVF embryos have been cryopreserved. This freezing technology in general has helped in keeping the calves in the can.

Cryopreservation of embryo produced in a complete *in-vitro* system

Embryos at the stage of morulae and beyond with good and fair morphology were selected for cryopreservation. Embryos were taken out from the culture dish and washed in modified PBS with 0.3% BSA at room temperature (25 °C). Then the embryos were equilibrated for 8–10 min with each of cryoprotectant solutions containing 2, 4, 6 and 8 and 10% glycerol in modified PBS with BSA in ascending order of concentration. After equilibration, each embryo was loaded into a separate 0.25 ml plastic straw (French mini straw; IMV, France). Freezing media was drawn into the straw followed by a small air-bubble and then media containing the embryo. A second air-bubble was drawn into the straw followed by freezing media. Thus, the embryo was retained in the middle of the straw surrounded by two air-bubbles on both the sides. The embryo was observed within the straw under the microscope. The loaded straws were fixed on the straw holder of the programmable freezing machine (Biocool; FTS system, USA and Hoxane, USA) and cooled as follows.

		Culling rate	
Ramp I	1.	20 °C 1.5 °C/min	-7 °C
	2.	10 min halt at	-7 °C
	3.	Seeding	
Ramp II	4.	-7 °C cooling rate (0.4 °C/min) Seeding (ice crystallization) was induced in the medium by touching the wall of straw with a pair of forceps cooled in liquid nitrogen.	-37 °C

The straw was taken out from liquid nitrogen container and transferred to a water-bath at -37 °C. The straw was gently agitated until the complete thawing of the media. The cryoprotectant of the freezing media was removed by equilibrating embryos for 8–10 min each of freezing media containing 10, 8, 6, 4 and 2 0% glycerol.

(e) Manipulation and bisection of embryos

Bisection of embryo has been successfully done and pregnancies established from bisected embryo. Twins produced through this technique is yet another step towards faster multiplication of superior germplasm. The technology can be used vary effectively for reducing the generation interval, particularly when using modern breeding techniques like Multiple Ovulation and Embryo Transfer (MOET) and open nucleus breeding system (ONBS).

(f) *In-vitro* fertilization

In order to bypass the problem of low embryo production rate, *in-vitro* fertilization technique has been developed. A method was developed for ovaries obtained from slaughterhouse. The recovery rate of good cumulus-complexes oocytes was 0.42. Only 40% of oocytes reached metaphase which is the maturation stage of the oocytes.

In-vitro maturation of buffalo oocytes

Among several commercially available culture media, the most commonly used medium for maturation seems to be TCM 199. However, for *in-vitro* culture, several different media such as TCM 199, CRI aa, mSOF, KSOM, and Menezo's B_2 with various modifications have been used. Very recently commercially prepared media for bovine IVF have been launched in the market. The culture conditions adopted during *in-vitro* maturation of oocytes significantly influence the fertilization rate and subsequent embryonic development which vary among different laboratories.

In brief, about 10 to 40 COCs are added to 50μl droplets of maturation medium. The droplets are covered with mineral oil and pre incubated under maturation condition for at least 2 hr (38.5 °C 5% CO_2 and 95% air atmosphere with 95–100% humidity) prior to incubation of COCs for 24 hr. At the end of incubation period, the oocytes emerge at the same stage of development as they are released at the time of ovulation in the live buffalo cow.

In-vitro capacitation of buffalo spermatozoa

In vitro capacitation of buffalo spermatozoa has been studied by incubation of ejaculated spermatozoa in Bo medium supplemented with 0.5% BSA and 10 μl/ml heparin and 5mM caffeine. The percentage of sperm motility (M) and acrosome reaction (AR) were determined morphologically by light and transmission electron microscopy. The effect of caffeine and heparin on these parameters were also examined. Addition of 5mM caffeine was effective for the temporary activation of the sperm motility but made difficult the maintenance of motility and viability after 3 hr of incubation. Calcium is an essential substance which is needed for capacitation and to induce the acrosome reaction. When sperm was incubated for 7 hr with 10 μg/ml heparin, exposed to 10 mM Ca for 20 min, about 80% sperms had undergone the acrosome reaction. The time course required for heparin to capacitate sperm was judged by *in-vitro* fertilization with buffalo *in-vitro* matured oocytes. This study suggests that the capacitation of buffalo spermatozoa by heparin requires at least 7 hr exposure of sperm to heparin for maximal response in fertilization.

In-vitro fertilization of buffalo oocytes

All mature oocytes with an expanded cumulus mass were introduced into 100μ 1 droplets of capacitated spermatozoa suspension for 6 hr in CO_2 incubator at 38.5 °C.

Initial experiments on the fertilization showed that out of 131 mature oocytes kept for fertilization, only 29.8% were fertilized. Only 0.76% reached up to the morula stage and 1.52% up to blastocyst stage. Most of the oocytes were arrested at pronuclei formation stage (11.45%). After making modifications in the technique it was possible to increase the fertilization and cleavage rates. Out of 174 mature oocytes about 55% oocytes were fertilized, 2.87% were at morula stage and 0.57% at blastocyst stage. The technique was further improved and about 80% oocytes were fertilized. Out of 185 mature oocytes, 14.6% reached morula stage and 8.10% blastocyst stage.

Reproduction management in buffalo

Reproductive efficiency is the primary factor affecting productivity and is hampered, in female buffalo, by the late attainment of puberty, seasonality of calving, long postpartum anoestrus and subsequent calving interval.

Proper oestrus management allows for more efficient planning of milk production, especially in larger herds and facilitates the use of artificial insemination. Essentially similar basic approaches are being used in buffaloes as there are in cattle. All of the pharmacological systems for oestrus management used nowadays in buffaloes have been adapted from the systems used in cattle through empirical approach and are supported with increasing number of controlled works reported in the literature. The frozen semen and the embryo transfer technologies have opened up immense possibilities for improving the productivity of the neglected farm animals in buffalo breeding.

CHAPTER 12

NUTRITION

The vast majority of buffaloes are reared in regions of the globe that are poorly developed. They are owned by farmers with hardly any capacity for financial investment. In the regions where buffaloes abound, there is considerable quantitative deficiency of feeds and fodders, and available foodstuffs are possibly qualitatively deficient. In such situations it becomes very necessary to search for non-conventional materials that can be utilized as feed to make good the quantitative deficiency and to evolve processes by which qualitative deficiencies of feeds can be rectified.

One basis for progress in livestock development is research, as this provides the necessary data for the formulation of effective developmental programmes. In order to prepare such a programme concerned with the nutrition of the buffalo and to ensure the supply of the optimal nutritional requirements of buffaloes of different ages and levels of production in different regions, two things are absolutely necessary information on the nutritional composition of the locally available feeds and fodders and knowledge of the alimentary physiology of the buffalo. As the nutritional physiology of cattle has been investigated in great detail, and as cattle and buffaloes are similar in many respects, it would be profitable to know if their alimentary physiology differs and, if so, in what details. Unfortunately, in many of the buffalo-rearing countries the organizational base for research in animal nutrition is weak and inadequate. As a result, very large gaps exist at present, in knowledge relating to different aspects of nutrition of buffaloes.

There has been a great scientific interest in studying the comparative performance in the utilization of nutrients in cattle and buffaloes, specially when there are distinct physiological differences between the two species in: (*i*) rumen movements, (*ii*) fluid volume in rumen, (*iii*) the rate of passage of digesta etc.

12.1 Feeding

Buffalo are, like cattle, ruminants. This means they utilize micro-organisms in the rumen to digest the feed. Feed eaten by ruminants is of vegetable origin. The ruminant is an expert in converting cellulose and other fibrous materials into high quality milk and meat. Their digestive capacity is greater than the non-ruminant. Ruminants "chew the cud", that is they regurgitate partly digested food to the mouth to chew it again, thus helping in breakdown of this plant material.

Feed enters the rumen when swallowed by the animal. The rumen is an anaerobic environment, *i.e.* no oxygen is present. The feed is exposed to microbes such as bacteria, protozoa and fungi. These microbes attack the feed particles and by enzymatic action the components are broken down and used for their own metabolism, growth and propagation. The feed is masticated, regurgitated and exposed to microbes in the rumen. Large particles will become smaller and eventually be transported to the reticulum and further on. How long a specific feed particle will stay in the rumen depends on size, palatability and the fibre content of the feed. Buffalo have slower rumen movement than cattle, which leads to a slower rate of ingesta outflow. The pH of the rumen content is similar to that of cattle, and it is affected in the same manner. Normal pH is between six and seven, depending on feed and time of feeding.

Feed components can be divided into protein, energy (carbohydrates), fat, minerals and water. The breakdown and utilization of the different feed components are reviewed below.

The waste end products of the microbial attack are methane and carbon dioxide that are eructated. Volatile fatty acids (VFA) of which acetic, propionic and butyric acids are the predominant ones, together with ammonia, absorbed through the rumen wall and transported via the blood to, *e.g.* the liver and udder where they serve as building material for chemical compounds such as glucose, protein and fat. Most ammonia is utilized directly by the rumen microbes to synthesize proteins.

Ruminants are dependent on the function of the rumen microbes. Therefore, it is important to keep the rumen environment healthy. The easiest and best way is to feed a high amount of good quality roughage and a smaller amount of good quality concentrate.

12.2 Composition of Nutrients

12.2.1 Protein

Almost all protein is attacked by the microbes and utilized in their metabolism and incorporated in the microbial mass. Microbial protein is of high quality and is absorbed as amino acids after being digested by gastric enzymes in the abomasum.

Ammonia which is absorbed by the rumen wall and transported by the blood to the liver, is converted to urea. When there is protein deficiency, urea can be utilized by the rumen microbes as a non-protein nitrogen source to build protein. In this way nitrogen is circulated and efficiently used by the animal.

Protein can be protected to withstand microbial attack. It is then called "by-pass protein". By-pass protein is only degraded in the abomasum and small intestine where it undergoes enzymatic attack similar to that of mono-gastric animals. By-pass protein is commercially available in some ready made concentrates and is usually given to high producers.

12.2.2 Carbohydrate

Carbohydrates are the main sources of energy for ruminants. Carbohydrates are the components of starch and fibres. Fibre is a common name for cell-wall components such as cellulose, hemi-cellulose and lignin. Starch can be degraded by animal gastric enzymes, whereas fibres cannot. Ruminants can utilize fibres to a larger extent than mono-gastric animals because of the ruminal microbes. However, lignin (wood-fibre) is not utilized. It is generally believed that buffalo utilize fibre more efficiently than cattle do. The efficiency of fibre digestion is five to eight percent higher in buffalo than in cattle.

12.2.3 Fat

Fat is required in small amounts for the ruminant. However, whatever fat is present in the feed undergoes microbial attack and degradation. Unsaturated fatty acids are to a large extent saturated. This is one of the reasons for the milk and body fat of the ruminant being of equal composition, largely independent of the type of feed given. If the fat can in some form, be protected from ruminal degradation, and instead be utilized in the lower intestinal tract, it may be used as an additional energy source. However, it may then unfavourably alter milk fat composition. Too much unprotected fat in the diet depresses the ability of the microbes to ferment fibres, negatively influencing energy utilization.

12.2.4 Vitamins and mineral supplements

This is the area of some significance since mineral (macro and micro) availability represents a limitation on productivity. Though their requirement is low, their deficiencies should be avoided if high performance in buffaloes is to be ensured.

Lall *et al.*, (2004) inferred that high plane of nutrition with proper mineral supplementation could bring majority of the anoestrous buffaloes into oestrus.

Kumar and Dass (2005) concluded that supplementation of niacin in the diet of buffaloes improved the rumen fermentation by decreasing the concentration of ammonia-N and increasing protein synthesis.

Bakshi and Wadhwa (2004) studied the effect of herbal feed additives on the nutrient utilization in buffalo calves and concluded the herbal combinations exhibited promising results for use in feeding buffalo calves.

12.3 Voluntary Feed Intake (VFI)

12.3.1 Dry matter intake (DMI)

A great deal of information is available on the DMI of buffaloes in Indian literature which varies from 56 to 120 g/kg $W^{0.75}$ depending upon the density of energy and physical nature of diet. Buffaloes consume 73.1 to 116.9% of dry matter consumed by cattle. Factors like various experimental treatments, composition of feed, frequency of feeding as well as processing of feeds are responsible for these differences. In lactating buffaloes (Sebastian *et al.*, 1970) consumption is 14% less than in lactating cattle. Mandal *et al.*, (2002) found that methionine and lysine supplementation had a significant and positive effect on dry matter intake and body weight gain in buffalo calves. Barman and Mohini (2002) studied the influence of rumensin supplementation and dietary level of concentrate mixture on the growth, nutrient digestibility, N balance, and cost of feeding in buffalo calves and concluded that supplementation of rumensin is cost effective at lower level of concentrate feeding in male buffalo calves.

The DMI in some of the studies conducted with buffaloes have been shown in (Table 12.1).

Ranjhan (1983) reported that the dry matter intake of buffaloes varies between 80 and 95 g per kg metabolic body weight and recommended a requirement of 92 g DM per kg metabolic body weight.

On conventional diets of wheat straw, concentrates and herd-grain mixture, an average DMI of 92.5 g per kg $W^{0.75}$ was observed from 18 feeding trials involving 120 buffalo male calves.

The dry matter intake by the lactating Murrah buffaloes was 132 g/d/kg $W^{0.75}$ when fed with different levels of concentrates (Shukla *et al.*, 1972).

Table 12.1 Dry Matter Intake in the Indian Water Buffaloes and Swamp Buffaloes

Particulars of diets	Particulars of animals		
	Live weight (kg)	% of live weight (kg)	DMI per kg $W^{0.75}$
Wheat straw + concentrate	Growing 200 kg	2.32–2.43	73.30–76.80[a]
Wheat straw	Adult	2.2	69.50[b]
Paddy straw	Adult	1.6	50.60[b]
Green fodder + concentrate	Lactating buffalo	2.54	80.30[c]
Wheat straw + concentrate	Growing 100 kg	1.97–03.1	62.3–58.0[d]
Berseem hay	Growing Surti calves 140 kg	2.35	74.30[e]
	Growing calves	2.17–2.37	68.6–74.9[f]
	Lactating buffaloes	2.05–2.22	64.8–70.2[g]
Wheat straw + concentrate	Surti heifers	1.40	44.2[h]
Maize straw + 10% molasses + 1% urea	Surti heifers	1.90	60.0[h]
Maize straw + 5% molasses + 1% urea	Surti heifers	2.03	64.1[h]
Wheat straw + concentrate mixture	Murrah calves	2.28	72.10[i]
Low	Growing	3.82	120.8[j]
Medium	Calves	3.60	113.8
High plane	–	3.59	113.5
Maize silage	Adult	–	81.6[k]
Sorghum stover	Adult	–	81.4
Corn silage	Swamp adult	–	65[l]

a, Sharma and Talapatra (1983); b, Jain *et al.* (1970); c, Sebastian *et al.* (1970); d, Ranjhan and Pathak (1979); e, Sharma and Mardia (1974); f, Sengar and Singh (1969); g, Shukla *et al.* (1972); h, Balasubramanium *et al.* (1979); i, Chabra and Arora (1979); j, Sehjpal *et al.* (1980); k, Taparia and Sharma (1980); I, Castillo *et al.* (1985).

The dry matter intake by Philippine Carabao (swamp buffaloes) varied from 67 to 116 g/ $W^{0.75}$ with different roughages. An intake of 96 g/$W^{0.75}$ was observed with mature Carabao of 405 kg with rice straw and 116 g with the chaffed straw (Castillo *et al.*, 1985).

The dry matter requirements worked out by various workers are given in (Table 12.2).

Table 12.2 Dry Matter Requirement for Maintenance and Growth

Body weight (kg)	Metabolic body weight (kg)	Daily gain (kg/day)	DM requirement (kg)				
			Ranjhan (1977)	Arora *et al.*, (978)	Ranjhan and Patak (1979)	Keral (1982)	Barush *et al.*, (1983)
–	–	0	2.5	–	2.5	2.4	–
–	–	0.5	2.4	–	2.5	2.8	3.24
100	31.62	0.6	–	2.65	–	–	3.48
–	–	0.7	–	–	3.0	2.8	3.72
–	–	0	3.0	–	3.0	3.3	–
–	–	0.5	3.6	–	–	–	3.84
150	42.86	0.6	–	3.78	–	–	4.08
–	–	0.7	–	–	–	–	4.32
–	–	0	3.5	–	3.5	4.1	–
–	–	0.5	4.8	–	5.0	5.1	4.44
200	53. 18	0.6	–	4.91	–	5.1	4.68
–	–	0.7	–	–	–	–	4.92
–	–	0	4.0	–	4.0	4.8	–
–	–	0.5	6.1	–	6.0	5.9	5.04
250	62.87	0.6	–	–	–	6.1	5.28
–	–	0.7	–	–	–	–	5.52
–	–	0	4.5	–	4.5	5.6	–
–	–	0.5	6.8	–	6.8	5.6	–
300	72.08	0.6	–	–	–	6.8	5.88
–	–	0.7	–	–	–	7.0	6.12

While calculating the requirement for dry matter, an intake of 95 g/d/kg $W^{0.75}$ for the dry adult animals, 99 g for pregnant animals (last quarter) and 132 g/d/kg $W^{0.75}$ for the lactating animals are recommended for both riverine and swamp buffaloes.

12.3.2 Water requirements

Water intake in buffaloes is affected by diet, the environment and physiological function. Buffalo calves weighing about 270 kg consume approximately 20 litres of water during winter and 36 litres during the heat of the summer. The losses through evaporation are about 5 and 19 litres during winter and summer, respectively.

12.4 Utilization of Nutrients

In India, crop residues like chopped wheat straw, paddy straw, stovers, etc., are the staple roughages fed to all categories of farm ruminants including buffaloes. The general belief that buffaloes are superior in utilizing the dietary fibre and non-protein nitrogen than cattle generated considerable amount of research on nutrient utilization.

There are many reports based on the concentration of the volatile fatty acids in the rumen and apparent digestibility coefficient of fibre in buffaloes which indicate that buffaloes utilize 5–8 units better fibre than cattle (Ichhponani and Sidhu, 1965).

12.4.1 Fibre utilization

The concentration of VFA in rumen liquor was higher on feeding wheat straw and concentrate mixture for maintenance of adult buffaloes, than in growing animals, (74.3 to 82.0 m mole/litre) (Ichhponani and Sidhu, 1965). Some improvement in fermentation was reported by chaffing and grinding of wheat straw (Chaturvedi *et al.,* 1973). Levels of energy in the rations do not affect the concentration of VFA in the rumen liquor but the proportion of acetic acid significantly decreases with a corres-ponding increase in propionic and butyric acids. Restricted or liberal feeding of green grass has no effect on the carbohydrate fermentation, but an increased concentration of VFA in rumen liquor has been reported on increasing the frequency of feeding from twice daily to four times a day.

12.4.2 Protein utilization

Like cattle, the digestion of protein in buffalo rumen is mostly microbial and depends on the solubility of the dietary proteins in the rumen liquor and rate of passage of proteins in lower gut. Proteins are hydrolysed to peptides and amino acid, which are deaminated to carbondioxide, ammonia and short chain fattyacids. The concentration of ammonia in the rumen liquor is dependent on the sources of dietary nitrogen, the solubility of intact proteins and the sources and the proportion of soluble carbohydrates. The major portion of the rumen ammonia is absorbed through the ruminal wall into the portal blood and is converted to urea in the liver. The greater part of this urea is excreted in the urine while part is returned to the rumen in saliva or through the ruminal wall by diffusion.

Feeding formaldehyde treated clusterbean meal improved growth and feed utilization in buffalo calves (Yadav *et al.,* 2000)

Kannan *et al.,* (2000) inferred that the mustard cake could be replaced completely with deoiled sunflower cake without affecting feed intake and digestibilities of nutrients in lactating Murrah buffaloes.

12.4.3 Utilization of non-protein nitrogenous (NPN) compounds

A good deal of work has been conducted in India where urea and biuret have been used as replacement for proteins. Satisfactory growth in young buffalo calves was observed by Pathak and Ranjhan (1976) on feeding a urea-molasses liquid diet *ad libidum* along with 200 g fish meal and a restricted amount of oat fodder. Supplementation of urea diets with soluble carbohydrates like starch, sugars and molasses enhances the utilization of urea nitrogen.

The Indian Standards Institution (ISI) has also recommended the inclusion of one percent urea in the concentrate mixture of ruminants.

Sihag *et al.,* (2003) found that feeding of UMMB (urea molasses mineral blocks) encourages optimal fermentation and microbial protein synthesis in the rumen of buffaloes maintained on wheat straw based diet.

12.4.4 Rate of passage of digesta

The use of a marker has shown that the complete expulsion of residues from food ingested by buffaloes requires a period of 4–5 days. Total retention time may be divided into two phases, namely; phase 1, which is the time required to reach the abomasum, and phase-2, which is the time to travel from the abomasum to the anus. Phase-1 is about three times longer than phase-2.

Using stained straw as a marker, Ponnappa *et al.* (1971) and Ashok Kumar (personal communication, 1971) found that 5% excretion time (which is supposed to be the transit time for digesta to travel from the abomasum to the anus) ranged from 26 to 35 hr between animals. Mean values for three levels of DCP and TDN feeding, *i.e.* NRC, 20% below NRC and 20% above NRC (NRC 1966) requirement, were 32.0, 29.7 and 32.3 hr respectively (Ashok Kumar, personal communication, 1971) and the values were close to the mean of 32.3 hr reported by Ponnappa *et al.,* (1971). The mean time spent in the reticulo rumen was 91.2 hr. By reducing the water intake of buffaloes to only three-fourths of *ad libidum* water intake, significantly reduced the DMI from 75 to 65 g per kg $W^{0.75}$ and the mean retention time of ingesta from 70.6 to 67.2 hr.

12.5 Nutrient Requirement

A good deal of work done on the energy and protein requirement in buffaloes had been reviewed by many authors (Ranjhan, 1977, 1983; Ranjhan and Pathak, 1979). In India feeding standards recommended by Sen *et al.* (1978) are being followed for dairy cattle and buffaloes. These are based on the mid Morrison values of 1959 with some modifications introduced from time to time. Recently, Kearl (1982) has also published the requirements for buffaloes.

12.5.1 Energy and protein requirements for maintenance

Different workers determined the energy requirements of buffalo for maintenance ranging from 103 to 130 Kcal/kg $W^{0.75}$. Values mostly used are 122, and 125 Kcal/kg $W^{0.75}$ (Agarawal, 1974; Ranjhan and Pathak, 1979; Sen *et al.,* 1978; Kurar and Mudgal, 1981; Kearl, 1982).

The DCP requirement for maintenance reported by various workers are 2.089 g (Singh, 1965), 1.819 (Negi *et al*., 1968) and 2.440 g (Kurar and Mudgal, 1981) per kg metabolic body weight. Ranjhan and Pathak (1979) recommended a value of 2.84 g/kg $W^{0.75}$/d, suggested by Sen *et al.* (1978). However, Kearl (1982) used a value of 2.54 g/kg $W^{0.75}$/d, about 11% less than the value recommended by Indian workers on the assumption that buffaloes are more efficient in utilizing protein. Requirements given for river buffaloes can be safely applied for the swamp buffaloes (Castillo *et al.,* 1985).

12.5.2 Energy and protein requirements for growth

The nutrient requirement in growing calves may be divided into two phases, namely pre-ruminant growth period and after the rumen is developed.

Feeding practices of pre-ruminant calves differ widely. The duration of milk/skim milk feeding from birth ranges from 45 days to 6 months of age. Similarly, the quantity of whole milk and skim milk fed to each calf varies from 100 to 300 kg.

Based on the work done at the Indian Veterinary Research Institute, Izatnagar, the proposed requirements are given in (Table 12.3. and Table 12.4).

Table 12.3 Digestible Protein Requirement for maintenance and Growth Recommended by Different Authors for Buffaloes

Body weight (kg)	Metabolic body weight (kg)	Daily gain (kg)	Digestible crude protein requirement (kg/d)					
			Ranjhan (1977)	Sen *et al.* (1978)	Arora *et al.* (1978)	Sivaiah and Mudgal (1978)	Kearl (1982)	Baruah *et al.* (1983)
100	31.6	0.0	0.150	–	–	–	0.080	–
		0.5	0.260	0.280	0.330	0.214	0.254	0.207
		0.6	–	–	–	0.214	–	0.235
		0.7	–	–	–	0.256	0.313	0.264
150	42.86	0.0	0.160	–	–	–	0.109	–
		0.5	0.300	0.300	0.398	0.253	0.319	0.240
		0.6	–	–	–	0.279	–	0.269
		0.7	–	–	–	0.294	0.378	0.298
200	53.18	0.0	0.170	–	–	–	0.135	–
		0.5	0.325	0.400	–	0.288	0.341	0.275
		0.6	–	–	–	0.314	–	0.303
		0.7	–	–	–	0.329	0.400	0.332
250	62.87	0.0	0.180	–	–	–	0.180	–
		0.5	0.350	0.435	–	0.321	0.374	0.309
		0.6	–	–	–	0.347	–	0.337
		0.7	–	–	–	0.362	0.493	0.366
300	72.08	0.0	0.200	–	–	–	0.183	–
		0.5	0.400	0.470	–	0.340	0.402	0.343
		0.6	–	–	–	0.364	–	0.371
		0.7	.–	–	–	0.388	0.461	0.400

Table 12.4 Daily Nutrient Requirements of Pre-ruminant Calves

Age	Body weight (kg)	Daily gain (g)	DCP (g)	TDN (kg)	ME (Mcal)	Ca (g)	P (g)	Vit A (1000 IU)	ViT D (IU)
Birth to 15 days	25	200	80	0.40	1.5	2.5	1.5	1.5	200
16 to 30 days	30	300	90	0.50	1.7	3.0	2.0	1.5	250
31 to 60 days	40	300	125	0.80	2.4	3.5	2.5	1.7	250
61 to 90 days	50	350	150	1.00	3.6	4.0	3.0	2.0	360

Source: Ranjhan, 1977

There is little information available about the nutrient requirements for the growing buffalo calves.

At Mathura (Sharma and Talapatra, 1963) male buffalo calves of 7–11 months of age were fed on 3 planes of nutrition: On a high plane of nutrition the calves received 0.45 to 0.74 kg DCP and 2.39 to 3.43 kg TDN. On a medium plane they received 0.20 to 0.32 kg DCP and 1.70 to 2.69 kg TDN. Under low plane of nutrition, calves received 0.10 to 0.20 kg DCP and 1.0 to 1:18 kg TDN. These levels were achieved by varying the ratio of concentrate and wheat straw offered to calves. The mean growth rates were 640, 400 and 0 g per day under high, medium and low plane of nutrition, respectively. In another study 490 g daily gain in body weight of 1.5 to 2 years old buffalo calves were reported when fed as per minimum level of Merrison's standard (Ichhoponani and Sidhu 1965).

Experiments at Izatnagar, have shown the DCP requirement as 0.26, 0.29 and 0.34 kg for growing buffalo weighing 150, 200 and 250 kg gaining at the rate of 350, 390 and 460 g per day, respectively (Kumar, 1971). On an all roughage ration a higher requirement of 0.32, 0.34 and 0.35 kg DCP for growing calves of 150, 200 and 250 kg body weight, growing at a lower rate of 260, 270 and 280 g per day was observed.

Based on the results reported so far it appears that ME requirement per g gain varies between 10 and 15.6 Kcal for animal weighing from 200 to 500 kg (Baruah *et al.,* 1983). The DCP requirement for maintenance and growth was 245.40 g/day for a calf weighing 102 kg body weight and gaining @ 461.84 g/day (Verma, 1998).

The protein requirement of growing buffalo calves for achieving a daily weight gain of 500 g is given in (Table 12.5) (Kakkar *et al.*, 1991). Buffalo calves weighing 100 to 300 kg required marginally lesser protein than that recommended by Kearl (1982).

Table 12.5 Estimated DCP and CP Requirements of Buffalo Calves

Body wt. of calves (kg)	DCP requirement (g/day)		CP requirement (g/day)
	Kearl (1982)	Estimated	
100	254	234	344
150	319	293	430
200	341	314	462
250	374	344	506
300	402	370	525

Source: Kakkar *et al.* (1991)

12.5.3 Energy and protein requirements for lactation

The energy requirements for maintenance of the lactating buffaloes are higher. The reported values are 32 Kcal/kg $W^{0.75}$ (Ranjhan and Pathak, 1979) and 137 Kcal/kg $W^{0.75}$ (Kearl, 1982).

Ranjhan and Pathak (1979) recommended a value of 1188 Kcal/kg of 4% fat-corrected milk, the same is recommended by Sen *et al.*, (1978). During the mid lactation period a value of 1003 Kcal ME/Kg rCM has been recommended (Sengar and Singh, 1969). Kearl (1982) suggested an average value of 1230 Kcal per kg of fat corrected milk.

The digestible protein requirement for milk production has been recommended as 126.03 g/100 g of protein secreted in milk.

12.5.4 Nutrient requirements for pregnancy

No systematic work has been conducted on the nutrient requirements of pregnant buffaloes during the last quarter (3 months) of pregnancy. However, as a general recommendation one and a half times of the maintenance requirements may be given during the last quarter of pregnancy (Ranjhan and Pathak, 1979). Chauhan *et al*., (2000) studied the of plane of nutrition on nutrient utilization in pregnant buffalo heifers and hence inferred that buffalo heifers may be fed 20% higher nutrients in the later part of pregnancy so that they can attain a body weight up to 500 kg before calving in their first lactation.

12.5.5 Nutrient requirements of working bullocks

In south east Asia and most of African countries, the buffalo are the main stay of agricultural operations. Systematic studies on the nutrient requirements of working buffaloes are lacking though very important.

For bullocks the work has been categorized into two types, normal and heavy work. The bullocks normally work for 2–4 hr of a farm chore of pulling a cart or thrashing, etc. For heavy work, they are supposed to plough or pull a loaded cart for at least 6 hr. The nutrients required for different types of work are shown in (Table 12.6).

Table 12.6 Daily Nutrient Requirement for Working Buffaloes

Live weight (kg)	Normal work				Heavy work			
	DM (kg)	DCP (g)	TDN (kg)	ME (meal)	DM (kg)	DCP (g)	TDN (kg)	ME (meal)
200	4.0	240	2.0	7.2	5.0	250	2.7	9.5
300	5.8	330	3.1	11.4	7.0	420	4.0	14.4
400	7.6	450	4.0	14.4	9.8	570	4.8	17.3
500	9.4	560	4.9	18.0	11.2	710	6.4	23.0
600	11.2	660	5.8	20.8	13.4	820	8.0	28.8

Source: (Sen *et al.*, 1978)

12.5.6 Nutrient requirements for breeding buffalo bulls

Properly reared bulls are normally fed on the maintenance requirement. Over feeding leads to excessive fattening which reduces libido. The nutrients required for breeding buffalo bulls have been shown in (Table 12.7).

Table 12.7 Daily Nutrient Requirements of Breeding Buffalo Bulls (Sen *et al.*, 1978)

Body wt. (kg)	Dry feed (kg)	DCP (g)	TDN (kg)	ME (meal)	Ca (g)	P (g)	Carotene (mg)	Vitamin (1000IU)
400	7.5	250	4.2	15.4	18	13	40	16
500	8.3	300	4.6	16.6	20	15	53	21
600	9.6	345	5.4	19.5	22	17	64	26
700	10.9	390	6.1	22.1	25	19	74	30
800	12.0	430	6.7	24.2	27	21	85	34
900	13.1	470	7.3	26.4	30	23	95	38
1000	14.1	500	7.9	28.6	32	25	105	42

Use of protected proteins

Ruminants in general obtain uniform quality of protein after microbial synthesis inside the rumen (Malik and Chopra, 1981). This microbial protein at the time of peak production, sometimes limits the availability of energy and/or protein. To overcome this situation, protected or bypass proteins have been recommended. One of the methods used is 1% formaldehyde treatment of vegetable proteins or other proteins which after treatment can escape microbial degradation and will be directly available to the animal. Ration formulations containing formaldehyde-treated GNC gave above 500 g daily weight gain in buffaloes (Malik and Chopra,1981). The use of higher level of formaldehyde or keeping the treated cake for longer period resulted in the over-treatment of protein. This caused nitrogen starvation in the rumen of buffaloes. Addition of 1 to 2% urea, mixed in concentrate mixture helped to increase rumen microbial population and ultimately improved cellulose digestion. (Table 12.8) shows the effect of adding urea on formaldehyde-treated cake-based ration. It was found that feeding rumen protected protein increased milk yield and fat percentage in lactating buffaloes.

Table 12.8. Weight Attained and Live Weight Gain in Periods I and II of Formaldehyde Treatment

Groups	Period I		Period II	
	Total wt. gain (kg)	Daily wt. gain (g)	Total wt. gain (kg)	Daily wt. gain (g)
GNC	73.3	479	75.7	501
TGNC	97.5	637	79.3	525
TGNCU	76.5	500	85.0	566

GNC = Groundnut cake, Control group; TGNC = GNC treated with formaldehyde (1% formaldehyde treatment in period I and with 2% in period II); TGNCU = Formaldehyde treated GNC + Urea.
Source: Malik *et al.* (1983).

12.6 Comparative Digestibility Coefficients

Buffaloes are superior to cattle in their ability to digest the organic nutrients. Generally buffaloes digest 2 to 5% more of each nutrient than cattle. In majority of the studies, however, the difference of 2 to 5% in the digestibility between species were not statistically significant (Chaturvedi *et al.*, 1973).

12.7 Fluid Volume in Rumen

Various *in vitro* experiments conducted with rumen liquor showed the superiority of buffaloes over cattle in utilizing cellulose (Ichhponani *et al.*, 1971a) and other cell-wall constituents. Cellulose digestion was faster when filter-paper was incubated with rumen liquor of buffalo than incubated with rumen liquor of cattle, showing thereby more cellulolytic activity of the rumen. This led many workers to conclude that buffaloes are capable of utilizing inferior quality roughages in a better way than cattle. The difference in the digestibility between the species was observed in all categories of animals such as growing, non-producing adult and lactating animals. Consumption of lesser quantity of dry matter by buffaloes when compared to cattle might be responsible for the slightly higher digestibility observed in buffaloes (Grant *et al.*, 1974). The slower rate of passage of food through the reticulo-rumen of buffaloes was also attributed to be one of the reasons for higher digestibility of crude fibre in buffaloes since this mechanism would provide for better and longer exposure of the ingesta to microbial digestion in the rumen.

The concentration of TVFA, NH_3 and bacterial N is higher in buffalo rumen liquor than in cattle rumen liquor (Ichhponani *et al.*, 1971a). It is however difficult to conclude the superiority of species from the mere knowledge of the concentration of metabolites in rumen liquor since the concentration at any time depends upon balance between production and removal. The removal may either be through absorption, onward passage to omasum or mutual use of metabolites by various species of rumen micro-organisms. It is therefore necessary that care is taken while interpreting the results of the concentration of rumen metabolites, since slower rate of passage of the ingesta in buffalo might be responsible for higher concentration of VFA, bacterial N or ammonia N, etc. The higher concentration of certain electrolytes in the rumen liquor of buffaloes such as potassium and calcium (Prasad and Raghavan, 1973) might also be due to the same reasons.

Meenakshi and Nangia (1998) determined the effect of varying rumen ammonia concentrations on the activity of cellulase and transaminases (alanine aminotransferase, GPT; and aspartate aminotransferase, GOT) by intraruminal infusion of 50, 100 and 150 g ammonium bicarbonate, respectively. They found the existence of a transaminase reaction for ammonia nitrogen utilization and hence concluded that high ammonia concentration coupled with higher microbial transaminase activity suggest a higher rate of ammonia nitrogen usage by rumen microbes.

The basic physiological difference in the flow rate of ingesta might also alter the rumen environment so as to influence the utilization of the food. For example, blood urea concentration of the growing buffalo calves below two years of age was almost twice (Mehra *et al.*, 1976) that of the cow calves. Higher blood urea was parallel to higher ruminal ammonia concentration in rumen of buffaloes. Higher ammonia level was either due to low rate of passage of ingesta or higher proteolytic activity of microbes. NH_3 concentration in cattle and buffaloes has always been more than 11 mg per 100 ml of SRL. It was suggested that higher blood urea concentration in buffaloes can be effectively recycled and used in times of deficit, though this has never been confirmed experimentally. One significant difference between species is a narrower ratio of acetate to propionate in rumen liquor of buffaloes than in cattle (Ichhponani and Sidhu, 1965 a and b; Singh and Ranhotra, 1970). This difference may be due to differences in the rate of passage, since actual production rates of propionate were lower or equal to that in cattle. Apart from the concentration of these metabolites the actual production rates of VFA determined by using isotope dilution techniques (Chaturvedi *et al.*, 1973) did not reveal any significant difference between species.

Buffaloes are superior to cattle in N retention (Sebastian *et al.*, 1970). When the intake of N is approximately the same in both the species, buffaloes show better retention of N than cattle. The reason for this is not obvious, since the digestibility of CP in many instances is only slightly higher in buffaloes. The higher balance of nitrogen has not been reflected in the higher growth rate in the buffaloes. This higher nitrogen balance may perhaps be due to the inherent capacity of buffaloes to hold more of NPN in their blood.

Rumen fermentation in buffaloes

The salient species variations between the fermentation processes of buffalo and cattle were investigated in the early stages (Ichhponani *et al.*, 1962; Singh *et al.*, 1968). Total volatile fatty acid concentration in rumen of buffaloes was significantly greater than of cattle on rations containing berseem + wheat straw + concentrate (Ichhponani and Sidhu, 1965a) and green pearlmillet and cowpea mixture (Ichhponani and Sidhu, 1965b). The buffalo rumen ecosystem was favourable for the establishment of significantly higher number of microiodophils (including *Oscillospira guillermondi*) than that of cattle (Langar *et al.*, 1968) and such types of micro-organisms have been considered as protein synthesizers from NPN. Confirmation of this variation in the rumen ecosystem was obtained by transferring buffalo digesta, rich in *Oscillospira*, into the emptied cattle rumen and vice-versa (Singh *et al.*, 1968).

In vivo rumen metabolism studies using straw-green fodder mix based roughages showed that the buffalo rumen liquor had high-protein synthesizing activity and total and individual volatile fatty acid concentration than that of cattle. Better utilization of crop residues and NPN by buffalo was confirmed by feeding straw, supplemented with an isocaloric concentrate mixture having 0, 17, 35 and 50% DCP substituted by urea-N (Langar *et al.*, 1984). The nitrogen fractions of strained rumen liquor (SRL) showed significant differences ($P > 0.05$) in the protein-synthesizing activities and microbial counts of buffalo and cattle. The cellulose and acid detergent fibre (ADF) digestibilities were also significantly higher in buffaloes than in cattle.

The amino acid profile of the bacterial and protozoal fractions separated from the strained rumen liquor showed quantitative differences in the proline, phenylalanine, diaminopimelic acid, cystine and methionine contents in the fractions collected from buffaloes and cattle. The buffalo protozoal fraction had also higher tyrosine content. With the increased urea intake, the sulphur amino acids content in the microbial fraction showed a decline. It suggested that NPN feeding affected the sulphur level of ration thereby limiting the methionine-cystine synthesis by the rumen micro-organisms (Multani and Ahuja, 1985). In the late seventies and early eighties a series of *in vitro* and *in vivo* experiments were conducted to compare the production of biochemical entities in the rumen (Ichhponani *et al.*, 1969a, 1972) and the utilization of nutrients (Ichhponani *et al.*, 1969b, 1971a, b) in buffalo and cattle. A comprehensive review (Ichhponani *et al.*, 1977) of the results of these studies established that buffalo was superior to cattle in many aspects like rumen fermentation pattern based on the type of micro-organisms, and in the digestion and utilization of various carbohydrate sources, from simple starch to very complex cellulosic materials.

One major observation which has immediate field application relates to studies on the effect of diets on microbial population and rumen metabolites in buffaloes. Pattnaik *et al.,* (2001) characterized a bacteriocin-like compound (Lichenin) produced anaerobically by *Bacillus licheniformis* isolated from water buffalo and concluded that lichenin could be a potential candidate for manipulating the rumen function at molecular level intended for improving the productivity of the ruminant. Puri and Garg (2001) found that adding 50 g sodium chloride/head daily in the feed mixture was optimum to promote better feed utilization in buffaloes as osmotic active agents like sodium bicarbonate and sodium chloride increased the dilution rate/outflow rate, which improves animal performance.

It revealed that the diet comprising wheat straw, green fodder, concentrates and mineral mixture sustained highest population of rumen bacteria and protozoa (Salim Iqbal *et al.*, 1992). Concentration of total volatile fatty acids, ammonia nitrogen and amino nitrogen attained their peak levels within 2 to 4 hr post-prandial period indicating optimal microbial activity in the rumen, whereas the diet containing only wheat straw exhibited least protozoal and bacterial count, decreased rumen metabolites and complete vanishing of *Epidenia* (Singh *et al.*, 1992). This could be attributed to differences in the protein: carbohydrate ratio and mineral supplementation. Wheat straw and green fodder diet showed highest percentage of *Holotrichs* and maximum concentration of free amino acids in the rumen liquor. Dietary differences had concentration of free amino acids in the rumen liquor. Dietary differences had highly significant effect on the population of rumen micro-organisms and percentage of *Holotrichs*, *Entodinia*, *Diplodinia* and *Epidinia* (Singh *et al.*, 1992).

The concentrations of iron, zinc, copper, cobalt and manganese were significantly higher in blood plasma, cerebrospinal fluid (CSF) and rumen fluids (RF) when animals were fed diet consisting of wheat straw, green fodder and concentrate. These elements were at low levels when wheat straw only was fed. Dietary differences had significant effect on the levels of trace elements. There was a positive correlation between blood and CSF. Interrelationship of these elements in the body fluids under different feeding regimens was highly correlated. A decreasing trend of trace elements in CSF revealed quadratic relationship with blood and plasma (Singh *et al.*, 1990).

Buffaloes thrive better than cattle on coarse fodders. This has given an impression that buffaloes are more efficient than cattle in digesting and utilizing crude fibre and cellulose. The present status of knowledge of the comparative alimentary physiology of the buffalo and cattle indicates that the differences exist but it is not possible to draw any definite conclusions as to whether one species differs from the other in any material manner in relation to digestion and utilization of nutrients in commonly used feeds and fodders, or if one species is superior to the other in digestive and metabolic functions.

On the basis of rumen studies, investigations were centered on the urea/uromol utilization, determination of energy values of forages, and energy and protein requirements of buffaloes.

Urea/uromol utilization

Experiments conducted on urea utilization with crop-residues as basal roughage showed that buffalo could tolerate higher levels of dietary urea than cattle (Kaushal *et al.*, 1972; Malik and Chopra, 1980). In spite of high capacity for dietary urea intake, chances of accidental ammonia toxicity were eliminated by developing a cooked urea-molasses complex named uromol (1:9 urea-molasses) reported first in 1974. This complex has above 50% bound urea and a slow rate of *in-vitro* release of ammonia (Chopra *et al.*, 1974). *In vivo* studies (Malik *et al.*, 1978a) showed slow rumen degradation of uromol in the rumen, thereby resulting in a steady ammonia concentration for better microbial protein synthesis and fibrolytic activity on crop residues.

Comparative rumen metabolism, growth and lactation studies on buffaloes with concentrate supplements having groundnut-cake (GNC), uromol or urea-molasses at iso-nitrogenous levels showed that uromol was as good as GNC (Malik *et al.*, 1978a,b). Use of uromol up to 65% of the total N in a concentrate supplement fed with straw resulted in higher CF and cellulose digestibilities and N-retention than that with urea-molasses. The steady release of rumen ammonia with uromol ration resulted in better synchronization on microbial protein synthesis and the nitrogen outflow, and the post-rumen availability of N was 95–93% of intake with GNC and uromol ration as compared to only 78% observed from urea-molasses ration (Malik *et al.*, 1978a; Malik and Chopra, 1978).

Feeding isonitrogenous levels of GNC or uromol-containing concentrate supplements resulted in comparable daily weight gains in buffalo calves as compared to low gains observed with urea-molasses (Malik *et al.*, 1978b). Feeding uromol-containing concentrate supplement for 120 days to lactating buffaloes (Malik and Chopra, 1977) did not have any adverse effect. A long-term study continued for two lactations with intervening dry pregnant periods revealed that as compared to GNC or uromol-fed group, the buffaloes fed on urea-molasses had high plasma urea, weight loss and drop in milk yield in the second lactation (Langar *et al.*, 1982). These results showed that urea fed alone or with molasses was inferior as compared to processed urea fed as uromol.

The problem of low shelf life of cold uromol and its subsequent mixing with other feed ingredients was overcome by developing uromol-bran product (Malik and Makkar, 1979). This product prepared by using urea, molasses and bran in the ratio of 1:3:5 had 36% digestible crude protein (DCP) and 72% total digestible nutrients (TDN) and could substitute (w/w) oilcakes in a concentrate supplement. Based on these results the recommendations regarding feeding of this product to buffaloes are: (*i*) uromol-bran mixture should form 15% of the concentrate ration (replacing 50% of the cake portion) for fast-growing (above 6 months of age) and high-milk-yielding buffaloes (peak yield 15 kg and above); and (*ii*) for slow-growing and low-milk-yielding (peakyield 8–10 kg) animals uromol-bran mixture can form 30% of the concentrate supplement (replacing 100% cake).

Uromol-impregnated wheat straw (UIWS) as maintenance ration with 0.5 kg concentrate mixture could form maintenance ration for a mature buffalo (Kakkar *et al.*, 1983).

Uromin lick

The advantage of solidification of uromol on cooling was taken for developing a uromin lick. This lick comprises 400 g urea, 1,200 g molasses (boiled for 30 minutes), 600 g mineral mixture, 400 g salt, 600 g starch or *maida*, 400 g deoiled groundnut-cake and 400 g deoiled rice bran, mixed thoroughly and pressed to form a brick of about 4 kg; 2% bentonite is added as binder. This brick can be kept in the feeding trough for licking. A growing calf licks about 0.5 kg of this brick daily and this provided a part of the crude protein and almost whole of the mineral requirements of the animals (Ahuja *et al.*, 1986; Makkar *et al.*, 1989).

Leucaena lick

A lick containing leucaena leaves (40%) has also been developed (Gupta and Malik, 1990) with the difference that it did not contain any conventional feed ingredients other than molasses. The composition of this lick on per cent basis is: urea 10, molasses 35, mineral mixture 14, leucaena leaf-meal 40 and acetic acid 1. These licks have been used in the Kandi area of Punjab in India and other such places where concentrates and good quality green fodders are in short supply.

12.8 Processing of Crop Residues

Cereal crop residues alone cannot meet the maintenance requirements of the animal because of high lignification, low nitrogen and mineral contents. Several physico-chemical and biological pre-processes have been tried and tested on buffaloes for upgrading the nutritive value of crop residues.

Chauhan (1999) studied growth production and nutrient utilization by buffalo calves offered ammoniated wheat straw and alfa-alfa hay *ad lib.* supplemented with available cereal energy sources and concluded that conventional concentrate mixture can be completely replaced with available cereal grains along with lucerne hay and ammoniated wheat straw for better efficiency of nutrient utilization in year-old growing buffalo calves.

Dayal *et al.,* (2002) studied the evaluation of urea-NH_3 treated palm press fibre by *in vitro*, in sacco and *in vivo* techniques. The DCP and TDN values of untreated and treated PPF were 3.70, 42.30 and 8.80, 52.7%, respectively, making treated PPF (palm press fibre) a potential substitute of conventional roughages for ruminants.

Physico-chemical treatments

Wheat straw or bagasse (sugarcane residue after extraction) after treatment with (w/w) 2 and 5% sodium hydroxide, when fed with uromol as N source results in increased availability of metabolizable energy (ME). But for meeting the total ME requirement for maintence of adult buffaloes (400–500 kg weight), per day 600 g of cereal or cereal by-product supplement is required with untreated straw-uromol ration. Processing using pure alkali is however uneconomical. An alkaline (0.8 to 1.0% alkalinity) keir boiling waste (KBW) effluent from textile industry when used for treating cereal crop residues (3 litres/kg crop residue) improved the digestibility of treated material to the same extent as by treating with 3.0 (w/w) sodium hydroxide (Nagra and Langar, 1984).

Natural fermentation

Fermenting straw only or stover (96.5%) with urea (3.5%) at 70% moisture could provide upgraded straw for animal feeding only in a 9 day period as compared to 30 day period used earlier. Further, using the fresh, naturally fermented material as inoculum in the subsequent batches hastened the process to provide upgraded straw preparation within 6 days only (Gupta *et al.*, 1984). Experiments on the 9 day natural fermentation when carried out with 40, 50, 60 and 70% moisture level, showed that 40% moisture level provided the straw preparation with nutritive value close to that available with 70% moisture (Bakshi *et al.*, 1987). These fermented straws provided enough nutrients for maintenance of buffalo calves.

12.9 Evaluation of Agro-industrial By-products

A large number of agro-industrial by-products have been evaluated as an alternate source of feed in buffaloes. Some of these by-products can be used with relative ease by the farmers themselves; whereas others have to be processed before use. The latter involves biotechnological approach to (*i*) degredation of plant cell-wall in crop residues with particular focus on lignocellulose breakdown using recombinant microbes to simple sugars; and (*ii*) detoxification of growth depressants, antimetabolites and incriminating toxins from oil-cakes, oil-meals and other seed products after deoiling.

Some of the technologies which are currently in use are given below:

Corn steep fluid

This is a by-product of starch industry. It has 40% CP and 23% soluble sugars. Compared to urea, corn steep fluid as nitrogen source in buffalo rations gives better rumen microbial protein yields (Virk *et al.*, 1981).

Spent brewers' grain

The fresh product from brewery has 75% moisture and on dry matter basis contains 18.8% CP, 14.6% DCP and 54.6% TDN. The dried brewers' grain can replace 50% concentrate supplement requirement of growing and milking buffaloes (Virk *et al.*, 1981).

Potato waste

There is about 7% wastage when the potatoes are stored in cold stores. These discarded potatoes can be used as alternate energy source in the rations of buffalo calves (Makkar *et al.*, 1984). The waste potatoes can be fed as such after chopping or with urea as a source of N. Potato waste contains 13% CP on dry matter basis.

Poultry excreta

Poultry droppings on wheat-straw-based poultry litter can be used in buffalo ration after processing (Malik *et al.,* 1980; Ahuja *et al.*, 1983). In a long-term experiment on buffalo heifers, poultry excreta with wheat straw was mixed at 27 and 37% levels in the concentrate mixtures after grinding. A groundnut-cake-based ration was used as control. In poultry litter ration, energy intake was made good by additional feeding of concentrates. The experiment was continued till the heifers conceived. There was no adverse effect of feeding poultry-litter-based rations, digestibility and N retention were similar to those of control group (Malik *et al.*, 1980).

Rumen fermentation pattern and nutrient requirements of buffaloes are not the same as those of cattle. The buffalo is a better converter of NPN compounds into protein and in the digestion of crude fibre. Adoption for the buffalo of the feeding standards as recommended by the National Research Council of the United States or the Agricultural Research Council of the United Kingdom for cattle appears to be a good base to follow, and a modified standard based on the research results be adjusted whenever considered appropriate.

12.10 Feeding Formulation

12.10.1 Practical feeding of the lactating buffalo

Lactating buffaloes should be given the best feed the farm can offer. Producing milk is one of the most energy demanding biological processes. Weight loss is common in high producing animals during the first month of lactation because they cannot consume a sufficient amount of energy. A popular term is that the animals are milking off the fat. It is therefore important that the buffalo is in good health status at partum. A well balanced ratio of protein, energy, vitamins and minerals in a palatable and tasty feed is the best way of increasing milk production and live weight, as well as improving health and fertility.

Traditional feeding patterns for buffaloes all over the world is subjected to forages and crop production of the season which affects the level of milk production. Forage is insufficient during the dry season and abundant during the rainy season. Shortages are overcome by conserving forages as hay or silage.Nutrient requirements for milk producing buffaloes is given in (Table 12.9).

Table 12.9 Nutrient Requirements for Milk Producing Buffaloes

Requirements for live wt.	Energy (ME in MCAL)	TDN (kg)	Total crude protein (g)	Calcium	Phosphours (g)
450 kg	13.0	3.4	341	18	13
500 kg	14.2	3.7	364	20	14
550 kg	15.3	4.0	386	22	16
600 kg	16.3	4.2	406	24	17
Requirements for milk yield per kg 4% fat corrected milk	1.24	0.32	90	2.73	1.68

12.10.2 Practical feeding of the calf

Calf mortality is very high, in India. It is often 30–40% before 3 months of age, and in Italy the figures may be higher. This is caused by malpractice such as negligence, limited milk feeding, injuries and diseases. By increasing the amount of feed to the calf's requirements and by practicing the following instructions the mortality can be decreased.

Colostrum is the most important and most suitable feed for the newborn calf. It contains all the nutrients needed (Table 12.10) along with the vital antibodies. It is crucial for the survival of the calf that it receives colostrum during the first 12 hours of its life, the earlier the better. The calves should be given colostrum as long as the mother provides it *e.g.* 3 to 4 days. Any surplus colostrum can be frozen and then thawed and carefully heated to 39 °C. If no freezing facilities are available colostrum can stay fresh for a couple of days if it is cooled in a hygienic container. Colostrum can be fermented with living lactic acid culture. Fermented colostrum can be kept for at least a week and up to two weeks if cooling facilities are available.

Table 12.10 Requirements of Nutrients of Calf up to 3 Months

Age (days)	Daily gain (kg)	DCP (g)	TDN (Mcal)	ME (g)	Ca (g)	p (g)	Vit-A (1000IU)	Vit-D (IU)
0–15	0.20	80	400	1.5	2.5	1.5	1.5	200
16–30	0.30	90	500	1.7	3.0	2.0	1.5	250
31–60	0.30	125	800	2.4	3.5	2.5	1.7	250
61–90	0.35	150	100	3.6	4.0	3.0	2.0	260

If the calf is not allowed to suckle its mother it should be provided with colostrum as soon as possible after birth. If it is not possible to feed the calf directly after milking the buffalo, colostrum should be cooled in order to maintain hygienic quality. When it is time to feed the calf, the milk should be carefully heated to no more than 39 °C. Colostrum must never be boiled. By boiling the milk the antibodies are destroyed and hence, cannot be utilized as such by the calf.

The natural eating behavior of the calf is to suckle its mother often and to consume a small amount of milk at each suckling period. It is best for the calves reared under artificial conditions if their eating behavior is as "natural" as possible. Colostrum should be fed to the calf at least twice daily with equal intervals.

The calf should be trained to drink from a bucket. The easiest way to do this is to dip clean fingers into the milk and then allow the calf to lick and suck the fingers. The hand is then gradually drawn into the milk in the bucket while the calf is still suckling. Once the calf has learn to drink it is easy to feed. The calf may need assistance for 5 days. There are special nipples which can be put in the bucket. The calf will suckle these, hence it will need less assistance from the trainer.

After the colostrum period, whole milk should be provided to the calf until 15 days of age @ a level of $^1/_8$th to $^1/_{10}$th of the calf's body weight. Milk replacer can be fed along with the whole milk provided that it has a certain composition of nutrients. It is not advisable to completely substitute whole milk with milk replacer. Milk and/or replacer should be offered to the calf on at least two occasions per day. The milk and/or replacer should be served at body temperature (38–39 °C).

At two weeks of age, the calf should be introduced to good quality green feed and concentrates, as a calf starter (Table 12.11). This stimulates the rumen to grow and function properly. By following the feeding schedule in Table 12.12 a daily gain of 0.35 kg can be expected in Murrah calves.

Table 12.11 Calf Starter Mixture

Feed source	Amount
Crushed barley	50 %
Groundnut cake	30 %
Wheat bran	8 %
Fish meal/skim milk powder/meat meal	10 %
Mineral mixture	2%
To increase acceptability, add, per 100 kg of starter	
Molasses	5–10 kg
Salt	500 g

Table 12.12 Feeding Schedule for Calves

Age (days)	Whole milk (l)	Skim milk (l)	Calf starter (g)	Hay (g)
0–14	4*	–	–	–
15–21	3.5	–	50	300
22–28**	3.0	–	300	500
29–35	1.5	1.0	400	550
36–42	–	2.5	600	600
43–49	–	2.0	700	700
50–55	–	1.0	800	800
57–63	–	1.0	1000	1000
64–70	–	–	1200	1100
70–77	–	–	1300	1200
78–84	–	–	1400	1400
85–91	–	–	1700	1900

* First 3 to 4 days food colostrum

** Ensure a smooth and gradual change to milk replacer

An alternative method is to rear calves with foster mothers. In Italy, 40% of the buffalo calves are reared by suckling an old and less productive buffalo or even a cow. This has several advantages, *e.g.* little labor is required concerning feeding of the calf and the calf will secure it's nutrient intake itself.

Buffalo calves fed with stovers of maize, *bajra* and oat cannot meet their nutrient requirements and are often in negative energy and protein balance. However, feeding the calves with treated stovers with a urea-molasses-salt complex both enhances the palatability of the stovers as well as the digestibility and nutrient value. Buffalo male calves weighing 150–200 kg has proven to increase the intake of treated stovers verses untreated ones and thereby increasing weight gain, nitrogen balance and health.

12.10.3 Practical feeding of the heifer

Until now farmers in general do not follow any systematic feeding standard or schedule for the feeding of the heifers, but from survey it has been observed that heifers received special attention of farmers regarding feeding and management as they are the future dairy animals and a potential source of income. Mostly heifers are allowed milk from one teat till 9 months of age along with good roughage and about 1 kg of rice bran/*chuni*, etc., On this system the Murrah heifers grow about 450 g per day. At organized farms (Central Government, State Governments and Military Dairy Farms), good feeding systems are followed which have been given in (Table 12.13) under these systems the Murrah heifers gain about 550 g per day.

Table 12.13 Feeding Schedules for Heifers at the Organized Dairy Farms

Age (months)	Schedule No.	Concentrate mixture (kg)	Roughage (kg)
3–6	1	1.2 to 1.5 1–2 kg	Green oat/maize/silage
	2	0.9 to 1.5	Berseem/Lucerne/cowpea 15–20 kg +wheat straw 2 kg
	3	1.4 to 2.0	Green fodder 3 kg + straw 2 kg
6–12	1	1.25	Green oat/maize 20–25 kg
	2	1.00	Berseem/Lucerne 25–30 kg + dry fodder 2 kg
	3	2.00	Straw/stover 2 kg + green 5 kg
1 year to age of conception	1	2.00	Green oat/maize 30–35kg
(140 to 300 kg)	2	1.00	Berseem/Lucerne 30 kg + straw 3 kg

She must have an average daily gain of at least 500 grams per day in order to reach the optimum size for calving within reasonable time (500 kg at 32 to 40 months). Unfortunately, many farmers consider heifers to be unproductive and hence they are not properly fed. Lack of feed is often a reality, it is therefore not possible to feed all animals in the herd with high quality feed. The following advice could be considered as a rule of thumb, bearing in mind that the quantity and quality of feed varies with the season. Furthermore, the condition and growth rate of the heifer should be checked regularly to see that she has the approximate growth rate and if not, adjust her feeding schedule accordingly.

The heifers should be fed green feed of the season of about 4–7 kg DM together with some straw and concentrate or grain per day. If the green feed is leguminous the ration of green feed and concentrate or grain can be reduced and the amount of straw increased. However, it is positive to feed the heifers a small amount of grain or concentrate (not less than 0, 5 kg per day) for making both them and their rumen accustomed to this type of feed, especially partus.

If available, ammonia treated straw could be given along with low quality green feed and concentrate. Silage could be given to heifers, but it is often a very valuable feed saved for milk producing animals. However, a few months before partus the heifer should slowly be introduced to the feed she will have as a milk producing buffalo.

Maximum voluntary intake of the heifer is obtained at approximately 1 to 1.5 kg dry matter of straw together with 3 kg (DM) of green feed and 1 kg concentrate.

Straw fed to growing stock should preferably be ammoniated and further supplemented with green feed or hay and some kind of concentrate to give the best result.

12.10.4 Practical feeding of the dry buffalo

Feeding the dry buffalo concerns preparing for partum and a high milk production. In the last two months of gestation the buffalo has increased requirements for nutrients for fetal growth. Experiments with Murrah buffaloes has shown that the best economical way of feeding dry buffaloes 2 months before calving is at 125% of the recommended level for cattle. By giving the dry buffalo a little more than she needs her chance to build up the body reserves and to be in good physical condition is improved. After calving, the buffalo can be fed at 100% of recommended level for cattle.

12.10.5 Feeding of pregnant buffaloes

During pregnancy 1.25 kg of concentrate mixture is fed during the last quarter of pregnancy along with the normal grazing/wheat straw feeding. The common practise with farmers is to feed wheat bran/rice bran/pulse chuni.

12.10.6 Feeding of breeding bulls

Breeding bulls are fed on good quality forages (sorghaum, oats, berseem), straw and concentrate at organized farms. One kg concentrate mixture over the maintenance ration is fed.

CHAPTER 13

BUFFALO HEALTH

By

M.C. Sharma and Mahesh Kumar

When compared with other domestic livestock, the water buffalo generally is a stardy animal. In spite of the fact that most of them live in hot, humid regions that are conducive to diseases. The buffalo is susceptible to most of the diseases and parasites that afflict cattle. Although the reasons are not specifically known, the effect of diseases on the buffalo and its productivity is often less serious than on cattle in the same ecosystem.

Antibiotics and vaccines developed for cattle work equally well on buffaloes. As a result, treatments are available for most of the serious diseases of buffaloes.

It is not possible to produce a disease prevalence map and pattern of disease occurrence in buffaloes for all the diseases. They share the same plethora of disease conditions it is susceptible to, excepting a few, which are peculiar to the buffalo and some others where there are differences. The manifestation of disease is still under investigation and it is still premature to say that it is resistant to certain disease due to absence of adequate challenge studies. There are ongoing studies to determine the susceptibility of the animal to various diseases, and its relation to age, environmental and management practises. For diseases like rinderpest, tuberculosis, pasteurellosis, foot-and-mouth disease, for which disease control programmes have been formulated. Buffalo health is a very vast topic, but some of the economically important infectious and non-infectious problems have been discussed here under.

The major cause of losses in buffaloes is due to calf mortality. Newborn buffalo calves, like other bovine calves, succumb in large numbers to viruses, bacteria and poor nutrition. This is largely due to poor management during the calf's first 2 months of life, *e.g.* depriving calves of mother's milk and proclivity for wallowing which exposes calves to water-borne diseases. In many cases a young one occasionally drowns when an adult rolls on the top of it.

In countries such as the Philippines, Vietnam, Kampuchea, Laos, Malaysia, Thailand and Myanmar, buffaloes are highly susceptible to rinderpest and may be even more so than the indigenous cattle managed under comparable conditions. In south-eastern Europe, however, the buffalo is relatively more resistant than the local cattle breeds. In the broadest sense it seems that, subject to local and regional variations, the susceptibility of the buffalo to this disease increases from western to eastern Asia. The clinical syndromes and autopsy findings are more or less similar to those observed in cattle. However, the body-temperature curve in the buffalo is quite often erratic. Clinical symptoms tend to be more acute and as a rule the conjunctiva is more severely congested.

In general, buffaloes are less susceptible than cattle to foot-and-mouth disease. Nevertheless, severe outbreaks among buffaloes are not uncommon, and in countries like Laos and the Philippines more buffaloes were affected than cattle in some of the past epidemics.

Water buffaloes are highly susceptible to pasteurellosis, (haemorrhagic septicaemia) caused by the bacterium *Pasteurella multocida*. Haemorrhagic septicaemia runs an acute course in buffaloes than in cattle and the oedematous form is more common. Buffaloes are more susceptible to it than cattle and die in large numbers when affected by pasteurellosis. A vaccine against pasteurellosis is effective in protecting both buffaloes and cattle; it is cheap and easily made.

Reports on the incidence of anthrax in buffaloes vary from country to country. In Egypt the disease appears to be rare in buffaloes. In India also, the disease is less common in buffaloes than in cattle. However, the situation is just the reverse in Myanmar where buffaloes frequently contract the infection; this poses a serious danger to the elephants working in adjacent forests. On some islands in Indonesia, the disease occurs in a severe form and affects buffaloes more than cattle.

Several workers have reported that the buffalo is relatively less resistant to tuberculosis. This is one of the most important diseases affecting buffaloes in Egypt, the incidence being higher in housed buffaloes in the southern region. Slaughterhouse examinations of carcasses in a Cairo abattoir showed that about 7% of the buffaloes slaughtered had tuberculous lesions. Tuberculosis tests carried out in four representative areas in India on 44519 cattle and 40201 buffaloes under comparable conditions of management showed 4.7 and 13.8% reactors, respectively, in the two species (Lall 1969).

The incidence of tuberculin reactors in India is higher in buffaloes than in cattle (Lall, 1969). Mammalian tuberculin evokes a stronger reaction in buffaloes on intradermal test as compared to that in cattle (Lall, 1946). Other strains of mycobacteria have been isolated from feral buffaloes and cattle in northern Australia but seem to have little effect on the animals. Tuberculosis occurs among buffalo herds of the world only because most of them are kept under unsanitary conditions.

Mastitis is one of the serious diseases of the buffalo, especially in countries where buffaloes are mainly kept for milk production. A mastitis survey carried out on 690 buffaloes in nine dairy herds in different parts of India showed an incidence of 20.7% of sub-clinical or insidious cases and 2.4% of clinical cases. In another random survey of 1193 buffaloes in the northern part of India, Katra and Dhanda (1964) found an incidence of infection of 11% in the urban and 9% in the rural areas. The corresponding data for 3097 cows were 10 and 7.5%. About 98% of the cases were caused by pathogenic staphylococci and streptococci while the remaining 2% of the cases were due to other organisms. In Egypt, incidence of mastitis in the two species was virtually the same (Wahby and Hilmy, 1946). In one survey 9% of 860 buffaloes brought to the Cairo clinic were affected with mastitis (El-Gindy *et al.*, 1964). Treatment and control programmes used for cattle are equally effective for buffaloes.

Brucellosis

Brucellosis is fairly common in buffaloes. In a survey of 13,565 animals in military dairy farms in India, 10% of the buffaloes and 13% of the cows recorded positive reactors on serological tests. In Turkey, 42% of the 31 buffaloes and 39% of the 285 cattle were positive.In Brazil 41% of buffalo serum samples gave a positive reaction on agglutination tests (Golem 1943, Santa Rosa *et al.,* 1961). In Egypt cultures of *Brucella abortus* were isolated from as many as 15% of 200 buffalo milk samples subjected to bacteriological examination. In Indonesia, however, brucellosis is sporadic in buffaloes, although it is endemic in cattle. Brucellosis in Venezuela is increasing more rapidly among buffaloes than among cattle. As many as 57% of some Venezuelan herds are infected with the disease. It is a frequent cause of abortion in buffaloes. Serologic procedures and measures developed for the control of the disease in cattle are also effective means of curbing this infection in buffaloes (consumption of raw milk or contact with aborted fetuses may cause undulant fever in humans).

Ticks

Buffaloes are notably resistant, although not immune, to tick bites. Buffaloes appear to be less affected by ticks than cattle. Warble infestation is also less prevalent in buffalo (Sen and Fletcher, 1962). In a tick-infested area of northern Australia only two engorged female ticks were found on 13 adult buffaloes during a 2 year test. Accordingly, healthy buffaloes are not commonly affected by diseases borne by ticks nor are the hides damaged by their bites. Since ticks are rarely found on buffaloes, anaplasmosis, theileriosis and babesiosis (these are tick-borne diseases) have little effect on buffaloes in the field buffaloes and cattle are equally susceptible. However, it inoculated with East cost fever, a form of theileriosis. This is important because tick infestations in cattle are particularly troublesome in the tropics and the pesticides used to control them are becoming ineffective as the ticks develop resistance. The pesticides are also becoming expensive. The basis of the buffalo's tick resistance is not known, but wallowing and rubbing may play a role in it; animals kept in experimental concrete pens in Australia have developed heavy tick infestation.

Screw-worm

Larvae of the screw-worm fly (*Callitroga* species), a major pest of livestock in Central and South America and some other tropical areas, do not affect adult water buffalo. In Venezuelan areas where cattle (zebu type) are severely infested, adult water buffaloes are virtually free of screw-worm larvae and the umbilicus of newborn calves seldom, if ever, becomes infected. The same is true in Papua New Guinea. It is thought that the mud plaster produced by wallowing suffocates the larvae. In India screw-worms do not affect water buffaloes either, and there they wallow in fairly clear water and the farmer usually washes them off.

Round-worm

Neoascaris vitulorum is a commonly occurring helminth parasite of buffalo calves and can cause considerable mortality in young stock. The heavy losses of young buffalo calves throughout the world are caused, in large measure, by the roundworm *Toxacara vitulorum.* The calves are more susceptible than mature animals and they become infected before birth or within 24 hr after birth through the mother's colostrum. The roundworm is the most serious buffalo parasite and in untreated calves the small intestine can get packed with worms to the point of complete occlusion. Anthelmintic drugs that control roundworms are highly effective and widely available.

The adult water buffalo appears to have a high degree of resistance to strongyloid nematodes. Being such excellent converters of ruffages, they do not suffer the nutritional deficiency and the resulting liability to these roundworms is experienced seasonally by cattle.

Liverfluke

Fascioliasis, or fluke infection, is another important disease of buffaloes. In India, *Fasciola gigantica* infestation is predominant in the plains while *F. hepatica* is mostly confined to the hilly tracts. In a survey lasting from 1949 to 1953 in Thailand, 7% of 29,421 buffalo livers were found to be cirrhotic due to fluke infestation. In Singapore, a large number of buffaloes imported from Thailand were found to be infected with liverfluke. Fascioliasis is also common in the Ararat valley in Azerbaijan and in the Philippines. During wallowing, water buffaloes can easily become infected with water-borne infective stages of liverfluke (*Fasciola hepatica*). Although the number of flukes in a buffalo may be phenomenally high, no clinical signs of the disease are usually evident. It seems likely that the resulting liver damage reduces the growth, the work and milk production of buffaloes more than is generally appreciated.

Trypanosomiasis

Among the protozoan parasites, *Trypanosoma evansi* appears to be the most important for the buffalo, especially in Southeast Asia. This microorganism generally causes a latent or subclinical form of trypanosomiasis, but severe outbreaks characterized by high temperature, signs of abdominal pain and mortality 6 to 12 hr after the onset of the symptoms have also been recorded. The water buffalo is susceptible to trypanosomiasis and is reportedly more susceptible than cattle to *Trypanosoma evansi*. Experience with this animal in Africa is limited, but trypanosomiasis may be the reason why Egypt is the only African country that has traditionally employed water buffalo.

Coccidiosis

Heavy mortality in buffalo calves due to coccidiosis has been reported from India and Sri Lanka.

Other Parasites

The walloing habit and its resulting mud-coat seem to protect water buffalo from many biting flies, but the main ectoparasite in Australia and Southeast Asia is the buffalo fly (*Siphona* sp). Pediculosis, caused by the sucking louse (*Hematopinus tuberculatus*), occurs widely among buffaloes. Sarcoptic mange (*Sarcoptes scabiei* Var *Bubalus)* is a serious disease, especially among calves and during dry seasons when wallowing opportunities are restricted. The lungworm *Dictyocaulus viviparus* thrives in warm, humid areas and sometimes infects buffaloes heavily, although its outward manifestations are rare.

A detailed description of major diseases their etiology, epidemiology, symptoms, diagnosis and treatment is described here.

13.1 Infectious Diseases

13.1.1 Bacterial diseases

13.1.1.1 *Anthrax*

Anthrax also referred as splenic fever occurs as a peracute or acute disease characterized by sudden death with the exudation of dark coloured blood from the natural orifices.

Etiology

Anthrax is caused by *Bacillus anthracis*, a large rod shaped gram positive spore forming organism with a long body and truncated ends. When exposed to air the organism forms spores within few hours, which are extremely resistant to damage and can survive for 15–20 years in soil. Hence anthrax infected carcasses must never be opened. The organisms possess a capsule and are capable of producing a toxin. The disease is of zoonotic importance as human beings suffer from Wool sorters disease.

Epidemiology

Though the occurrence of the disease varies with the soil and climate, it has a worldwide distribution. However, most of the outbreaks are restricted to 'anthrax belts' and originate from contaminated soil, which remains infective almost indefinitely. The outbreaks are more common in warm and humid conditions like a rain after drought. The disease results in heavy economic losses due to mortality as Mansjoer (1961) estimated that in Indonesia it caused an annual loss of 6.5 million US dollars. Buffaloes are considered to be more susceptible than cattle. Infection is through ingestion, inhalation or entry of organism or spores through the skin. Direct spread from animal to animal occurs rarely. Biting flies and insects act as mechanical carriers.

Symptoms

In buffaloes, the disease is manifested in peracute or acute forms. At the beginning of an outbreak, the buffaloes commonly suffer from the peracute form. The infected animals die suddenly within two hours without showing any symptoms, but escape of blood from nostrils, anus and mouth usually occurs before death. Before death, fever, muscle tremors, dyspnoea and congestion of the mucosa may be observed with collapse following terminal convulsions.

In acute form, the animals survive for 3–4 days as the course of disease is of about 48 hr. Body temperatures of 42 °C, depression and deep and rapid respiration are seen. Pregnant animals abort, and milch animals show reduction in milk yield and milk is usually blood tinged. There may be diarrhoea, dysentery, and edema of the tongue, throat, sternum and perineum. An outbreak of a cutaneous form has also been reported in India (Dabbas and Sharma, 1981). Respiration and heart rates are elevated while ruminal movements are absent. The visible mucous membranes are congested with haemorrhagic spots on them.

Necropsy

Anthrax suspected carcasses should not be opened. If at all it is to be opened it should be only after laboratory tests and only in segregated premises, which can be disinfected thoroughly. The anthrax-infected carcass putrefies rapidly and no rigor mortis is seen. The abdomen is invariably distended with gases, the anus and vulva are everted and there is a tarry black discharge of blood, which does not clot. There is intense hemorrhage of the gut, and all other mucous membranes. Echymotic haemorrhage is noticed on the whole body. The spleen is soft, liquefied and enlarged while other lymph nodes are also enlarged. The organisms can be seen in the blood or edematous fluid.

Diagnosis

History and typical symptoms can detect anthrax. It can be confirmed by isolation of the organisms from blood and edematous fluid. Peripheral blood or edema fluid smears will reveal the organism, in positive cases. For ascoli test, a small piece of the muzzle or ear should be collected for preparing antigen for conducting the precipitation test. The disease being zoonotic, all precautions must be observed in handling the carcass and by the handlers themselves.The disease should be differentiated from peractue black quarter, lead poisoning, acute leptospirosis and bacillary haemoglobinuria.

Treatment and control

Treatment is of little importance in both peracute and acute forms of the disease. Anti-anthrax serum @ 100–150 ml I/V can be given to valuable animals in conjunction with antibiotics considering the high cost of the product. Though penicillin @ 10000 IU/kg body weight twice daily, had been in considerable vogue, streptomycin @ 8–10 g/day in two doses I/M has been found to be much more effective. Oxytetracycline @ 5 mg/kg body weight per day parentally is also efficacious. It is desirable to prolong the treatment up to a minimum of 5 days to prevent recrudescence of the disease.

For the control of disease, live spore vaccine is available which is derived from organisms of low virulence. This vaccine provides immunity for one year. It is recommended that after an outbreak, annual vaccination should be carried out for at least 3 years and in endemic areas annual vaccination is recommended. When outbreaks occur in clean areas for the first time, all incontact animals must be vaccinated. Entry or exit of animals should be restricted from the known infected areas.

13.1.1.2 *Black quarter*

Black quarter, also known as emphysematous gangrene or black leg, is an acute disease characterized by inflammation, severe toxaemia, emphysematous swelling of the muscles and high mortality.

Etiology

The disease is caused by *Clostridium chauvoei*, which are gram positive, spore forming, and rod shaped organisms. The spores of this organism are highly resistant to disinfectants and adverse environmental conditions and can persist in the soil or dried infected materials for many years. Its vegetative forms can easily be killed by heat or common disinfectants.

Epidemiology

The disease has a worldwide distribution and is enzootic in certain areas particularly those affected with floods quite often. It is more prevalent in low altitude hilly regions. In India, the disease is milder than that in cattle (Dhanda, 1977). The disease is a soil-borne infection and spreads mainly through intestinal mucosa after the ingestion of contaminated feed. The organism may also gain entry through cuts and wounds on the skin/mucosa. It may be found in the spleen, liver and alimentary tract of normal animals.

True black leg develops when the spores, which are lodged in normal tissues, are caused to proliferate by mechanisms such as trauma and anorexia. The disease is confined to young stock between 6–24 months of age and occurs more frequently in animals raised on a high plane of nutrition. The morbidity and mortality rates are high and most of the outbreaks occur during early rains.

Symptoms

The disease usually occurs in acute form. The symptoms are mainly produced by the toxaemia liberated from the germination of spores of the organisms. There is marked swelling in the muscles of shoulder, hip, chest, back or flank region. The swelling that is hot and painful to touch initially, becomes very extensive, cold and painless later on. On palpation of swollen area, crepitating sounds can be heard. The affected area later on, turns black and dry and froth mixed foul smelling fluid is released by puncturing it. There is high fever (41 °C), depression, anorexia, and ruminal stasis with increased heart and pulse rates. In terminal stage, diaphragm, tongue and heart muscles are also involved and the animal dies within 12–36 hr after the appearance of clinical symptoms.

Necropsy

In dead animal, putrefaction occurs very rapidly, the animal reveals very characteristic posture, with the animal lying on one side with affected limb stuck out stiffly. There is severe bloat and blood mixed frothy exudate comes out from the nostrils and anus. Peritoneum is full of haemorrhagic fibrinous exudate. Incision of the affected part shows the release of dark coloured sanguineous thin fluid containing gas bubbles and rancid odour. A metallic sheen is noticed on the cut surface.

Diagnosis

In typical cases of black quarter, a definitive diagnosis can be made on the clinical signs and the necropsy findings. However, positive identification on gross post-mortem findings of which one of the clostridial myositides is present, requires the services of an experienced pathologist. Isolation of the causal agents and serological tests are also helpful in the confirmation of the disease. Necropsy findings are useful only in fresh cadavers.

The disease must be differentiated from anthrax, lightening strike, lactation tetany, bacillary haemoglobinuria, haemorrhagic septicemia and acute lead poisoning.

Treatment and control

The severe cases must be treated immediately with BQ antiserum and high doses of antibiotics. The antiserum can be given @ 200–400 ml I/V and repeated after 24 hr. Penicillin @ 1000 units/kg body weight is highly effective. Treatment should commence with crystalline penicillin I/V, followed by long acting preparations, some of which must be injected into the affected tissue. Oxytetracycline has been found to be effective in experimental infection (Afzal *et al.*, 1966). Ciprofloxacin and norfloxacin are also effective against this disease. Treatment with antibiotics must be continued for 5–7 days.

In areas where the disease is enzootic, annual vaccination should be carried out of all animals between the age group 6–24 months, just before the anticipated danger period. Vaccination of calves at 3 weeks of age is recommended when the incidence of disease is very high. Of the various preparations available, the formalin-killed alum precipitated bacterin is the most satisfactory. Quarantine procedures and vaccination of all unaffected cattle is advised in an outbreak, along with injections of benzathine penicillin @ 6000 units/kg body weight. All carcasses must be destroyed by burning or be disposed off by deep burial to limit contamination of soil with spores of the organism.

13.1.1.3 *Haemorrhagic septicaemia (Pasteurellosis)*

Haemorrhagic septicemia also known as shipping fever or locally in India as 'galghontu' and 'ghurrha' is one of the most serious diseases affecting buffaloes due to its acute manifestation with resultant high mortality. It was first described in Italy in 1887 as 'Barbone' by Oreste and Armamni (Bain, 1963). High fever, excessive salivation and petechial haemorrhages on submucosae characterize the disease.

Etiology

Pasteurella multocida types 6B or 6E cause the disease. The causal organism is a small gram negative, bipolar, coccobacillus, which is not resistant to heat and adverse environs. These organisms are capsulated and take characteristic bipolar staining by methylene blue, which can be noticed easily in smears of blood or tissues from dead animals.

Epidemiology

The disease had worldwide prevalence in buffaloes except in Australia, Oceania and Japan. In the tropics, incidence is higher in the rainy season due to high humidity and changes in nutrition and management during this period. Immunity to the disease increases with age due to exposure to natural infection and vaccination (Adlakha and Sharma, 1992). An age-mortality relationship has been reported in enzootic areas (De Alwis, 1981). High mortality rates have been attributed to less immunization and stress related management and environmental problems leading to greater susceptibility to the disease. Molina *et al.* (1994) noticed the outbreak of disease in Philippines and suggested that work stress, adverse weather and poor nutrition played a role in the spread of disease. In Pakistan Yaqub *et al.* (1997) noticed an outbreak of the disease in buffalo calves.

In Asia alone, more than one lakh bovines die of the disease (Bain, 1957) while in India 33000 animals die annually (Dhandha *et al.*, 1956). In Nepal also, the disease is highly prevalent (Thakuri *et al.*, 1992). Malaysia suffered a loss of 2 lakh Malaysian dollars in 1976 as the disease resulted in 73% mortality in buffaloes (Joseph, 1979). Similarly Verma (1998) reported that in Malaysia 72% of all losses in livestock sector were due to HS in buffaloes. The death rates in buffaloes are three times more than cattle due to less immunization and stress factors leading to higher susceptibility to disease (Bain *et al.,* 1982). In India also about 30000–50000 deaths result from this disease annually (Verma, 1998).

The tonsillar region of the nasopharynx has been identified to be the main portal of entry of the organism whether through inhalation or ingestion. In favorable surroundings the organism can survive in the environment for a week. Carriers include cattle and buffaloes while reservoir hosts include pigs, sheep, goat and horses. In India, no carriers were detected in areas where its occurrence was low but in Sudan, a carrier rate of 3% was found in area free of the disease (Mustafa *et al.,* 1978). Naturally acquired immunity following sub-clinical infection has been reported to be both strong and long lasting. Bain *et al.* (1982) suggested that 10% buffaloes in field have naturally acquired immunity.

Symptoms

Clinically, the disease is characterized by sudden rise in temperature up to 42 °C, profuse salivation, severe depression, sub mucosal petechiae and death in about 24 hr. In the common throat form, there is a hot painful swelling of the throat, brisket or perineum and severe dyspnoea may occur. In the later stages of the outbreak some animals may develop pulmonary and alimentary involvement. A predominantly pneumonic form has been described in young buffaloes with partial immunity (De Alwis *et al.,* 1975). The animals die as a result of respiratory distress. In an outbreak in calves (Yaqub *et al.,* 1997) noticed high fever, respiratory distress, nasal discharge, excessive salivation, cyanotic conjunctiva and hot painful swellings of the throat.

Necropsy

Buffaloes show less variable pathological features due to the shorter duration of disease before death. The main lesions include edema of lungs, lymph nodes, glottis and peri-tracheal tissue. Generalized petechiae on the mucous membranes and the auricles may be noticed. Congestion with pneumonic changes is seen in the lungs and congestion from abomasum to large intestine is also seen. Pasteurella lipo-polysaccharide has been attributed to be the cause of intestinal lesions and clinical symptoms in natural infections (Bain *et al.,* 1982). In intestinal involvement, haemorrhagic gastroenteritis is commonly noticed.

Diagnosis

It is made based on clinical symptoms, demonstration of the causal organism in smears of blood/edema fluid, made immediately after death, since the organism disappears from dead animals fast. Sometimes negative smears may be obtained, so all suspicious cases must be confirmed to be negative only after cultural and biological tests. Rabbits are very susceptible, with death occurring in 24 hr of infection, showing haemorrhagic tracheitis and septicemia. Mice can also be used effectively for the same purpose. Biochemical studies of blood taken in the febrile stage show increased bilirubin, and other bile salts, decreased cholesterol level and changes in serum protein, all of which are indicative of hepatic insufficiency (Gupta *et al.,* 1977). A rapid ELISA has been used to diagnose specific serotypes responsible for the disease (Dawkins *et al.,* 1990). It is necessary to differentiate the disease from anthrax, leptospirosis and black quarter.

Treatment and control

Sulphadimidine is highly effective and earlier used to be the drug of choice and given at the dose rate of 0.5–1.0 gm/10 kg body weight S/C for 3–4 days. Combination of trimithoprim and sulphamethroxazole is highly effective when used @ 3–5 ml/50 kg body weight I/M for 4–5days. Chloramphenicol or oxytetracycline may also be used but these should be continued for a period of one week (Patil *et al.,* 1994). In severe cases, anti HS serum is also recommended. The cases can be treated with steroids or non-steroid anti-inflammatory agents and broad spectrum antibiotics as the treated animals show 100% recovery (Yaqub *et al.,* 1997).

For the control of HS, three types of vaccines are commonly in use, broth bacterin, alum precipitated and oil adjuvant vaccines. Killed vaccines with adjuvant are the vaccines of choice now a days. Alum precipitated vaccines are given S/C, which provide immunity for up to 5 months. Oil adjuvant vaccines are favored for systematic vaccination programs because they afford protection for over 6 months and may be up to a year. Vaccination is advisable a month before the onset of the rainy season. Primary vaccination is recommended at 4 months of age with a booster after 3 months and thereafter annual vaccination. A live streptomycin dependent mutant. *Past multocida* vaccine, which provides good protection with very few side effects, has also been developed. Recently, a live vaccine developed from a fallow deer strain (B:3, 4) is being used in Myanmar which provides immunity for more than a year (Verma, 1998).

13.1.1.4 *Brucellosis*

The disease also known as contagious abortion or Bang's disease, is characterized by abortion in late pregnancy and a high rate of infertility.

Etiology

Brucellosis is caused by *Brucella abortus* that are small gram negative, non-motile, non-sporulating coccobacillus organisms. They can be easily killed by heat, sunlight or common disinfectants. These organisms have a predilection for the gravid uterus, testicles, accessory male sex glands, lymph nodes, joint capsules and bursa. Prasad (1967) has also isolated *Brucella melitensis* from the genital tract of female buffaloes. These organisms remain in greatest concentration in the contents of the pregnant uterus, foetus and foetal membranes (Rajendran *et al.*, 1998).

Epidemiology

The disease is prevalent in buffaloes throughout the world and in India, its prevalence is high. In Punjab, its incidence has been reported to be 3.17% (Saini *et al.*, 1992) while in Bareilly of Uttar Pradesh its incidence was 5.41% (Gupta, 1997). In Murrah buffaloes, 16.25% serum samples were found positive by dot-ELISA for brucella antibodies by Rao *et al.* (1999). However in Egypt, its incidence was recorded as 19.9% in female buffaloes and 25% in male animals (Ahmed and Abd-El-Aal, 1996) but Ahmed and Abd-El-Aal (1997) noticed 2.61% incidence in females. In Iraq its incidence has been 4.6% (Hussain *et al.*, 1996) while in Bangladesh its incidence was 6.9% (Rahman *et al.*, 1997). It has also been reported in buffaloes in Vietnam (Sharma *et al.*, 1982 a). Infection occurs in animals of all age groups but persists only in sexually mature animals. Rahman *et al.* (1997) found no difference in incidence due to age or pregnancy status of animal as it was almost same in pregnant or non-pregnant animals. Infection is through ingestion of contaminated materials, which are usually from uterine discharge of an infected animal or its foetus. The organism can also penetrate intact skin or mucous membranes. Congenital infection can also occur but is seen in calves. The semen from infected bulls is a good source of infection. Contamination of udder during milking may also be a source of spread. The disease is of zoonotic importance as dairy workers, veterinarians or butchers may pick up infection and suffer from undulant fever.

Symptoms

The symptoms are primarily based on the immune status of buffaloes. Highly susceptible pregnant animals suffer from abortions after 6 months, retained placenta and catarrhal metritis. After one or two abortions, the animal may give birth to full term calves. In herds there is usually a 'storm' of abortions. In bulls there is epididymitis and orchitis involving one or both scrotal sacs. The testicles are enlarged and reveal painful swellings. In mild cases, sinovitis and painful swelling of affected joints are noticed.

Necropsy

Though the mortality rates are low, but on post-mortem examination edematous placenta, leathery plaques on the external surface of the chorion and necrosis of the cotyledons are observed. The lesions of primary pneumonia may be noticed in aborted foetus. These lesions are less severe than in cattle (Mamatelashvili, 1973).

Diagnosis

The cases can be diagnosed by history of abortions in late pregnancy and can be confirmed by isolation of the organisms from uterine discharges of an infected animal or the stomach contents and heart blood of the aborted foetus. Ziehl Neelsens staining is specific for confirmation of the organisms. Serological tests include the Rose Bengal test for herd testing and the standard tube agglutination test, spot agglutination test, ELISA, and CFT in individual cases. An intradermal test has been developed which has a sensitivity equivalent to the CFT. The Rose Bengal test is recommended for initial screening and the CFT or indirect ELISA for confirmation. Agarwal and Batra (1999) found ELISA more specific and superior test for its diagnosis while Rao *et al.* (1999) recommended the use of dot-ELISA for specific diagnosis of brucellosis in buffaloes. Batra *et al.* (1995) detected the brucella antigens in foetal stomach contents of aborted buffaloes by coagglutination and counter-immunoelectrophoresis tests. The disease should be differentiated from infectious bovine rhinotracheitis, listeriosis, leptospirosis, vibriosis etc.

Treatment and control

Treatment is difficult and not undertaken since organisms can multiply and remain alive in the cells of macrophage system. Sulpha drugs, penicillin or streptomycin alone are of little value. Chloramphenicol or combination of long acting oxytetracycline and streptomycin are useful in its treatment.

The method of test, slaughter of positive animals and vaccination of healthy animals are the best ways to achieve control of the disease. Cooperation at all levels is pre-requisite for its eradication (Rajendran *et al.*, 1998). However due to the persistence of agglutinating antibodies which interfere with serological testing of animals, vaccination is not resorted to except when the incidence of the disease is more than 20% in the herd. Calfhood vaccination with *Br. abortus* Strain 19 vaccine, between 4–8 months is carried out when the incidence is 5–20%. The only other vaccine with some currency is Strain 45/20 in adjuvant. The performance of a commercially available strain S45/30 bacterin is variable. A live vaccine (strain 82) has also been used in Russia (Aliev and Sultanova, 1980).

Provision of floor space, running space, lighting, ventilation and sanitation are important for control of the disease (Saini *et al.*, 1992). Greatest care must be taken in handling and disposal of aborted foetus, foetal membranes and uterine discharges etc. as it may serve as source of infection to other animals and human beings.

13.1.1.5 *Tuberculosis*

Tuberculosis is known as a chronic infection, which is characterized by the progressive development of tubercles in various organs of the body.

Etiology

The causative organism in buffaloes is Mycobacterium bovis, which shows minor cultural differences from cattle isolates (Mohan, 1968, Lall, 1969). However Australian workers (Hein and Tomasovik, 1981) have isolated *M. avium, M. fortuii* and *M. flavescens* also in addition to *M. bovis* from buffaloes. *M. bovis* are acid fast, non-sporulating and non motile, long slender rod shaped organisms, which may be slightly curved and may occur in small groups. These are moderately resistant to heat, desiccation and many disinfectants but can be killed easily by 2–3% cresol. Pasteurization also kills the organisms.

Epidemiology

The disease occurs world wide and is of major importance in dairy cattle. Buffaloes have been reported to be more susceptible with an incidence of 6.8% in comparison to 2.4% in cattle (Lall *et al.,* 1967). This has been contradicted by some workers, who said that the incidence of the disease is low in buffaloes when compared to cattle and Zebu cattle had a higher rate of incidence than European breeds (Bala and Sidhu, 1981). In Haryana and Punjab, its incidence varied from 3.8–4.4% (Kulshreshtha *et al.,* 1980) while in Australia, it was 8–25% (Mc Cool and Newton Tabret, 1979). It is seen in swamp buffaloes in Thailand also (Kanameda *et al.,* 1997).

The infected animals serve as the main source of infection to other animals since the organisms are excreted in the sputum, faeces, milk, urine, vaginal and uterine discharges and open peripheral lymph nodes. The organisms enter via inhalation or ingestion but inhalation is the common route of spread (Mc Cool and Newton Tabret, 1979). The newly born calves usually pickup infection by suckling of milk from infected dam. The morbidity rates are 20–50% but mortality rates are 60–80%. The infected buffaloes also serve as a source of infection to human beings.

Symptoms

Tuberculosis is a chronic disease that runs a course of few months to years. General malaise is the first observable sign in affected buffaloes. In the pulmonary form, low grade fever, inappetance, chronic dry husky coughing, progressive weakness and dryness of skin are observed. The coughing is pronounced after exercise, in cold weather or pharyngeal palpation. In later stages pleurisy and dyspnoea are also noticed. In the intestinal form, there is persistent diarrhoea. The affected mammary glands are painlessly enlarged and the supramammary lymph node may also be enlarged. The milk becomes watery and large numbers of organisms are present in it. Such animals show symptoms of mastitis also.

Necropsy

As pulmonary form is more common, there is involvement of the bronchial and mediastinal lymph nodes. Tuberculous granulomas are noticed in lymph nodes in the majority of cases. In the lung, miliary abscesses may extend to cause suppurative bronchopneumonia. Calcification of the lesions is less pronounced in buffaloes, which show a greater tendency of encapsulation.

Diagnosis

The cases can be detected on the basis of clinical symptoms and post-mortem lesions. It can be confirmed by the isolation of the organisms from tissues or excretions, or by animal inoculation test performed into laboratory animals like guinea pig. In live animals delayed hypersensitivity reactions like the single intradermal test (SID), short thermal test, Stormont test and the comparative test with tuberculin of various origins, are used for diagnosis. The reaction to this test is more pronounced in buffaloes than in cattle, and may last for more than 10 days. Stormont test is superior since it can be employed for initial herd screening, in advanced cases of tuberculosis, recently parturited animals, suspicious reactors as well as in differentiation from paratuberculosis and avian tuberculosis. The indirect hemagglutination test is reported to be a very sensitive test to detect early and advanced cases of tuberculosis, in which the tuberculin test failed (Shukla and Singh, 1972). However, these results have been contradicted to show that the tuberculin test is more effective by studies undertaken in Egyptian buffaloes (Salim *et al.*, 1978). An *in vitro* assay of cell mediated reactivity known as the 'Interferon-gamma assay' has been found to be superior to the SID test (Wood *et al.*, 1991). It should be differentiated from traumatic reticulitis, aspiratory pneumonia and contagious pleuropneumonia

Treatment and control

Treatment is of no value since it is very costly and to be continued for a long period. However, valuable animals may be treated with a combination of streptomycin and para-amino salicylic acid as a long-term treatment measure. Isonicotinic acid and dihydrostreptomycin treatment for 90 days has also been found to be efficacious (Portngal *et al.*, 1972). Isoniazid has also been reported to have some efficacy. However, a combination of 2.5 g streptomycin I/M, 1.5 g rifampin and 2.0 gm isoniazid given orally daily for 4–6 months has been reported highly effective.

Effective eradication of the disease is based on removal of infected animals, prevention of spread of infection and avoidance of further introduction of disease. So the test and slaughter technique, of all clinically positive and tuberculin positive cases is best policy for its control. A modification involves the slaughter of clinically positive cases and the separation of the rest into, tuberculin reactors and non-reactors. All the segregated animals are tested with tuberculin testing every 6 months and the reactor herd is reduced in size by rearing their offspring in clean premises, and following strict hygiene and sanitation practises.

13.1.1.6 *Johne's disease*

The disease is also known as paratuberculosis and is characterized by progressive weakness, enteritis leading to chronic diarrhoea and thickening and corrugation of the intestinal walls.

Etiology

It is caused by *Mycobacterium paratuberculosis*, which are acid fast and moderately resistant bacilli as they can survive up to one year in pasture. They can be killed easily by common disinfectants, heat and light.

Epidemiology

The disease has been reported from many countries in buffaloes all over the world. In Haryana, its incidence has been reported to be 3.6% (Kulshreshtha *et al.*, 1980). It runs a chronic course with high morbidity rate but less than 1% mortality rate. Economic losses mainly occur due to ill health, poor growth and reduced productivity and working efficiency. Though young animals are more susceptible, the clinical symptoms are seen mainly by 3–6 years of age since organisms grow very slowly. When adults pick up infection, they do not show symptoms. Stress and nutritional deficiencies influence the development of clinical signs. The disease is transmitted by the ingestion of contaminated food and water or by intra-uterine route. The organisms in the faeces of infected animals are excreted 15–18 months before the onset of clinical signs.

Symptoms

In affected buffaloes reduced working efficiency and productivity, submandibular edema and progressive weight loss are noticed in spite of normal appetite. There is chronic diarrhoea and faeces resembles like pea-soup but is without any odour. The disease runs a protracted course and terminates in death.

Necropsy

The lesions are mainly confined to the posterior part of the intestine. There are corrugations of the lining membrane and the associated lymph nodes are edematous and enlarged.

Diagnosis

Due to non-specific symptoms and non-availability of dependable tests, diagnosis is difficult. However history of persistent diarrhoea and continuous weight loss give an indication of the infection. Demonstration of the organisms by microscopic examination of the faeces or of the rectal mucosa also helps in its detection. However repeated examinations are necessary before declaring the animal to be negative. It can be confirmed by performing intradermal Johnin testing, but is not very reliable.

A CFT having a sensitivity of 90% and specificity of 70% and an AGID that has 96% sensitivity and 94% specificity are the most widely used modes of diagnosis of this disease (Sherman *et al.,* 1984, 1990). A *Mycobacterium phlei*–adsorbed ELISA, a lipoarabinomannan–antigen ELISA and a dot-ELISA have also been developed with good sensitivity and specificity (Tsai *et al.,* 1989; Mc Nab *et al.,* 1991). A molecular diagnostic probe using restriction endonucleases has been developed which is faster in diagnosis and is more accurate. However, no single test can detect all the cases of paratuberculosis. So for its accurate diagnosis, two to three tests should be performed. Lymphocyte transformation test is of value in detecting resistant animals.

Treatment and control

The treatment is difficult as organisms are mostly resistant to antibiotics and it requires prolonged treatment. Streptomycin has the maximum activity against the organism, but treatment of affected animals with daily doses of 50 mg/kg body weight causes only a transient improvement in clinical signs. A combination of dihydro streptomycin, rifampin and isoniazid has been found effective when used for 3 months.

Control of the disease is difficult due to long incubation period and lack of diagnostic test for field application. Eradication of infected animals and carriers and segregation and proper faeces disposal using biodigester technology can help to control the disease. A live attenuated vaccine has been used successfully in calves of less than a month of age (Wilesmith, 1982).

13.1.1.7 *Leptospirosis*

It is a disease of zoonotic importance and is characterized by interstitial nephritis, haemolytic anaemia and abortions.

Etiology

The pathogenic leptospira are classified into one species *Leptospira interrogans* containing over 100 serovars arranged in 23 serogroups. The organisms are tightly coiled rod shaped and very fragile spirochetes.

Epidemiology

The disease is prevalent worldwide. In Bulgaria, its incidence has been reported as 10% (Khalacheva and Sherkov, 1981) while in Vietnam, it has been found to cause abortions (Sharma *et al.,* 1982b). The wallowing habit of buffaloes and the fact that water sources are easily contaminated through the urine of rodents and wild life, which act as natural carriers, helps in perpetuating a transmission cycle. Calves are severely affected while adults show less severe forms. The organism gains entry through mucosal and cutaneous abrasions, from contaminated pasture and water by infected urine, aborted fetuses, and infected uterine discharges. In human beings it spreads through skin cuts or consumption of raw milk.

Symptoms

Buffaloes suffer from acute, sub acute or chronic form of disease. In acute form, fever, anorexia, acute haemolytic anaemia, haemoglobinuria, jaundice, petechial haemorrhages on mucosae, dyspnoea, abortions and blood in milk are noticed. In sub acute form, moderate fever, anorexia, dyspnoea and haemoglobinuria are observed. Abortions occur after 3–4 months of infection. In chronic form of disease, mild symptoms and abortions are recorded.

Necropsy

In the acute form, anemia, jaundice, haemoglobinuria and sub-serous and sub-mucosal hemorrhages are constant. There may be ulcers and haemorrhages in the abomasal mucosa. If haemoglobinuria has been severe then there may be associated pulmonary edema and emphysema. In the later stages there is progressive interstitial nephritis.

Diagnosis

The cases can be detected by clinical symptoms and confirmed by the isolation of organisms. Positive diagnosis in individual animals is difficult due to the rapidity with which the organism dies in specimens once they are collected and their transient appearance in various tissues. So specimen must be processed immediately after collection. Acute and convalescent sera taken 7–10 days apart should be submitted from each clinically affected animal or from those with a history of abortion. Sera should also be taken from 15–25% apparently normal animals. Herd based diagnosis is much more easier, since the probability of some animals having a higher titer of particular animals having the organism, are much higher.

Of all the laboratory tests the examination of urine samples for the organism probably offers the most profitable opportunity of demonstrating the presence of infection. Culture of the organism from urine, blood or milk by inoculating into special media or hamsters/guinea pigs can also be attempted. This method is slow and expensive. Of the serological tests, the microscopic agglutination test (MAT), is the most commonly used one. Other tests include ELISA and FAT.

Treatment and control

The treatment should be initiated before liver or kidney damage. Streptomycin @12–15 mg/kg body weight twice daily for 3 days is effective when given as soon as the signs appear. If organisms are present in urine, it should be given in double doses. Tetracyclines are also effective. Elimination of infection in carriers can be effected by a single dose of streptomycin @ 25mg/kg body weight. Animals which are severely affected with hemolytic anemia can be given blood transfusion @ 5–10 liters/450 kg body weight. As supportive therapy, liver tonics and haematinics are recommended for early recovery.

Though control is difficult due to involvement of wild animals and rodents, it can be attempted by observing strict hygiene and sanitation and by periodic testing in endemic areas and the treatment of the clinically infected and carrier animals. Vaccination with formalin inactivated bacterin with either aluminum hydroxide or Freund's complete adjuvant can be used The latter gives a better serological response. Vaccination should start at 4–6 months of age followed by annual revaccination. A live vaccine prepared from avirulent *L. interrogans* serovars *pomona* gives good immunity and is quite safe (Stahlheim, 1973). There is no cross immunity between *L. interrogans* serovars *pomona* and *hardjo* and in endemic areas, a bivalent vaccine should be used.

13.1.1.8 *Listeriosis*

It is an infectious disease of some zoonotic importance and characterized by encephalitis, abortion and septicemia or occasionally by mastitis or keratoconjunctivitis.

Etiology

The disease is caused by *Listeria monocytogenes* that has 13 serovars. However, serovars 4b, $^{1}/_{2}$a, $^{1}/_{2}$b and 3 are commonly isolated (Bosgiraud *et al.,* 1991). Virulent strains can be identified by their ability to multiply in macrophages and monocytes and their ability to produce a hemolysin, listeriolysin 'O'. The organisms are resistant to adverse climatic circumstances and can survive in pond water, dried faeces, damp soil, and contaminated grain for periods up to over one year. They are however susceptible to common disinfectants.

Epidemiology

The disease has a worldwide incidence. In India, its seroprevalence has been reported to be 2.04% in buffaloes (Adlakha and Sharma, 1992). The organism has an ubiquitous occurrence and is excreted in the faeces, nasal secretions, urine, uterine discharges, milk etc. and so can be isolated from farm slurry, sewerage, soil, surface water, plants, spoilt silage, fomites etc. The animals of all age group may be affected but adult buffaloes are more susceptible than young ones and buffaloes are less resistant than cattle (Khan, 1964). The occurrence of clinical disease is related to conditions of stress, which include nutritional, environmental, reproductive stress etc. Transmission can occur by several routes, which include ingestion, through intact mucosa, venereal transmission, inhalation or conjunctival contamination (Farber and Peterkin, 1991).

In human beings, it occurs as a sporadic infection or as a food-borne outbreak resulting in septicemia, meningoencephalitis and infection to other organs or abortions. Milk products or raw milk are incriminated for the disease in man (Thapliyal, 1998).

Symptoms

Among buffaloes, it produces two types of manifestations–listerial encephalitis or listerial abortions. The course of listerial encephalitis in adults is 1–2 weeks while in calves, death occur in 3–4 days. The affected buffaloes develop cranial nerve dysfunction and press their head onto fixed objects and there is unilateral facial paralysis. The animals are dull and there is deviation of the head to one side with poll-nose relationship undisturbed. The gait is unsteady and animals often exhibit blindness and sometimes show temporary excitement. In later stages, the animal becomes recumbent and die due to respiratory failure.

Mostly sporadic abortions occur due to the disease in buffaloes during the late 3rd stage of pregnancy. Retention of afterbirths is commonly noticed and in such cases temperature rises up to 40.5 °C. In the calves, septicemic listeriosis may occur in which case, emaciation, depression, fever and diarrhoea are noticed. In 3–7 days old calf, corneal opacity, nystagmus and dyspnoea have been observed.

Necropsy

On post-mortem examination, some degree of congestion of meningeal vessels and cloudiness of cerebrospinal fluid are recorded. Microabscess are noticed in brain stem. There is focal necrosis of the liver, spleen, endocardium and myocardium. In the aborted foetus, small yellow foci of necrosis in the liver, small abomasal erosions and yellow-orange meconium can be observed. The aborted foetuses are usually edematous and autolyzed. In aborted animals, placentitis and endometritis are also noticed.

Diagnosis

The diagnosis of disease becomes difficult as clinical symptoms or necropsy findings are not very characteristic. However, the isolation and identification of the causal organism are important in confirming the disease. Examination of CSF for presence of inflammatory cells and increased protein content are also helpful in the diagnosis. ELISA test developed for its diagnosis has also been found satisfactory (Miettinen *et al.*, 1990). The disease must be differentiated from other conditions in which nervous symptoms are manifested like rabies, acetonemia, brain abscesses etc. and those which cause abortion like brucellosis, vibriosis etc.

Treatment and control

Though the organism are resistant to many drugs, these are sensitive to chlortetracycline when given @ 10 mg/kg body weight I/V for 5 days. Penicillin when used @ 44000 units/kg body weight for 10–14 days has also been proved effective. The response of treatment is primarily dependent on its speedy commencement.

Avoiding stress especially in endemic areas effects control. The feeding of excess/spoilt silage must be avoided in these areas. Killed vaccines can be used to control the disease. Live attenuated vaccines have also been tried, however the results from both these vaccines are not very encouraging to adopt them on a regular and wider scale (Vagsholm *et al.,* 1991).

13.1.1.9 *Vibriosis*

The disease vibriosis in buffaloes may be wide spread as it has been recorded in India. It is caused by *Campylobacter fetus* (*Vibrio fetus*) and its three strains have been isolated (Prasad, 1967; Mohan, 1968). Infection is spread by the use of infected bulls for artificial insemination/natural service. The disease causes abortion between 4 and 6 month of gestation. Other features of the disease are temporary infertility characterized by repeat breeding and a prolonged dioestral phase. Isolating the organism from the uterine exudates and stomach contents of aborted fetus can make diagnosis. A mucous agglutination test has been developed and is used as a reliable diagnostic tool. The disease is usually self-limiting, with the infected animal developing immunity after one abortion. Adoption of good hygienic measures and screening of bulls can help to control the disease.

13.1.1.10 *Tail necrosis*

The drying of tail due to gangrenous necrosis is quite common in buffaloes. Though many factors like deficiency of essential fatty acids, microfilarial infection or degnala disease has been identified, but this condition is caused by *Corynebacterium bovis*. It begins with a swelling and inflammation at the tip of the tail, which gradually extends to the entire tail if not checked by amputation. The primary lesion becomes necrosed and a part of the tail sloughs off. The whole of the tail may also be lost as it is a chronic disease. Treatment usually consists of amputation of the affected part.

Gangrenous syndrome involving tail, ear tips and extremities has been reported to be caused by mycotoxins. Various fungi isolated are *Fusarium equiseti*, *Aspergillus flavus*, *A. terreus*, *A. ochraeus*, *A. niger*, *Penicilium citrinum*, *Curvuleria lunata* and *Botryodiplodia theobromae* (Kwatra, 1980; Manikam *et al.,* 1985) and mycotoxins causing the syndrome are aflatoxin, citrinin, ochratoxins and penicillic acid (Manickam *et al.*, 1985).

13.1.1.11 *Calf diarrhoea*

It is commonly seen in newly born calves and is recognized as one of the major causes of neonatal fatality.

Etiology

Though several bacteria, viruses and fungi have been isolated from the cases of calf diarrhoea, but the major causes are colibacillis and salmonella. The bacteria usually found associated with this condition include strains of *Escherichia coli,* Salmonella Typhimurium, *S.* Dublin, *S.* Newport, *S.* Bovis morbificans, *S.* Weltevreden, etc. Haemorrhagic diarrhoea has also been reported in buffalo calves due to *Clostridium perfringens* types A, C, and D (Acone *et al.*, 1970; Award *et al.*, 1979). In addition to it, *Pasturella*, *Klebsiella*, *Proteus vulgaris*, *Citrobacter* etc. have also been recorded (Farid *et al.*, 1976; Kaura and Sharma, 1981). Multiple infections due to enteropathogenic viruses and bacteria are more common than a single infection in a calf or a group of calves (Acres *et al.*, 1977). However Singh *et al.* (1993), Sunil-Chandra *et al.* (1994) and Sunil-Chandra and Mahalingham (1994 b) reported it to be caused by rotavirus while Iovane *et al.* (1995) isolated four type A strain of rotavirus from a severe outbreak of diarrhoea in buffalo calves. Byomi *et al.* (1996) isolated rotavirus, coronavirus and calcivirus from the cases of neonatal diarrhoea in buffalo calves.

Epidemiology

Mortality from salmonellosis ranges from 40 to 72% and that from colibacillosis 47% (Srivastava and Sharma, 1980). Overall 9.84% mortality in Surti buffalo calves was recorded by Patil *et al.* (1992). In Pakistan Khan and Khan (1996) reported 7.1% mortality while Khan and Khan (1997) isolated *E. coli* from 54–58% and salmonella from 13–14% cases of diarrhoea. *E. coli* incidence was significantly higher in the first week and salmonellosis in third week. Sharma *et al.* (1982c) and Singh *et al.* (1998) also found *E. coli* as a major cause of diarrhoea in calves. Maximum deaths were seen in third week of age and in summer. In Surti buffalo calves overall mortality was observed as 33.97% up to the age of one-year (Patil and Appannavar, 1998). Among these deaths the major cause was enteritis.

Symptoms

The affected calves suffer from profuse diarrhoea, sometimes with dysentery, straining during defaecation, cyanotic mucous membranes, depression, weakness, and incoordination of gait and severe dehydration. Sometimes, lungs may also be involved leading to development of pneumoenteritis. Usually there is increase in Hb, PCV and TEC suggesting haemoconcentration (Kumar *et al.*, 1981; Deshpande *et al.*, 1993) and lymphopenia, neutrophilia and hyperproteinemia (Kumar *et al.*, 1981).

Diagnosis

The cases can be detected by clinical symptoms and confirmed by isolation and identification of the causal organisms. Rotavirus induced calf diarrhoea can be detected by dot immunobinding assay (Kapoor *et al.*, 1996). However, Sunil-Chandra and Mahalingam (1994a) found that ELISA technique can be profitably used in the diagnosis of rotavirus infection in buffalo calves. Blood examination may also help in its diagnosis.

Treatment and control

The management of calf diarrhoea is based on alteration of diet, replacement of lost fluid and electrolytes and use of antibiotics and intestinal protactants. Effective control can be achieved by reduction of the exposure of the neonate to infectious agents, providing adequate colostrum and by increasing non-specific resistance, and increasing specific resistance by vaccination of the dam or the neonate. El-Said *et al.* (1996) recommended that buffalo dams should be vaccinated with ultracorn immunopotentiated K99 vaccine before calving to protect newborns against *E. coli*

13.1.1.12 *Mastitis*

It is referred as the inflammation of mammary gland resulting in physical, chemical or bacteriological changes in the milk of lactating animals.

Etiology

In buffaloes, large number of infectious agents causes the disease. The bacteria commonly associated with mastitis are different species of *Streptococcus* including *Str. agalactiae, Str. dysaglactiae, Str. faecalis,* and *Str. uberis*. The other bacterial organisms involved are *Staphylococcus aureus*, *E. coli*, *Corynebacterium pyogenes*, *Klebsiella*, *Mycobacterium tuberculosis*, *Pseudomonas*, *Mycoplasma*, *Bacillus* sp. etc. The incidence of streptococcal mastitis is higher than staphylococcal mastitis in buffaloes (Jafery and Rizvi, 1975; Misra, 1976; Kapur and Singh, 1978). However, Shukla *et al.* (1998) found *Staph. aureus* as major pathogen in buffaloes. *Mycoplasma laidlawii* has also been isolated from a buffalo suffering from mastitis (Rahman *et al.*, 1981). Cases of chronic mastitis have also been diagnosed in buffaloes which are caused due to botrymycosis, tuberculosis, actinomycosis or candidiosis (Nighot *et al.*, 1996).

Some of the fungi have also been isolated from mastitic milk of buffaloes. It includes *Candida krusei* (Monga and Kalra, 1971), *Candida tropicalis* (Pal 1997), *Cryptococcus ater*, *Candida stellatoidea*, *C. parapsilosis*, *C. quillermondi*, *Rhodotorula* sp., *Alternaria* sp., (Jand and Dhillon,1975), *Cryptococcus neoformans*, *Cryptococcus albidus*, *Candida albicans*, *Aspergillus fumigatus*, *Rhodotorula* sp., *Trichosporan* sp., *Torulopsis* sp., *Saccharomyces* sp., (Award *et al.*, 1980) and *Rhizopus nigricans*.

Epidemiology

Though lot of research work has been conducted on mastitis throughout the world, it still remains an important problem of dairy animals resulting in heavy losses to the livestock farmers due to reduced production and cost of management of disease. Buffaloes are susceptible to most of the organisms which cause mastitis in cattle. The disease is seen more in dairy buffaloes than swamp buffaloes. The main sources of infection are the infected udder and the environment, with infection taking place through milkers' hands, contaminated litter or milking machines. In Nepal its prevalence has been reported as 17.1% in Murrah crossbreds and 8.8% in local breeds (Joshi and Shrestha, 1995). In India, Chand *et al.* (1995) noticed 4.2% incidence of bacterial mastitis while in Pakistan, Chaudhury *et al.* (1994) found 36.3% incidence of streptococcal mastitis. Muhammad *et al.* (1995) found 10.34% incidence of *E. coli* mastitis. Youkhana (1997) noticed cases of acute non-suppurative, acute suppurative, subacute suppurative, chronic non-suppurative and chronic suppurative mastitis in buffaloes in Iraq.

The disease is more prevalent in animals that are housed in unhygienic places and in hot and humid environmental conditions. Mastitis due to trauma is less frequent although the udder and teats are pendulous in buffaloes. It is due to the fact that streak canal epithelium in buffaloes is thicker and compact and they have better developed smooth muscle sphincter around the streak canal which helps in its tighter closure (Uppal *et al.*, 1994). The neutrophil counts in buffalo milk is high due to which a large number of cells are available for phagocytosis and destruction of the invading pathogens (Silva and Silva, 1994). It may also be a probable cause of its low incidence in buffaloes. The rear quarters are more frequently affected (Chand *et al.*, 1995) and clinical mastitis is most frequent in the 4th lactation (Kapur and Singh, 1978). However Badran (1985) noticed it more in first lactation and Rasool *et al.* (1985) in sixth lactation. Joshi *et al.* (1997) noticed that incidence of clinical mastitis decreases with the advancing lactations and it usually affects rear right quarters. Haematogenous infection can also occur specially in case of involvement of mycobacterium organism. The prevalence of sub clinical and clinical mycotic mastitis in buffaloes has been reported in India as 3.7 % (Monga and Kalra, 1971) and 8.5 to 9.9% (Sharma *et al.*, 1977) respectively, while Mahapatra *et al.* (1996) noticed 8.1% incidence of clinical mycotic mastitis. The morbidity rates are quite variable but mortality rates are absent or very negligible.

Symptoms

From outer surface of the teat or from environment, the organisms penetrate inside teat canal and multiply there to produce the symptoms. The disease usually occurs in clinical or sub-clinical forms and shows all degrees of variation in signs, from the gradual onset of fibrosis, through acute inflammation. The type of inflammation and signs are dependent on the causative organism. In acute inflammatory reaction, significant alterations in milk quality occur as it contains large number of clots or flakes and there is change in its colour which may be even blood mixed sometimes. Udder reveals hot and painful diffuse swelling. In sub-clinical form, apparent symptoms are not seen and there is increase in leukocyte count in the milk. In coliform mastitis, the buffaloes reveal sudden onset of pyrexia, muscular tremors of head and upper hind legs, rumen atony and diffuse local swelling of the udder with straw-coloured watery non-odourous flaky secretion (Muhammad *et al.*, 1995).

In mycotic mastitis, the affected animals show reduced milk yield, loss of appetite, a mild rise in temperature, sub cutaneous edema of the udder, perineum and the sternal regions, swelling of the supramammary lymph nodes and a mucilaginous grayish secretion from the teats.

Diagnosis

Clinical mastitis can be easily detected by clinical symptoms and palpation of udder but sub clinical form is difficult to diagnose. An early diagnosis is very important for better management of cases. So the periodical examination of udder, strip cup testing, use of indirect tests like white side and California mastitis tests give an indication of the presence of disease. Dhakal *et al.* (1991) recommended somatic and viable cell counts and CMT for its diagnosis in buffaloes. Direct examination of milk for leukocyte count and bacteriological culture for identification of the causal organisms can also be undertaken. Measurement of electric conductivity by a conductivity meter or analysis of lactose content, N-acetyl-beta-D-glucosaminidase and lactate dehydrogenase in milk is reliable tests for detecting the disease. However ELISA has been found useful in detecting the cases of mycoplasmal mastitis (El-Shabiny, 1994). The cases of sub clinical mastitis can be detected successfully with trypsin inhibitor test as it can correctly diagnose

95% cases of sub clinical mastitis (Abdul-Samad *et al.,* 1997). The sensitivity and specificity of this test were 88.2 and 89.8%, respectively.

Treatment and control

The success of treatment is primarily based on the etiological agent involved, severity of disease and degree of fibrosis. Therefore, prompt treatment with antibiotics after a sensitivity test is indicated. For treatment of bacterial mastitis, large number of antibiotics are available. It should be selected on the basis of its activity against infective agent, diffusibility in mammary gland and cost. Though the treatment with antibiotics is recommended after sensitivity testing, in emergency cases broad-spectrum antibiotics like cephalosporins or penicillin G with other drugs may be infused. Muhammad *et al.* (1997) reported 97.4% efficacy of cephalexin in the treatment of clinical mastitis while Shukla *et al.* (1998) found 97.3% efficacy of aureomycin against mastitis in buffaloes. In cases of fibrosed udders, use of fibrinolytic agent along with antibiotics is advisable.

None of the antibiotic has been found effective against all cases of mastitis (Dahiya and Kapur, 1984; Paranjape and Das, 1986). Combination of ampicillin and cloxacillin has also been found highly effective (Uppal *et al.,* 1995). Gentamycin and cephalosporin were found highly effective against clinical and subclinical mastitis (Uppal *et al.,* 1998). However, Dhillon *et al.* (1995) treated the cases of mastitis with trisodium citrate given at the dose rate of 12 g/day in 250 ml water per buffalo. They suggested that citrate was essential for milk synthesis and calcium ions for sequestration. Tiamutin 10% oily injectable solution given @ 6 ml per quarter four times a day has been found 76% effective in its treatment (Khanna *et al.,* 1995). However, Dubal *et al.* (1999) noticed the residue of tiamutin in buffalo milk up to 48 hr after therapy. So the milk of buffaloes treated with tiamutin should not be used for 48 hr post-therapy. A combination of pefloxacin and dextrose has been recommended for its treatment as dextrose helps in preventing the physiological disturbances of carbohydrate metabolism and thereby helps in early recovery (Bagherwal and Shukla, 1996). The cases of coliform mastitis can be successfully treated with parenteral injection of norfloxacin, frequent stripping of affected quarter and oral administration of sodium salicylate and potassium nitrate. A herbal gel preparation has also been reported to be effective against subclinical mastitis (Joshi *et al.,* 1996).

Establishment of infection is encouraged by damage to the mammary gland and stimulated by antibiotic therapy. For example, *Candida* sp. utilize penicillin and tetracycline as a source for nitrogen. Hence the use of antibiotics in the treatment of mycotic mastitis is contraindicated. Treatment with iodides, either sodium iodide I/V organic iodides orally or iodine in oil as intra-mammary infusions show good results. Other drugs, which have efficacy in treatment, include, cyclohexamide, nystatin, polymixin B, neomycin, micanozole, clotrimazole and isoniazid.

Maintenance of strict hygiene in the milking sheds, proper sterilization of milking machines and other dairy utensils and adoption of other sanitation procedures along with the judicious use of antibiotics are of help in controlling the disease. Uppal *et al.* (1996) recommended the use of two teat dip treatments post-milking and dry herd therapy with ampiclox-DC for the prevention of mastitis in buffaloes. Though vaccination has given inconclusive results, it may be attempted to control streptococcal mastitis.

13.1.1.13 *Contagious bovine pleuropneumonia*

It is a highly contagious and septicemic disease characterized by high rise of body temperature and respiratory grunts.

Etiology

The disease is caused by *Mycoplasma mycoides* var *mycoides* (small colony type) which are pleomorphic, gram negative and highly fragile organisms. They cannot survive for more than 3–4 hr outside the host and are easily killed by heat treatment or by common disinfectants.

Epidemiology

The disease is a serious problem in large areas of eastern Europe, Asia and Africa. Buffaloes of all age group are equally susceptible but once infected, they become immune for subsequent infections. They are more resistant than cattle and show milder symptoms and have a higher rate of recovery than cattle (Sabry *et al.*, 1978). The spread of the disease is directly from infected animals by inhalation of droplet infection (Pathak, 1998). Outbreaks tend to occur more in housed animals and those in transit during humid conditions. Carrier animals also serve to perpetuate infection. The morbidity rates are comparatively high but mortality is almost negligible.

Symptoms

Incubation period varies from 3–6 weeks and occasionally up to 6 months. There is a sudden rise of temperature, up to 40 °C, anorexia, absence of ruminal movement, dullness, depression and a fall in milk yield. Coughing is observed first on exercise and then even at rest. There is pain in chest region due to which the animal stands with elbows abducted, back arched and head extended. The respiration becomes shallow and accompanied by grunting which is pronounced during expiration. Edema of the dewlap and throat is also seen. The auscultation of chest area reveals pleuritic friction sounds and moist rales. In recovered animals a sequestrum develops with a necrotic center, which commonly breaks down at times of stress resulting in an acute attack of the disease.

Necropsy

Animals die due to anorexia and lesions are primarily noticed in thoracic cavity. The lungs are consolidated giving a marbled appearance and may be red, brown or grey in colour. There is thickening and inflammation of the pleura with heavy deposits of fibrin and large amounts of clear serous fluid in pleural cavity. Adhesions between pleural surfaces may also be found.

Diagnosis

Diagnosis can be made based on clinical symptoms, anamnesis and necropsy findings. Isolation of organisms from serous fluid in thoracic cavity and identification of the organism can confirm this. Complement fixation and serum agglutination tests are highly useful in its diagnosis. Early cases may show a negative CFT and some positive reactors show no lesions on necropsy. The CFT is effective in detecting carriers also. Coloured antigens are commercially available which can be used for agglutination test. Intradermal allergic test has also been developed for its confirmation.

Treatment and control

Treatment is undertaken only in endemic area. Sulphadimidine and organic arsenicals have been used extensively and appear to reduce the mortality rate. Penicillin, streptomycin and tertacyclines are ineffective. Tylosin, @ 10 mg/kg body weight every 12 hr for 6 injections and spiramycin, @10–50 mg/kg body weight for 3 days are effective. Erythromycin given @ 20 mg/kg body weight twice daily for 4 days is also effective. However, maximum response occurs with tiamulin given @ 10 mg/kg body weight for four days.

Good hygiene and sanitation practises and removal of sources of infection effect control. Vaccination by a live attenuated vaccine administered at the tip of the tail can be given to calves after the age of 2 months. Vaccination reactions can be controlled with tylosin and spiramycin.

13.1.2 Viral diseases

13.1.2.1 *Ephemeral fever*

The disease is also referred as 'three days sickness' and affected buffaloes reveal hyperpyrexia, lameness and muscular stiffness.

Etiology

The ephemeral fever virus, which has four serovars and belongs to the family rhabdovirus causes it. This virus has affinity for the leukocytes-platelet fraction of the blood.

Epidemiology

The disease has been reported from many countries of the world and is endemic in Gujarat (Patel *et al.*, 1993). It primarily affects the adult buffaloes and calves below 6 months of age are not affected by it. Most of the cases occur in hot and humid environmental conditions (Patel *et al.,* 1993) when the mosquito population is comparatively high as the blood-sucking flies or mosquitoes primarily transmit the disease. It can also spread through blood transfusion. The morbidity rates are up to 80–90% while mortality rates are almost absent. Patel *et al.* (1993) recorded 1–2% mortality in affected buffaloes in Gujarat.

Symptoms

The virus after entry, localizes in the mesodermal tissue in the muscles and joints and affected animals reveal sudden rise in body temperature (41 °C), anorexia and reduction in milk yield (Patel *et al.,* 1993). Muscular area over shoulder, back and neck regions show swelling. Shivering, stiffness and clonic muscular movements are also noticed. Lameness becomes very prominent and typical posture of laminitis may be observed. Sometimes abortions in pregnant buffaloes may also occur. There is increase in heart and respiration rates with presence of nasal and ocular discharge. After three days, the animal becomes almost normal and start eating and body temperature becomes almost normal.

Necropsy

There is excess accumulation of serofibrinous exudate in synovial, pericardial, pleural and peritoneal cavities and the fluid has high number of neutrophils. The lymph nodes are swollen and enlarged and patchy congestion occurs on serous membranes. Abomasal mucosa reveals congestion and there is pulmonary emphysema and bronchitis. Degenerative changes in joints are also noticed.

Diagnosis

The cases can be easily diagnosed by clinical symptoms and can be confirmed by blood analysis, which reveals leukocytosis, neutrophilia, lymphopenia and increased fibrinogen. The disease can also be confirmed by agar gel precipitation, complement fixation, ELISA and fluorescent antibody tests (Gard *et al.,* 1984; Zakrzewski *et al.,* 1992). It is necessary to differentiate the disease from laminitis, parturient paresis and traumatic reticulitis.

Treatment and control

Usually the symptoms disappear in 3 days, so supportive treatment is recommended for affected animals. These are given the drugs to relieve the temperature and muscular stiffness. So paracetamol and phenyl butazone are given by parental route. To prevent secondary bacterial complications, antibiotics like strepto penicillin or tetracyclines should be used. Malviya and Prasad (1977) recommended the use of oxytetracycline, sodium salicylate and sodium bicarbonate for its treatment.

As no vaccine is available commercially against the disease, the only way to reduce the occurrence of disease is by adopting hygienic measures and reducing the vector population.

13.1.2.2 *Foot-and-mouth disease*

Foot-and-mouth disease (FMD) or aphthous fever is a contagious acute viral disease characterized by fever and formation of vesicles in the mouth and on feet.

Etiology

FMD is caused by an aphtho virus (family picornaviridae), which occurs in seven major serotypes A, O, C, SAT 1, SAT 2, SAT 3 and Asia 1. Of these, the predominant ones in India are, the serotypes O, A, Asia 1 and C. Dutta *et al.* (1983) isolated Asia 1, O and C while Hajela and Sharma (1978) reported that Asia 1 is more severe in buffaloes than in cattle. In Bangladesh, major outbreaks occurred due to A22 (Rahman and Prasad, 1985). Each serotype has several immunologically virulent subtypes of varying virulence. The virus is quite resistant to common disinfectants and heat treatment. However changes in pH, boiling at 100 °C and treatment with 1–2% NaOH kill the organizms.

Epidemiology

The disease is prevalent worldwide except Australia, New Zealand, Papua New Guinea and a number of European countries. The buffaloes of all age group are equally affected but it is more severe in calves. It is commonly noticed in buffaloes in Socialist Republic of Vietnam (Sharma *et al.,* 1985), Nepal (Thakuri *et al.,* 1992), Thailand (Chamnanpood *et al.,* 1994) and Egypt (Hosney *et al.,* 1996). The virus remains present in excretions and secretions as well as in blood of infected animals. It spreads by inhalation due to aerosol transmission (wind-borne infection can take place even up to distances of 250 km), ingestion of contaminated material or by direct contact. The pharynx is the first site of infection and replication. Carriers exist after recovery and vaccinated buffaloes may also serve as carriers if exposed to infection (Jarez *et al.*, 1979). The morbidity rate is up to 100% while mortality in adults is very low but in calves, mortality may be up to 20–25%.

Apart from the milk losses, there are losses like loss of draught power, loss due to permanent disability, loss due to death and loss due to cost of medical treatment. All such losses accounts for Rs 180 million each year and on per buffalo basis, the loss is of Rs 250 in India (Saxena, 1994). Of the losses, maximum loss occurs due to reduced draught power.

Symptoms

In adult buffaloes, the symptoms are mild in the majority of cases and they recover rapidly. Lesions occur mainly in the mouth and foot lesions are rare and mild in nature. Foot lesions in buffaloes have a scaly appearance initially which become vesicular later on (Gomes *et al.,* 1997). Occasionally in contact buffaloes are found to be resistant even when in direct contact with affected cattle. Sometimes a high rate of mortality is seen in suckling calves, which is however lower than in cattle. Milk production is affected up to 30% (Kassem and Soliman, 1966). The animals show high rise of body temperature, excess salivation and careful chewing movements. Small vesicle formation is noticed on buccal mucosa, dental pad and tongue while sometimes these may be observed in inter digital space also. Such lesions are seen on coronet also resulting in lameness (Dabas *et al.,* 1993).

Necropsy

The post-mortem lesions are relatively mild and are characterized by vesicular eruptions and erosions. The young calves affected with the malignant form of the disease, reveal extensive hyaline degeneration of the myocardium resulting in the characteristic 'tiger heart' appearance.

Diagnosis

Diagnosis can be made from clinical symptoms and confirmation can be done by the isolation of virus in cell culture and laboratory animals (mice and guinea pig) and then typed. Serological methods which can be used for diagnosis include CFT, ELISA, plaque reduction assay, virus neutralization, radial immunodiffusion and the virus infection associated antigen test. ELISA has been found to be more sensitive for demonstration of antibody levels than microserum-neutralization test

(Rahman *et al.,* 1988). FMD should be differentiated from vesicular stomatitis, mucosal disease, rinder pest and buffalo pox.

Treatment and control

Treatment is directed to the use of mild disinfectants and protective emollients on the lesions, administration of antibiotics to prevent secondary bacterial infection and the use of NSAIDS like flunixin meglumine to reduce inflammation. Use of non-specific immunomodulators like levamisole has been found useful as these agents stimulate the host immune response. Washing of foot and mouth lesions with 2% alum solution twice daily and use of long acting tetracycline @ 20 mg/kg body weight I/M once only have been found effective for its early recovery (Dabas *et al.,* 1993).

Slaughter and destruction are best achieved control (by burning or deep burial) of all affected and in contact cloven-footed animals. Chamnanpood *et al.* (1994) stated that simple quarantine of early cases during outbreaks is likely to be effective in reducing the spread of disease from one place to other. The thorough disinfection of all premises, equipment, etc. should also be done with 1–2% formalin or sodium hydroxide or 4% sodium carbonate solution. Vaccination should be done regularly with multivalent vaccines against the serotypes prevalent in that area. Vaccination twice in a year is recommended. The calves born from non-immunized dams are vaccinated first at 4 months and then at 8 months. However the calves born from immunized mothers are vaccinated at 6 and 10 months of age. In newly introduced adult buffaloes, vaccination is done immediately and then after 2 months followed by revaccination after every 6 months. The results of vaccination with virus of cattle origin are however, not as satisfactory as in cattle (Girard *et al.,* 1959). Allergic reactions have been reported with multivalent strains (Sharma *et al.,* 1979). The use of cell cultures of buffalo origin can be used to produce vaccine that may give better results.

Presently tetravalent oil adjuvant binary ethylenimine inactivated vaccine is available which can be given @ 3 ml I/M or S/C in buffaloes. This vaccine is to be repeated after 44–48 weeks and can be used in endemic areas as well as in containing the spread of disease during outbreak. In calves it is given at 1 month, 4 months and 1 year of age and then repeated after every 44–48 weeks.

13.1.2.3 *Rinderpest*

Rinderpest, also known as cattle plague, is an acute highly contagious viral disease of ruminants and is characterized by high fever, necrotic stomatitis, diarrhoea and high mortality rates.

Etiology

Rinderpest virus, which is a RNA agent, is a morbilli virus belonging to family paramyxoviridae causes the disease. It has several strains of varying virulence and almost identical immunological properties. The virus is very fragile and is readily killed by the use of common disinfectants, heat treatment or by drying.

Epidemiology

The disease has been eradicated in most parts of the world with the exception of certain African and Asian countries. A national eradication program implemented in India (1992–1996) lead to the declaration of India as rinderpest free. OIE has been monitering the status and has decleared India free from rinderpest. The natural hosts of the disease include cattle, buffaloes, sheep and goat. Susceptibility of buffalo varies from the highly susceptible species of the far East and the relatively resistant species of Egypt and Turkey. In India, the buffalo is thrice as susceptible to the disease as cattle (Edwards, 1927). All age groups are equally susceptible. In Pakistan, Asif and Akhtar (1997) recorded rinderpest antigens in 37.6% buffaloes and advised that there is an urgent need of mass vaccination of all ruminants against RP.

The primary mode of transmission is through direct contact, between sick and healthy animals, The greatest danger is the movement of ruminants and pigs out of the enzootic area. Virus excretion is limited to 2–3 weeks after infection and the virus is excreted through faeces, urine, conjunctival and oral secretions. The primary site of entry in natural infections is the nasopharyngeal mucosa. Sheep, goats and wild ruminants serve as reservoirs of infection. The morbidity and mortality rates are quite high.

Symptoms

The clinical signs depend on the virulence of the strain of the infecting virus and the resistance of the animal infected. After penetrating the upper respiratory epithelium, organisms multiply in the regional lymph nodes and disseminate to other lymph nodes, lungs and epithelial cells of mucous membranes. These organisms destroy lymphocytes and symptoms of the disease develop. There is sudden onset with high fever, anorexia, rough hair coat and depression. Mucosal lesions appear 2–3 days after the onset of fever. Diarrhoea, often haemorrhagic, sets in after the fever subsides. These animals become prostrated 7–8 days after the onset of these symptoms with temperature becoming subnormal a few hours prior to death.

There is high mortality, with some animals dying even without exhibiting any symptoms. The mortality rate recorded in India is about 77.5% (Mohan, 1968).

Cutaneous eruptions have also been recorded under the tail, on the perineum, around the udder, scrotum and between the legs (Joshi *et al.,* 1977). Abortions in the pregnant animals have also been noticed.

Necropsy

Characteristic lesions are seen around the mouth, in the intestines and the abomasum. The mucous membrane is extremely congested and the folds of the abomasum are thickened and become dark chocolate in colour. The lips and ventral surface of tongue have yellowish grey necrotic patches and erosive ulcers. Mucosa of colon has transverse zones of haemorrhage and erythema giving characteristic 'zebra shaped markings'.

Diagnosis

Apart from clinical signs and the nature of outbreak of disease, post-mortem lesions and detection of specific viral antigens in the tissues and isolation and identification of the virus can do confirmation. Serological techniques that can be used include, AGID (serum must be taken 3–5 days after the onset of fever for accuracy), CIEP, CFT, FAT, ELISA, immunoperoxidase, serum neutralization and the use of specific cDNA probes. Guha and Chatterjee (1999) found virus neutralization and indirect-ELISA tests quite sensitive in detecting the disease. The disease should be differentiated from FMD, haemorrhagic septicemia, bovine malignant catarrhal fever and bovine viral diarrhoea.

Treatment and control

Since it is a viral disease, there is no specific treatment and it should not be undertaken because of the danger of dissemination of the infection. However attempts should be made to replace lost fluid during diarrhoea by saline or lactated ringer's solution.

Disease can be controlled easily as virus cannot survive outside the host for longer period and various strains of virus have immunological identity. It is effected by good quarantine measures and the culling and slaughter of all infected and in-contact animals. For proper control, wild ruminants, which serve as reservoir, should not be allowed to come in contact of buffaloes and sheep and goat should be separated from buffaloes. In enzootic areas, control depends on the use of an effective vaccine and regular disease surveillance. The vaccine used should not cause the triggering off of latent infection and cause adverse reactions in the vaccinated animals. Tissue culture vaccine has been found to be suitable in buffaloes and the duration of immunity is for the whole productive life of the animal. Recombinant vaccines are also being developed and will replace the tissue culture vaccines now *in vogue.*

13.1.2.4 *Malignant catarrhal fever*

It is a highly fatal acute disease of buffaloes characterized by gastroenteritis, lymph node enlargement and encephalitis.

Etiology

There are 2 forms of the disease, one of which is caused by Alcelaphine herpesvirus-1 (AHV-1) which is a herpes virus and transmitted from blue wild beast. The virus has been named as bovid herpesvirus 3 also. The other form is associated with the domestic sheep and is caused by ovine herpes virus-2 (OHV-2).

Epidemiology

The disease has been reported from most of the countries where buffaloes are reared (Singh *et al.,* 1979; Hoffmann *et al.*, 1984; Wiyono *et al.,* 1994) and is mostly fatal in unattended cases. The morbidity may vary from 20 to 50% in susceptible herds and most of the cases are seen in late winter or spring months. All age groups are equally susceptible and buffaloes are more susceptible than cattle (Liggitt *et al.,* 1980). Inhalation of aerosol or ingestion of pasture contaminated with nasal or ocular discharge transmits the disease. Spread by direct contact is not possible. Wild beasts, wild ruminants and sheep are the reservoirs of the disease.

Symptoms

The disease is seen in head and eye or alimentary forms. In head and eye form, the course of disease is comparatively longer and buffaloes are suddenly infected and reveal high fever (41°–42 °C), anorexia, rapid pulse rate, reduction in milk yield, profuse purulent nasal discharge and severe dyspnoea. Discrete local area of necrosis are seen on the hard palate, gums and gingivae due to which mouth is painful and animal moves jaw carefully. The skin of muzzle is extensively involved and there is excess salivation. Nervous signs like weakness of a leg, incoordination, muscle tremors, nystagmus, head pressing, paralysis and convulsions may also be seen.

The alimentary form is more severe and occurs in peracute form for a period of 1–3 days. The animals reveal high fever, dyspnoea and acute gastroenteritis. There is marked diarrhoea, haemoconcentration and loss of weight. In an outbreak among buffaloes, Hoffmann *et al.* (1984) noticed hyperemia of skin, enlargement of lymph nodes and depression.

Necropsy

Haemorrhages and inflammation are noticed in mouth, pharynx and nasal cavity. In oesophagus longitudinal, shallow erosions are observed while intestines reveal catarrhal enteritis. The liver is swollen and shows evidence of degeneration. All the lymph nodes are swollen, edematous and often haemorrhagic. Serofibrinous epipericarditis and myocarditis were also observed by Hoffmann *et al.* (984) in an outbreak in buffaloes. Pop (1990) recorded congestion, haemorrhages, oedema and ulceration with suppurative, fibrinous and fibrinonecrotic exudate and sub-mucosal and glandular infiltration by lymphocytes and plasmocytes.

Diagnosis

The cases can be diagnosed on the basis of clinical symptoms and can be confirmed by necropsy lesions in dead animals. The isolation of the virus is difficult so the immunological tests could not be developed. It can be isolated from the buffy coat of infected blood in calf thyroid tissue culture or by transmission trials in buffalo calves. The disease can also be confirmed by polymerase chain reaction test (Wiyono *et al.*, 1994). DNA extracted from the formalin-fixed, paraffin-embedded tissue blocks can also be used for confirmation of the disease by PCR (Tham, 1997). Magnetic particles have been used to extract DNA from brain tissue of buffalo to confirm the disease by PCR by Ellis and Masters (1997). It should be differentiated from mucosal disease, rinderpest, infectious bovine rhinotracheitis and viral encephalitis or haemorrhagic septicemia.

Treatment and control

Though there is no specific treatment, use of non-steroid anti-inflammatory drugs may be given to relieve from discomfort. Supportive treatment is given in the form of broad-spectrum antibiotics and fluid therapy are also of some value. For the control of disease, buffaloes should be isolated from sheep flock.

13.1.2.5 *Rabies*

Rabies also known as lyssa, is a highly fatal viral infection of the central nervous system resulting in deranged consciousness, inability to swallow and paralysis.

Etiology

Rabies virus which is a rhabdovirus (genus lyssavirus) causes the disease. It is a truly neutrophilic virus and causes lesions only in the nervous tissue. It is relatively fragile and can be killed by standard disinfectants or heat treatment and dies in dried saliva in a few hr.

Epidemiology

It occurs in most countries of the world except islands, which are able to exclude it by rigid quarantine measures. Salem *et al.* (1996) diagnosed 11 cases of rabies in Egypt. Buffaloes like other mammals are susceptible to the disease (Srinunthapanth *et al.,* 1985). The source of infection is always an infected animal and the spread is invariably through a bite, though transmission can take place by contamination of a break in the skin with infected saliva. Ingestion of very large quantities of saliva may also produce disease. The virus is excreted in milk in low quantities and therefore, it may not produce disease. Though its occurrence in buffaloes is low but mortality is 100%.

Symptoms

Reports of the disease affecting buffaloes, are not numerous since they are more adopt at defending themselves from rabid animals than cattle. Incubation period is almost 3 weeks. The affected animals either show paralytic or furious form. There is drooling of saliva, eructation, grinding of teeth, continous tail movements, anorexia, stiffness of hind limbs, paralysis and recumbency. These recumbent animals die in 2–3 days in paralytic form. However in furious form, the animals become alert but tense and hypersensitive. They show sexual excitement, apparent inability to swallow, violent ramming of head on fixed objects and loud bellowing. Such symptoms persist for 36–48 hr and then animal collapses and die. Salem *et al.* (1996) noticed anorexia, ptylism, agalactia, hydrophobia, dilatation of pupils, frequent loud bellowing, restlessness, aggression, excitability, hypersensitivity to sound or touch, aimless running, incoordination, prostration, recumbency and death. In addition to these symptoms, Singh *et al.* (1999) noticed kinked tail, constipation, posterior paralysis and passing of wind from rectum with loosened sphincter. The skin from forehead was peeled off and serum oozed from that area.

Necropsy

Animals, which have died of suspected rabies, should have the following tests done on the brain tissue, FAT of impression smears of the hippocampus, medulla oblongata, cerebellum or gasserian ganglion, inoculation of infected material into mice and a histological search for negri bodies is to be made.

Diagnosis

The diagnosis of rabies is the most difficult but important duty of any veterinarian. Yang (1983) stressed that in places where laboratory diagnostic facilities are not available, an early diagnosis can be made from clinical signs. Impression smears prepared from brain can be tested by FAT for confirmation. CFT or ELISA can also confirm it. Animal inoculation test using mice is also of greater importance. Singh and Grewal (1998) found modified counter immuno-electrophoresis as most sensitive test for detecting neutralizing antibodies of rabies followed by direct immunofluorescence test. The disease must be differentiated from other diseases showing nervous symptoms like subacute lead poisoning, polioencephalomalacia, listeriosis, lactation tetany and deficiency of vitamin A.

Treatment and control

No treatment must be attempted after the onset of clinical signs. Immediately after exposure the wound should be irrigated with soap solution and water. Post exposure vaccination then can be given, depending on the nature and extent of risk involved. Euthanasia must be avoided and suspected animals must be kept under close observation. Control is effected by the destruction of wild fauna in and around animal holdings and the vaccination of all domestic cats and dogs being maintained at the premises. Live and inactivated vaccines both of chick embryo origin and tissue culture origin are available and can be used.

13.1.2.6 *Buffalo pox*

It is a mild viral disease of buffaloes and is characterized by the presence of typical pox lesions on the teats and udder. It was described under various names like variola-vaccinia, variola and vaccinia previously, but now it is named as buffalo pox.

Etiology

The disease is caused by buffalo pox virus of the genus orthopox of the family pox viridae. It resembles with vaccinia virus in structure (Bloch and Lal, 1975) and is resistant to inactivation by ether but sensitive to change of pH and action of chloroform and bile salts (Singh and Singh, 1967).

Epidemiology

Pox in buffaloes has been reported from many states of India, Indonesia, Italy, Pakistan, Russia and Egypt (Chandra *et al.,* 1987; Dumbell and Richardson, 1993; Muhammad *et al.*, 1998). In buffalo tracts, the disease usually occurs in endemic form though both generalized and localized forms have been recognized (Lal and Singh, 1973). Morbidity up to 70% has been reported but mortality rates are comparatively low. However, Ghosh *et al.* (1977) recorded 20% mortality in affected buffalo calves. The disease usually spreads through direct contact or by intradermal inoculation.

It is recognized as a disease of zoonotic importance since lesions can be seen on hands, wrist and thumb of human beings. Such human beings may show fever and involvement of regional lymph nodes and they pick up the infection during milking or handling of the infected case (Muhammad *et al.,* 1998).

Symptoms

The affected buffaloes suffer from fever (40 °C), anorexia, dullness, depression and congestion of conjunctivae. These animals show typical pock lesions mainly involving teats, udder and medial aspect of thighs. However in some cases, the lesions are noticed on lips and around eyes and nostrils also. The lesions are noticed either in localized form or in generalized form. In localized form, these are present on udder, teats and inside of thighs whereas in generalized form, these are seen on various parts of the body. In all such cases, there is development of mastitis due to secondary bacterial infection in teat. In young calves which suckle infected udder, purulent gingivitis may also develop.

Diagnosis

The cases can be detected by the clinical symptoms and can be confirmed by isolation of the virus from the lesions present on the teats or udder. It can also be confirmed by precipitation, complement fixation and neutralization tests but ELISA has been found to be more sensitive (Ghildyal *et al.,* 1986). For its diagnosis, buffalo pox virus antigens can be demonstrated by immunoperoxidase technique in cell culture and formalin fixed paraffin embedded tissue sections (Ghildyal *et al.,* 1986). Counter immuno electrophoresis has also been found quite sensitive, specific and rapid for its diagnosis (Mohanty and Rai, 1990).

Treatment and control

As it is a viral infection, so there is no specific treatment. However, to check the secondary bacterial contamination, antibiotic therapy with broad spectrum antibiotic must be initiated as early as possible. Care must also be taken to prevent the occurrence of mastitis in affected animals.

Presently no suitable vaccine is available against the disease so adopting strict hygienic measures can only control it. The human beings handling the infected animals, must properly clean their hands with quaternary ammonium compound or other strong disinfectant solution.

13.1.3 Mycotic diseases

13.1.3.1 *Dermatophytosis*

It is also referred as ringworm and characterized by alopecia, scab formation and pityriasis.

Etiology

Dermatophytosis in buffaloes is mainly caused by the fungus *Trichophyton verrucosum* (Azimov, 1963; Sharma *et al.,* 1980 and 1981). Other species, which are also associated with the disease in buffaloes, include *T. mentagrophytes* (El-Sherif *et al.,* 1981), *T. terrestre* (Gupta *et al.,* 1970) and *T. ajelloi* (Saxena and Mehra, 1973). The fungus can grow on hairs and produce spores which are resistant but moderate heat and drying can kill the spores.

Epidemiology

The disease occurs all over the world (Kamel *et al.,* 1977; El-Refai *etal.,*1979; Torky and Hammad, 1981; Pal and Singh, 1983). The infection is more severe in young animals especially those less than one-year old and kept in ill-ventilated, dark and humid pens. The incidence of the disease is higher in winter than in summer. The spread occurs primarily with direct contact on infected animals. Slightly alkaline pH of skin helps in rapid multiplication of the organisms. The infection has zoonotic importance.

Symptoms

The affected animals exhibit small circular, raised, greyish white patches of about one inch diameter on the face, neck, abdomen, limbs and other part of the body. The surface below scabs remains moist initially but is dried up later on and removed easily leading to development of alopecia. There is no itching.

Diagnosis

The disease can be diagnosed by clinical symptoms and can be confirmed by the isolation of causal organisms from periphery of lesions. Skin scraping examination also helps in its confirmation.

Treatment and control

The local application of tincture of iodine, 3% salicylic acid, 5% benzoic acid, Whitefield's ointment, 10% ammoniated mercury ointment, antibiotics like natamycin and nanomycin A as topical preparations and the oral administration of griseofulvin @ 15 mg/kg body weight, are effective treatment measures. Another systemic treatment measure, which is advocated in farm animal, is sodium iodide @ 1g/14 kg body weight as a 10% solution I/V 2 to 3 times in a week. The individual animal recovers quickly.

Control can be effected by isolation of infected cases, use of good hygiene and sanitation and application of disinfectants like 2.5–5 % phenolic detergents or 0.25% sodium hypochlorite solution. A Russian vaccine LTF-130 has been used successfully to control the incidence of bovine ringworm (Gudding and Naess, 1986). A live attenuated *T. verrucosum* vaccine can also be used to achieve similar results (Gudding *et al.,* 1991).

13.1.3.2 *Degnala disease*

It is an important fungal infection resulting in reduced functional or productive capacity of the affected animals. It has been described under different names in different regions of India like buffalo leprosy in Tamil Nadu, Kerala and Maharashtra; gangrenous syndrome in Punjab and grass poisoning in Assam.

Etiology

Though many workers have claimed that different etiological agents cause the disease, it has now been established that the toxins of fungus *Fusarium* sp. cause it. The disease occurs due to consumption of rice straw, as the affected animals recover quickly with the change of fodder. Apart from *Fusarium* sp. the fungi of other genera *viz Penicillium, Aspergillus, Alternaria* and *Curvularia* were also isolated from rice straw causing the disease. Some of these isolates like *F. equiseti* were found toxic by rabbit skin bioassay (Dandapat and Verma, 1998). In the affected rice straw, Bhatia *et al.* (1983) found different quantities of mycotoxins T-2, diacetoxyscirpenol, fusarenone-X and deoxynivalenol. Mouli (1991) also stated that the disease occurs due to feeding of either spoiled concentrate or paddy straw.

Epidemiology

The disease occurs commonly among buffaloes in India. It is also prevalent in Pakistan (Irfan, 1971; Khan *et al.,* 1981). The buffaloes are more susceptible and the disease is noticed in rice-growing areas mainly (Hussan *et al.,* 1994). A similar type of disease has been reported from China also (Wu *et al.*, 1991). From India, the cases have frequently been reported from Andhra Pradesh, Gujarat, Haryana, Kerala, Punjab, Uttar Pradesh and West Bengal (Dandapat and Verma,

1998). As rice straw is the major source of fodder in these states, the prevalence of disease is more in buffaloes in these areas. The disease results in serious economic losses due to mortality, decreased productivity and functional capacity.

Symptoms

The symptoms are mainly produced due to the toxins of the fungal agent. The affected buffaloes reveal diarrhoea or dysentery, ataxia and mucosal haemorrhages. At times emesis, posterior paresis, lethargy and hunger are also noticed. There is necrosis of muzzle, oral mucosa and extremities and gangrene and necrosis of limbs. Coagulative deficiency marked by internal haemorrhages and haemopoeitic suppression is frequent sequlae to ingestion of contaminated paddy straw. The growth rate of young calves is suppressed and milk yield in lactating animals is also reduced. Intermittent salivation and depression are also noticed in the affected animals.

Diagnosis

The disease can be diagnosed on the basis of clinical symptoms. However, its confirmation is primarily based on the demonstration of the causal agent or the toxins in the paddy straw. Tracheal organ culture and cell culture techniques are of value in demonstrating the causal factor while the toxins can be demonstrated by thin layer chromatography, liquid chromatography, gas chromatography, immunofluorescence or ELISA tests. The specific toxins can also be studied by nuclear magnetic resonance, mass spectrometry, UV spectroscopy, molar absorption coefficient, fluorescence spectra, melting point determination and species specific rotation data (Sydenham *et al.*, 1996). It should be differentiated from selenium toxicity that has some common symptoms.

The degnala disease is different from selenium toxicity because of low selenium content in paddy straw. The paddy straw has only 0.11 to 0.20 ppm selenium, which is insufficient to produce the disease. It is a disease of cattle and buffaloes only while selenium toxicity occurs in all the animals and birds. Selenium toxicity is a chronic disease while degnala is having an incubation period of up to two weeks only. Moreover selenium contents in the tissues of buffaloes died of degnala disease were found very low while in animals that died of selenium toxicity, had high selenium levels in tissues (Dandapat and Verma, 1998).

Treatment and control

If the disease is in the primary stage, the change of fodder and supplementation of green fodder can cure the sick buffaloes (Mouli, 1991). Dhillon (1973) recommended the use of arsenicals while for preventing secondary bacterial infection, broad spectrum antibiotics can be given. Degeure has also been found effective in its management. Of late, pentasulphate mixture given orally and 2% nitroglycerine ointment applied topically have been found highly effective (Maqbool *et al.,* 1996). The treatment of infected straw with 4% NaOH solution also helps in neutralization of the toxins and increases the palatability of straw (Maqbool *et al.,* 1996, 1998).

For the control of disease it is essential that the paddy straw should be free from the toxic fungi. So it must be stored in dry place after proper drying. The paddy straw suspected to be infected with fungi should not be fed to the animals and if it is to be used, it should be treated with 4% NaOH solution.

13.1.3.3 *Mycotic abortion*

Aspergillus fumigatus has been identified to be the main fungus causing mycotic abortions (Ainsworth and Austwick, 1973). Abortions are usually seen in the 6–7 month of gestation. The aborted foetus shows, discrete, raised ringworm like lesions on the skin of the head and back. The placenta is found to be thickened, haemorrhagic, edematous and necrotic. The diagnosis can be confirmed by the isolation of the pathogen from the stomach of the aborted foetus and placenta, demonstration of the fungal hyphae in foetal tissues and the detection of *A. fumigatus* precipitins in the sera of the dam. In the absence of any effective treatment, the housing of the affected animals in warm, dry, and well-ventilated sheds and the avoidance of moldy feeds for pregnant animals will help to reduce incidence of the disease.

13.1.3.4 *Pulmonary mycoses*

The condition is not very common among buffaloes. *Aspergillus* sp. has been recorded to be the cause of pneumonia in cases which had been extensively treated with cortisone and antibiotics (Gill *et al.,* 1977). The affected animals show dullness, labored breathing, anorexia, fever, nasal discharge, and weight loss. *Absidia corymibifera* has been identified to be the etiological agent in a case of mycotic pneumonia (Mandal *et al.,* 1981) and coccidiodomycosis has been identified to be the cause in some other cases (Ilahi *et al.,* 1966).

13.1.3.5 *Mycotic gastroenteritis*

Mycotic ulcerative gastritis due to *Rhizopus* sp. and *Candida* sp. has been reported (Kharole *et al.,* 1976). In it the gastric mucosa reveals severe congestion, haemorrhages, thrombosis and necrosis, besides infiltration of neutrophils and lymphocytes. *A. fumigatus* has also been isolated from a case of chronic diarrhoea, which was unresponsive to conventional therapy.

13.1.4 Parasitic diseases

13.1.4.1 *Fasciolosis*

It is commonly referred as liverfluke infection, distomatosis, liver rot or water poke and characterized by hepatic insufficiency and obstruction of the bile duct.

Etiology

Fasciola hepatica and *F.gigantica* parasites in buffaloes cause thedisease. These parasites are leaf-shaped and greyish-brown and remain in bile ducts.

Epidemiology

The disease is commonly seen in Myanmar, Egypt, India, Iraq, Japan, Pakistan, Philippines, Taiwan, Turkey, Vietnam as well as in Europe (FAO, 1977). Its incidence has been found up to 71% in Iraq (Griffiths, 1974), 79% in Bangladesh (Shaikh *et al.,* 1983), 68.5% in Thailand (Sukhapensa, 1980), 15.3% in Pakistan (Maqbool *et al.,* 1994), 22% in Vietnam (Ludng *et al.,* 1997) and 11.7% in Bangkok (Chaisrisongkarm, 1973). In Europe and higher altitude districts of India and Pakistan, the disease is caused by *F. hepatica* while in south-east Asia, in plains of India and the Indian subcontinent, it is caused by *F. gigantica*. Mixed infection with both the species has been observed on the boundaries of highland areas of Pakistan (Kendall, 1954). The incidence of disease varies and is dependent on many factors like age, location of raising the animals, management practises and sanitation (Sukhapensa, 1992). Two annual peaks of infection in September and February are usually noticed in (Maqbool *et al.*, 1994). Lymnaeid snails serve as intermediate host for these parasites. So spread of disease occurs through snails and its occurrence is more in rainy season. The animals of all age group are equally susceptible but Sukhapensa (1992) failed to observe any mature fluke in the buffalo calves of less than one year of age.

Symptoms

The buffaloes may suffer from acute as well as chronic form of the disease though chronic form occurs commonly. The young animals may pick up infection in early winters and show acute form of the disease. In such cases, anorexia, poor weight gain, pale mucous membrane, lack of vigour, debility and emaciation, constipation, dullness and weakness leading to prostration are seen. In advance stage, animals show diarrhoea and may die in 3–5 days if not treated (Kendall, 1953).

The chronic form of disease occurs due to ingestion of small number of metacercariae over a long period. Such buffaloes suffer from inappetance, moderate weight loss, sub mandibular edema, mild to moderate anaemia, pale mucous membrane, and loss of milk yield and chronic diarrhoea. There is progressive biliary cirrhosis and hardness of the liver area. It is recognized as a disease of great economic importance due to reduced working ability, decreased production and condemnation of infected liver.

Necropsy

Lesions are seen in the liver and bile duct. There is fibrosis and shrinkage of liver and biliary cirrhosis. Bile ducts become prominent, thickened, fibrous and calcified. If the liver is cut, a large number of flukes can be seen in the bile ducts.

Diagnosis

Clinical symptoms help in the diagnosis of disease, which can be confirmed by faecal examination for the presence of liverfluke eggs. However, in acute disease, eggs are not seen in faeces. Such cases can be diagnosed on blood examination, which reveals normocytic anaemia, low Hb and TEC and eosinophilia (Nguyen *et al.*, 1996), low albumin level and high activity of liver enzymes like OCT, GD, SD and ICD (Kumar *et al.,* 1982) . Blood examination also helps in diagnosis of the chronic cases.

Treatment and control

For its treatment, large numbers of drugs are available. However, some of the drugs are active against mature flukes while others are effective against immature and mature flukes both and such drugs should be used. For complete elimination of infection, Griffiths (1974) recommended use of 12-ml carbon tetra chloride in milk @ 0.04 ml/kg body weight to be given by stomach tube into the rumen. Three treatments at the interval of 2–2.5 months each were recommended. Kumar *et al.* (1983) found high efficacy of disophenol when used @ 1.2 mg/kg body weight S/C. Maqbool and Ifran (1983) reported efficacy of hexachlorophene, nitroxynil as well as oxyclozanide. Niclofolan @ 3 mg/kg, triclobendazole @ 12 mg/kg, oxyclozanide @ 10 mg/kg and closantal @ 7.5 mg/kg body weight given orally have been found highly effective (Sukhapensa, 1983; Maqbool *et al.,* 1994; Ludng *et al.,* 1997). Sanyal and Gupta (1996) found high efficacy of triclabendazole in buffaloes when given @ 1.5 mg/kg body weight intraruminally but it was ineffective @ 12 mg/kg body weight when given orally (Sanyal, 1996). Supportive treatment in the form of liver tonics and good nutritious diet helps in early recovery.

Control of the disease is possible either by killing the snail population or by preventing buffaloes from picking up the infective stage. Snail population can be controlled by using molluscicides or through destroying their habitats (Sukhapensa, 1992). For preventing picking up of infection, clean and hygienic water should be supplied and they should be allowed to graze only on higher ground. Buffaloes should be dewormed regularly after every 6 months with broad-spectrum anthelmintic. However, underdosing and repeated use of same anthelmintic must be avoided.

13.1.4.2 *Paramphistomosis*

Etiology

It is also known as stomach fluke disease and there are about 18 species of amphistomes, which occur in fore-stomach of buffaloes. Of the different stomach flukes, *Paramphistomum, Fischoederius* and *Cotylophoron* sp. occur in rumen; *Gastrothylax, Carmyerius, Calicophoron* and *Oliveria* sp. in reticulum and abomasum and *Gigantocotyle* in bile ducts (Sukhapensa, 1992).

Epidemiology

The disease has been reported from Myanmar, China, Egypt, India, Indonesia, Malaysia, Philippines and Sri Lanka (FAO, 1977), Pakistan (Khan and Anjum, 1994) and Afghanistan (Tenora *et al.,* 1974). Life cycle of these flukes is complex and requires water snails for the development of larval stages. The infection is transmitted by ingestion of metacercariae which are attached to the grasses around water sources. Metacercariae encyst on grasses and other objects in water and remain viable for considerable time. Though buffaloes of all age groups are infected but Sukhapensa (1986) could not record any adult flukes in calves less then one year of age. The number of adult flukes varies in individual animals.

Symptoms

Though a large number of adult flukes may be present in rumen or reticulum, they remain non-pathogenic. The disease is primarily produced by immature flukes, which remain in duodenum or other parts of small intestine or bile duct and produce necrosis and haemorrhage. Such animals reveal edema, profuse watery diarrhoea, anorexia, emaciation, anaemia, marked weakness and even death. In calves, Tenora *et al.* (1974) recorded dark-coloured, foetid diarrhoea, edema under jaws and anaemia. These calves suddenly collapsed due to dehydration.

Diagnosis

The cases can be detected by clinical symptoms and can be confirmed by seeing the eggs of flukes in faeces. Serological tests like rocket immuno electrophoresis (Manji *et al.*, 1998) or immuno diffusion test can also confirm it.

Treatment and control

Hexachlorophene given @ 10 mg/kg body weight orally has been found 89% effective against the disease (Panda and Misra, 1980) while niclosamide @ 50–90 mg/kg body weight is highly effective against immature flukes. Varshney and Singh (1975) found 87.5–100% efficacy of nitroxynil against clinical cases of paramphistomiasis when used @ 12.5 mg/kg body weight. By checking the snail population and preventing the buffaloes from grazing on infected grasses are helpful in controlling the disease. To control snail population, molluscicides should be sprayed or broadcast. The buffaloes should be grazed on higher, well-drained areas and grasses from marshy, low-lying areas should not be fed.

13.1.4.3 *Schistosomiasis*

It is caused by blood flukes and two types of the disease visceral and nasal schistosomiasis have been reported in buffaloes.

a) *Visceral Schistosomiasis*

The disease is caused by *Schistosoma spindale*, which is elongated, unisexual and dimorphic trematode. It occurs commonly in buffaloes in Myanmar, Cambodia, India, Malaysia, Pakistan, Thailand and Vietnam (Griffiths, 1974). The adult parasites are commonly seen in buffaloes of 4–10 years of age and Sukhapensa (1992) failed to record it in calves of less than one-year age. However, Pande *et al.* (1968) considered it as an important cause of enteritis in buffalo calves up to 2 years of age. Incidence of infection varies from 20–50% in different countries (Sukhapensa, 1992). Water snails serve as intermediate host. The animals pickup infection through penetration of infective stage larvae cercariae into skin or through intake of water containing cercariae.

These flukes occur in mesenteric and portal blood vessels and may reach the liver in heavy infection. Usually apparent clinical symptoms are absent but in heavy infection, anorexia, loss of body weight, weakness and anaemia are noticed. Cercariae can penetrate the skin of man also resulting into development of cercarial dermatitis, clam digger itch or swimmer's itch. At the site, there is mild erythema, edema and itching followed by pruritis and vesicle formation.

b) *Nasal Schistosomosis*

The disease is caused by *Schistosoma nasalis* and is widely prevalent in India while in Pakistan its incidence has been reported as 49% (Sukhapensa, 1992). The parasite inhabits the veins and venules of nasal mucosa resulting in development of nasal granuloma or snoring disease. Spread of the disease mainly occurs through snails (Sreetamulu, 1994). It causes inflammation of mucosa and there is mucopurulent discharge and proliferation of fibrous tissue and epithelial cells. Such animals usually show sneezing, snoring and dyspnoea and presence of pin head sized eruption and congestion of nasal mucosa (Rajamohan and Peter, 1975).

Schistosomiasis can be detected by seeing the eggs in faeces. For the treatment of nasal granuloma, antimosan given @ 1.7 mg/kg b wt daily for 6 days or sodium antimony tartrate @ 2 mg/kg b wt twice daily for two days have been found effective (Sukhapensa, 1992; Sreetamulu, 1994).

13.1.4.4 *Other fluke infections*

In addition to the *Fasciola* species, buffaloes have been reported to carry *Dicrocoelium dendriticum* infection in hills of India, Italy, Turkey and Russia and *Eurytrema pancreaticum* infection in pancreatic ducts in south-east Asia and Brazil (FAO, 1977). From the bile duct of buffaloes, *Gigantocotyle* species have also been reported from India, Iraq, Pakistan, Sri Lanka and some south-east Asian countries (Sukhapensa, 1992; Rao *et al.,* 1993). It produces lesions in bile duct and in heavy infection, thickening and cirrhosis of bile duct are noticed (Seneviratna, 1967).

13.1.4.5 *Toxocariasis*

It is primarily a disease of young calves resulting in poor growth and digestive disturbances.

Etiology

The disease is caused by *Toxocara vitulorum*, which was previously referred as *Neoascaris vitulorum* also.

Epidemiology

It is recognized as number one problem of buffalo calves, and results in heavy mortality in India, Brazil, Malaysia, Russia, Pakistan, Sri Lanka and Egypt (Griffiths,

1974). Though the buffaloes of all age group are infected yet the symptoms are seen in calves below three months of age. Le *et al.* (1997) found cases in calves below three months of age but not in animals below three weeks of age. They can be infected prenatally (Fernando *et al.*, 1981) or can acquire it by ingestion of colostrum containing larvae in large numbers (Tongson, 1971). The larvae are passed in milk of dam up to a period of 40 days after calving. In 74% of one-and half-month-old calves, worms were seen by Selim and Tewfic (1966) while Samizadeh-Yard *et al.* (1981) reported 23% incidence in Iran. The disease is more prevalent during summer in India as Chauhan *et al.* (1973) reported 55% occurrence from March to June.

Symptoms

The affected young calves reveal poor weight gain, rough and dry hair coat, anaemia, inactivity, exhaustion, diarrhoea and steatorrhoea. Usually the faeces are foul smelling and white or clay coloured. There is characteristic butyric odour in the breath. If the number of parasites is too large, they may block the intestinal passage and such calves suffer from constipation. Larvae may migrate to different tissues like liver and lungs also resulting in hepatic insufficiency, coughing or respiratory distress.

Diagnosis

The disease can be easily diagnosed by clinical symptoms and can be confirmed by faecal examination for the presence of eggs. However the eggs are seen only when the worms become mature. So for early detection of disease, serological tests have been developed and found satisfactory.

Treatment and control

Though a large number of drugs are known for the treatment of ascariosis, piperazine salts are in common use. Piperazine adipate when given @ 0.25 g/kg body weight, is 100% effective against immature and mature worms both. The drug is recommended to be given initially at two weeks of age and repeated twice after one month each. Apart from piperazine, pyrental @ 5 mg/kg body weight (Singh, 1974), fenbendazole @ 7.5 mg/kg body weight (Gautam *et al.*, 1976; Sukhapesna, 1992), levamisole (Singh, 1982), tetramisole, mebendazole (Le *et al.*, 1997), morantal tartrate (Gupta *et al.*, 1983) and thiophanate @ 70 mg/kg body weight (Sukhapensa, 1992) have also been found 100% effective. Knox *et al.* (1994) found considerably reduced systemic availability of fenbendazole in buffaloes and recommended that a higher dose rate of the drug should be used in buffalo than that in cattle. Kasarilikar *et al.* (1999) compared therapeutic efficacy of levamisole, piperazine and albendazole and found that all the drugs are effective. However levamisole hydrochloride was recommended @ 7.5 mg/kg body weight orally as it helped in early elimination of worms and quick recovery. Its cost is also low as compared to other drugs. Buffalo calves over 5 months of age are immune to the infection, these animals can eliminate their worms spontaneously (Sukhapensa, 1992).

For the control of disease, effective management practises should be adopted. It is essential that the eggs passed in faeces should be properly disposed off since they serve as source of infection to other calves or dams. The eggs can be killed by 2% lysol solution in 2 hr, by boiling water instantly or by exposure to direct sunlight in 1–3 hr (Chauhan *et al.*, 1978). The dams should not be allowed to lick the soiled body of their calves since they may serve as source of infection to calf in next pregnancy.

13.1.4.6 *Thelaziasis*

Etiology

It is also known as eye worm infection and caused by *Thelazia* parasite. From India, Pande *et al.* (1970) recovered *T. gulosa, T. rhodesii* and *T. skrjabini* from lachrymal duct of buffaloes. These are slender milky-white roundworms and males are 7–13 mm and females 12–16 mm long. The parasites are seen in conjunctival sac, lachrymal ducts, third eye-lid and corneal surface.

Epidemiology

Infection with *T. rhodesii* has been seen in countries of south east Asia and in Thailand, its prevalence has been found 12% (De Jesus, 1962) and 5–15% in Romania and Bulgaria (Griffiths, 1974). From India, Pande *et al.* (1970) found 35% buffaloes infected with either of 3 species. There is seasonal variation of disease as 59% cases are seen in November-February (winter), 44% in July-October (rains) and 41% in March-June (summer) and mature worms were greater in winter and immature during rains (Arora *et al.,* 1975). In India, *T. skrjabini* is the most frequently observed species as it has been seen in 97% cases followed by *T. rhodesii* in 8% and *T. gulosa* from 3% only (Arora *et al.,* 1975). The life cycle of parasite is complex and requires insects like *Musca* sp. as intermediate host for larval development.

Symptoms

Usually one eye of the buffalo is involved but infection of both eyes is also seen in some cases (Sukhapensa, 1992). Lacrimation, conjunctivitis and photophobia are common symptoms noticed in affected buffaloes (Udupa *et al.*, 1995). In acute infection, swelling of the conjunctiva, corneal opacity and mucopurulent ocular discharge followed by keratitis and corneal opacity are seen (Sukhapensa, 1992).

Diagnosis

The cases can be detected by seeing the worms in eye-wash or by removing the worm from the eyes.

Treatment and control

The disease can be treated by flushing the conjunctival sac repeatedly with clean water or 2–3% aqueous solution of boric acid. For flushing, 50 ml of aqueous iodine solution (0.5% iodine and 0.75% potassium iodide) or 0.05% lysol may also be used. Sukhapensa (1992) recommended use of several drops of 10% levamisole solution in eye for killing the parasite. Application of 20% butacaine sulphate ointment to the conjunctival surface relieves pain. Udupa *et al.* (1995) treated a case of *T. rhodesii* in buffalo with a single dose of ivermectin given S/C. For the control of infection, general hygienic measures and control of insects may be adopted.

13.1.4.7 *Filariasis*

Filariasis is an important infection among buffaloes and is caused by *Setaria*, *Stephanofilaria*, *Parafilaria* and *Onchocerca* parasites.

a) *Setariosis*

Etiology

The disease is caused by *Setaria* parasites. Though from buffaloes *S. labiatopapillosa, S. digitata, S. marshalli, S. marshalli pandei* and *S. cervi* have been isolated but *S. cervi* occurs most commonly (Kumar *et al.*, 1986; 1987; 1993; Kumar and Sharma, 1994). Adult setarial worms remain in peritoneal cavity and are non-pathogenic to buffaloes. These are several cm long, milky white in colour and taper towards the hind end. The larvae circulate in the blood and damage the tissues. So the disease is mainly produced by larvae, which is referred as microfilariasis (Kumar *et al.*, 1991).

Epidemiology

It occurs commonly in India and countries of south-east Asia (Griffiths, 1974) and

Vietnam (Sharma *et al.,* 1985 a) in adult animals of 3–10 years of age. Sukhapensa (1992) failed to notice and parasite in the buffalo calves of less than one year of age. Its prevalence depends on the ecological condition of the area since geographical regions with long, hot and humid seasons are conducive for the propagation of the parasite (Kumar *et al.,* 1993). The disease is transmitted through biting flies and mosquitoes.

Symptoms

There are no pathognomonic symptoms of the disease and affected buffaloes reveal inappetence, slight rise in body temperature, weakness and continuous weight loss. Such animals become debilitated, lean and thin and have rough and dry skin coat. There is watery discharge from nostrils and conjunctivae become congested and had mucopurulent discharge. The milk yield of lactating animals is continuously reduced and in some buffaloes, swellings of dependent parts and joints may be observed (Kumar and Sharma, 1994). The heart and respiration rates are elevated while rumen motility is reduced. In few cases, circling head pressing against fixed objects and other nervous symptoms may be seen (Pachauri, 1972).

Diagnosis

Clinical symptoms may give some idea about the disease and it can be confirmed by seeing the larvae in circulation by blood examination. The number of larvae remains variable in circulation so it is not possible always to get them on direct blood smear examination. The examination of buffy coat or filtration through nitrocellulose paper or membrane filters help in its diagnosis. Blood analysis reveals anaemia, high erythrocyte sedimentation rate, lymphocytosis and eosinophilia. Blood urea nitrogen, creatinine and activities of various liver enzymes are elevated. Serological tests like complement fixation, indirect fluorescent antibody, ELISA and blast transformation also help in the diagnosis of the disease (Kumar, 1985).

Treatment and control

For effective treatment of disease, it is necessary to remove adult and larvae both. Any of the broad-spectrum anthelmintic like mebendazole, tetramisole or levamisole is effective against the adult parasite (Verma *et al.,* 1994). However against the microfilariae, diethyl carbamazine citrate (10–25 mg/kg body weight for 10–15 days), lithium antimony thiomalate (0.3–0.4 mg/kg body weight I/M on alternate days for 10–12 days) or combination of both alternately for 10–15 days may be used. Ivermectin when given @ 0.2 mg/kg body weight S/C, was found highly effective (Sharma, 1991) both against adult and larval stages of parasite. Supportive treatments given in the form of iron, vitamin B complex and liver tonics help in early recovery.

Since the disease is primarily transmitted by flies and mosquito, the measures taken to check the spread and breeding of these vectors, will help in control of the disease.

b) *Stephanofilariosis*

Etiology

The disease is caused by *Stephanofilaria* parasite, which has three species *S. dedoesi, S. zaheeri* and *S. assamensis*. In addition, *S. andamani* has also been isolated from the naval-flap of buffaloes (Sinha and Das, 1958).

Epidemiology

It is prevalent in many countries in the buffaloes like in Myanmar, India, Indonesia and Thialand (Sukhapensa, 1992). The incidence in Thailand varies from 20–80%. In India, it is mainly prevalent in the eastern regions as majority of cases have been reported from Andaman Island, Assam, Bengal, Bihar, Madhya Pradesh and Orissa (Malviya, 1972). In Jabalpur, India, the incidence of *S. zaheeri* infection has been reported as 79% (Agrawal and Dutt, 1978). Maximum cases are seen in buffaloes of 2–4 years of age while calves below 5 months of age do not show ear sore lesions (Agrawal and Dutt, 1978; Sukhapensa, 1992). Poor body condition and high rainfall are predisposing factors in increasing the infection rate in buffaloes (Sukhapensa, 1992). The disease occurs during rainy season, and in winter the lesion become quiescent (Das *et al.*, 1975). Flies mainly *Musca*, *Stomoxys* or *Haematobia* are mainly responsible for the spread of disease.

Symptoms

In this disease, mainly skin lesions are noticed which involve inner surface of the pinna of the ear leading to development of dermatitis and the infected area is cyanosed. The disease caused by the parasite is commonly referred as ear sore and lesions have been classified as moderate, chronic and atypical (Agrawal *et al.*, 1978). The parasites mainly localize in and around hair follicles and sebaceous glands. In moderate disease, there is thickening and cracks in skin with or without exudation while in chronic cases, skin of the whole of the inner surface of pinna becomes granulated, dry and hard. Scattered black spots over the skin are seen in atypical lesions. In later stage of the disease, exudation is absent as lesions reveal encrustation, granulation and dryness.

Diagnosis

The cases can easily be diagnosed on the basis of clinical symptoms. It can be confirmed by demonstration of adult worms in lesions or microfilariae in the circulation. Like microfilariae of setarial worms, the larvae of these parasites are also not always present in sufficient number in circulation. So for the demonstration of the larvae, concentration methods may be used.

Treatment and control

A large number of drugs have been tried for the treatment of ear sore with variable efficacy. Parenteral injection of lithium antimony thiomalate and topical application of tartar emetic ointment helped in temporary relief (Ahmad, 1961) while two applications of 5% formalin or one application of 10% formalin have been claimed highly effective (Misra, 1969). Use of 4% sumithion ointment for 30 days has also been found effective in its treatment (Das *et al.*, 1975). With the application of 6% melathion ointment Agrawal and Dutt (1977) could cure 58% cases completely while Patnaik and Khan (1980) found trichlorophon as most effective treatment against the otitis external. By controlling the flies or mosquito population, the disease can be controlled.

c) *Onchocercosis*

The disease is caused by *Onchocerca* parasite and its four species have been reported. *O. armillata* is present in the intima of the thoracic aorta, *O. gutturosa* in ligamentum nuchae, *O. gibsoni* in subcutaneous nodules and *O. sweetae* in intrademal nodules in pectoral region. It is commonly seen in India and some countries of southeast Asia including Malaysia and Australia (FAO, 1977). Mosquitoes especially the *Culicoides* sp. serve as intermediate host (Spratt *et al.*, 1978). In *O. armillata* infection deeply corroded areas on the intima of thoracic aorta exhibiting complex sinuous tunnels and bluish elevated nodules are noticed while in *O gutturosa* infection, nodules of different sizes are seen in ligamentum nuchae. Some of these nodules are calcified also (Chauhan and Pande, 1972).

d) *Parafilariosis*

The disease is caused by *Parafilaria bovicola* and occurs in India (Patnaik and Pande, 1963). It has also been reported from the Philippines. The cases are mostly seen during summer and rainy season *i.e.* from April to October in adult buffaloes. The nodules of 2–3 cm diameter develop on the skin of neck and other body parts. Blood oozes from these nodules in bright sunlight and there is release of female worms from nodules. In the blood coming from the nodules, microfilariae can also be detected which helps in the diagnosis of the disease.

13.1.4.8 *Trypanosomiasis*

It is also referred as surra and results into development of intermittent fever, loss of condition and weakness of hind quarter.

Etiology

The disease is caused by *Trypanosoma evansi* in most of countries though *T. vivax* has also been reported from river buffaloes in America.

Epidemiology

The disease is widely prevalent in buffaloes in the Indian subcontinent and in a number of countries in south east Asia (Loses, 1980) with considerable economic losses in the form of morbidity, weight loss, reduced milk production and mortality (Juyal *et al.,* 1998). The parasite is transmitted mechanically by biting flies like *Tabanus*, *Stomoxys* and *Haematobia* and most of the cases are seen in rainy season when vector population is high. Singh and Joshi (1991) also recordedhigh incidence of disease from August to November in recently calved buffaloes. The mortality rate varies among buffaloes though acute form of surra becomes fatal in buffaloes (Anon, 1977). Outbreaks of disease have been reported from India (Chhabra *et al.,* 1978) while in Thialand, its prevalence varied from 2–13% (Sukhapensa, 1992). It occurs in acute form in Indonesia and Philippines also(Sukhapensa, 1992).

Symptoms

The buffaloes are usually symptomless reservoirs of the parasite or carry latent infection (Patchimasiri *et al.,* 1983). However, the clinical symptoms flare up when these are put to stress of adverse climatic conditions, vaccination or other infections including fascioliasis, filariasis, and haemorrhagic septicemia or foot-and-mouth disease. The affected animals reveal hyperthermia or intermittent fever (up to 41 °C), emaciation, conjunctivitis, anaemia, nasal discharge, loss of milk, dyspnoea and weakness of muscles of back and hind-quarter resulting in difficulty in movement and staggering gait (Lohr *et al.,* 1985; Singh and Joshi, 1991; Partoutomo *et al.,* 1994). Bilateral mucous discharge from the eyes, rough hair coat, enlargement of the lymph nodes, weakness of hind quarter and recumbency have also been noticed (Verma and Gautam, 1978). The disease may also occur in acute or per acute form resulting in hyperthermia and death within few hours. Such cases reveal very high degree of parasitaemia and petechial haemorrhages on the serous membranes of thorax and abdomen. In the buffaloes suffering from acute form of the disease, intermittent fever, anaemia, edema of the legs and other lower parts of the body, inappetence, loss of condition and weakness have been reported (Sukhapensa, 1992).

Necropsy

Emaciation, excess fluid accumulation in body cavity, anaemic changes, enlargement and congestion of spleen and liver, lymph nodes and gelatinization of heart and kidney are noticed in buffaloes died of the disease (Singh and Joshi, 1991).

Diagnosis

The clinical symptoms help in detection of the disease, which can be confirmed by seeing the organisms in blood smears. It can also be confirmed by animal inoculation test by injecting blood from suspected buffalo into rat or mice I/V. Serological tests like capillary agglutination, indirect fluorescent antibody and ELISA also confirm the disease. There is increase in IgG and circulating immune complexes in affected buffaloes (Juyal *et al.*, 1998).

Treatment and control

The cases can be treated either with suramin (3 mg/kg), quinapyramine sulphate (5 mg/kg) or samorin (0.25 mg/kg body weight) as recommended by Sukhapensa (1992). Of these drugs, samorin prevents reinfection also for a considerable time following treatment. Diminazine aceturate @ 10–15 mg/kg body weight (Mallick and Dwivedi, 1981) and isometamidium @ 0.5 mg/kg body weight (Singh and Joshi, 1991) were also found effective.Presently quinapyramine prosalt containing quinapyramine methyl sulphate and chloride salts is commonly used and is very effective @ 2.5 g I/M as single dose. The sulphate salt present in it, has therapeutic efficacy while chloride salt has prophylacting action. Supportive treatment given in the form of haematinics and liver tonics help in early recovery of the cases.

The disease can be controlled by checking the vector population and by preventing the buffaloes from getting bitten by *Tabanus* or *Stomoxys*.

13.1.4.9 *Babesiosis*

The disease is also known as red water disease and caused by *Babesia* parasite. It occurs rarely in buffaloes in China (Liu *et al.,* 1997) and in India, Tonkin, West Malaysia, Italy and Thialand (Griffiths, 1974; Sukhapensa, 1992), Vietnam (Sharma *et al.,* 1982 f; 1985 b) and Pakistan (Shabbir *et al.*, 1995). Callow *et al.* (1976) found antibodies against *B. argentina* in buffaloes by indirect fluorescent test while

Dwivedi *et al.* (1979) reported two cases of *B. bovis* infection among buffaloes. The affected buffaloes revealed anaemia, anorexia, dullness, depression, disinclination to move, severe haemoglobinuria and rise in body temperature up to 41 °C. Dyspnoea, reduced milk yield and low PCV, Hb, TEC and TLC are also noticed (Shabbir *et al.*, 1995). In China, the disease is reported to be caused by a new species *B. orientalis* (Liu *et al.,* 1997). Blood smear examination and serological tests like card agglutination, slide agglutination, ELISA, indirect fluorescent antibody and passive haemagglutination help in confirmation of the disease. Dwivedi *et al.* (1979) successfully treated the cases with diaminazine aceturate @ 6 mg/kg body weight. Imidocarb has also been found highly effective @ 1 mg/kg body weight S/C. As the disease is transmitted by ticks, control of tick population can help in controlling the disease.

13.1.4.10 *Theileriosis*

The disease is caused by *Theileria* parasite, which has many species. In East African countries, the buffaloes are highly susceptible to *T. parva* infection while in India, Egypt and Russia, it is caused by *T. annulata* and in these countries buffaloes are less susceptible than cattle and occurrence of sub clinical form is more common in buffaloes. In countries like Indonesia, Malaysia and southeast Asia, it is caused by *T. mutans* (Griffiths, 1974). In Thailand, no case could be detected by Sukhapensa (1992) while Burridge and Odeke (1973) successfully established *T. lawrenci* infection in 2 buffaloes experimentally. These buffaloes revealed marked fever and leukopenia and thrombocytopenia before death, which occurred by day 16. A clinical case reported by Chhabra *et al.* (1972) also revealed high fever (41.5 °C), dullness, emaciation and ocular discharge. In the absence of any treatment, the case died. Use of berenil and tetracycline has been reported to have some efficacy. Buparvaquone which is highly effective against theileriasis in cattle, can be used in buffaloes also @ 2.5 mg/kg body weight I/M. However, controlled studies are to be made. Supportive treatments given in the form of liver tonics and fluid administration are advantageous.

13.1.4.11 *Coccidiosis*

It is a protozoan infection characterized by debility, anaemia and enteritis.

Etiology

The disease is caused by *Eimeria* and its 12 species *E. alabamensis, E. auburnensis, E. bareillyi, E. bovis, E. brasiliensis, E. bukidnonensis, E. canadensis, E. cylindrica, E. ellipsoidalis, E. subsphaerica, E.wyomingensis* and *E. zurnii* have been isolated from buffaloes in India (Griffiths, 1974). However, one more species *E. ankarensis* was isolated from buffaloes in Turkey. Of these species, *E. bareillyi* has been found in buffaloes only (Sanyal *et al.,* 1985). These are intracellular parasites of epithelial cells of intestine and complete sexual and asexual multiplication in single host.

Epidemiology

The disease is commonly noticed in all the countries that have buffalo population. Sanyal *et al.* (1985) reported that in Hisar, India, 51% buffalo calves harbour coccidia while Sharma *et al.* (1982d) found it as a common problem in buffalo calves in Vietnam. The calves below 6 months of age are primarily affected and may occur in epizootic forms also sometimes (Biswal, 1948). It is more prevalent in humid wet climate and overcrowded stock. So November to February are highly favourable months for its occurrence. The disease spreads either by ingestion of contaminated feed and water or by licking of contaminated and soiled skin. The infected animals pass oocysts in faeces, which sporulate in favourable temperature and humidity and require oxygen. High environmental temperature and dryness are detrimental for the oocysts.

Symptoms

The disease results in heavy economic losses due to calf mortality and occurs in acute, sub-acute and chronic forms. Initially there is moderate fever but with the onset of diarrhoea, body temperature becomes normal or subnormal. There is sudden onset of foul smelling diarrhoea, which contains excess of mucus and blood. It may appear as a dark tarry staining of faeces or red streaks of blood or clots. Later on, anorexia, weakness, anaemia and tenesmus develop. In acute form of the disease, nervous symptoms like muscular tremors, hyperasthesia and convulsions are also recorded. If such cases are not treated in time, death occurs.

Necropsy

In calves died of the disease, lesions are mainly seen in intestines. In small intestine, congested and oedematous areas and irregularly margined white areas are noticed due to microgametocytes and epithelial schizonts. White pin-point to pin-head size cystic bodies representing schizonts and white raised opalescent areas or slightly raised whitish patches developed due to aggregation of forming gametocytes and oocytes are also seen. In large intestine, catarrhal enteritis is noticed and there is accumulation of blood mixed fluid and fibrinous clots.

Diagnosis

The cases can be diagnosed on the basis of symptoms and can be confirmed by necropsy findings in dead animals. Demonstration of oocysts on faecal examination also confirms the disease.

Treatment and control

The treatment recommended for cattle is usually followed. Sulphadimidine given @ 10 g as single dose, resulted in decreased severity of intestinal lesions (Patnaik, 1965). Sulphamenomethoxine, nitrofurazone and chlortetracycline have also been found highly effective (Li, 1981; Sharma *et al.,* 1982e). Amprolium when given @ 10 mg/kg body weight for 3 days was also effective (Sanyal *et al.,* 1985).

For proper control of the disease, strict hygienic measures, isolation of infected calves and avoidance of overcrowding are necessary. The use of coccidiostats at lower dose rate is also recommended but such treatment is continued for longer periods. Sulphadimidine @ 35 mg/kg body weight for 2 weeks, amprolium @ 5 mg/kg weight for three weeks or monensin @ 33 g/ton of feed for a month may be used to prevent the disease.

13.1.4.12 *Sarcocystosis*

Etiology

It is caused by protozoan parasite *Sarcocystis*, which has three species in buffaloes *S. levinei, S. fusiformis* and *S. buffalonis* (Dubey *et al.,* 1989; Huong *et al.,* 1997).

Epidemiology

S. levinei and *S. fusiformis* have been reported from Ankara (Dundar and Ozer, 1996) and Brazil, China, Egypt, India, Malaysia, Philippines, Romania and Turkey (Dubey *et al.,* 1989) while *S. buffalonis* is a new species reported from Vietnam (Huong *et al.,* 1997). In China and Taiwan, its incidence is 50% while in Hongkong it is 60% (FAO, 1977), in India 70–80% (Juyal *et al.,* 1982; Sherikar *et al.,* 1998) and in Thialand 56% (Sukhapensa, 1992). However Reddy *et al.* (1990) reported 40% occurrence of *S. fusiformis*. Dog is the definitive host of *S. levinei* while cat is of other 2 species. Parasite produces 2 types of cysts, microscopic sarcocysts are produced by *S. levinei* and is transmitted by dogs while *S. fusiformis* and *S. buffalonis* produce macroscopic sarcocysts and transmitted via faeces of cats. Sex has no influence on the incidence and cases are seen more in buffaloes over 5 years of age (Reddy *et al.*, 1990).

Symptoms

These parasites are known as non-pathogenic and no clinical signs of sickness have been observed in infected buffaloes. However, it is important since during meat inspection carcass parts infected with the cysts are condemned for human consumption (Sukhapensa, 1992). Cysts of *S. buffalonis* varies from 1–8 mm in size and present in skeletal muscles and esophagus while *S. fusiformis* cysts are 3–38 mm long and appear commonly in esophagus and less common in skeletal muscles. The cysts can also be found in tongue, diaphragm, ocular musculature and heart. The highest percentage of infection by Sherikar *et al.* (1998) was found in esophageal muscle followed by shoulder and diaphragmatic muscle and lowest in heart muscle.

Diagnosis

Diagnosis of the disease is difficult since it does not produce any clinical signs. However, sacocystis antibodies can be detected by IHA, ELISA, dot-ELISA and IFA tests (Dubey *et al.,* 1989; Sherikar *et al.,* 1998). Digestion technique is more frequently used for surveillance and diagnosis.

Treatment and control

Anticoccidials have been reported effective against the disease. Amprolium @ 100 mg/kg, salinomycin @ 1 mg/kg or halofuginone @ 0.22 mg/kg body weight have been found effective in other hosts (Dubey *et al.,* 1989). For the control of disease, it is essential to break the transmission cycle. So, carnivores should be excluded from livestock. Good management practices also help in its control.

13.1.4.13 *Anaplasmosis*

The disease is caused by *Anaplasma marginale* and occurs in China, Egypt, India, Indonesia, Philippines, Taiwan and Thailand (Griffiths, 1974). In Haryana and Punjab, its incidence was reported as 33% (Gautam *et al.*, 1970) while in Thialand it was 2.3% (Dissamarn and Thirapat, 1965) and 1.8% (Sukhapensa, 1992). Direct blood smear examination and serological tests help in confirmation of the disease. The symptomatology and line of treatment are not different from cattle; as Sharma *et al.* (1984) found single dose of berenil and 5 doses of oxytetracycline effective.

13.1.4.14 *Mange*

Mange is recognized as a serious skin disease of buffaloes and all the three types *i.e.* sarcoptic, psoroptic and demodectic mange have been noticed in buffaloes.

(a) *Sarcoptic mange*

It is commonly observed in buffaloes and caused by *Sarcoptic scabiei,* which is a minute parasite, roughly circular in outline and has very short legs. The parasite is also referred as sarcoptic itch mite or scabies mite. The disease occurs in all the countries in hot dry months, where buffaloes are raised (Sukhapensa, 1992). Though it does not produce generalized lesions, it may become fatal in young stock. Hazarika *et al.* (1995) found 14.8% incidence of the disease in calves. Calves below one year of age and male calves are likely to be infected more (Hazarika *et al.,* 1995). It is present where the hairs are short and skin is thinner like on the neck, brisket, axilla, abdomen, and inner surface of thighs and base of tail.

Skin of affected area becomes wrinkled and thickened and hairs fall resulting in itching. The animals try to rub the area with solid objects to relieve itching due to which wounds or bruises may be produced and secondary infection sets up. In heavy infestation of young calves, progressive emaciation, restlessness, weakness and death may be noticed. The cases can be diagnosed by skin scraping examination for the presence of mites. The disease can be transmitted to human beings also which show intense itching (Chakrabarti *et al.,* 1981).

(b) *Psoroptic mange*

In buffaloes, it is caused by *Psoroptes communis* and occurs commonly in Bangladesh (Mondal, 1994) and Myanmar, Egypt, India, Indonesia, Pakistan, Philippines and Thailand (FAO, 1977). In Italy, the disease is caused by *P. natalensis* (Manfredini and Marangon, 1992) while Puccini *et al.* (1986) recorded an outbreak due to *P. ovis*. The disease occurs in buffaloes of all age group and its incidence is higher during rainy season in India as Maske and Ruprah (1981) found 71% cases in July. These mites attack the skin at the root of horn and lesions are noticed on horn core. Due to its presence, animal feels irritation in horn and strikes it frequently with manger. Lesions can be seen on shoulder region and base of tail also. The disease can be detected by examination of skin scraping taken from the base of horn.

(c) *Demodectic mange*

The disease is caused by *Demodex bovis* and its incidence is low. The cases have been reported from Egypt, India, Indonesia and Malaysia (Griffiths, 1974). It produces pinpoint to 5 cm in diameter nodular lesions over the whole body.

Sarcoptic mange can be treated with 0.1% dieldrin, 0.25–0.50% asuntol, 0.15% neguvon or 0.05% lindane (FAO, 1977). Application of 10% sulphur suspension in liquid paraffin on clean affected skin has also been found very effective against it (Sukhapensa, 1992). This therapy is cheap and safe. Topical application of amitraz daily for 14 days, topical spray of 0.6% butox daily for 14 days or injection of ivermectin @ 1 ml/50 kg body weight S/C 3 times at weekly interval have also been found effective against sarcoptic mange (Kumar and Suryanarayana, 1995). Ivermectin given @ 0.5 ml/25 kg body weight S/C has been found effective against it (Purohit *et al.*, 1997). Against psoroptic mange, Ruprah *et al.* (1980) found 1% malathion or 1% carbaryl effective when applied at 5-days interval for a month. Ismail and Amer (1978) found high efficacy of 0.012–0.025% chlorpyrifos or 0.025–0.05% diazinon against both types of mange. Aqueous 0.1% suspension of coumaphos when applied once in a week for a month, cures both psoroptic and sarcoptic mange (Pholpark *et al.*, 1985). Two to three sprays of 0.03% gamma BHC suspension at 10 days interval have also been found effective against both types of mange in adults and calves (Griffiths, 1974). Pathak and Sharma (1988) reported efficacy of topical application of tetmosol for 2–3 days along with injection of *scabiezma* and a preparation of *karanj* oil in engine lubricating oil or other mineral oil.

13.2 Non-infectious Diseases

13.2.1 Indigestion

Difficulty or incompatibility in digesting food is referred as indigestion. It is of four types *viz.* simple, acid, alkaline and vagus indigestion.

13.2.1.1 *Simple indigestion*

Consumption of indigestible food material or minor alteration in gastrointestinal tract function results in simple indigestion, which is characterized by inappetance or anorexia, atony of forestomach and abnormal faeces.

Etiology

The disease is commonly noticed due to sudden change in diet or environmental condition. Buffaloes with history of depraved appetite consume indigestible materials like sand, soil, wood pieces etc. that result in indigestion. Poor quality roughage having low protein content when fed during draught along with restricted water intake results in indigestion (Akhtar and Anjum, 1997). The animals that have been under stress or fatigue for 12–16 hr and then offered concentrate or roughage usually suffer from the disease (Joshi *et al.*, 1999). Spoiled, damaged or mouldy feed or prolonged use of sulphonamides or antibiotics orally results in it. Excess feeding of sorghum or pearl millet stovers during lean period is also responsible for the disease (Ranjhan and Pathak, 1992).

Symptoms

Accumulation of indigestible food material results in abnormality of rumen motility. Any change in rumen pH also results in atony. Initially buffaloes show in appetance and reduction in milk yield. Subsequently they become dull and depressed and there is absence of rumination. The ruminal movements are depressed and low in amplitude while rumen becomes firm, doughy and without distension. Faeces is reduced in frequency and dried. However, after 24–48 hr, there may be diarrhoea. Though animals do not show any systemic reaction, there is moderate tympany. Most of the cases recover spontaneously or with simple treatment within 2–3 days.

Diagnosis

History of feeding and clinical symptoms helps in detection of the disease. It can be confirmed by examination of rumen pH. Rumen liquor can be examined for sedimentation activity and cellulose digestion tests. In indigestion, cellulose digestion time is increased and there is prolongation of floatation time. Blood examination of affected buffaloes reveals neutrophilia, lymphopenia and eosinopenia along with high glucose urea and total bilirubin (Akhtar and Anjum, 1997). Since regurgitation does not occur in indigestion, the increase in blood urea may be due to the failure of urea recycling through salivary glands and its non-utilization by microbes in the rumen (Singh *et al.,* 1989). Simple indigestion should be differentiated from ketosis, traumatic reticulo-peritonitis, acute grain engorgement, abomasal displacement, vagus indigestion and secondary ruminal atony due to hypocalcemia, allergy and anaphylactic states.

Treatment

Treatment is usually symptomatic and is attempted to restore the normal ruminal function. One of the most common lines of treatment is the use of saline purgative to evacuate the rumen. For it 400–800 g magnesium sulphate or magnesium oxide is dissolved in 3–4 litre of water and given orally as a single dose. Tartar emetic @ 10–12 g given orally is also beneficial. Alternately a mixture of 150 g magnesium sulphate, 150 g sodium sulphate and 15 g ginger powder dissolved in 750 ml water can be given once orally (Ranjhan and Pathak, 1992). Parasympathetic stimulants like physostignine or neostignine given @ 5 mg/100 kg body weight help in restoring rumen motility (Radostits *et al.,* 1994). Depending on the ruminal pH, acidifier or alkalizer may be given to restore the pH. Reconstitution of the rumen flora helps in early recovery of the case.

13.2.1.2 *Acid indigestion*

It is also referred as acute rumen engorgement, lactic acidosis, ruminal acidosis, founder, and grain overload or carbohydrate engorgement. The disease occurs due to excess intake of highly fermentable carbohydrate and characterized by toxaemia, ruminal stasis, recumbency and severe dehydration.

Etiology

The disease occurs due to accidental ingestion of highly fermentable carbohydrate diet such as wheat, barley or maize. It can be caused due to bread, sugarcane tops, potatoes etc. Wheat, barley and maize are most toxic while oat and sorghum grains are least toxic. The grains become more toxic when crushed or ground finely as starch content is exposed to ruminal flora (Radostits *et al.,* 1994). The morbidity may be 10–40% but mortality is around 90% in untreated cases.

Symptoms

Due to ingestion of highly fermentable carbohydrate, the population of ruminal microbes is changed and *Streptococcus bovis* population is increased. These bacteria produce large quantities of lactic acid resulting in low ruminal pH. When pH becomes 5 or low, cellulolytic bacteria and protozoa are destroyed and concentration of volatile fatty acids is increased resulting in further low in later stages, laminitis and lesions in brain are also developed.

Severity of the clinical symptoms in affected buffaloes is based on the amount and nature of grains consumed. Within few hours of ingestion, rumen is distended and there is abdominal pain and restlessness. It is followed by anorexia and soft faeces diarrhoea (Prasad *et al.,* 1972). Ruminal movements are absent and animal becomes recumbent. Temperature becomes subnormal while respiration and heart rates are elevated. If heart rate is increased to 120/min along with constipation, prognosis of such cases is poor. Severe dehydration, staggering gait, impairment of eyesight, anuria and acute laminitis develop in later stage. If the cases are not treated properly, they die within 72 hr.

Diagnosis

Proper history recording and clinical symptoms help in diagnosis of the disease. Examination of rumen fluid reveals acidic pH and absence of microflora help in its confirmation (Prasad, 1979). Urine is concentrated and has acidic pH. Blood examination reveals haemoconcentration, low blood pressure, high lactate and inorganic phosphorus and low bicarbonate levels. The disease should be differentiated from hypocalcemia and acute diffuse peritonitis.

Treatment

There are three principles for the treatment of disease *viz.* correction of ruminal and systemic acidosis and prevention of further production of lactic acid, restoration of fluid and electrolyte loss and restoration of ruminal and intestinal motility (Radostits *et al.,* 1994). In severe cases rumenotomy is required while mild cases recover spontaneously (Joshi *et al.*, 1999). For neutralizing lactic acid, 5 litre of 5% sodium bicarbonate solution is given I/V within 30 min followed by its 1.3% solution @ 150ml/kg body weight for next 6–12 hr. Alkaline aperient like magnesium hydroxide or magnesium carbonate can be used @ 300–500 g orally followed by 100–150 g after 12 hr. If evacuation of rumen contents is not possible, giving few litre of water and 20–50 million units of penicillin or 2–4 g chloramphenicol orally can decrease rumen hypertonicity. Use of antihistaminic and fluid therapy is also recommended in such cases. For early recovery of the case, liver tonics, rumenotorics and thiamin may be given as supportive treatment. During therapy, supply of water to the animal should be restricted.

13.2.1.3 *Alkaline indigestion*

It is also known as rumen alkalosis and is characterized by excess ammonia production in rumen, depression and decreased ruminal contractions along with nervous disturbances.

Etiology

The disease is caused due to feeding of urea or other non-protein nitrogenous substances along with inadequate supply of easily digestible carbohydrate. The disease develops when young forage grown with excess use of urea and ammonium fertilizers is fed to buffaloes. Overfeeding of wheat straw, putrefied fodder or urea treated paddy straw can also lead to it. Drinking of water from sewerage source or fertilizer industrial waste also result in it (Venkateswarlu *et al.,* 1998).

Symptoms

There is excessive ammonia production in the rumen which enters into systemic circulation resulting in hepatic, renal, circulatory and nervous disturbances (Venkateswarlu *et al.,* 1998). Affected buffaloes reveal inappetance, dullness, and distension of rumen and reduced ruminal contractions. The ruminal pH becomes alkaline and affected animals are reluctant to move and show dehydration and congestion of mucous membranes. Respiration becomes shallow and slow and there are muscular tremors. In terminal stage, there is dyspnoea.

Diagnosis

Proper history recording and clinical symptoms help in diagnosis of the case, which can be confirmed by examination of rumen fluid. It becomes brownish yellow to deep brown with watery consistency and alkaline pH. The number of large sized protozoa is increased and cellulolytic and sedimentation activities are absent. Amount of total volatile fatty acid also decreases.

Treatment

The diet responsible for alkaline indigestion should be changed and ruminal pH should be corrected. For it, 2–4 litre of 5% acetic acid or vinegar mixed in water may be given orally as a single dose. Rumenotoric helps in early restoration of ruminal motility while thiamin given intramuscularly also helps. If there is ruminal impaction, laxative or saline purgative may be used. Intravenous administration of saline solution helps in early recovery.

13.2.1.4 *Vagus indigestion*

Lesion involving vagus supply to the forestomach and abomasum are characterized by delayed passage of ingesta and ruminal distention.

Etiology

TRP and adhesions are the most common cause of disease. It can also be caused due to actinobacillosis of the rumen and reticulum, enlargement of lymph nodes due to tuberculosis or lymphomatosis and diaphragmatic hernia.

Symptoms

The major effects are due to failure of relaxation of reticulo-omasal and pyloric sphincters. Buffaloes suffer from inappetance, chronic indigestion, and ruminal atony and gradual reduction in milk yield. Scanty pasty faeces and gradual abdominal distention are seen. Due to accumulation of gas in rumen, it gives apple shaped appearance on the left side and pear shaped abdomen on the right side. The animals become progressively weak, eyes are sunken and progressive dehydration develops. Hair coat becomes rough and skin elasticity is lost. In such cases, temperature and respiration remain within normal physiological range.

Diagnosis

History and clinical symptoms help in the diagnosis of disease, which can be confirmed on blood analysis. As there is progressive dehydration, PCV, total plasma protein, blood urea nitrogen and creatinine levels are elevated. Moderate hypocalcemia, hypochloremia, hypokalaemia and alkalosis are seen. Neutrophilia with a shift to left are indicative of TRP. In the rumen fluid, chloride level is elevated.

The disease should be differentiated from abomasal displacement, chronic reticulo-peritonitis, ascites, rupture of the urinary bladder, peritonitis, intestinal obstruction and subacute abomasal torsion (Radostits *et al.,* 1994).

Treatment

Prognosis in most of the cases is unfavourable, as the animal does not respond to any treatment. Rumen fistula or rumenotomy for emptying the rumen contents be done. Balanced electrolyte solution @ 3–4 litre per day I/V and 1–1.5 litre mineral oil orally per day are helpful if given for 3–4 days. Use of 120–150 ml dioctyl sodium sulphosuccinate as 25% solution given I/M for 3–5 days also help in its recovery. The valuable pregnant animals should be maintained on electrolyte solution till parturition as they may recover spontaneously after parturition.

13.2.2 Ruminal tympany

It is also known as bloat, tympanitis, hoven or meteorism and refers to the over distension of rumen with gases and characterized by ruminal distention, abdominal pain and marked dyspnoea.

Etiology

Gas in the rumen may be present either in the free form (free gas bloat or secondary bloat) or trapped form (frothy bloat or primary bloat). Primary bloat is caused due to excessive intake of leguminous feed. These leguminous feeds have quality of foaming due to soluble leaf proteins present. Ingestion of finely ground grains also promote frothiness of rumen contents. Buffaloes that secrete less amount of saliva, are more susceptible to bloat (Joshi *et al.,* 1999). Heavy application of nitrogenous fertilizers to forage crops and pastures is also associated with the occurrence of tympany (Ranjhan and Pathak, 1992).

Secondary bloat occurs as a result of failure of eructation of free gas due to any physical interference. Choking of oesophagus by a foreign solid object, external pressure on oesophagus due to enlargement of mediastinal lymph node, spasm of oesophageal musculature, muscular atony or disturbance in nerve supply are the common cause of free gas accumulation in rumen (Radostits *et al.,* 1994).

Symptoms

In primary tympany, the gas bubbles remain dispersed in rumen inhibiting eructation. Foaming of ruminal contents and atony of rumen are important factors for its development. Fast growing legumes are easily digested and leaf mesophyll cells are ruptured resulting in release of chloroplast particles. Rumen microbes leading to trapping of air bubbles colonize these particles. In secondary tympany, gas bubbles coalesce and separate from rumen contents. However due to narrowing of oesophageal lumen or reticulo-rumen portion, these gases do not escape and rumen is distended.

The severe cases usually die suddenly without showing symptoms. In other cases after 8–12 hr of feeding, there is anorexia, dullness and signs of indigestion. Rumen is distended and ruminal movements present initially, are completely absent afterwards. Abdominal pain, restlessness, kicking at belly, rolling of animal and frequent lying and getting up are noticed. Dyspnoea and open mouth breathing, protrusion of tongue, extension of head, cyanosis and salivation are common signs. Heart and respiration rates are elevated.

Diagnosis

History of the case and clinical symptoms help in diagnosis of the disease. Passage of stomach tube can give some idea about the presence of an obstruction in esophagus. Exploratory puncture of left paralumber fossa also helps in its confirmation. However it is necessary to differentiate primary bloat from secondary bloat for proper treatment of the case. In primary bloat, ruminal distension is not highly prominent, rumen yields to pressure on palpation, highly resonant sounds are heard on percussion and gases come out intermittently on exploratory puncture. In secondary bloat, ruminal distension is very prominent, rumen is dense on palpation and drum like sounds are heard on percussion and gas comes out freely and continuously on exploratory puncture. The disease should be differentiated from oesophageal choke, diaphragmatic hernia, vagus indigestion and tetanus.

Treatment

The treatment depends either it is primary or secondary bloat. The feed responsible for it should be withdrawn. Acute secondary bloat can be relieved by use of trochar and cannula on left paralumber fossa while acute primary bloat can be relieved by rumenotomy. Running of animal and rubbing of tongue help in eructation. The drugs having carminative, antispasmodic, antiflatulent and stomachic properties are used for its treatment (Bhardwaj, 1998). To reduce the foam, antifoaming agents like vegetable oil may be used. Peanut, soybean, linseed, mustard or groundnut oils can be given @ 300–400 ml orally. For denaturation of stable foam, 200–250 g sodium bicarbonate should be used orally. Synthetic non-ionic surfactants like poloxane or dimethion silicon can be given orally once or twice daily depending on the severity of case. A mixture of 25 ml turpentine oil, 5 g *asafoetida* and 500 ml linseed oil or a mixture of 30 ml cresol and 500 ml of any vegetable oil given as a single dose is also very effective (Ranjhan and Pathak, 1992). Antizymotic treatment followed by rumenotoric for 2–3 days should be used for restoring the ruminal activity.

13.2.3 Traumatic reticulo-peritonitis (TRP)

It is also referred to as hardware disease, traumatic reticulitis or traumatic gastritis and characterized by anorexia, ruminal stasis, abdominal pain and sudden reduction in milk yield.

Etiology

The disease occurs commonly in adult buffaloes due to ingestion of sharp foreign objects like wire piece, nails or other sharp metallic object. Non metallic objects like sharp bones or wood pieces may also produce the disease (Kumar *et al.,* 1999). Plastic bags, nylon ropes and rags may also result in the similar type of disease.

Symptoms

Metallic or non-metallic foreign object may lodge in the reticulum and remain there without producing harmful effects. Reticular contractions may push these objects further resulting in abdominal pain, ruminal atony and local peritonitis. These foreign objects may penetrate the pericardium, pleura, spleen or liver also or there is rupture of gastro epipolic artery resulting in sudden death due to internal haemorrhage. The plastic bags, nylon ropes or rags are lodged in the rumen and block the flow of ingesta (Joshi *et al.,* 1999). Reticulum is more vulnerable for the settling of foreign bodies followed by rumen (Kumar *et al.,* 1999).

There is sudden onset followed by anorexia, decrease in milk yield, arching of back, abduction of forelegs, abdominal pain and reluctance in movement. Rumination is suspended and rumen contractions are either decreased or absent. Body temperature and pulse rate are elevated while reticular grunt can be heard on auscultation. Pinching of withers causes depression of the back and deep palpation of xiphoid area incites pain.

Diagnosis

Recording of history and clinical symptoms help in diagnosis of disease while metallic objects can be confirmed by metal detector or by X-ray examination of xiphoid region. Blood analysis reveals leukocytosis, neutrophilia and shift to left, monocytosis and increase in fibrinogen level.

The disease should be differentiated from simple indigestion, rumen impaction, tympany, abomasal displacement, pyelonephritis, ketosis and ephemeral fever (Radostits *et al.,* 1994).

Treatment

Surgical removal of the metallic or non-metallic object help in the management of disease. Use of high doses of antibiotics like ampicillin @ 5–10 mg/kg body weight or streptopenicillin @ 2.5 g should be given for a week. Corticosteroids like dexamethasone @ 20–25 m and antihistaminic like promethazine hydrochloride @ 0.3–0.6 mg/kg body weight are also useful. Feeding of cylindrical or bar magnet helps in prevention of the disease as these magnets are lodged in rumen and attach metallic objects to themselves preventing the object to pass to reticulum or other organ.

13.2.4 Foreign body syndrome in calves

The buffalo calves when weaned at birth and maintained on pail milk feeding and housed in groups, they develop the vice of licking each other. Such habit is usually developed as a result of depraved appetite caused by the mineral deficiency. Due to licking of each other, the hairs are ingested by the calf and these hairs start coiling into round structures usually referred as hair balls (Pathak and Sharma, 1988). Formation of hair balls is enhanced when calf suffers from gastric disorders such as impaction and stasis of stomach. In the initial stage, these balls are small and loosely knit but subsequently, they become larger in size and hard due to deposition of calcium and other minerals. There may be more than one ball in the rumen or reticulum and it results in recurrent digestive abnormality in calves. When such balls are lodged in omasum or abomasum, the condition becomes more serious. Sometimes they may choke pylorus also resulting in retention of food material in digestive tract. Such food is later on putrefied and polygastric mucosa reveals ulceration. The affected calves die due to starvation and toxaemia (Sharma *et al.,* 1983a). These calves can only be treated by removing of hair balls by rumenotomy.

13.2.5 Abomasal displacement

Abomasum of buffalo is a sac like elongated organ, which remains on the lower right quadrant of the abdominal cavity. It can move either in the left side or right side resulting in development of either left side or right side displacement of abomasum. In left side abomasal displacement, the abomasum is trapped in between rumen and left abdominal wall and is characterized by inappetance, reduction in milk yield and abdominal distension in left side. Right side abomasal displacement is also known as dilatation of abomasum and characterized by depression, dehydration and right side abdominal distension.

Etiology

Left side abomasal displacement is common in large sized, high producing buffaloes immediately after parturition, though occasional cases are seen before parturition also. About 91% cases have been seen within 6 weeks following parturition. Feeding of low roughage and high concentrate diet is associated with the disease or it can also occur due to rolling of the animal. In 88% of the cases, abomasum is displaced to the left side while right side displacement is seen in 12% cases only (Radostits *et al.,* 1994). The disease is common in adult buffaloes than the younger ones.

Right side displacement of abomasum is a sub acute disease, which occurs 3–6 weeks after parturition. Grain feeding and stress of parturition result in atony of abomasum. It can also occur either due to obstruction of pylorus or primary atony of abomasal musculature.

Symptoms

Due to feeding of concentrates in large amount, rumen ingesta flow to abomasum is increased resulting in abomasal distension and left side displacement. Feeding of excessive concentrate also results in increased amount of volatile fatty acid in rumen and decreased smooth muscle motility and excess gas production which further helps in the displacement of abomasum in left side. Affected buffaloes reveal inappetance or anorexia, sharp decline in milk yield, left abdominal distension and secondary ketosis. Scanty soft faeces are noticed while rectum looks empty on rectal examination. Temperature, respiration and pulse rates remain almost normal while ruminal movements are decreased both in intensity and frequency. Auscultation of left paralumber fossa reveals splashing sounds.

Due to atony of the abdominal musculature, gases and fluid are accumulated in abomasum resulting in gradual distension and displacement in caudal direction in right side. The buffaloes show inappetance, poor milk yield, depression, dehydration and polydipsia. Temperature, heart and respiration rates remain normal while mucous membranes become pale and muddy. Faeces are soft, scanty and dark coloured. Rumen becomes atonic while distended abomasum may be palpated behind and below the right coastal arch. In terminal stage, animal becomes recumbent and dies due to shock and dehydration and rupture of abomasum (Shelke *et al.,* 1996).

Diagnosis

History and clinical symptoms help in diagnosis. Palpation, percussion and auscultation of left or right side of flank can confirm it. Exploratory puncture of the distended portion also helps in its confirmation since the abomasal fluid is highly acidic (p^H 2–3) and has no protozoa. Blood analysis reveals increased ketone bodies while glucose remains normal. There is increased Hb, PCV and total protein, decreased potassium and chloride and metabolic alkalosis. The disease should be differentiated from ketosis, TRP and vagus indigestion.

Treatment

Abomasum can be put to normal position by surgicaloperation. For non-surgical correction, animal is cast on left side and then by tying the feet, this is rolled from side to side in 70° angle by putting on dorsal recumbency. Such rolling and manipulation brings abomasum in normal position specially if it is done after keeping buffalo off feed for 2 days to allow rumen to shrink. Secondary ketosis can be cured by giving 180 ml propylene glycol orally or glucose intravenously. To evacuate abomasal contents, 2–3 litre of mineral oil or 500 g magnesium sulphate can be used orally for 2 days. Alkalosis can be corrected by using potassium chloride (108 g), ammonium chloride (80 g) and water (20 litre). Up to 10 litre of this is given in 4 hr followed by 100–150 ml/kg body weight over 24 hr I/V. Potassium chloride @ 50 g/day should be used orally until buffalo becomes normal. Feeding of grains should be avoided for 3–5 days.

13.2.6 Impaction of omasum and abomasum

Since buffaloes are commonly fed large quantities of poor quality dry roughage, they usually suffer from impaction of omasum and abomasum. The affected buffaloes reveal anorexia, dullness, depression and passage of scanty faeces (Ranjhan and Pathak, 1992). Buffaloes become progressively weak and die. There is no satisfactory treatment for such cases as the impaction cannot be relieved by use of purgatives.

13.2.7 Abomasal milk impaction in calves

In buffalo calves the rumen is not developed in the early stage of life and milk reaches directly into abomasum through reticular abomasal opening. Sometimes in these calves, the milk coagulates and become rubber-like elastic material and cause abomasal impaction. It is a serious gastric disorder noticed in buffalo calves up to 5 weeks of age. In affected calves, initially there is inappetance followed by anorexia. There is scanty, white yellow coloured foul smelling faeces which resembles like scours. After 2–3 days, there is absence of defecation. The milk, whey and coagulum are accumulated in the abomasum due to which abdomen becomes tense and pendulous (Pathak and Sharma, 1988). Palpation of abdomen yields doughy sounds. Such calves usually die due to circulatory failure, starvation and dehydration as there is no response of sulphonamide or antibiotics, stomachics or parenteral administration of glucose saline treatments (Sharma *et al.,* 1983b)

13.2.8 Hepatitis

Liver is involved in the various metabolic and detoxification processes and 75% of its cells are damaged without producing apparent clinical signs. It is referred as degenerative and inflammatory disease of liver characterized by anorexia, jaundice and depression.

Etiology

The disease is caused by number of agents and depending on the etiological factor, it can be of different types. Toxic hepatitis can be caused by inorganic substances (copper, arsenic, phosphorus), organic compounds (carbon tetrachloride, chloroform, hexachloroethane, plants (*Lantana*, *Crotalaria*) and fungi (*Aspergillus flavus*, *Fusarium*). Infectious hepatitis occurs due to leptospirosis, listeriosis, chlamydiosis, salmonellosis or histoplasmosis. Nutritional hepatitis is caused due to deficiency of cobalt, vitamin E, selenium or methionine. Parasitic hepatitis occurs in liverfluke infection or migratory ascaris larvae while congestive hepatitis occurs due to congestive heart failure. It can also occur in neonatal buffalo calves (Keskar *et al.,* 1995)

Symptoms

Any of these agents can damage the liver parenchyma resulting in development of hepatitis. Initially there is inappetance followed by anorexia. Most of the buffaloes reveal mental depression while in few cases excitement may be noticed. Jaundice, diarrhoea or constipation, muscular weakness, abdominal pain, haemorrhagic diathesis, photosensitization and recumbency are observed. There is comma with intermittent convulsions, sub-acute abdominal pain, arching of back, clay-coloured faeces and edema. In addition, nervous symptoms like head pressing, aimless walking etc. are also noticed.

Diagnosis

The disease can be diagnosed by history and clinical symptoms and can be confirmed by palpation, percussion, biopsy and X-ray examinations. Liver function tests also help in its confirmation. Bile pigment metabolism in urine and serum,

BSP clearance test, galactose tolerance test, cholesterol metabolism and serum enzymes like estimation of OCT, SD, GGT and arginase are of great value in confirming the liver damage. Severity of jaundice can be known by measuring icterus index. It should be differentiated from acute grain engorgement or encephalomyelitis.

Treatment

Dietary management helps in early recovery of the case. Protein and protein hydrolysate should be avoided and diet should be rich in carbohydrate and calcium and low in fat. To treat putrefaction, antibiotics like neomycin @ 10 mg/kg or chlortetracycline @ 4–8 mg/kg body weight I/M should be given. Laxative like mineral oil @ 500–750 ml helps in evacuation of gastro intestinal tract. For early recovery, liver tonics should be used continuously for one week by oral or I/M route. To prevent hypoglycemia, glucose solution should by used I/V.

13.2.9 Pneumonia

It refers to the inflammation of the pulmonary parenchyma and bronchioles and characterized by coughing, increased respiration rate and abnormal lung sounds on auscultation. It may be classified as lobular, lobar, focal or interstitial pneumonia and can be caused by various factors. However, stress factors precipitate the disease and different infectious agents found associated with it include *Pasteurella multocida*, *P. haemolytica,* and *Mycoplasma* and *Chlamydia* organisms. The disease occurs in young calves and results in 10–40% mortality (Tomar and Tripathi, 1987). Cases are usually noticed in cold and hot-humid environmental conditions and they become more severe when calves are kept in unhygienic, poorly ventilated houses. In addition to it, poor health, low birth weight, wet bedding and sudden change in climatic conditions also exposes the calves to pickup infection (Pathak and Sharma, 1988).

The affected calves reveal high rise of body temperature, bilateral mucopurulent nasal discharge, rapid shallow breathing, coughing, dyspnoea, anorexia, dry muzzle and rough hair coat. Such cases can be detected by symptoms and for isolation of the causal agent, nasal swabs can be taken. The calves usually respond to oral or parenteral administration of sulphonamides or other antibiotics. Antiseptic inhalations or massage of turpentine liniment over chest area also help in relieving of the symptoms (Joshi *et al.,* 1999). These calves should be kept in well-ventilated warm houses and should be provided good nutritious diet.

13.2.10 Nephrosis

True cases of nephritis are rarely noticed in buffaloes and they suffer from nephrosis, which refers to the degenerative and inflammatory lesions of the renal tubules and characterized by uremia, dehydration and polyurea. The disease can be caused by haemodynamic factors like renal ischaemia, haemoglobinuria and dehydration or by different toxins. Various toxins causing nephrosis are oxalates, benzimidazole compounds, highly chlorinated naphthalenes, organic copper compounds, selenium, mercury, cadmium and arsenic (Radostits *et al.,* 1994). In affected buffaloes, oliguria, proteinuria and uremia develop. Subsequently there is anorexia, hypothermia, depression, slow heart rate and weak pulse. The urine has high specific gravity and reveals presence of protein while blood has high urea nitrogen level. Such cases can be cured by correcting the primary cause.

13.2.11 Urolithiasis

It is referred to the presence of urolith or calculi in the urinary system and characterized by distension of bladder, retention of urine, unsuccessful attempts for urination and abdominal pain.

Etiology

The disease is caused due to formation of calculi in the urinary tract. It forms when urinary solutes are precipitated and deposited around the nidus. Urinary casts, bacteria, leukocytes, degenerated or desquamated cells and mucoproteins help in nidus formation while change in pH, lack of water, dehydration and metabolic defects help in precipitation of the solutes over nidus. Mostly inorganic salts are precipitated over the nidus and mucopolysaccharides serve as cementing agent and help in calculi formation (Radostits *et al.,* 1994). Disease is more common in buffalo bulls because of more length of urethra and presence of curve in it due to sigmoid flexure. It occurs where fodder crops have high quantity of phosphates like in paddy growing areas and hills. Feeding of cotton seed cake has also been associated to the disease (Wang *et al.*, 1997). Calculi are formed in urethra in males and urinary bladder in females. Chances of calculi formation are increased when there is deficiency of vitamin A, presence of urinary tract infection, high salt concentration, deficiency of green fodder or excess use of sulphonamides and hormone therapy. Composition of the calculi is affected due to pH of urine.

Symptoms

Formation of calculi results in partial or complete blockade of urinary passage and distension of bladder due to urine accumulation. If distension persists for 24–48 hr, there is uremia and toxaemia followed by rupture of urinary bladder resulting in urine accumulation in pelvic cavity.

The buffaloes reveal acute abdominal pain, signs of discomfortness, kicking at belly and stiff gait. Frequent attempts for urination is made and few drops of blood mixed urine pass with grunting. There is bilateral abdominal distension and rectal palpation reveals distended bladder. If bladder ruptures, it will be palpated empty and urine may accumulate in body cavity.

Diagnosis

The disease can be diagnosed by clinical symptoms and confirmed by rectal palpation of the bladder and urethra. Passage of a catheter through urethral opening also helps in its confirmation if calculi is located in urethra. Radiological examination can confirm presence of calculi. Blood analysis reveals increased levels of urea nitrogen and creatinine while on urine analysis, RBCs, pus cells, epithelial cells and crystals can be seen. It should be differentiated from cystitis and pyelonephritis.

Treatment

The disease can be managed by removing the calculi through surgical operation. Calculi located in the upper part of urethra can be pushed into bladder by a catheter. Muscle relaxants like aminopromazine also helps in such cases. Some of the ayurvedic preparations help in dissolution of small calculi. To prevent ascending infection, broad-spectrum antibiotics like streptopenicillin or cephalosporins may be used along with urinary antiseptics like hexamine @ 5–6 g.

13.2.12 Parturient paresis

It is also referred as hypocalcemia or milk fever and characterized by general muscular weakness, depression and low blood calcium level.

Etiology

The primary cause of disease is the deficiency of calcium in the circulation. Ionic calcium remains in circulation while non-ionic calcium is present in bones and calcium stores. Deficiency of calcium may occur due to defective calcium absorption from intestine during parturition, insufficient mobilization of calcium from skeletal reserve sources during terminal stage of gestation to meet out the demand or due to excess drainage of calcium in cholostrum as a result of higher concentration in other blood vessels than the normal capacity for its absorption from the intestine and mobilization from the long bones. The mobilization of calcium is also affected by insufficiency of parathyroid gland. Requirements of calcium, which are low during dry period are, increased three folds following parturition. If these requirements are not met, disease occurs. Low levels of phosphorus during gestation and lactation predispose the animal for milk fever (Ranjhan and Pathak, 1992) so calcium and phosphorus should be in appropriate ratio in diet. Some antibiotics like streptomycin, gentamycin and neomycin reduce calcium ionization (Radostits *et al.,* 1994). So their use during pregnancy should be avoided.

Symptoms

The disease is commonly seen in high milk producing dairy buffaloes (Sane *et al.*, 1957) in third to sixth lactation. It occurs more in dry areas where straws and stovers are the staple roughage (Ranjhan and Pathak, 1992). Disease is noticed within 72 hr after parturition (Sharma, 1982) though cases have been seen up to 3–8 weeks after parturition (Joshi *et al.*, 1999) or during last week of pregnancy (Radostits *et al.*, 1994). Hypocalcaemia results in reduction in muscular tone, cardiac output, arterial blood pressure and rumen and abdominal motility. Due to it, hypothermia, muscular weakness and paresis develop.

The affected buffaloes initially reveal excitement and tetany. They go off feed, disinclined to move and hypersensitive. Later on muscle tremors, shaking of head, protrusion of tongue, stiffness of hindlegs and grinding of teeth are noticed. Such animals fall easily and remain in sternal recumbency in semiconscious or unconscious stage. They look depressed. Drowsy and has dry muzzle, cold extremities, subnormal temperature, static rumen, relaxed anus, constipation and dilated pupils. Intensity of heart sounds is reduced while pulse become weak. Afterwards, the animal goes in lateral recumbency and reveals flaccidity of muscles, marked depression, bloat, impalpable pulse and weak and inaudible heart sound. If the animals are not treated in time, condition deteriorates rapidly in 12–24 hr and they die due to respiratory failure.

Diagnosis

History and clinical symptoms can diagnose the cases. Blood examination confirms the disease as it reveals eosinopenia, lymphopenia, neutropenia, hypocalcaemia, and hypophosphatemia and increased activity of AST and creatine phosphokinase enzymes. Urine can be examined for presence of calcium, which also helps in the confirmation of disease. The disease should be differentiated from hypomagnesemia, ketosis, Downer's syndrome and ephemeral fever.

Treatment

The animals recover quickly with the parental administration of calcium. A 25% solution of calcium boro gluconate given 300–350 ml I/V and 250–300 ml S/C provides immediate response. The drug should be given slowly I/V as it may cause acute heart block if given rapidly. In recurrence of disease, solution containing 20.8% calcium gluconate, 4.4% boric acid, 5% magnesium hypophosphite and 20% dextrose should be given @ 350 ml I/V and 200 ml S/C after 24 hr (Singh *et al.*, 1974). Aged buffaloes respond slowly to calcium therapy, so calcium boro gluconate should be repeated 2–3 times after every 12 hr. Mild cases may be treated by giving calcium boro gluconate orally @ 60–100 g daily for 3–4 days (Ranjhan and Pathak, 1992).

Maintaining proper calcium and phosphorus ratio in diet (2.3:1) can prevent the disease. There should be low sodium and potassium and high sulphur and chloride contents to maintain anion-cation balance and reduce the occurrence of disease.

13.2.13 Hypomagnesemia

It is also referred as grass tetany, grass staggers or lactation tetany and characterized by clonic and tonic convulsions and muscular spasm.

Etiology

The disease occurs due to deficiency of magnesium in the body, which is primarily present inside the cells. Magnesium level in blood circulation or extracellular fluid is dependent on magnesium intake, loss in faeces and milk and its homeostasis by kidney. High producing buffaloes excrete more magnesium in milk and if it is not replaced by dietary source, hypomagnesemia develops. Excess protein in diet results in excess ammonia production, which chelates available magnesium causing its deficiency. Excessive use of nitrogenous fertilizers in soil results in magnesium deficiency in plants. When such plants are fed continuously, buffaloes suffer from the disease. Lush green pastures are deficient in magnesium and their prolonged use also results in development of the disease (Dabash, 1976). The calves maintained on magnesium deficient milk or milk replacers, also suffer from it. The buffalo calves reared on milk replacer containing 27% protein developed the disease within 6 months (Awad *et al.,* 1979).

Symptoms

The disease occurs in calves as well as adult animals. However, in calves, there is mortality up to 33% and magnesium level may fall up to 0.54–0.55 mg/dl (Awad *et al.,* 1979). In adults, it occurs commonly in second or third lactation. Though the disease can be recorded in lactating as well as non-lactating pregnant buffaloes, most of the cases are seen within 2 months of calving. Magnesium deficiency alters impulse transmission and sensitivity of motor end plate resulting in muscular irritability.

It can be observed in acute, subacute or chronic forms. In acute form, buffaloes stop feeding and ruminating suddenly and become alert and uncomfortable. Staggering gait, twitching of muscles, clonic convulsions for short duration, tetany of limbs and hyperasthesia are noticed. During convulsions, clamping of jaw, retraction of eyelids and excess salivation are observed. Body temperature, pulse and respiration rates are elevated and such buffaloes die due to respiratory failure.

There is gradual onset in sub acute form of disease. The animals show mild symptoms and they have reduced appetite and milk yield. Mild tetany, muscle tremors, spasmodic urination and frequent defecation are also noticed. In less severe form, buffaloes become dull and there is gradual loss of appetite resulting in unthriftiness. Dabash (1976) recorded 22–55 kg loss in body weight of pregnant buffaloes and about 27% drop in milk yield of lactating buffaloes.

In buffalo calves, retraction of head, kicking at belly, tonic and clonic convulsions and clamping of jaws are observed (Awad *et al.,* 1979). Pulse rate becomes very high and in terminal stage convulsion disappear, pulse becomes impalpable and mucous membranes look cyanotic. The calves died due to the disease reveal pale mucous membrane, ascites, varying degrees of impaction, pericardial and thoracic effusion and degeneration of adipose tissue. Emaciation, dehydration, pulmonary lesions and enteritis may also be recorded (Awad *et al.,* 1979).

Diagnosis

History and clinical symptoms help in detection of the disease. It can be confirmed by measuring the level of magnesium in serum, cerebrospinal fluid or urine. Early detection of the disease is possible by measuring magnesium level in CSF as it is reduced significantly in CSF even before tetanic symptoms become visible (Radostits *et al.,* 1994). The blood calcium level is also reduced in affected buffaloes. Hypomagnesemia should be differentiated from nervous form of ketosis, acute lead toxicity or rabies.

Treatment

The animals respond quickly if calcium and magnesium salts are given. A mixture containing 25% calcium boro gluconate and 5% magnesium hypophosphite is used @ 500 ml I/V followed by 200 ml S/C after 12 hr. If it is not available, 400–500 ml of either 20% magnesium sulphate solution or 6.5% magnesium borogluconate solution can be given S/C (Ranjhan and Pathak, 1992). To relieve convulsions, sedatives should be used. Giving 50–60 g magnesium oxide for a period of 7–10 days can treat the less severe cases by oral route. For the control of disease, diets should be supplemented with magnesium carbonate, magnesium oxide or magnesium sulphate or salt licks containing magnesium salts should be freely available to the animals.

13.2.14 Phosphorus deficiency haemoglobinuria

It is also known as haemoglobinuria and is a haemolytic syndrome of buffaloes associated with severe hypophosphatemia (Samad, 1997) and characterized by intravascular haemolysis, haemoglobinuria and anaemia.

Etiology

The disease is related to the deficiency of phosphorus and can occur due to feeding of low phosphorus containing feeds for long periods (Nagpal *et al.,* 1968). During pregnancy, phosphorus demand is increased and if it is not met, animal suffers from the disease (Pirzada and Hussain, 1998). Metallic ions of high molecular weight are present in sugarcane tops, mustard, cabbage and lucerne. If such fodders are given in excess amount, these ions interfere with the absorption and assimilation of phosphorus in body (Ranjhan and Pathak, 1992). Molybdenum and copper present in some feeds compete with the absorption of phosphorus and thereby the disease may occur (Radostits *et al.,* 1994).

The disease is noticed in adult buffaloes in advanced pregnancy (Samad *et al.,* 1979) probably due to high demand of phosphorus for the growth of foetus during last trimester of pregnancy (Samad, 1997). The cases are seen during the puerperal period also in India and other countries (Mc Williams *et al.,* 1982). Most of the cases occur in third pregnancy (Nagpal *et al.,* 1968). It is seen commonly in sugarcane growing areas where sugarcane tops alone or mixed with mustard forage are fed to buffaloes for long periods (Ranjhan and Pathak, 1992). Cases are seen in summer when animals are fed wheat, paddy and stovers of maize or sorghum, which are poor sources of phosphorus. However, Bhikane *et al.* (1999) found more number of cases in December followed by January. The disease results in 20–80% mortality in affected animals (Gautam *et al.,* 1972).

Symptoms

Due to phosphorus deficiency, glycolysis and ATP synthesis in erythrocytes are reduced due to which fragility of erythrocytes is increased. The decreased activity of glucose-6-phosphate dehydrogenase was thought to be a possible factor in causing intravascular haemolysis by indirectly increasing the oxidative stress on the erythrocytes (Singari *et al.,* 1991). The increased fragility of RBCs results in development of haemoglobinuria. One of the most consistent findings in affected buffaloes is the coffee-colour urination. These buffaloes have almost normal appetite but their milk yields are reduced significantly and have constipation with hard and black tinged faeces (Pirzada and Hussain, 1998). Dehydration, pale mucous membranes, dyspnoea, tachycardia with normal or slightly subnormal body temperature develop subsequently (Bhikane *et al.,* 1999). In terminal stage, jaundice and jugular pulsation are also recorded. All these clinical symptoms are mostly observed after 4 months of pregnancy (Nagpal *et al.,* 1968; Gautam *et al.,* 1972).

Diagnosis

The cases can be diagnosed by proper history recording and clinical symptoms. Blood analysis helps in the confirmation of the disease. Hb content and RBCs are slightly elevated and they may have Heinz bodies (Gautam *et al.,* 1972). Serum inorganic phosphorus is significantly decreased while calcium remains normal or slightly decreased (Nagpal *et al.*, 1968). Samad (1997) suggested that serum inorganic phosphorus level of 2.0 mg/dl could be regarded as optimum cut off level. Glucose-6-phosphate dehydrogenase activity was significantly low in these buffaloes (Singari *et al.*, 1991). Chemical analysis of urine reveals presence of albumin and haemoglobin while ketone bodies, glucose, bile and erythrocytes are absent (Ranjhan and Pathak, 1992). The disease should be differentiated from leptospirosis, bacillary haemoglobinuria, babesiosis and chronic copper poisoning.

Treatment

Since the disease is caused due to deficiency of phosphorus, replacement of phosphorus is necessary. For it, 120 g sodium acid phosphate can be dissolved in 600 ml water and half of it is given I/V and half S/C. Phosphorus deficiency can also be overcome by oral administration of 125 g bone meal or dicalcium phosphate twice daily for 5 days or 60 g monobasic sodium phosphate twice daily for 3 days. Randhawa *et al.* (1994) recommended the use of copper glycinate I/V and dicalcium phosphate orally for effective treatment of the condition. The severe cases should also be transfused with whole blood @ 4–6 litre as supportive therapy (Pirzada and Hussain, 1998). Use of haematinic mixture containing copper, iron and cobalt salts and liver tonic also helps in early recovery of the case. In sugarcane growing or phosphorus deficient areas, the diet should be supplemented daily with about 100 g steamed bone meal or 30 g monobasic sodium phosphate (Ranjhan and Pathak, 1992). Botrophase, a blood coagulant prepared from the venom of snake *Bothrops jararaca*, seems to have antifibrinolytic activity and can be used for the treatment of disease in buffaloes (Pirzada and Hussain, 1998).

13.2.15 Ketosis

It is also referred as acetonaemia and characterized by hypoglycaemia, ketonuria and moderate loss of appetite and milk yield.

Etiology

The disease occurs as a result of defective metabolism of carbohydrate and volatile fatty acids or dysfunction of adrenal gland (Radostits *et al.,* 1994). Main cause of disease is hypoglycaemia, which is produced as a result of impaired carbohydrate metabolism. Feeds that have excess butyrate or ketogenic substances may produce the disease. Feeding of poor quality fodders, which are deficient in protein and propionate, also result in it. During early stage of lactation if buffaloes are provided low energy ration or given inadequate exercise, they suffer from disease. It can also occur due to cobalt deficiency or in well fed animals due to feeding of excessive protein as there is more production of butyrate in rumen by protein metabolism (Joshi *et al.,* 1999). It is commonly recorded in high producing dairy buffaloes in early and at peak lactation. Sporadic occurrence of disease has been reported from buffaloes throughout the world (Ranjhan and Pathak, 1992).

Symptoms

Due to hypoglycaemia, liver glycogen level is also reduced and ketonemia develops. Wasting and nervous forms of ketosis as recorded in cows, are seen in buffaloes also. Wasting form of ketosis is more common and animals show gradual decrease in feed intake and fall in milk production. Usually buffaloes take dry roughage and refuse concentrates (Ranjhan and Pathak, 1992). Subsequently animals have depraved appetite and reveal reduction in body weight and subcutaneous fat. There is excessive urination of watery consistency and urine and milk have ketotic smell. The animals become stiff, dull and reluctant to move.

Nervous form of ketosis may also be recorded in buffaloes and they reveal excess salivation, false chewing movements, hyperaesthesia and abnormal movements. Such animals move aimlessly and may press their head against a wall and move in circles (Ranjhan and Pathak, 1992). Staggering gait, moderate tetany and trembling may also be observed.

Diagnosis

Recording of history and clinical symptoms help in diagnosis of the disease. Blood and urine examinations can confirm the disease. Ketone bodies in urine can be detected by Rothera's test. Blood examination reveals eosinophilia, lymphocytosis, and neutropenia, reduction in glucose and calcium and increase in ketone bodies. The disease should be differentiated from parturient paresis, abomasal displacement, TRP, pyelonephritis, metritis or rabies.

Treatment

The drugs that maintain blood glucose level, are highly effective. For it, 500 ml of 50% glucose solution can be given by slow I/V injection and repeated daily for 3–5 days. Propylene glycol or glycerol can also be given @ 250 g twice daily for 2 days followed by 125 g once daily for 2 days. To enhance gluconeogenesis, dexamethasone can be given @ 15–25 mg I/M once and may be repeated after

24 hr if necessary. For proper utilization of glucose inside the body, insulin can also be given @ 0.5 IU/kg body weight along with cortisone or glucose therapy (Purohit, 1997). Sodium propionate given @ 120 g in 250 ml water twice daily for one week also helps in recovery of the case. Anantwar and Singh (1998) found high efficacy of 2–3 injections of triamcinolone acetamide given @ 0.05 mg/kg body weight I/M. Proper feeding and exercise help in prevention of the disease (Joshi *et al.,* 1999). Providing proper balance of cobalt, phosphorus and iodine in the diet can also prevent it.

13.2.16 Rickets

It is a generalized metabolic disease of bones which is noticed in young calves and characterized by defective bone calcification and overgrowth of joints. The disease is primarily caused by deficiency of calcium and phosphorus in diet or due to defective absorption of calcium from gastrointestinal tract or deficiency of vitamin D (Radostits *et al.,* 1994). In affected calves, costochondral junctions and joints are enlarged, long bones are curved and there is stiff gait. Such calves feel pain during walking and remain in recumbency. The cases can be detected by clinical symptoms and confirmed by radiological examination. Blood analysis reveals hypocalcaemia and hypophosphataemia. The affected calves can be treated by giving calcium tablets or syrup and cod liver oil. The similar condition in adult animals is referred as osteomalacia and can be treated by giving sodium acid phosphate @ 30 g in 300 ml water I/V and vitamin D @ 11000 IU/kg I/M.

13.2.17 Deficiency of vitamin A

Deficiency of vitamin A is commonly noticed in calves though it may occur in adult animals also when they are fed wheat or rice straw for a long period. The disease occurs due to deficiency of vitamin A or carotene in diet for a long period. The affected calves reveal lacrimation, night blindness, complete blindness and degeneration of kidneys (Ranjhan and Pathak, 1992). Depressed growth, papillary edema, pityriasis, respiratory disturbances and enteritis are also seen (Amer *et al.,* 1982). Such cases can be treated by providing vitamin A.

13.2.18 Poisoning

13.2.18.1 *Fluorosis*

If 300–400 ppm of fluorine is present in drinking water or feed, it may produce toxic effects. The severity of toxicity depends upon amount of fluorine ingested, its solubility and compound of fluorine taken (Radostits *et al.,* 1994). Some soils are having excess of fluorine. Similarly smoke vapours or dust from industries involved in production of copper, aluminum, glass, glazed tiles, bricks and enamels are also potent source of fluorine. The ingested fluorine is converted into hydrofluoric acid resulting in irritation of mucosae. It also chelates calcium ions due to which nervous symptoms develop and blood does not clot. Affected buffaloes reveal anorexia, ruminal stasis, diarrhoea, weakness, muscle tremors, constant chewing movements, hyperaesthesia and dilatation of pupil (Joshi *et al.,* 1999). In chronic fluorosis, mottling and pigmentation of teeth may occur. Such teeth become brittle, have pits and can break easily. Bones of these buffaloes are enlarged and joints are swollen. They reveal lameness, stiffness and painful gait due to fluorosis.

The cases can be easily diagnosed by radiological examination and clinical symptoms. Blood analysis reveals high fluorine content. The affected animals should be removed from the places having high fluorine content and be given 40–50 gm of aluminum sulphate orally daily for 5–7 days. Calcium preparation and glucose solution given intravenously and gastric sedatives given orally also help in early recovery of the case.

13.2.18.2 *HCN poisoning*

It may occur due to accidental ingestion of cyanide-containing fertilizers, rodenticides or fumigants. It is however, commonly caused due to excessive feeding of fodders having cyanogenic glucosides like *Sorghum vulgare* (jowar), *S. sudanensis* (sudan grass), *Zea mays* (maize) etc. Drought affected fodder cut after first rain has high content of hydrocyanic acid. The oxidative enzyme system is impaired due to ingestion of HCN and as a result of it, various symptoms are seen (Ranjhan and Pathak, 1992).

In acute poisoning, buffaloes die suddenly without showing symptoms. In less severe cases, depression, dyspnoea, staggering gait and muscle tremors are noticed. Such animals become progressively weak and mucosae of natural orifices become bright red (Rao *et al.*, 1991). The pulse rate is weak and rapid and in terminal stage, dilatation of pupils and cyanosis may be observed (Radostits *et al.,* 1994). Such cases can be effectively treated with a mixture of 3 g sodium nitrite and 15 g sodium thiosulphate dissolved in 200 ml water and injected I/V. For the control of poisoning, drought stricken fodder should not be given to animals in large quantities.

13.2.18.3 *Nitrate and nitrite poisoning*

Young cereal forage like oat, barley, wheat or mustard when grown with heavy dosing of nitrogenous fertilizers, they contain toxic levels of nitrate and nitrite. When such fodders are ingested in large amounts, symptoms of toxicity are noticed. Nitrate present in fodder crops causes gastroenteritis while nitrite formed due to reduction of nitrate, causes respiratory distress (Radostits *et al.,* 1994). The affected buffaloes reveal profuse salivation, abdominal pain, diarrhoea, muscular weakness and incoordination. There is difficult and rapid breathing due to anoxia (Ranjhan and Pathak, 1992). These cases can be treated with methylene blue given @ 1–2 mg/kg body weight as 1% aqueous solution I/V. It can be repeated after 24 hr in severe poisoning.

13.2.18.4 *Lantana poisoning*

Lantana camara plants grow abundantly in India and other countries. Excessive intake of leaves of the plant produces toxic substances. The plant is not very palatable, so animal consumes it only during fodder scarcity particularly in dry season (Radostits *et al.,* 1994). Its excessive intake produces acute symptoms while chronic ingestion over a long period, produces chronic symptoms. The affected buffaloes reveal anorexia, severe constipation, depressed gastrointestinal motility, photosensitization, and icteric discolouration of conjunctiva, weakness of hindlegs and edematous swelling of face and ears (Dwivedi *et al.*, 1971). The acute cases usually die and reveal cloudy swelling, fatty degeneration and focal areas of necrosis in liver and kidneys. The affected buffaloes recover spontaneously if they are kept in dark and intake of lantana is stopped. Wounds developed due to photosensitization, are treated with antiseptic dressings (Ranjhan and Pathak, 1992). For early recovery, use of liver tonics, vitamin B complex and glucose saline are recommended.

13.2.18.5 *Ergot poisoning*

It occurs due to intake of large quantities of fodders infected with naturally occurring ergots. Pearl millet is a good host for the ergot-producing fungi (*Claviceps purpurea*). When buffaloes ingest ergot produced on the heads of pearl millet or other cereal plants, they suffer from ergotism (Ranjhan and Pathak, 1992). The affected buffaloes reveal persistent diarrhoea, lameness, stiffness of lower joints, oldness and loss of sensation in extremities and dry gangrenous inflammation of the feet, ears and tail (Kalra *et al.,* 1973). Though there is no specific treatment, complete rest and withdrawal of infected fodder will help in recovery of less severe cases.

13.2.18.6 *Bracken fern poisoning*

Bracken fern (*Pteris aquilina*) grows abundantly in sub-Himalayan region of India and is used as a fodder in hills during fodder scarcity or even as a routine. Large intake of bracken fern or its chronic ingestion over a long period produces the bracken poisoning (Radostits *et al.,* 1994) and is noticed in buffaloes also (Pachauri and Joshi, 1977). The affected buffaloes, though have almost normal appetite, become progressively weak and have rough and dry skin coat. There is excess mucus discharge from mouth and nostrils. Later on, animals suffer from haematuria and enteritis and there are blood clots in faeces. Clotting time of blood is prolonged and occasionally there may be edema in throat region. Such cases can be detected by urine examination. Though there is no specific treatment of bracken fern poisoning, withdrawal of bracken fern from fodder and symptomatic treatment help in recovery of the cases. Intravenous injections of toluidine blue (250 mg in 250 ml normal saline) or butyl alcohol (Ranjhan and Pathak 1992) have effect against bracken poisoning but controlled studies need to be done for field application of such treatments.

13.2.19 Miscellaneous diseases

Lapping or inability to drink water, known as Arai locally in Haryana, was noticed in buffaloes initially by Bhardwaj (1991). Water trickles from the nostrils when buffaloes attempt to drink water. It occurs due to inflammation of mucous membrane in the area of incisive papilla as it prevents closure or the opening of the incisive duct, and due to it, water is sucked into nares. The disease can be cured temporarily by plugging the opening on incisive duct with gelatin powder. Local application of lugol's iodine with glycerine twice daily relieves the symptoms in 3 days (Bhardwaj, 1991).

Hereditary suprabasilar acantholytic mechanobullous dermatosis characterized by trauma induced sloughing of haired skin, hooves and horns was described in Murrah buffalo calves by Riet-Correa *et al.* (1994). The lesions were noticed shortly after birth on the distal extremities, scapular and gluteal regions and tip of tail. It appears to be inherited as a recessive trait.

CHAPTER 14

BUFFALO MANAGEMENT

Water buffaloes are easily adaptable and are managed in many ways. Given the system of husbandry followed in a country, the system varies from an intensive feed lot system of Asia and Middle East to that of ranching and herding in Latin America, parts of southeast Asia (Malaysia, Indonesia and Philippine and Australia). In general, they are raised like cattle. But in some operations they must be handled differently. This section highlights these differences.

Water buffaloes are managed in "backyards" in Asia under intensive management system in herds of 2–50 animals by small and marginal farmers mostly by women, children and old folks, and around urban centers in herds of 20–50. They exist on the resources of small holdings. Management expenditures are generally minimal. Care of the family buffalo is usually entrusted to children, old people or women not engaged in other farm duties; the buffalo allows them to be useful and productive.

The buffalo fits the resources available on the farm, but it is also an urban animal. Thousands of herds of 2–20 buffaloes may be found in the cities and towns of India, Pakistan and Egypt all fed, managed, and milked in the streets, generally called *Khatals*.

The buffalo has important qualities as a feed lot animal; it can be herded and handled with relative ease because of its placid nature. The buffaloes of the small and marginal farmers of India, Pakistan, Srilanka and Egypt are fed and milked by their owners under feedlot-like conditions in their villages. Many of Italy's 100,000 buffaloes are maintained under similar conditions. Wherever buffaloes are raised as dairy animals for milk production, they are housed, managed and fed individually with great care.

As thermo-regulatory mechanisms are poorly developed in buffaloes, they suffer thermal stress during summer. Therefore, they need wallowing or splashing of water in order to keep their bodies cool. If this is not done, they show summer infertility or anoestrum. It has been found by researchers that buffaloes show normal heat and normal oestrus cycle during summer due to water treatment. Generally for minimizing thermal stress during hot summer, proper shelter/housing is provided to avoid direct solar radiation. Washing/wallowing/sprinkling/splashing or showering of water is done to lower temperature. Cool drinking water is provided along with cooling devices e.g. wet curtains or panels of air coolers or fans. Trees or landscaping around the shelter cuts down temperature. Feeding green fodder/silage/hay/is generally helpful and should be done at relatively cooler times *e.g.* morning or late evenings. Providing concentrate feed at night, grazing only if green pastures/grasslands are available, regular mineral mixture supplementation also helps.

14.1 Housing

Shelter Space

Buffaloes need more standing space than cows. A shed measuring 24 sq. m should be able to comfortably house six buffaloes. The walls of the shelter should be about 1.5–2 m high and cement plastered. This will make them damp-proof. The roof should be 3–4 m high to allow good ventilation. The roof can be supported on pillars. The floor should be hard, impervious and well drained to remain dry and clean. The manger (feeding space) should be 1 m long, 0.5 m high with a depth of 0.25 m. The corners of the manger or feeding trough should be rounded to avoid injury and for easy cleaning. A common sprinkler should be provided in the loafing area and/or if possible a wallowing tank big enough to accommodate 6 buffaloes. The water in the tank should be clean. There should be regular and abundant supply of clean drinking water as well as enough water for splashing/ sprinkling, showering if wallowing tank is not provided. This should be done 3–4 times a day in summer; overcrowding of animals should be avoided.

Water requirement

Buffaloes require more water than cows. Water consumption is maximum in summer, much leser in winter and rainy reason. Normally a buffalo consumes about 75 liters of water per day, in summer as compared to 60 liters in rainy/winter season. The consumption of water has a strong relation with dry matter consumption. On an average each kilogram of dry matter consumed requires 5.5 liters of water during rainy season when humidity is high.

14.1.1 Housing and management in backyard farms

In India most buffalo are reared in the backyard of the home, so much so that the buffalo is like a family member and is very well taken care of. The buffalo in turn has a close affinity for people. It is usually the lady of the house who takes care of these animals. The buffalo are tied up for the night in small shelters very close to the farmer's house. The shelters are cleaned early in the morning and the buffalo are then fed concentrates: home-made mixes of oil cakes and wheat bran. The calf is allowed to suckle to stimulate milk let down. Buffalo are hand milked and towards the end of milking some milk is left behind for the calf, which is then allowed to suckle again. There is very little awareness of pre-and post-milking hygiene. The animals are then fed around 8 to 10 kg of green fodder and in seasons of scarcity they are fed the same quantity of wheat straw. Around 09:00 to 10:00 am they are set free to go and drink water in the common village ponds, and they wallow in these tanks. For a couple of hours during midday and the afternoon they are tied up under shady trees and are again fed 8 to 10 kg of forage. Towards evening they are again taken to the shelters, fed concentrates, milked and then fed some roughage towards the late evening.

14.1.2 Housing and management in organized buffalo farms in the Indian subcontinent

In all the major cities of north India where large-scale buffalo population exists, owners of these farms have been in this business for the past two to three generations. In these organized dairy farms the routines are similar to the backyard farms, except the animals are not let out for wallowing. Usually one farm hand takes care of about ten buffalo. These animals are usually housed in, tied up housing, in a head-to-head or tail-to-tail system, with a high raised manger in which the animals are fed. There would be a centrally located water trough where the animals are brought to drink water, and they will also be washed here once or twice a day. Animals are fed roughage two or three times a day and usually all the concentrate in the ration is fed twice a day during milking. The barns are washed and cleaned with water twice a day and manure is picked up and dumped into manure pits outside the barns. All these farm activities are very labour intensive.

14.1.3 Improved housing and management of buffalo

Production performances of Murrah buffalo in tied up housing and in loose housing were studied, and the result proved beyond doubt that loose housing system was more productive and economical, with increased yields. Giving the animals protection from hot sun and cold has provided to be very useful. In winter, curtains helped lactating buffalo to produce about 500 g more milk daily than animals kept in an open shed. A higher conception rate of 80% was obtained in animals given showers in addition to wallowing facilities. This may also prevent early calf mortality. In areas where loose housing cannot be practised, buffalo should be tied up in a conventional half-walled sheds through the daytime (after milking) from April to June. Over-herding of buffalo in the shed should be avoided, with a maximum 25 buffalo in a floor space of 25 ft × 50 ft. The animals should be let out into an open paddock or yard overnight, for exercise and to provide opportunity for natural breeding. They also need to be able to wallow for half an hour daily in clean water. Care should be taken to empty and disinfect the wallowing tanks at least once every week, otherwise they can spread a variety of contagious diseases. With proper management buffalo farming is indeed profitable. By deciding at birth whether a calf should be a milk producer or not, proper care of the calves is easier and less costly. The farmer can then focus on the future milk producers and cull the others. No matter how good the genetic potential, no animal will perform well if it is not cared for and fed properly. Bull calves from high yielding dams can be kept at the farm for future breeding. They can also be sold to breeding stations for progeny testing.

14.1.4 Housing in warm and temperate regions

As discussed, housing for water buffalo should protect against thermal stress particularly from direct exposure to sun, heavy rains and cold weather. It must allow good ventilation. Housing may therefore be different in different areas of the world, due to differences in climate. But all housing should allow enough space for each buffalo. The outdoor yard should preferably be covered with grass or maybe concrete, to prevent it from becoming an unhygienic mud hole in rainy periods.

Buffalo may appear to be misplaced in a hot and humid environment as they are more or less dependent on water for their cooling. This is not entirely true. Buffalo protected from direct sunlight do very well even during hot and humid days, partly because of their ability to lose heat through the respiratory tract. But note that high milk production requires a high feed intake, and that leads to higher metabolic heat production. High yielding buffalo thus have a disadvantage over lower yielding animals, and need more cooling facilities. If buffaloes are not provided with proper shelters, wallows or cool showers, their feed intake and growth rate decline; there could even be loss of body weight; water intake increases; and in lactating buffalo there could be a drop in milk production. There is also a marked reduction in fertility.

Cattle have much more efficient thermo regulatory mechanisms with greater density of sweat glands which enable much more heat dissipation through sweating. In spite of their limitations, buffalo adapt and thrive in hot and humid tropical and sub-tropical climates, principally due to the semi-aquatic behaviour by which a buffalo seeks water to immerse its body as a means of reducing the heat load (Mahadevan, 1992). Buffaloes are known to have a higher water turnover rate than both *Bos taurus* and *Bos indicus* cattle (Siebert and MacFarlane, 1969). They are also less efficient users of water per unit of dry matter intake, have higher urine outputs and a lower percentage of kidney reabsorption of water (Moran *et al.*, 1979).

Buffalo becomes more restless, nervous and aggressive during hot-dry and hot-humid climatic conditions. The percentage of restless, nervous and aggressive buffaloes increases with increasing atmospheric temperature. During dry, hot and humid seasons almost all nervous and aggressive buffaloes and more of the docile buffaloes need oxytocin injections for milk letdown (Pathak, 1992). Buffalo seems to tolerate cold better than is commonly believed. However, cold winds and rapid drops in temperature appear to have caused illness, pneumonia and even death (BSTID, 1981).

A study on the effect of certain summer management practises on lactating Murrah buffaloes indicates that there is a definite increase in yield of about 20 to 25% by providing cooled drinking water and showering the animals during the afternoons (Radadia *et al.,* 1980a,b). A distinct improvement in the summer breeding of buffalo following managerial changes in farm practises has been reported (Roy *et al.,* 1968). A higher conception rate of 80%, was obtained in animals given showers in addition to wallowing facilities. Showers may prevent early embryonic mortality. This study further established that there is no quiescence of reproduction rhythm during summer. Buffalo heifers whose age at puberty coincides with onset of summer can also be located in heat and can conceive during summer. Resting animals under a tree instead of the hot sun could also prevent pre-natal mortality.

14.1.5 Thermal ameliorative measures to improve comfort levels of buffalo

The comfort or thermo neutral zone is described as the environmental temperature range in which no apparent demands are made upon physiological thermoregulatory mechanisms. This temperature range is from 2 to 21 °C for *Bos taurus* and 10 to 27 °C for *Bos indicus*. But buffalo are more sensitive than cattle to direct solar radiation and high ambient temperatures. There are several reasons for this.

1. The dark body colour which absorbs heat well when the animals are exposed to sunlight.
2. The relatively fewer number of sweat glands per unit area of skin, which is unfavourable for high heat loss by sweating.

3. The thick epidermal layer of skin, which protects against heat loss by conduction and radiation.

Practises for minimizing thermal stress should include

- Proper sheltering/housing to avoid direct sun radiation
- Washing/wallowing/sprinkling/splashing or showering of water
- Cool drinking water
- Cooling devices *e.g.* wet curtains or panels, air coolers or fans
- Trees or landscaping around the shelter

Considerations for improving nutritional status

- Feeding green fodder/silage/hay
- Providing feeding at night
- Grazing only if green pastures/grasslands are available
- Mineral mixture supplementation.

14.2 Calf Management

14.2.1 Care of calves at birth and early age

The following management practices should be followed at birth and during the early growth period to reduce the mortality.

1. Immediately after birth the nose and mouth of calf must be cleaned of membranes to facilitate easy breathing.
2. After cleaning the nostrils and mouth, calf should be left with the mother to lick it dry.
3. Tincture of iodine should be applied at the end of broken umblical cord in order to avoid the risk of infection which may cause naval ill. In case of intact umblicus, it should be severed with a sterile knife at a distance of about 6–8 cm from the umblicus before applying the tincture of iodine.
4. In emergency cases due to arrested breathing, measures should be taken to induce breathing by artificial means, *i.e.*, by blowing air in the mouth of calf or by lifting the calf from its hind limbs with a sudden jerk, etc.
5. In case of early weaning, calf should be removed immediately after birth to spare the dam from any distress at parting.
6. A light bedding of straw, dry grass, wood savings, etc., free from nails and

sharp materials should be provided to dam and calf. Dry ash is also used for the purpose in villages during rainy season to keep the *Kuchha* floor dry.

7. Neonatal ascariasis is common in buffalo calves and arrangement for deworming should be made as early as possible, preferably in first week of life. A single oral dose of adequate piperazin hexahydrate is recommended for the calves.
8. New born calf should void meconium in 4 to 6 hours of first colostrum feeding and first faeces is tarry in colour and consistency. This is followed by subsequent yellowish faeces at normal intervals of 6–8 hours.
9. It is advantageous to keep new born calf in individual pen for the first 3–4 weeks of age. This allows more attention to individuals and the chances of naval sucking of each other are minimized. At about one month of age calves may be kept in a group of 20 to 30.
10. At large dairy farms mark of identification should be put on the calf as early as possible. The colour of body which is mostly black is a great restraint for the identification. Tatooing is done in the interior surface of the ear or on the ventral surface of tail because these places are less pigmented which lasts for 8–10 years.
11. Body weight of calf is recorded on a balance along with length, breadth and height for the computation of milk allowance.
12. Generally buffaloes do not attack human being. Even then, as a measure of safety, dehorning may be done during first 2 weeks of life. At present it is not in common practice.
13. Vaccination against haemorrhagic septicaemia (HS), black quarter (BQ) and anthrax should be arranged as per the advice of Veterinary Doctor of the farm or village.
14. Supernumerary teats are not uncommon in buffalo heifers and such teats should be amputated during early life before wasteful development starts.
15. For feeding of early weaned calves, clean and sterile utensils should be used.
16. Antibiotic feed supplements may be advantageous in early weeks of pail feeding.

14.2.2 Some important principles to be observed with pail feeding of young calf

The principles are given as follows:

1. Treat each calf as an individual and calculate its daily feed allowance on the basis of birth weight, growth response and health.
2. In order to avoid gastric disorders causing indigestion and diarrhoea, daily feed should be divided into 2 or more meals as per convenience and should be fed at equal intervals.

3. Strict hygienic measures should be adopted. Milk containers, feeders, calf pans and other appliances should be clean and dry.
4. Boil milk and then cool to body temperature (39 °C) before feeding each time.
5. Feed 3 to 4 times in first one week of age, reduce gradually to 2 meals per day by the end of second week and continue this up to 90 days of age.
6. Feed correctly calculated allowance, avoid over-feeding during first month of age.
7. If the calf is off feed, cut out next feed and dose with 30–50 cc castor oil.
8. Plenty of fresh drinking water should be provided to calf in the pen at the sloppy end of the slanted floor to avoid moistening of bedding and floor which may provide favorable environment for maggots, etc. and may also cause chilbrain and pneumonia.
9. Calves should be provided good quality hay, preferably a legume hay in racks.
10. Deworm calf in first week of age for ascariasis, which has been reported even in newly born calves of less than 7 days age.
11. Mix an antibiotic feed additive either in milk or in concentrate mixture
12. Maximum calving take place in rainy season when it is hot and humid in India. Care should be taken to keep calf shed well ventilated, lighted and dry.

At 3 months of age a calf will eat 1.5 to 2.0 kg calf starter daily. In order to minimize the effect of weaning (removal of milk from the ration) milk/skim milk is gradually reduced to zero level by the 60th day of age. Feeding schedule of buffalo calves at birth is given in (Table 14.1) and alternative feeding schedule for buffalo calves is given in (Table 14.2).

Table 14.1 Feeding of Buffalo Calves at birth

Age (days)	Colostrum/milk (liters)	Medication	Prevention against
1	2	Aureomycin or Terramycin or Steclin or nitrofurazone or Terramycin with antigen 77	Scour
		Sealing of Naval cord with Tr. Iodine	Naval ill
2	2	Vitamin A supplement (Revisol, Rovimix, Vitablend, Fish liver oil)	Night blindness
3	2	Piperazine adepate	Ascariasis
		Enterovioform	Dysentery
7	2	Piperazine adepate (if not given on 3rd day)	Ascariasis
8 to 11	2	Sulmet	Coccidiosis

Source: Bhosrekar, 1993.

Table 14.2 Alternative Feeding Schedule for Buffalo Calves

Age (days)	Body weight	Colostrum (kg)	Milk (kg)	Milk replacer
1–5	Up to 25	1 to 1.5	–	–
6–15	26 to 35	–	1.0	100 g
16–25	36 to 40	–	2.0	250 g
25 and above	41 to 45	–	2.5	325 g
	46 to 50	–	3.0	500 g
	51 to 55	–	3.0	600 g
	56 to 60	–	2.0	1000 g
	60 and above up to	–	1.0	1000 g

14.2.3 Teaching the calf to feed

Suckling calf hardly needs any assistance. In case of early weaning at birth or within a week of birth, it becomes necessary to teach the calves pail feeding. Usually buffalo calves are lazy and slow in learning to drink milk from the pail, basin or bucket. The scheduled quantity of boiled milk should be poured in the clean or sterile feeder after cooling to body temperature (38–40°C) and then container should be moved to calf so that the milk touches the muzzle of the calf. The calf is trained by allowing him to suck the clean finger of the operator so that he may be gradually guided towards milk in the container (pail/basin). Once calf sucks a little milk, it becomes easier to guide him to feeding. It takes 2 to 7 days to teach a buffalo calf pail feeding.

14.2.4 Exercise

Weaned calves under loose housing get sufficient opportunity for exercise. When calves are housed in calf pens individually, provision should be made for a common paddock in which calves may be let loose after milk feeding. Hay may be offered in paddock. Arrangements for sufficient drinking water and shelter from sun rays are important for buffalo calves because of their highly pigmented skin and coat. Buffalo calves, like their parents, are very fond of wallowing and they enjoy it for hours without any adverse effect on health.

14.2.5 Training to lead

Buffaloes are tied only in stalls and they need little training for roping. Training a calf for leading should be started at about 2–3 months of age. In villages, calves are tied with a rope noose in legs during first month of age. Rope is changed to all the four legs alternatively. After about 3–4 weeks of age they are allowed to follow the herd for grazing. Rope halter training helps in controlling, and restraining of animals.

14.2.6 Housing of calves up to 3 months of age

Houses of young calves should be clean, dry, well ventilated and comfortable during all seasons besides being economic. Housing is as important as feeding during early age of artificial rearing of calf. Inferior housing may prove disastrous causing heavy mortality by infectious diseases under unhygienic environment. Mud plastered roof improved the intake and utilization of nutrients in buffalo calves (Jat *et al.*, 2005) There should be adequate light and ventilation to prevent dampness in the calf pen. In villages calves are tied in a corner at some distance from the dam during first 2–3 weeks of age and then moved to manger where they learn to eat fodder in company of other animals. At organized farms specially designed houses exclusively for calves are built. Calves are kept either individually in small pens of 1.0 m × 1.5 m size or in stalls in single or double rows with partitions for individual calf. Housing of calf in individual pens during first 3–4 weeks of life provides opportunity for individual care and an illness can be detected easily. The chance of naval sucking, a vice of young calves, is also minimized. After about one month of age, calves of both sexes may be kept together in groups of 10, 15, 20 or more calves in single or double row stalls.

Pens for calves

Calves should be kept in individual pens for the first month. The pens should be easy to keep clean, with shelter from direct sunlight, rain, snow and draught. Keeping the calves in separate pens makes it easier to check what they eat, that they are growing properly, and to detect illnesses. Also, naval suckling is avoided and diseases are less likely to spread. The calves should have access to fresh and clean water at all times.

Individual pens are recommended for young buffalo calves

Preferably, the buckets for milk and water should be outside the pen, in a steady holder within easy reach for the calf, so that calves cannot splash liquid on the bedding. Humid bedding will facilitate growth of germs and parasites. The pen should contain a holder for hay and concentrate. These holders should be placed above the ground so that the calf cannot step or defecate on them.

14.2.7 Dehorning of young calves

The process by which horns of animals are not allowed to grow is called dehorning. Clinical dehorning is also done in case of broken horn and diseased horn. The latter is generally called as amputation of horn.

Methods of dehorning

Dehorning is usually not in practice for buffalo calves, however it may be done by any of the devices like hot iron, caustic stick and electrical dehorning cone. With the help of these devices, horn buds are destroyed at the very early age. Dehorning, when required, should be done within one month of age in order to avoid excessive bleeding, labour and inconvenience to calf. Horns are very helpful for the identification of breeds of buffaloes and dehorning is not normally practised.

Merits of dehorning

1. The danger of injury by sharp horns is avoided and handling needs less care.
2. Less space is required by the animals and they may be kept in groups under loose housing system.
3. During fight between them, chances of bruising and tearing do not arise, which cause economic loses and reduce the quality of hides.
4. Chances of horn diseases due to fracture of horns and consequent infection are minimized.

Demerits of dehorning

1. Animals loose their natural appearance.
2. Horns of buffaloes are specific breed characteristics in most of the cases and dehorning will vanish this important character of Murrah, Zaffarabadi, etc.
3. Animals loose a defensive mechanism provided by nature.

14.2.8 Identification

Small farmers do not practise a specific marking system for their animals. They know their animals and can identify them individually by their age, nature, gait, structure and production, etc. At large farms identification is practised. Limited methods of marking can be used for buffaloes because of pigmented skin.

Ear notching

Notches cut in the pina of ear represent a number depending on its location and side. It is an easy method of identification.

Number tags

Large numbers allotted to animals and engraved on a smooth metal piece is fastened in the neck collar like a locket. Bright contrast colours are used for pinning the number plate so that it will be visible from a reasonable distance.

Ear tags

Metalic and plastic ear tags having numbers in series or as per requirement of the farm are being used for the identification of animals.

Tattooing

It may be followed for buffaloes also but sites are limited, *i.e.* inner surface of pinna and under surface of tail. This is a permanent marking and lasts for 8–10 years.

Branding

Dorsal surface of hoof or horn may be used for the branding of number with red hot metalic marks. Hot branding is done generally on hind quarters.

14.2.9 Removal of supernumerary teats from the udder of female calves

Occurrence of supernumerary teats in buffaloes is not uncommon and it has been observed in about 30% buffaloes (Dwivedi, 1966). Usually supernumerary teats occur on the posterior surface of one or both rear teats but they are reported from other teats or sides of the udder also. The extra teats are of no use, besides they disfigure the normal udder, hence such teat (teats) should be extirpated surgically at an early age (1 to 3 months). It is very easy to cut off extra teat with the help of a sharp pair of sterile scissors as there is no bleeding in most of the cases and if it bleeds it can be controlled easily by applying some pressure for 3–5 minutes. Tincture of iodine or any antiseptic lotion or cream should be applied on the wound after the removal of extra teats. In case of a double teat where base of the normal teat is involved, assistance of a veterinary surgeon should be taken so that there is no risk to damage the teat canal which may be blocked during the process of healing of the wound.

14.2.10 Calf diseases

Mortality in young buffalo calves is a serious problem under village conditions as well as organized farms. Buffalo calves when young are lazy and slow and require more attention of the herdsman. Any negligence in feeding and management results in heavy loss of animals. About one-fourth of calves are lost in first 3 month period in India. Mortality is more in male calves and that is because they are neglected by the owner from the very first day of birth. Pneumonia and gastroenteritis have been found to be a serious cause of calf mortality.

Pneumonia

It is one of the major causes of mortality of young calves specially during rainy season and winter. The mortality due to pneumonia has been reported to be 33.8 to 41.3% of total calf mortality (Verma and Kalra, 1974; Sharma *et al.*, 1975). The predisposing factors are poor health, exposure to draft, wet bedding, dampness in the house and weakness due to other diseases. The infection may be viral or bacterial. High temperature (103°–106 °C) associated with copious nasal discharge, coughing and rapid breathing are the apparent symptoms. The muzzle becomes dry and dull. Infected calf should be immediately removed to isolation house and pen/stall should be thoroughly cleaned and disinfected. Sick calf should be kept in a warm, dry house and veterinary doctor must be consulted for the treatment.

Gastroenteritis (Diarrhoea, Scour, Enteritis, etc.)

This is another and equally fatal cause of neonatal mortality in buffalo calves. The incidence at organized farms has been found to be 32–35% (Verma and Kalra, 1974; Sharma *et al.*, 1975). It may be non-infectious due to indigestion following overfeeding or due to feeding of spoiled grain mixture and hay, or it may be infectious scour caused by virus, bacteria or parasites. Symptoms are severe diarrhoea, yellowish white faeces with very offensive smell. In severe cases blood and mucus also appears in the faeces. These are followed by prostration, extreme weakness, dehydration and death. Treatment should be done as per the advice of veterinarian.

Ascariasis

Death due to heavy load of round worms is very common in young buffalo calves in villages. The incidence of ascariasis has been reported in neonatal calves. Deworming at about one weak of age has been found to be very effective.

14.2.11 Calf mortality

The calf mortality in early age is very high under village conditions and at private dairies in towns and cities. Both diseases and negligence on the part of farmers are factors responsible for the heavy deaths in neonatal calves. Under artificial rearing, mortality can be reduced significantly. In urban centres of production, mortality is induced in calves in India, Pakistan, Bangladesh and Sri Lanka. This is motivated by economy of avoiding calf raising. They are deleberately starved of milk and die due to starvation so that owners save the expenditure on calf raising for which they neither have space nor would like to invest in feeding. In buffalo calves, the highest mortality was observed in calves born in summer, followed by those born in winter, rainy season and autumn (Kumar *et al.,* 2002).

14.2.12 Records to be maintained

The following records should be maintained for having a complete account of the calf.

Calf register

The calf at birth should be entered in the register giving the details of birth, date, month, sex, body weight, identification number, dam number and sire number. This register should also contain complete information regarding the fate of calf and utilization or disposal.

Animal history card

A history card of calf should be prepared on the date of birth and it should contain complete details, *viz.,* number, name, date of birth, sex, pre-medication, vaccination, etc and subsequent history.

Health card

It may be maintained either at the farm or in the Veterinary hospital depending upon the convenience and approved practices. This card contains detailed information of sickness and treatment given to calf from birth onwards.

14.3 Heifer Management

Buffalo heifers are the valuable replacement animals and receive special attention of farmers right from the date of birth. Care and management significantly influence the production performances of heifers. Under village conditions, the age of heifers at first conception varies widely depending upon the feeding and management practices. The average age at first conception is 18–20 months under good management. Buffalo heifers under good management conditions calve at about 30–36 months of age and 300–350 kg body weight. Mean daily gain in body weight from 12 weeks of age to 12 months of age is around 500 g/day. The average body weight of farm bred buffalo heifers at one year of age ranges from 200 to 300 kg depending on the breed, climatic conditions and region.

14.3.1 Age at first calving

The age of buffalo at first calving varies from 30 months to over 60 months. The average age of water buffaloes at first calving is higher in tropics than in temperate climate of former Soviet Union, Bulgaria and other countries.

14.3.2 Housing of Heifers

At about 6 months of age heifers should be moved to more spacious sheds meant for growing animals. In villages and at small farms it is a prevalent practices to tie the heifer (other animals) with rope. The house is generally made of mud walls with *kacha* floor and tiled roof, it could be made of thatch or bamboo, in some cases it may be *pucca* brick work with concrete floor.

Loose housing

At organized dairy farms heifers are usually kept under loose housing system with a shed for feeding and shelter during extreme climates. Gupta *et al.,* (2004) found that buffalo heifers could be reared in loose housing in the northern part of India without any adverse affect on performance. The floor space requirement of growing heifers at various physiological stages is given in (Table 14.3).

Table 14.3 Space Requirements of Heifers

Physiological stage	Floor space (m^2)		Manager length (cm/animal)
	Open area	Covered area	
Young heifers	5–6	1.0–1.5	40–50
Pregnant heifers (early pregnancy)	8–10	3–4	50–75
Pregnant heifers in (late pregnancy)	Kept in calving pen during the last fortnight and after parturation should be moved to lactating animals shed.		

Source: Bhosrekar, 1993.

Conventional houses

They are standard stalls with facilities for feeding, watering and grooming of individual animals. The shed is partitioned at about 1.5 m distance with galvanized iron pipe of 5 cm diameter into several compartments in order to keep the animal confined to his own area and to avoid any injury specially to udder and teats from the neighboring animals. This also facilitates the management of individual and special feeding without much investment on labour. This type of stall is also harnessed with a stanchion for each animal to fasten their neck. In a second type called 'Tie stall' animals get more comfort as the floor space is not partitioned. In this type of stall a little more floor space has to be provided to each animal.

14.3.3 Factors affecting growth rate of heifers

Water buffaloes of all breeds, when let loose for grazing, like to wallow in water for a long time specially during summer. They wish that they should be provided maximum time for wallowing or arrangement for water sprinkling over their body should be adequate.

Fodder

On *ad libitum* feeding of legume forage like berseem, lucerne, cowpea and weed like Ghiabati (*Ipomoea pestigridis*), growing heifers may gain about 370 g daily. Supplementation of a small quantity of energy rich concentrate mixture significantly improves the rate of daily gain in body weight.

Effect of shelter and water sprinkling

Buffaloes are very fond of wallowing. However, under limited conditions of intensive rearing or keeping of small herds in towns and a few head (1 to 5 or more) by farmers, it is not practicable to provide wallowing ponds. Under such conditions animals are washed once or twice a day depending upon the atmospheric conditions or water is sprinkled on the body. Water sprinkling has a significant effect on the growth of heifers. Similarly, if the shelter is provided against direct solar heat, growth performance is dramatically improved, particularly if shade is provided by thick foliage of large trees.

14.3.4 Feeding of pregnant heifers in the last month of gestation

The requirement of feed and nutrients of heifer increases significantly depending upon the milk production capacity. When production potential of dam of the heifer concerned is not known, average of the breed is used for the calculation of rations of heifers in the last month of pregnancy. Usually, when straws and poor quality hay are the roughage source, pregnant heifers are fed 1.5 to 2.0 kg concentrate mixture depending upon the availability of small amount of greens. The concentrate feeding should be increased gradually during 15–20 days before calving to 4–5 kg per day for a milch breed of buffalo.

14.3.5 Care of heifer at parturition

The mean gestation length in Murrah buffaloes has been reported to be 306 days. Remove the heifer to disinfected loose box 2 weeks before calving. Three days before calving, feed a laxative and light ration, increase proportion of bran and molasses in the diet. Monitor 4 hourly before 2 days of calving. The actual calving takes 2 hr, watch it from a distance as heifers tend to be nervous. In case the foetal bag is not reptured, tear it open; in case of delay in delivery call a veterinarian immediately. After the delivery, give the heifer a bucket full of fresh tap water followed by warm mash made from bran, cracked wheat or corn, molasses, salt, ginger and mineral supplement. This will restore vigour and help in cleaning the bowels. The placenta is expelled 4 hr after delivery, it should be removed and immediately burried, otherwise heifers are known to eat it and thus suffer from impactation of rumen. If the placenta is retained for more than 24 hr call a veterinarian.

14.3.6 Milking a first calver

Usually the milking of buffaloes which have calved for the first time needs a little special attention of the milker in handling and approaching. Buffaloes are easily trained in comparison to cows and they can be milked in a milking shed or elsewhere. It is a common scene in Indian towns to see the door to door milking of buffaloes by the small dairy keepers. At large farms the following training should be given to all pregnant heifers during calving.

1. Heifers should be taught to enter milking stall in a queue. For this purpose they should be led by 3–4 experienced buffaloes.
2. In the milking stall each heifer should be assigned a definite place and she should be brought every day to her assigned place only.
3. About 15–20 days before calving the milker is required to spend a few minutes with each heifer and heifer should be gently approached by patting on the back or loin. In due course her udder and teats should be gently rubbed and pulled.
4. Pregnant heifers and buffaloes after calving should be handled gently otherwise it may be difficult to milk a hard milking animal easily. Shouting or striking the

cow is always harmful for the farmer as such treatments lead to holding of the milk and thus the lowered yield on that day.

5. Udder is very tender at first calving and it should be approached gradually and gently so that the buffalo is not hurt or annoyed at milking.
6. After milking buffaloes should be taken out of the milking stall in an orderly manner.

The problem of short teats is rare in buffalo heifers and if there is a case of short teats it can be corrected to some extent by regular pulling and stripping every day for few minutes about 3 months before calving. However, this practise should be stopped about 2–3 weeks before calving to avoid milk formation in the udder.

14.4 Care and Management of Bull Calves

In India male buffalo calves are neglected from the very first day of their life and those who survive the stress of bad management, insufficient feeding and neglected conditions are used either for drought purposes or slaughtered for meat. Normal male buffalo calves are allowed full feeding on udder only for 3–5 days after birth and thereafter they are just allowed to suckle for 2–3 minutes to stimulate the let down of milk. This practice of milk feeding to male calves is followed as long as they survive. This practice is common in small private dairies of towns. The average birth weight of male calves of Indian buffaloes is 30 kg and they reach hardly 100 kg of body weight at about one year of age under neglected feeding and management at most of the farms whereas heifers maintained at organized dairy farms reach body weight of 200–250 kg at the same age. The wide difference in the body weight at one year of age is clear identification of negligence in the case of management of male calves. Studies on rearing of male calves for producing breeding bulls under standard feeding and management practices show that they attain a body weight of 250–300 kg at about 24 months of age and can be given training for semen collection. However, actual collections are taken at about 30 months of age.

14.4.1 Purpose of rearing bull calves

Bull calves of buffaloes are reared mainly for the following 3 purposes :

1. Selected bull calves are reared for future breeding bulls to be used for natural service or artificial collection at the insemination centres and semen banks,
2. Bull calves are reared for draught purpose mainly in humid paddy zone,
3. The third purpose is for the production of meat.

The selected bull calves to be reared for breeding purposes need special housing and management. The other bull calves may be reared similar to that of heifer calves after 3 months of age. The bull calves to be used for ploughing and other traction works may be castrated at about one year of age and those to be fattened for slaughter should be fed liberally good quality feeds and fodders for faster growth and earlier finishing. Economy of production and availability of feeding stuffs should always be kept in mind. It is economical to house them in groups in loose housing system with adequate provision of shade for extreme summer and winter seasons.

14.4.2 Care of the breeding bull calves

At about 5–6 months of age bull calves should be separated from the heifer calves and then should not be allowed to mix with the heifers and non-pregnant buffaloes. At this stage bulls require special training so that they can be led to dummy for semen collection or may be moved conveniently in a show ground if needed. Head collar or neck roping should be used for a short time in the beginning to lead on a street or road etc., at short intervals of 3–45 days so that they can move in freely and normally. After 5–6 months of age bull calves should be exercised for 30 minutes in a group in a bull exerciser or at a suitable place meant for exercise. In the beginning, a seasoned bull should be used to lead the young ones for few days. The period of exercise should be increased gradually to about 1 hour per day which is adequate for the bulls in regular use. A sheltered area of 5–6 square meters after one year of age is ideal for diversified climatic conditions of India. Bull calves can be kept in groups of 10–15 animals up to about 2 years of age in loose houses with adequate open space.

Bull calves should be fed liberally a fairly rich ration consisting of grain mixture and good fodder to ensure a daily gain of 500–700 g in body weight so that they may attain a body weight of 300–350 kg at about 2 years of age to be used for breeding purposes.

14.4.3 Insertion of nose ring/nose string

For easy control and handling of bulls, a metallic ring or rope string is inserted in the nose through the cartilage of nasal septum. In India metallic rings are inserted in the nose of breeding bulls and rope strings are used for working bullocks/bulls. The ring should be placed conveniently in the young bull calves of 8–12 months of age. The nose ring is made of non-rusting metals like copper and brass for young bulls and gun metal for adult and unruly bulls. A light or medium weight ring of 5–6 cm diameter is inserted in the young bulls, which is replaced by a heavier ring of 6–8 cm diameter at about 2 years of age. Extra rings should be kept in hand for immediate replacement in case of breakage or loss.

Ringing of bulls is not difficult and a ring can be inserted by a skilled farmer. The bull calf is secured in a stanchion and the sterilized trocar and cannula is used to put the ring in place. The one open end of nose ring is put in the left side hole of cannula and pushed forward so that the attached end of ring comes out to other side. After this cannula is withdrawn and ring ends are closed and fastened tightly with the help of a screw. Usually ends of ring and screw are smooth but if they are observed to be rough after insertion then they are smoothed with a file or sand paper. After putting the nose ring it takes about 8–10 days for healing and bull should be tied or led only after complete healing, in order to avoid further injury and delay in scar formation. In village condition large sized needle with wide eye at distal end is used for inserting rope string in the nose of working buffaloes. Rings are also available with sharp pointed end which pierce the cartilage when pushed at the time of insertion.

Occasionally an unruly bull pulls the ring out of nose tearing the nasal septum. There is no perfect remedy for such accidents, however, a gentle bull can be controlled and led with the help of a head collar. In some cases rope string passing a little high in the nasal septum and tied behind the horns may serve the purpose of handling and leading.

14.5 Management of Pregnant Cows and the Down-Calvers

Proper care and management of the herd are the key of success in dairy industry. Special attention of farmer is required for the watchful care of pregnant animals during the last quarter and at calving time so that animals should be in comfortable condition.

14.5.1 Care of pregnant buffaloes

- The average gestation period in buffaloes is 310 days.
- Pregnant animals should be fed well during last 2 to 3 months of pregnancy as maximum foetal growth takes place during this period.
- About three days prior to calving, the buffaloes should be fed a laxative and light rations. The proportion of bran and gur in the diet should also be increased.
- On an average Indian buffaloes carry their calves for a period of 305 days but gestation period may range from 290 to 320 days from the date of service (conception). At organized farms where large number of animals are maintained it is very essential to keep records of each female individually. The following care should be taken for pregnant cows and down calvers.
- Calculate the approximate date of calving from the breeding record of the cow and move her to calving box about 3–5 days in advance.

- The calving box should be thoroughly washed and disinfected.
- Buffaloes in second and higher parity should be dried at least 6–8 weeks before calving. For drying off the buffaloes quantity of concentrate mixture should be reduced to zero and poor quality roughage should be fed for 4–5 days. Simultaneously milking should be reduced from twice a day to once a day which should be stopped in 7–10 days period. This is required only in case of good yielders.
- Buffaloes should be fed liberally rich ration during the rest period of 6–8 weeks to bring them in a good state of flesh at freshening. About 2–3 kg concentrate mixture should be fed to good yielders with good quality fodders.
- During the last 10–15 days before calving buffaloes should be fed a mild laxative diet consisting of plenty of green fodder, hay or silage. In case of dry fodder feeding, a concentrate mixture containing wheat bran and groundnut/ linseed cake in a ratio of 2:1 or crushed barley should be fed @ 1 kg per head.

14.5.2 Caring for the cow at calving

Calving in general sheds and milking parlour is undesirable. At large farms special sheds of 9–10 square meters closed area known as calving boxes are made to put cow a few days before parturition. About 7–10 days before the expected day of calving, a calving box is cleaned thoroughly after removing previous bedding and manure, if any. The house is disinfected with a suitable disinfectant and sufficient quantity of slaked lime is spread at the entrance and exit. A light bedding of straw or wood savings is provided during summer and rainy season and a little thick bedding is required in winter. The calving box should be kept dry specially during winter and rainy season in order to avoid chilling. Direct air/draft should be avoided in cold weather. If calving box is not warm enough, cover the body of buffalo with a rug. Little attention is required in summer months.

Usually parturition is normal in buffaloes and they should not be disturbed they should be left alone for calving. However, buffalo should be observed from a distance at intervals of 30–60 minutes. In cases of unusual delay in parturition and suspected dystocia, a veterinary surgeon should be called without any delay for examining and assisting the buffalo.

A buffalo is generally thirsty after parturition and she should drink a large quantity of water if given free choice. In hot weather fresh drinking water should be provided liberally but in cold weather it should be warmed to body temperature before offering so that chilling is avoided.

The feeding of laxative diet which started a few days before freshening should be continued for few days and feeding schedule should be changed gradually in 7–10 day period.

Usually placenta is expelled within 6 hours of calving but in few cases it may take a little more time. If it is not expelled after waiting overnight, say 20–24 hours, a veterinary surgeon should be called for necessary assistance. Any negligence may prove very harmful as longer delay in explusion will cause decomposition and poisoning of the entire system resulting in drastic fall in milk production and other breeding problems.

It may be ensured that the calf has received colostrum within 4 hr after birth. Weak calf should be assisted in nursing.

Care during calving

The process of calving takes about 2 hours. It is advisable to watch the calving from a distance as buffaloes tend to be nervous animals and may get disturbed. It has been observed that maximum calving (60%) takes place in the late night and early morning. In a study, out of the 46 buffalo calving recorded, 27 buffaloes calved between 12 midnight to 8 am, while only 19 buffaloes calved during the day.

Some considerations:

- Tear open the foetal bag if not ruptured
- In case of delay in delivery after appearance of foetal bag as an implication of wrong presentation, call a veterinarian immediately
- After the deliver (normal or otherwise), give the animal a bucketful of fresh tap water followed by warm mash made from bran, cracked wheat, bajra, jaggery, common salt, ginger and mineral supplement. This will restore the vigour and help cleaning.
- The placenta is normally expelled shortly after birth. In cows this is done after about 3.78 hr after delivery. The buffalo foetal placenta is heavier by 1 kg than that of cows. It weighs 3.27 kg immediately on expulsion.
- The calf at birth weighs significantly higher as compared to cow calf. The average weight is 30.37 kg.

14.5.3 Management after reproduction

Buffaloes demonstrate certain behavioral changes when in heat. These include bellowing, uneasiness, raising of head and tail, frequent micturition and staying away from rest of the herd. Withholding of milk, swelling of vulva and reddening of the nictating membrane are important sings confirming heat or oestrus. The buffaloes are more vulnerable in this stage and therefore sufficient care should be taken. Some tips of proper management of buffaloes during and after reproduction are given below:

- Protect all buffaloes, heifers and dry buffaloes against thermal stress
- Heat detection should be carried out by close observation of buffaloes
- Systematic scheduling of the heat detection programme *e.g.* records of calving and expected date of oestrus etc., will be useful information while determining buffaloes which are open. This will also help identifying buffaloes showing silent heat.
- Inseminating during the cooler hours of the day.
- Testing for confirmation of oestrus, Mucus arborization pattern, cow-side ELISA test and Estron probe will help in confirming the oestrus and the optimal time for AI.
- The placenta should be removed immediately after expulsion and buried. Otherwise the buffalo is likely to eat it and might suffer from impaction of rumen. If the after-birth is not expelled within 24 hr, terramycin tablets or furea boluses should be placed in the horns two in each horn using clean gloves.
- The placenta should never be removed manually. It will fall on its own.

14.6 Management of Dry Herd

The length of lactation period in buffaloes varies from 8 to 10 months and calving interval from 11 to 18 months. Thus, a dry buffalo has to be maintained for an average period of about 5 months (2–8 months) before the next calving. On good quality pasture no extra feeding of concentrate is required. Good grazing for 6–8 hours will be sufficient to meet the requirement of dry pregnant buffaloes. In rainfed areas supplement feeding is not needed from July to October. November and December are the lean periods in Central and Western India and during this period wheat *bhusa*, paddy straw and stovers of sorghum, maize or pearl-millet form the main roughage part of the ration. Under such conditions about 2–3 kg concentrate mixture should be fed along with *ad libitum* roughage. From January berseem and oat fodders become ready for harvesting. About 20–35 kg green berseem and 3–4 kg wheat *bhusa* or other poor quality roughage should be fed to dry and pregnant buffaloes.

In village condition autumn is a natural breeding season, however by providing shelter and water sprinkling/wallowing facility normal breeding performances and fertility have been observed around the year. Hence, non-pregnant dry buffaloes and breedable heifers should preferably be provided protection from solar radiation by providing shelter and water sprinkling/wallowing so that unnecessary delay in breeding can be avoided.

Buffaloes in their last 6–8 weeks of gestation period should be separated from the dry herd and should be fed on the basis of their previous lactation performance or expected yield. This feeding is known as "streaming up" which is followed by a "lead feeding" of 7–10 days period. Lead feeding is required only for high yielding females producing 20 liters or more milk per day.

14.7 Management of Bulls

A breeding bull is said to be half of the herd as it sires the whole herd. The bull should be reared under optimum management conditions from the early age so that it is gentle in behaviour. The training to lead the bull is given from the young age of 8–10 months by a skilled bull attendant. After about 9–10 months of age selected bulls are housed individually in bull sheds. The following steps are needed under standard management practices.

14.7.1 Housing and exercise of the bull

By nature buffalo bulls are gentle in behaviour but sometimes a few may become vicious. Vicious bull should be kept under proper scrutiny in order to avoid injury to some one. A sufficient large-sized shed with 10–12 square meters covered area is ideal for mature buffalo bull. The shed should open in an adjoining paddock of 15–20 square meters in size fenced with galvanised iron pipe, wooden logs (*balli*), barbed wire or a wall of bricks and cement to a height of 2–2.5 meters. The stall should be fitted with a stanchion for holding the bull during cleaning of the stall and washing of the bull. However, for most of the time a bull should be let loose to move freely in the paddock and stall. If it is necessary to tie the bull, arrangement of a heavy neck strap should be made and use of nose ring for daily tying should be avoided. Arrangement of a shower bath in the stall will be good for convenient washing of the bull. The floor of stall is made of bricks or roughened concrete work with sufficient slope for efficient drainage. Paddock should have a good levelled ground. In Indian climatic conditions a *Kacha* ground is economical and quite suitable provided a proper drainage is provided to avoid mud and filth in the barn. Brick paving is desired at places where soil is not suitable. This also reduces labour on cleaning and maintenance. Usually heavy bedding is not required except for 3–4 months of winter during which direct wind and draft should be avoided. Bull should be let loose on pasture or in paddock to have sufficient exposure to sunlight.

A bull should be given regular exercise at a fixed time on an open ground or bull exerciser. All bulls of a farm except the unruly bulls, if any, should be led in a queue in orderly manner. Fast walking, jumping and running for about an hour in open ground or fast movement in circle in a bull exerciser for 50–60 minutes every day is enough to keep the bulls active and healthy.

14.7.2 Feeding and watering the bull

An adult bull should be fed as per requirement only. Grazing of 6–8 hours on good pasture is adequate to maintain the health and vigour. A bull on stall feeding should be fed 15–20 kg green fodder or silage, 6–8 kg wheat straw/paddy straw and 2–2.5 kg concentrate mixture. Buffalo bull should be provided liberal amount of drinking water and it should be washed at least twice daily in summer.

14.7.3 Controlled use of the bull

Age and body size both are important points to be considered before putting the buffalo bull to service. A bull should not be allowed to serve a buffalo or used for semen collection unless it is about two years of age and 300–350 kg body weight. Use of young bull should be limited to twice a week for semen collection or mating two buffaloes per week. However, mature bulls of over three years of age may serve one female a day during the breeding season and collections may be made on alternative days. It is better to use the bulls on alternate days for natural service also. Too frequent use of bulls for longer period is often responsible for aspermia or nonmotile semen which leads to non-fertile service. During breeding season bulls should be kept away from the females under strict security. Movement of bull with the herd reduces its vigour and it is difficult to record the date of service of the buffalo and her identification. In India, two types of bulls are found, one donated by the public and others maintained at the artificial insemination (AI) centres of animal husbandry departments. The former type of bulls generally move along with the buffalo herds in the villages for breeding purposes.

14.7.4 Management of the unruly bulls

Almost all good breeds of Indian buffaloes are gentle and docile. However, sometimes a bull becomes vicious. A vicious buffalo bull is troublesome and difficult to control. Usually they are disposed of to be slaughtered or used for traction purpose after castration. Sometimes a well-bred bull is retained under strict security. Unruly bulls are kept in a strongly built stall and mostly they are kept tied in the heavy chain of iron. Two experienced and strong attendants are needed to lead such bulls for exercise and semen collection. Sometimes large size paddock is provided adjacent to bull stall with breeding facility and such bulls are used only for natural service.

14.7.5 Rotation of the bulls

In villages one to five female buffaloes are kept by the farmers and in milk colonies of towns usually 20–30 lactating buffaloes are maintained. Mostly breeding facilities through AI have been provided by the Animal Hubandry departments of the States. Indeed dairy farmers having more than 20 heads of females maintained a bull for breeding. Departmental bulls at AI centres are shifted from one centre to another after about 3 years of services at one centre. This practice is followed for avoiding inbreeding. At larger dairy farms several bulls are maintained and breeding is recorded, hence chances of inbreeding are only rare (accidental). At such farms bulls are retained for many years. A buffalo bull can be used up to 10 years of age with fair potency. Muscular strength and desire continue up to 15 years of age but the number of unsuccessful services increases frequently after 6–7 years of age (Cockrill, 1974).

14.7.6 Management of bullocks

Castrated as well as entire male buffaloes are used for draught purposes in India. Buffalo bullocks are used for ploughing the land, thrashing the grains, pulling the cart and other traction purpose.

14.7.7 Training

Buffalo males of 24–30 months of age and about 300 kg body weight are yoked with an old and experienced bullock. The yoke is connected with a fairly heavy wooded log with the help of a rope or iron chain and then the pair of bullocks are driven in a field. While training, the owner talks and pats the animals and at the end of the exercise they are hand fed. After a few days of empty yolk training when bullock becomes easy to be led, it is introduced to plough or cart yoke again with a trained bullock. Bullocks are normally trained in 3–4 week period for most of the agricultural operations but it is advantageous to take light work in the first year of training. Cockrill (1974) has referred two instances of buffaloes continuing to work until the age of 40 years. On an average buffalo working life is 20 years but it is not uncommon to see older animals at work.

Buffaloes are used to work for 6–10 hours a day. During summer solar radiation is avoided because it is hard for black coloured animals to work and stay in the sun for longer period. After work bullocks should be given rest for ½-1 hour before watering and feeding. Liberal quantity of fresh drinking water should be offered to the animals after return from the work. The ration of bullocks should be adequate both quantitatively and qualitatively depending upon the body size and work load. After working, bullock should be given at least equal number of hours for eating, drinking and resting.

14.7.8 Shoeing of bullocks

Unshoed bullocks should not be put to work specially on gravelled and metalled roads for carrying heavy loads otherwise severe injury to sole will render them unfit to work for several days. This is uneconomical for the owner and painful for the bullocks. Various types of iron shoes are used in different parts of world and local farmers put shoes to animals in the markets and shoeing shops. In villages mobile farrier's services are utilised for this purpose. In several countries shoes made of wood are also used.

14.7.9 General management practises

Grooming

Grooming keeps the animal clean and helps in the maintenance of their health. Daily grooming helps in detecting illness and injury in the buffaloes. The face, eyes and nostrils of all the animals should be cleaned in the morning with the help of a clean cloth or towel and body should be brushed gently with a dandy brush and curry comb or straw/hay pad. Angular parts of the body, inguinal region, thigh, udder and grooves of hooves should be cleaned thoroughly.

Regularity in daily operations

As far as possible grooming, feeding, watering, washing, exercise, etc., should be scheduled and followed.

Kindness in handling

Domesticated farm animals are not beasts and they should be handled gently. Rough behaviour, shouting and beating of lactating buffaloes cause lowered milk production on the day which is an economic loss to the owner.

Wallowing/bath

Buffaloes like wallowing in freshwater and they wallow in herds, in rivers, ponds and other waters. One can see buffaloe herd wallowing in a group of 5–10 animals clustered together. When sufficient water is not available specially in summer months, buffaloes lie in the muddy water to keep their body cool.

At large farms wallowing tanks are provided and animals in small number are washed once or twice a day depending upon the atmospheric temperature. At some farms shower bath facility has also been provided.

Shelter

Buffaloes dislike hot sun and severe cold and they should be protected from the extreme climatic conditions. Protection of animals from direct solar radiation has been found to increase the breeding performance of female buffaloes and *libido* in bulls.

Culling

Uneconomical and surplus buffaloes should be identified and disposed of. Sterile females and non-working bullocks find their way to slaughterhouse. Young breedable surplus animals are sold in weekly *bazars* and cattle fairs.

Control of flies

Flies in stall are a serious nuisance to buffaloes. Cleanliness and hygienic maintenance of stalls keep the flies away.

Heat detection in females

The oestrus in buffaloes is detected with the help of a teaser bull which is paraded from one buffalo to another in the stall once in the morning and again in afternoon or a teaser bull is let loose in the herd of females in loose housing for about an hour. Buffaloes come in heat all round the year but the occurrence is highest during August to March.

Most of the buffaloes show the sign of heat during night and it has been found to be maximum between 0 and 1 hour followed by 20 and 21 hours. In more than 60% buffaloes heat has been detected at night. The signs of normal heat is also weaker in buffaloes. The duration of heat has been observed to be 48 hours or more in 94% Murrah buffaloes, 87% in nondescript buffaloes and 91% in Murrah heifers. Timely detection of heat and insemination of animals at right stage is very important for maximum conception. On the basis of the cervical mucus crystallisation pattern the breeding performance has been found to be highest when animals were inseminated in the heat stage.

Clipping

The coat (hairs) of buffaloes is clipped in the month of April-May in order to keep the ticks and lice away. Any infestation of ectoparasite is easily detected.

Control of vices

The common vices like kicking, sucking their own teats or teats of other cows, licking and naval sucking, observed frequently in cattle, are rare in buffaloes. If a buffalo has developed some bad habit it should be corrected.

(i) *Kicking*

It has been observed that some times first calvers do not allow milker to touch and milk the udder. Such animals are trained for milking by a skilled milker who easily controls the kicking by fastening the hind legs above hock with a cord in a shape of 8. Sometimes due to soreness, teats and udder become sensitive and painful to touch and such conditions should be treated properly by a qualified veterinary doctor.

Unusual noise, presence of strangers or dogs/cats are a few other causes which frighten the animal. A frightened buffalo should not be beaten and should be dealt gently.

(ii) *Sucking*

This is also rare in buffaloes but sometimes one may come across such animal which sucks her own milk or the milk from other lactating animals. This causes reduced milk production which is an economic loss. This habit is controlled by the continuous use of anti-sucking devices like nose-plate or a cradle.

(iii) *Licking*

Licking is a natural phenomenon of dams which lick their calves at birth and during milking. However, an animal of either sex may develop the habit of licking their other companions. This causes ingestion of hair which may accumulate in the stomach and form hair balls. This may cause indigestion and other gastric disorders. Licking can be prevented by using a pinch of salt on the tongue continuously for several days.

(iv) *Naval sucking*

It frequently occurs in young calves weaned at an early age and housed in groups. Naval sucking leads to naval ill, and it should be prevented by applying some repulsive paste on the naval of calves.

(v) *Chewing of rope*

In villages ropes are used for tying the animals with a peg. Sometimes animals develop a habit to chew their rope. This is controlled by using an iron chain.

Spraying

Lice infestation is frequently observed during winter season in stall fed buffaloes. The incidence of ectoparasitic infestation is light in buffaloes on pasture. Spray with 0.5% DDT or 0.1% lindane efficiently destroys buffalo louse. Dipping is generally not practised in India for buffaloes. When wallowing is available, ectoparasitic infestation is rare.

Transportation

In India, mostly high-yielding buffaloes in early lactation are transported from rural areas to urban stalls exclusively for milk production. The transportation is done by truck or train depending on the number of animals and distance to be travelled. A truck carries 3 buffalo cows with calves along with feed and straw for 3–5 days period. In a goods train wagon 8 buffaloes with their calves, one attendant and sufficient quantity of hay and water, which are replenished in route, are carried. Four buffaloes in each half of the wagon are placed face to face. The floor is covered with saw dust or straw. No significant loss in body weight and other signs of stress has been observed in buffaloes even after 10–15 days of train journey, however, lactating animals may reduce some milk. Patience and gentle behaviour with the animals in transportation specially at the time of loading and unloading is very important. A terrified animal can become violent which may cause serious injury to animals itself or to attendant or to both of them.

Collection, storage and utilisation of manure

The quantity and composition of dung is dependent on the age of buffalo, physiological state and sources of feeds and fodders. An adult buffalo voids every day 3–5 kg dry matter and 50 to 100 g nitrogen besides potassium and other minerals of high manurial value. The dung is collected either once or twice a day depending upon the weather condition and stored at a distant place on a high ground or in a pit. During dry season greater proportion of dung mixed with waste straw and chaff is used for the preparation of dung cakes which are dried in sun. Dung cake is used as fuel for cooking food in villages. The other important use is the preparation of compost which is used for manuring the fields. Now dung is being used more and more in the bio-gas plant where gas is used for fuel and the digested slurry is used for manure.

Vaccination programmes

There are a number of vaccines available for common diseases. Most vaccination programmes are more efficient if applied to young calves, with boosters given at regular intervals. Buffalo are sensitive to the same diseases as cattle. Diseases strike harder on animals in poor condition. In order to protect the animals, they should be properly vaccinated and de-wormed at regular intervals. It is important to include all animals at the farm in a veterinary control programme in order to minimize risks of disease outbreaks. Vaccination programme for buffaloes against contagious diseases is given in (Table 14.4).

Table 14.4 Programme for Vaccination of Farm Animals Against Contagious Diseases

Name of disease	Type of vaccine	Type of vaccination	Duration of immunity	Remarks
Anthrax (Gorhi)	Spore	Once in a year pre monsoon vaccination	One season	–
Black quarter (Sujab)	Killed	-do-	-do-	–
Haemorrhagic septicaemia (Galghotu)	Oil adjuvant	-do-	-do-	–
Brucellosis (contagious abortion)	Strain 19 (live bacteria)	At about 6 months of age	3 or 4	To be done only in infected herds
Foot-and-mouth disease (Muhkhar)	Polyvalent tissue culture	At about 6 months of age with booster dose 4 months later	One season	Repeat vaccination every year in Oct/Nov

Record keeping

- Complete and accurate herd records are a valuable asset to the management of cattle, buffalo and other livestock.
- Herd records are essential in the operation of purebred herds when the management expects to register the animals in the herd-books and for other purposes. The progressive farmer must therefore maintain information on date of birth, sex, colour, tattoo and other individual identification marks of the animals.
- In addition to these, records of breeding performance including productivity of all animals in the herd should be maintained. These should include date/ dates of services, dates of calving, calves born, number weaned, weaning

weight, mothering ability of cows, efficiency of gain in body weight and production, such as lactation yield, lactation length, dry periods, diseases and treatment given including diseases resulting in regard to breeding sires, should be maintained.

- These particulars provide valuable information when selecting herd replacements and aid in culling the animals. They are important in determining the net income from livestock enterprise.

Range management

Water buffaloes can also be managed on rangelands. In Brazil, Venezuela, Trinidad, the United States, Australia, Papua, New Guinea, Malaysia, Indonesia, the Philippines, and elsewhere there is rising interest in raising buffalo beef on the range. The production practices for raising them are similar to those used for range cattle. Water buffaloes in the humid tropics must be able to cool off. Shade trees are desirable, and although a wallow is not essential, it is probably the most effective way the animal has of coping with heat.

Water buffaloes are intelligent animals. Young ones learn patterns quickly and often are reluctant to change their habits. Feral animals even those born in the wild tame down after a week or two in a fenced enclosure.

Provision of adequate fencing is one of the great problems of buffalo management. The animals have strong survival instincts and if feed runs short, such as in the dry season, they will break through fences that would deter cattle who would remain and starve. They will also break through fences if their family unit is split up. Barriers must be stronger than those used for cattle and the wires closer together and lower to the ground because buffaloes lift fences up with their horns rather than trample them down. In northern Australia, Papua New Guinea, and Costa Rica it has been found that buffaloes are particularly sensitive to electric fences (a single wire is all that is needed), and in Brazil a special suspension fence has been devised. Both of these seem to be cheap and efficient answers to the fencing problem.

Water buffaloes are easily handled from horseback and easily worked through a corral. Actually, because of their docility they can be mastered on foot, even on ranges where cattle require horses. Unless they come from different area they tend to herd together and can be mastered like sheep.

One of the major management adjustments to be made by cattlemen is understanding and capitalizing on the buffalo's placid nature. Buffaloes are naturally timid and startle easily; they must be handled quietly and calmly. Rough handling, wild riding and loud shouting make their handling more difficult and training much harder.

The identification of individual buffaloes is difficult. Fire brands do not remain legible on the skin for long. Cryobranding (freezing branding) is more durable. Most types of ear tags are not very successful; the numbers wear off and mud covers up the tag's color. In northern Australia ear tatooing has been the most successful identification technique, with tatoos remaining legible for at least 8 years.

When they pasture together, cattle and buffaloes coexist satisfactorily. They segregate themselves into their own groups and do not interfere with one another. The buffaloes, however, usually dominate the cattle and tend to monopolize the areas with the best feed supply.

Feed troughs and mineral boxes used for cattle are suitable for buffaloes, but chutes and crushes must be widened to accommodate the buffalo's broader body and, when necessary, the Swamp buffalo's greater horn spread.

Water buffaloes are powerful swimmers. In Brazil they have been known to escape by swimming down the Amazon River. An unusual management difficulty is caused by piranha in the rivers and swamps of Venezuela. In one herd of 100 heifer buffaloes, 40 have lost all or part of a teat to these voracious fish.

CHAPTER 15

MEAT PRODUCTION

15.1 Buffalo Meat Production in India and the World

India has 52% (98 million) of the world's buffalo population. As per the latest census of 2003 there was a growth of 7.5% in buffalo livestock during the previous 5 years. Contrary to popular belief, India does display a preference for buffalo meat. According to the USDA, though the demand for poultry is fast eroding in the global market, there is still substantial production of buffalo meat in India.

The growth of buffalo meat consumption has given an impetus to exports. One of the strengths of Indian buffalo meat is its competitive price as compared to other sources of animal protein. The meat industry in India is an off shoot of the dairy industry. Meat Production from Registered Sector for the Year 2002–2003 (State wise) is provided in (Table 16.1).

Buffaloes, particularly those which have completed their lactation cycles, are used for meat, hence, the cost is competitive.

Buffalo meat is usually eaten in all countries of Far East and near East where buffaloes exist. It generally commands a ready sale, though sometimes it is regarded as inferior to beef and is sold at a lower price. Buffalo meat in most of the countries is produced and handled under very unsatisfactory conditions and mostly old buffaloes, unfit for economic milk production or work, are slaughtered. The dressing percentage in such cases is low, being about 43–47%. Even as such buffalo meat is considered a delicacy by people used to eating it. The meat of the buffalo is used with advantage for preparation of hamburgers, meat rolls, kababs and sausages. It also makes good biltong and tasty soup. Like the milk and butterfat, the body fat of the buffalo is white on account of absence of carotene. Buffalo meat, like the milk is lower in cholesterol and higher in mineral content than that of cows. Lean buffalo meat has less than half (44%) the total fat content of lean beef and has proportionately less saturated fat. According to Bhat (1994) in India 10.52 million buffaloes are slaughtered to produce 2.62 million metric tonnes of buffalo meat with average carcass weight of 146 kg. Of the total buffaloes, 89% were slaughtered in Asia to produce 1.60 million metric tonnes of meat. Pakistan and China alone contributed close to 79000 metric tonnes of meat.

Certain European countries, most notably Italy, Bulgaria, Czechoslovakia and the CIS countries have done valuable research in buffalo meat production in recent years. Buffaloes in these countries are fed for slaughter at 16 to 20 months when they weigh 350 to 400 kg. The dressed carcass percentage is 51–60. Meat quality is good and flavour is indistinguishable from that of beef (Alim, 1967; Dzafraov, 1968). In Trinidad, Indian breeds like Murrah, Surti, Jaffarabadi, Nili-Ravi and Bhadawari have been crossed with native buffaloes; crossbred progeny is raised for beef production (Rastogi *et al.*, 1978). An average group weight of 476 kg at 2 years of age was attained in these crosses giving a daily weight gain of 512 g. Higher rates of gain between 900 and 1150 g per day have been achieved in China and the Europe. The carcass dressed out between 52 and 62%. Buffalo meat was as palatable as beef sold in the market (Dzafraov, 1968; Pei-Chien, 1978).

In buffalo meat the fibres are not interspersed with fat and hence the tenderness produced by ‘marbling’ is absent. This, however, does not appear to be a characteristic of meat from this species. It is more likely that the coarseness observed in buffalo meat by Hafez was not due to any intrinsic defect but to the fact that the animals were usually old and in poor condition when slaughtered. The dressed carcass of the buffalo generally weighs 45 to 47% of the live weight of the animal. In Yugoslavia, Bulgaria and Italy, where rearing of buffaloes for meat is common, a dressing percentage of 51 is not unusual. Caucasian buffaloes yield carcass with a dressing percentage varying from 48 to 53. The protein content of the meat ranges between 19.2 and 22.4% and that of the fat between 11.4 and 22.2%. In some markets such as Hong Kong, buffalo meat constitutes more than half of all the beef sold.

15.2 Meat Export

The buffalo meat export industry has grown from 167291 MT (FOB value Rs 797 crores) in 1999–2000 to 343817 MT (FOB value Rs 1647 crores) in 2003–2004. This growth has been achieved due to a growing demand from the traditional markets. The traditional importing countries include, Malaysia, Philippines, Mauritius, UAE and other GCC countries and Jordan.

Table 15.1 Meat Production from Registered Sector for the Year 2002–2003 (State wise)

State	No. of animals slaughtered (in 000 nos.)	Average yield per animal (in kg)	Meat production (in 000 tonnes)
Andhra Pradesh	445	103.50	46.10
Arunachal Pradesh	38	180.00	6.84
Assam	9	39.25	0.36
Bihar	577	62.13	35.85
Chattisgarh	–	–	–
Goa	–	–	–
Gujarat	36	91.97	3.28
Haryana	–	–	–
Himachal Pradesh	–	–	–
Jammu and Kashmir	–	–	–
Jharkhand	–	–	–
Karnataka	89	95.95	8.56
Kerala	162	56.17	9.10
Madhya Pradesh	45	112.12	5.05
Maharashtra	610	134.23	81.88
Manipur	24	130.00	3.10
Meghalaya	218	120.00	26.10
Mizoram	–	139.29	–
Nagaland	10	175.00	1.75
Orissa	–	–	–
Punjab	–	–	–
Rajasthan	153	56.13	8.59
Sikkim*	–	–	–
Tamilnadu	72	99.97	7.19
Tripura*	–	–	–
Uttar Pradesh	1120	120.78	135.23
Uttarakhand	19	124.73	2.41
West Bengal	72	137.90	9.93
Union Territories			
A and N Islands	–	–	0.03
Chandigarh	–	–	–
D and N Havel*	–	–	–
Daman and Diu*	–	–	–
Delhi	143	193.05	27.61
Lakshadweep	–	–	–
Pondicherry	3	37.33	0.11
All India	**3845**	**2209.50**	**419.07**

*denotes no data reported for some states, goat and sheep are combined.

Source: State Animal Husbandry Departments.

In the last few years, there have been increased sales to countries in Africa (including Gabon and Angola), some CIS countries and intermittent supplies to

Iran and Egypt. In the last few years, newer markets which have emerged and are growing include Afghanistan, Iraq (through Jordan and Kuwait) and most importantly Saudi Arabia.

The water buffalo offers promise as a major source of meat, and the production of buffaloes solely for meat is now expanding. Because buffaloes have been used as draught animals for centuries, they have evolved with exceptional muscular development; some weigh 1000 kg or more. Until recently, however, little thought was given to using them exclusively for meat production. Most buffalo meat was generally, and still is, derived from old animals slaughtered at the end of their productive life. As a result, much of the buffalo meat sold is of poor quality. But when buffaloes are properly reared and fed, their meat is tender and palatable. Water buffaloes are exported for slaughter from India and Pakistan to the Middle East and from Thailand and Australia to Hong Kong. Demand for meat is so great that Thailand's buffalo population has dropped from 7 million to 5.7 million head in the last 20 years, a period in which the human population has more than doubled.

15.3 Carcass Characteristics

All buffalo breeds, even the milking ones, produce heavy animals whose carcass characteristics are similar to those of cattle. Despite heavy hide and head, the amount of useful meat (dressing percentage) from buffaloes is almost the same as in cattle. In Mediterranean type buffalo in Brazil dressing percentage is 55.5. Dressing percentages in swamp buffalo in Australia is 53.

Buffaloes are lean animals. Although a layer of subcutaneous fat covers the carcass, it is usually thinner than that on comparably fed cattle. Even animals that appear to be fat proved to be largely muscular. Research in Australia revealed that it is difficult to produce swamp buffaloes with more than 25% fat. An average choice-grade beef carcass may contain about 35% fat. This lower level of fat is sometimes seen even under feed lot conditions, although animals liberally fed with concentrated rations will eventually fatten. Castrated males have a reasonably even layer of subcutaneous fat.

In general, the buffalo carcass has rounder ribs, a higher proportion of muscle, and a lower proportion of bone and fat than beef.

Buffalo meat and beef are basically similar. The muscle pH (5.4), shrinkage on chilling (2%), moisture (76.6%), protein (19%) and ash (1%) are all about the same in buffalo meat and beef. Buffalo fat, however, is always white and buffalo meat is darker than beef because of more pigmentation or less intramuscular fat (2–3% "marbling", compared with the 3–4% in beef) and is low on cholesterol.

15.4 Eating Quality

Taste panel tests and tenderness measurements conducted by research teams in a number of countries have shown that the meat of water buffalo is as acceptable as that of cattle. Buffalo-steaks have been rated higher than beef-steaks in some taste tests in Australia, Malaysia, Venezuela and Trinadad.

There is some evidence that buffaloes may retain meat tenderness to a more advanced age than cattle because the connective tissue hardens at a later age or because the diameter of muscle fibres in the buffalo increases more slowly than in cattle. In one test the tenderness (measured by shearing force) of muscle samples from carcasses of buffalo steers, 16–30 months old, was the same as that from feedlot Angus, Hereford and Friesian steers 12–18 months old. This gives farmers more flexibility in meeting fluctuating markets while still providing tender meat.

Buffalo hide makes excellent, thick, tough leather much valued for making shoe soles, belts and many other leather articles requiring this kind of material. Strips of buffalo hide softened with fat are woven to make very strong, attractive reins, lassos or ropes. Very good and durable water buckets are made out of buffalo hide for lift irrigation purposes. Buffalo hide can be split with modern machinery to make thin strong sheets, which after processing and dyeing make as good a leather as that obtained from other animals.

Tasty 'buffalo chips' are made from buffalo hide in parts of Thailand, Nepal and Indonesia. The hide is cut into small strips, boiled in water for a long period and dried under the sun. The pieces are fried in deep fat to make crisp delicious chips. With adequate publicity it should be possible to create an export market for buffalo chips.

Massive horns of the buffalo support a cottage industry. Based on them a wide range of useful as well as fancy and decorative horn articles are manufactured. The ingenuity, workmanship and beauty of many of these products are really fascinating. In Malaysia, a musical instrument known as tetuag is made out of buffalo horn (Cockrill, 1966).

The potentiality of buffaloes as a meat animal to meet the present and future requirements of rapidly growing population in Asian countries deserve special attention, because of its capacity to convert coarse roughages and other cereal by-products into meat economically. Although buffaloes are being used for meat in these countries, there are no feeder/slaughter grades. There is an urgent need to produce these types.

15.5 Rearing of Male Buffalo Calves for Meat

The animal nutritionists of this country are exploring the male buffalo calves as a valuable source of meat. Concerted efforts are being made to develop appropriate feeding regimes for this class of animal. The earliest work was by Sarma and Talapatra (1963) which was followed by Icchponani and Siddhu (1966) and Ichhponani *et al.* (1972). The classical studies include three doctoral programmes (Agarwal, 1974; Baruah, 1982; Sarma, 1984) and two long term research projects (Pathak *et al.*, 1980; Senger *et al.*, 1985). A number of other studies were also conducted (Ranjhan *et al.*, 1977; Pande and Shukla, 1979; Sarma *et al.*, 1979; Joshi *et al.*, 1984; Gupta *et al.*, 1984; Sarma and Sharma, 1985). Generally, feed formulations based on NRC (1976) requirements or Morrison's standards for beef cattle, growth rate, feed efficiency and carcass evaluation at a predetermined slaughter weight (usually about 300–350 kg) formed the parameters of these studies. Variable growth rates were reported. A daily gain of about 500 g and a dressing yield of about 50–55% were obtained.

The planes of nutrition did not significantly affect the dressing percentage, carcass composition and meat quality (Anjaneyulu *et al.*, 1985). Ranjhan and Pathak (I983) have proposed an optimum slaughter weight of 350 kg for male buffalo calves. Sharma *et al.* (2005) also studied the effect of plane of nutrition on carcass composition of buffaloes. He found that the level of feeding or age of slaughter did not have any significant effect on shear force value, chemical composition or trace element composition of buffalo meat except that triglycerides and cholesterol increased in the meat of calves fed at higher plane of nutrition. Sharma *et.al.* (2005) studied the effect of roughage to concentrate ratio on buffalo meat production. He inferred that the efficiency of utilization of nutrients was low after 18 months resulting in low daily live weight gain at 22 months. Maximum daily live weight gain was observed between 12 to 15 months of age with the best feed conversion ratio.

Agarwal (1974) has attempted to study the economics of meat production. However he has only given the feeding costs per kg of dressed carcass and lean meat; the corresponding values were Rs 7.07 and 9.40 and 6.10 and 8.12 on high and low planes of nutrition respectively.

15.6 Meat Traits of Slaughter Buffaloes

Valuable information has been collected on the quality characteristics of market slaughter buffaloes through two elaborate research projects launched by IVRI, India. In the first project (Kondaiah *et al.*, 1981, 1982 and 1983), a score card system was developed to grade live animals and carcasses. The overall conformation of animals was between fair and poor grades. The females were better in conformation, bigger in size and fatter than males. Live animal conformation score was significantly correlated with carcass conformation score, fat score, fat thickness and eye muscle area. The dressing yield was about 46.4%. carcasses showed practically no marbling. Meat from males was darker. In the second project (Lakshmanan *et al.,* 1985 and 1986), the mean carcass weight of animals at Kolkata, Mumbai, Chennai and Hyderabad were found to be 265, 236, 100 and 88 kg respectively. Carcasses were evaluated for cuts, bruises and contamination. Usually poor grade carcasses alone had cuts indicating negligence in handling them by butchers. Bruises and contamination were prevalent in majority of the carcasses and the latter was more pronounced with irregular floor slaughter practices as in Delhi and Hyderabad.

Pillai *et al.* (1987) reported a mean carcass yield of 148 and 187 kg for buffaloes of Marathwada and Vidarba regions respectively. They also reported a better conformation for females. Lakhsmanan *et al.* (1984) proposed a novel classification for variety meats of buffaloes and reported their yields in market animals.

Buffalo farmers do not attend to the male calves. Those, which escape death due to malnutrition or other causes, grow by grazing on scrub land for 1–2 years of age and attain a body weight of 70 to 130 kg. These calves are usually poor in condition. Ramamohana Rao (1978) conducted slaughter studies on such animals and reported a dressing yield of 39%. Lakshmanan *et al.* (1986) reported a higher yield of 43.6 and 46.7% for calves slaughtered at Mumbai and Delhi abattoirs respectively. However, the same type of calves when raised on adequate nutrition and grown up to a body weight of about 300 kg gave consistently a dressing yield of 50% and more (Lakshmanan and Pathak 1979).

Adopting proper hygienic and sanitary practices during processing of buffalo meat resulted in meat with acceptable microbiological quality and extended shelf-life (Yashoda *et al.*, 2000).

15.7 Quality Attributes of Buffalo Meat

The first Indian study on quality characteristics of buffalo meat as initiated by Yadav and Singh (1974) who reported the cholesterol content of different muscles. The same authors, in 1985, investigated the effect of age on muscle proteins and showed that the total and stromal proteins increased with increasing age. Rao *et al.* (1985) observed that muscle proteins differed significantly between muscles. The amino acid composition, essential amino acid index and biological value were worked out by Madhawi *et al.* (1982). Their results indicate that buffalo meat is nutritionally better than beef. The intramuscular fat and energy value were comparatively low in buffalo meat (Anjaneyulu *et al.,* 1985), while the saponification value of internal fat was significantly higher than external fat, the iodine values did not differ much (Rao *et al.,* 1985). The fatty acid composition of intramuscular lipid of muscles and liver tissue was determined by GLC analysis (Sharma *et al.,* 1986). A doctoral work was completed on the effect of processing and storage on buffalo meat lipids (Kesava Rao, 1988).

Kondaiah *et al.* (1986b) conducted a series of studies on the functional properties of buffalo meat. The normal values for carcass meat, head meat, tripe and heart were estimated and it was served that head meat is qualitatively superior. Incorporation of salt (2%) and tetra sodium pyrophosphate (1%) significantly increased the pH, water holding capacity, emulsifying capacity and decreased the cooking loss (Kondaiah *et al.* 1985). A polyphosphate blend which is equally effective as sodium pyrophosphate in improving the functional properties of ground buffalo meat and containing 3% less sodium has been developed. The effect of handling conditions on quality was investigated and it was concluded that frozen meat was better for use in products (Kondaiah *et al.,* 1986a). Two doctoral works were carried out on the microbiological quality of buffalo meat, one on qualitative and quantitative assessment of microbial load on market meat (Bachhil, 1985) and, the other on keeping quality and spoilage pattern (Agnihotri, 1988). The microbial profiles of buffalo meat cuts and the effect of chilling and freezing on them have been investigated by Bhadekar *et al.* (1987) and Kulkarni *et al.* (1987).

Babu *et al.,* (2002) observed that the buffalo meat can be stored safely with high consumer's acceptability up to 15 days in the modified atmosphere of 80% oxygen+20% carbon dioxide in multi layer pouches at 4 ± 1 °C. Sen and Sharma (2003) studied the quality changes in buffalo meat during storage in dry ice pack. They observed that mean pH values remained almost unchanged throughout the storage period, indicating minimal chemical changes due to better temperature control in the dry ice pack method. A significant increase in tyrosine value during storage in dry ice was observed; it was more pronounced ($P<0.05$) after 12 hr of storage. Microbial counts exhibited increasing patterns over time. However, the counts never reached the maximum allowable level for fresh meat, throughout the storage period. Up to 36 hr of storage did not show significant changes in colour; however, a significant decrease in colour score was observed on further storage. No undesirable odour was observed throughout the entire storage period; odour score did not decrease below 2.

Sen and Sharma (2002) suggested that the measurement of LPDH, LDH and SDH could be used to develop a simple method which could distinguish the fresh and frozen-thawed buffalo muscle.

15.8 Processed Buffalo Meats

India is a vast country with maximum number of ethnic groups, each group having its own taste profiles of foods. Consequently, this country has inherited a rich culture of traditional meat products. *Goshtaba, Rista* and *Nate-Yagni* are the world renowned mutton products from Kashmir Valley.

Chander *et al.* (2003) studied the role of irradiation on microbiological safety and shelf-life extension of non-sterile and sterile convenience meat products stored at ambient temperatures and found that the products subjected to irradiation at 10 kGy showed an absence of viable microorganisms and also had high sensory acceptability up to 9 months at ambient temperature. Curing and PP (polyphosphate) treatment further improved the tenderness of the meat products (Hoda *et al.*, 2002). Better scores for appearance, flavour, tenderness and overall acceptability were observed in ginger extract treated meat samples (Naveena *et al.*, 2004). Sachindra *et al.* (2005) studied the microbial profile of buffalo sausage during processing and storage and inferred that measures such as low initial microbial counts, hygienic precautions during preparation of sausage, steam cooking for 45 min, vacuum or CO_2 packing and storage at 4 ± °C would control the microbial growth and provide wholesomeness and safety to the buffalo sausage. Muthukumar *et al.*, (2005) assessed the quality attributes and shelf-life of buffalo *haleem* under refrigerated storage (4 ± °C). He inferred that the product was physico-chemically, microbiologically and sensorily acceptable up to 10 days in aerobic package under refrigerated storage (4 ± °C).

Some of the popular products made from buffalo meat are described here.

15.8.1 Kababs

Kababs also called as *seekh kabab,* are ready-to-serve meat products made from buffalo or sheep or goat meats. A number of roadside stalls in metropolitan and other cities are engaged in exclusive production and sale of buffalo kababs About 30–40 g of minced buffalo meat, containing lean, fat, salt, spices, condiments, green curries, herbs etc. in appropriate proportion, is moulded by hand into hollow cylindrical structures around the skewers; grilled and smoked for 2–3 minutes on burnt charcoal placed in a specially designed charbroiler. Buffalo *kababs* are very cheap and consumers enjoy eating *kababs* with *roti/naan* together with salad and chutneys as a lunch/dinner on stalls.

15.8.2 Restaurant dishes

Hotels and restaurants serve a number of delicacies made from buffalo meat. These dishes, can also be prepared at home ; but, some consumers usually buy/

eat them at restaurants and special eating houses who specialize in these preparations, *e.g. kofias, tikkis, butt,* paya etc.

15.8.3 Koftas and tikkis

Carcass and/or offal meats are made into *keema* by chopping, salting and mincing. *Keema* and bengal gram are mixed in the ratio of 4 : 1. The material is cooked and ground into a paste; mixed with spices and green curries; either rolled into balls of 3–4 cm diameter or flattened to round dishes for production of *koftas* and *tikkis* respectively ; pan fried in oil and served hot. *Shammi kabab is* the other name for *tikki.*

15.8.4 Butt

Rumen meat, very tough in texture, is processed into a fairly crispy delicacy called *butt.* The meat is cut into small pieces of 2–4 cm^2 size; cooked in water for about 30 minutes; water is decanted; the pieces are pan-fried in oil with salt, grated onion, garlic and condiment mixture re-cooked for 20–30 minutes after adding little water. *Butt* looks like "fried fish in sauce".

15.8.5 Paya

Paya is prepared from flat bones, shank bones, and large chunks of meat. The bones are split ; salt and turmeric powder are applied; water is added to immerse the materials ; heated for prolonged period of 2–6 hr at simmering temperature to produce a fine broth of meat, tendons, ligaments and marrow contents; the long bones are removed and the remainder is pan fried with spices and condiments; finally, the material is re-cooked in water till a viscous, soupy and flavourful product is formed. *Paya is* always served hot. *Paya* with rice gives high satiety. It is a special dish of die winter season. It is strongly recommended for recouping die health for convalescent people.

15.8.6 Bheja

The processed brain is called *bheja.* It is cooked in water for 5–8 minutes; outer membrane is peeled off; cut into small pieces and pan-fried in oil with grated onion, cut coriander leaves, salt and spice mixture to taste.

15.8.7 Buffalo curries

Carcass meat, offal meats (which are used as chunks or *keerrea)* and organ meats (like liver, kidney, heart, tongue, ears, udder etc.) can be made into respective curries either individually or in combination; the name of the curry will signify the kind of meat. The culinary technique generally consists of frying or cooking of meat, adding salt, spices, vegetables etc., and further cooking to produce a meaty product in gravy, die quantity and consistency of which may vary considerably from one curry to another.

15.8.8 Corned beef

A private firm was engaged in the manufacture of corned beef for export to Middle East and EEC countries. The corning process consisted of mixing of cooked diced meat, minced raw meat, salt and other ingredients and canning. The venture did not succeed for reasons of low capacity utilization and the Governmental regulation on labeling of the products as containing "buffalo meat", which adversely affected its marketing.

15.9 Challenges and Opportunities for Indian Meat Industry

(a) Setting up of the State of the art-abattoir-cum-meat processing plants

(b) Packaging of technologies to raise male buffalo calves for meat production

(c) Buffalo rearing under contractual farming as backward integration to the modern abattoirs for meat production

(d) Establishing disease-free zones for rearing animals

India can also achieve the Number one position in meat production. This could be achieved by reducing the mortality rate in male buffalo calves (80%), and rearing the animals scientifically for quality meat production, For example, about ten million buffalo calves, which were otherwise eliminated in their very infancy, would become available for quality meat production. This will raise the standard of living of small and marginal farmers in the long run. Meat production has been neglected, and has not been given adequate attention by the scientists, policy makers, and entrepreneurs, coupled with lack of political will. If all these are combined, the meat production can be greatly enhanced. Meat production is intimately linked to quality leather production in which India has acquired number two position in the world after Italy. If substantive support is given by the Government, both meat and leather can also achieve Number One position in the world, like milk.

In order to achieve the *Pink Revolution,* the following steps have been taken by the Government and the entrepreneurs in India. The Government and the Private Sector have already initiated many of the steps.

(a) Setting up of the State-of-the-art-abattoir-cum-meat processing plants

The recent trend in India is to establish latest state-of-the-art abattoirs-cum-meat processing plants. India has already established 10 most modern state-of-art mechanized abattoirs-cum-meat processing plants in various States based on slaughtering buffaloes and sheep. These plants are environment friendly, where all the slaughterhouse by-products are utilized in production of meat-cum-bone meal, tallow, bone chips etc. They are also adopting appropriate technologies to obtain value added products. These plants have effluent treatment devices which treat all the washings of abattoirs, lairage etc. to safe water discharge having 30 PPM of BOD. A few more (eight) are under construction.

The plants follow all the sanitary and phytosanitary (SPS) measures required by the International Animal Health code of OIE. Having no social taboo, like the cow in India, with buffalo slaughter, these plants mostly produce buffalo meat for export. India is becoming a major buffalo meat producing country and will be a main player in the international market with additional establishment of the state-of-art-abattoirs cum meat processing plants and control of FMD in three zones in few years from now.

(b) Packaging of technologies to raise male buffalo calves for meat production

In India, every year, about 10 million male calves are removed from the buffalo production system due to intentional killing by the farmers to save dam's milk due to non-remunerative cost of raising male animals, thus incurring a loss of about US $ 11 million per annum. These calves could be salvaged for meat production, which will not only improve the economic condition of the farmers but also would increase meat production for domestic consumption and export market.

In India intensive feeding of male buffalo calves has started for meat production. The male calves at the age of 6–8 months, purchased from the farmers, are quarantined for 15 days during which vaccination and deworming are provided. Thereafter, they are fed on high protein/high energy diet to put on a weight of 120 kg in 4 months to produce quality meat. They are never fed on antibiotics, hormones and growth promoters. They are raised in organic farming. Meat from such animals is tender, lean and juicy and goes to the wet market.

(c) Buffalo rearing under contractual farming as backward integration to the modern abattoirs for meat production

A strong need has been felt to establish a production base around each modern abattoir to produce quality disease-free animals as per the sanitary and phytosanitary

(SPS) requirements of O.I.E. Hind Livestock Development Foundation has established a model backward integration with 110000 farmers who are raising more than half a million buffaloes in 2200 villages under contractual farming system. The Foundation is providing animal health, animal feeding and extension management services to the farmers at their doorsteps. The marketing of the animals to the meat plant is organized by the Foundation to pay them remunerative prices. This has reduced the mortality in the male calves as inputs for animal rearing are provided by the Company in vaccination, deworming and feeds.

(d) Establishing disease-free zones for rearing animals

India is now fortunately free from most of the trade related diseases listed at List 'A' of the Office International des Epizooties (OIE), namely, rinderpest, contagious bovine pleuropneumonia (CBPP), etc. India has also not reported bovine spongiform encephalapathy (BSE–mad cow disease). However, foot and mouth-disease (FMD) is still prevalent in an endemic form in some States in India. The Government of India has established 3 Zones with 56 Districts to control FMD in the Tenth Five Year Plan. These Zones are in the North, Central and Southern zone where most of the EOU plants are located. Hopefully, in another 2–3 years, OIE recognized FMD free zones with vaccination would be established in the country, which will further boost meat export. The Project has started with 100% financial assistance from the Central Government. Training of the Veterinarians and purchase of vaccines had been completed. The mass vaccination has been started since October 2003.

CHAPTER 16

MILK PRODUCTION

By

P. N. Bhat and A. J. Pandya

India which has 66% of economically active population, engaged in agriculture, derives 17.62% of gross domestic product (GDP) from agriculture. The share of livestock product is estimated to be about 21% of total agricultural sector (National accounts statistics, 2004).

The dairy sector in India has shown remarkable development in the past decade and India has now become one of the largest producers of milk and value added milk products in the world. Production of buffalo milk by countries of the world between 1999–2000 (Table 16.1).

There are 174 million buffaloes in the world, and India possess the highest buffalo population of 98 million (FAOSTAT website year 2006). FAO data also showed that the buffalo milk production in India has increased from 43000 thousand tones (1999) to 50740 tones (2004). India has the highest volume of buffalo milk production followed by Pakistan, China and Egypt with contribution of 66.3, 25.1, 3.6 and 3.0%, respectively.

Latin America is emerging as a predominant area for buffaloes with largest concentration in Brazil.

16.1 Milk Composition

Buffalo milk is much richer than cow milk with respect to butter-fat content that may be as high as 15% under good feeding and management. The average fat

Table 16.1 Buffalo Milk Production by Countries of the World 1999–2004

Country	Years						Percentage share to world total					
	1999	2000	2001	2002	2003	2004	1999	2000	2001	2002	2003	2004
Albania	0.01	0.01	0.01	0.01	0.01	0.01	0.0	0.0	0.0	0.0	0.0	0.0
Bangladesh	22.40	22.40	22.80	22.80	22.80	22.80	0.0	0.0	0.0	0.0	0.0	0.0
Bhutan	0.42	0.39	0.35	0.32	0.32	0.32	0.0	0.0	0.0	0.0	0.0	0.0
Brunei Darussalam	0.04	0.04	0.04	0.04	0.04	0.04	0.0	0.0	0.0	0.0	0.0	0.0
Bulgaria	10.77	9.21	5.19	4.41	5.28	6.23	0.0	0.0	0.0	0.0	0.0	0.0
China	2600	2650	2680	2700	2750	2750	4.0	3.9	3.8	3.7	3.6	3.6
Egypt	2018	2030	2213	2087	2550	2267	3.1	3.0	3.1	2.9	3.4	3.0
Greece	0.03	0.03	0.04	0.03	0.06	0.305	0.0	0.0	0.0	0.0	0.0	0.0
India	43000	44400	46600	48000	50100	50740	65.7	65.9	.66.2	66.4	66.3	66.3
Iran, Islamic Rep. of	214.00	216.19	220.99	225.95	230.00	235.00	0.3	0.3	0.3	0.3	0.3	0.3
Iraq	26.40	26.60	27.30	27.60	27.60	27.60	0.0	0.0	0.0	0.0	0.0	0.0
Italy	133.20	135.10	153.76	147.26	171.87	167.05	0.2	0.2	0.2	0.2	0.2	0.2
Malaysia	7.25	6.79	7.02	7.36	7.48	7.48	0.0	0.0	0.0	0.0	0.0	0.0
Myanmar	108.70	110.98	113.66	116.02	120.00	124.00	0.2	0.2	0.2	0.2	0.2	0.2
Nepal	744.03	759.57	781.39	806.69	834.38	863.32	1.1	1.1	1.1	1.1	1.1	1.1
Pakistan	16391	16910	17454	18022	18617	19240	25.1	25.1	24.8	24.9	24.6	25.1
Philippines	0.00	0.00	0.00	0.00	0.00	0.00	0.0	0.0	0.0	0.0	0.0	0.0
Sri Lanka	25.50	25.52	25.56	25.64	25.56	25.84	0.0	0.0	0.0	0.0	0.0	0.0
Syrian Arab Republic	0.90	0.93	0.94	1.52	1.52	1.52	0.0	0.0	0.0	0.0	0.0	0.0
Turkey	75.00	67.33	63.33	50.93	48.78	39.28	0.1	0.1	0.1	0.1	0.1	0.1
Vietnam	30.00	30.00	30.00	31.00	31.00	31.00	0.0	0.0	0.0	0.0	0.0	0.0
World	**65408**	**67401**	**70399**	**72277**	**75543**	**76548**	**100.0**	**100.0**	**100.0**	**100.0**	**100.0**	**100.0**

content is higher than that of cow milk, possibly a little over 7% except in Egyptain buffaloes (6.6%) (Alim, 1978). The solids-not-fat content is around 9 to 10.5% and is generally slightly higher than that of cow milk (Table 16.2). The smooth creamy texture of buffalo milk makes it ideal for many types of dairy product. The high level of solids make processing very much more cost effective when compared to cow milk.

Protein, lactose and mineral together constituted 9.6% in Indian buffalo (Pal *et al*., 1971), whereas almost similar contribution by these factors was in Egyptain buffaloes (Alim, 1978). The average composition of buffalo milk and its comparison with different species is given in (Tables 16.2) and that for different breeds of buffaloes from other countries is given in (Table 16.3).

Table 16.2 Average Composition (%) of Milk from Different Species

Milk nutrient	Buffalo	Cow	Sheep	Goat	Human
Fat	7.0	4.3	6.0	4.5	3.5
Protein	4.0	3.4	4.8	3.8	1.9
Lactose	5.1	4.8	5.0	4.7	6.5
Minerals	0.8	0.7	0.9	0.8	0.2
SNF	9.8	9.0	10.3	9.0	7.3
Total solids	16.7	13.3	16.3	13.5	12.1

Pandya and Khan (2006).

Table 16.3 Composition of buffalo milk from different countries (all values in %)

Breed	Water	Fat	Proteins	Lactose	Minerals	SNF
Bulgarian	82.56	7.50	4.33	4.80	0.81	9.94
Carabaos	78.46	10.35	5.98	4.32	0.84	11.19
Caucasian	82.66	7.58	4.05	5.22	0.74	9.76
Chinese	76.80	12.60	6.04	3.70	0.86	10.60
Egyptian	82.09	7.96	4.16	4.86	0.78	9.95
Hungarian	83.79	7.22	3.65	4.56	0.78	8.99
Italian	81.94	7.85	4.28	4.97	0.75	10.21
Murrah	81.80	8.03	4.51	4.75	0.91	10.17
Murrah	82.76	7.38	3.60	5.48	0.78	9.86
Rumanian	81.75	8.23	4.79	4.48	0.76	10.02
Russian	81.04	8.56	4.76	4.78	0.86	10.46

Rao and Nagarcenkar (1977).

16.1.1 Milk fat

As stated earlier, buffalo milk contains much higher amount of fat than cow milk. It is nearly twice as rich in fat as cow milk. The average size of fat globule of buffalo milk ranges from 5.4 to 5.7 μ. Buffalo milk fat globules are coarser and one ml of buffalo milk contains roughly 2.7 million fat globules with 60 % fat globules having size between 3.5 and 7.5 μ. Numerous factors influence the size, composition and no of fat globules in buffalo milk. Although the buffalo milk fat contains higher concentration of total glycerides due to high fat content, the proportions of the 3 types of glycerides (mono, di and tri) are similar in buffalo milk and cow milk.

The fatty acid composition of buffalo milk fat is distinctly different from that of cow milk fat. The detailed fatty acid composition of buffalo and cow milk fat is given in is given in (Table 16.4). Fatty acid profile of other lipid components has also been studied in details.

Table 16.4 Fatty Acid Composition (w/w %) of Buffalo and Cow Milk Fat

Fatty acid	Buffalo milk fat	Cow milk fat
C 4.0 butyric	4.36	3.20
C 6.0 caproic	1.51	2.11
C 8.0 caprylic	0.78	1.16
C 10.0 capric	1.28	2.57
C 10.1 decinoic	Trace	0.31
C 12.0 lauric	1.78	2.78
C 14.0 myristic	10.81	11.93
C 14.1	1.27	2.12
C 15.0	1.29	1.23
C 16.0 Br.	0.18	0.30
C 16.0 palmitic	33.08	29.95
C 16.1 palmitoleic	1.99	2.16
C 17.0	0.58	0.34
C 18.0 Br.	0.24	0.35
C 18.0 stearic	11.97	10.07
C 18.1 oleic	27.15	27.42
C 18.2 linoleic	1.51	1.49
C 18.3 linolenic	0.47	0.59

Ramamurthy and Narayanan (1971).

The values of physico-chemical constants of buffalo milk fat obtained by various workers are given in (Table 16.5). Buffalo milk fat has higher melting point (MP), density, specific gravity and saponification value and larger grain size, but lower BR reading, refractive index, acid value, iodine value, Richert Missel value and Polenske value than cow milk fat. These constants are affected due to the stage of lactation, season, feed and thermal oxidation of fat.

Table 16.5 Physico-chemical Constants of Buffalo and Cow Milk Fat

Physico-chemical constants	Buffaloes	Cows
Softening point °C	34.3–36.3	33.5–35.9
Melting point °C	33.4–46.4	31.5–35.2
Acid value µm	0.17–0.352	0.26
Refractive index	1.4515–1.4533	1.4498–1.4530
BR reading	41.00–43.50	41.05–42.40
Saponification value	218.23–236.10	221.0–238.0
Iodine value	27.00–33.90	27.70–37.32
RM value	27.83–35.50	24.6–29.7
Polenske value	0.7–1.6	1.3–1.8
Density g/ml	0.905–0.917	0.888–0.911
Grain size, mm	0.20–0.41	0.098–0.190

Compiled from many sources.

Buffalo milk and ghee contain significantly lower free fatty acids than cow milk and ghee. Skim milk contains higher amount of free fatty acids than whole milk and its concentration increases on souring. In initial first to third lactation the content is higher, which decreases during third to sixth lactation and again increases in subsequent lactations.

A significantly lower concentration of total and free cholesterols (275 and 212 mg/100 g respectively) in buffalo milk fat has been found than that in cow milk fat (330 and 280 mg/100 g respectively). The phospholipids content of buffalo milk per unit weight of fat is much lower than in cow milk fat.

The buffalo fat contains more squalene and ubiquinone, less leutin, lanosterol, ethers and alkanoles (ethanol, methanol and butanol) than cow milk fat. Buffalo milk has less unsaponifiable matter than cow milk.

16.1.2 Milk proteins

The protein content of buffalo milk has been reported to range between 3.87 and 4.32 %. Although buffalo milk has higher amount of total proteins, caseins and whey proteins than cow milk the proportions of different fractions are similar in both the milks. Detailed distribution of nitrogen in different fractions of buffalo and cow milk (Table 16.6) revealed that out of the total proteins about 80 % are caseins and 20 % whey proteins. Colostrum has greater proportions of whey proteins but caseins are lower than in normal milk.

Table 16.6 Distribution of Nitrogen in Different Protein Fractions in Buffalo and Cow Milk

Nitrogen fraction	Buffalo milk			Cow milk		
	Nitrogen mg/100	% of total N	% of total protein	Nitrogen mg/100	% of total N	% of total protein
Total nitrogen	600.33	100.00	–	573.33	100.00	–
Protein nitrogen	573.67	94.15	100.00	542.33	94.59	100.00
Casein nitrogen	460.67	75.60	80.30	437.00	76.22	80.58
á lactalbumin	48.33	7.93	8.43	39.00	6.80	7.19
â lactoglobulin	37.00	6.07	6.45	36.33	6.37	6.70
Proteose-peptone	31.00	5.09	5.46	29.67	5.18	5.47
Non-protein	35.00	5.74	–	31.00	5.41	–

Sindhu and Singhal (1988).

Almost all the casein of buffalo milk is present in the micellar form. In cow milk 90–95 % is in the micellar form and the rest is present in serum. The ratio of micellar and soluble casein for buffalo milk is 91 and that for cow milk is 21. The size of the micelle in buffalo milk ranges from 80–250 nm with majority of them having 110–160 nm size compared to 70–110 nm in cow milk. The voluminosity of buffalo casein micelle is 2.68–3.72 ml/g in the temperature range of 25–37 °C. Solvation of casein micelle as calculated from voluminosity is 2.6–2.9 g water/g casein. This is much lower than the solvation of cow milk caseins. Opacity of buffalo casein micelle is higher that of cow casein micelle.

The amino acid composition of casein micelle from buffalo milk has been studied by many workers. The amino acid composition of buffalo casein and cow casein is almost alike.

The relative concentration of main three fractions (α_s, β and κ) is 4.4, 52.4 and 3.1 5 for buffalo and 54.4, 39.1 and 6.4 % cow milk caseins respectively. All the three fractions of buffalo milk casein have slower mobility than cow milk casein.

Buffalo milk contains higher proteose peptone than cow milk. Its proportion is 4.7 % of the total caseins in buffalo milk.

The proportions of different fractions of whey proteins in buffalo milk are similar to those in cow milk. Amino acid composition of buffalo β-lg is identical to that of cow β-lg. However, buffalo β-lg does not exhibit genetic polymorphism and is identical to cow β-lg in electrophoretic mobility, sedimentation and titration behaviour.

Buffalo and cow ά la have same crystalline form, similar nitrogen content, specific extinction coefficient at 280 nm, and tyrosine and tryptophane contents.

The concentrations of Igs is very high in colostrum compared to those in normal milk. Four classes of these proteins, *viz.* IgGa, IgA_1, IgA_2 and IgM, have been identified in buffalo milk and colostrum. Lactoferrin content of buffalo milk is much higher than that in cow milk. Its content in buffalo colostrum is still higher.

16.1.3 Minerals

Buffalo milk contains more minerals than cow milk. The values (mg/100 ml) related to different mineral constituents are: total minerals (700–900), calcium (177–224), magnesium (16.03–29.94), phosphorous (102–137), sodium (44.75–137), Potassium (43–119.75), chloride (63.82–106), citrates (157.7–217.97) and iron (0.19).

There was a slight positive relationship between specific gravity and ash content of buffalo milk, which was absent in cow milk. Between ash and crude protein content a high linear relation was found.

The partitioning of minerals, particularly Ca, Mg, P and citrates in two phases of milk has been studied by many workers because of influence of these minerals on many technological properties of milk.

The macro and micro elements in buffalo milk are shown in (Table 16.7).

Table 16.7 Macro and Micro Elements (ppm) in Buffalo Milk

Sodium	750	317
Potassium	1390	908
Calcium	2030	1880
Magnesium	200	91.9
Iron	–	0.325
Phosphorus	1290	–
Zinc	–	6.26
Copper	–	0.303

16.1.4 Vitamins

In general the vitamin A content in buffalo milk is higher than that in cow milk. However, due to the absence of carotenoids and high fat content, its total vitamin A potency per unit weight of fat is lower than in cow milk fat. Buffalo milk fat contains lower tocopherol than of cow. However, due to higher fat content buffalo milk contains higher than cow milk. Its content is several times higher in colostrum. The ascorbic acid content of buffalo milk was found to be 25–30 mg/kg.

The concentrations of various vitamins of group B in buffalo milk as per Boman (1953) (in μg/ml) are thiamin, 0.5; riboflavin, 1.02; nicotinic acid, 2.60; biotin, 26.75; folic acid, 0.13; pantothenic acid, 1.5%, pyridoxine, 3.75 and p-aminobenzoic acid 26.75. Similar concentrations for thiamin (0.5 μg/ml) and riboflavin (1.07 μg/ml) but lower concentration for nicotinic acid have been reported in another study.

16.1.5 Enzymes

Like cow milk, buffalo milk also contains a variety of enzymes. Of these some 20 enzymes have been isolated, purified and identified. Only some of the buffalo milk enzymes like lipases, alkaline phosphatases and xanthine oxidases have been studied in detail.

Alkaline phosphatase activity in buffalo milk is also about two-thirds of its activity in cow milk. The activity of lipase enzyme is also lower in buffalo milk. The activity of xanthine oxidase enzyme in buffalo milk is similar to that in cow milk according to some, whereas others have reported it to be lower than cow milk.

The protease activity of buffalo milk is slightly higher than of cow milk. The lysozyme content of buffalo milk is lower than of cow milk. The ribonuclease is present in higher concentration in buffalo milk. The concentration of A protein of lactose synthetase is higher but of β protein lower in buffalo milk than in cow milk. The concentration of y-glutamyl transpeptidase is less in buffalo milk than in cow milk.

16.1.6 Composition of colostrum

Approximately during the first three days of lactation the buffalo secretes colostrum. Colostrum is vital for the newborn calf and its composition reflects the calf's need (Table 16.7). Colostrum contains the important proteins: the immunoglobulins, which are the newborn calf's source of antibodies. The content of iron and copper is markedly higher in the colostrum as compared to normal milk.

Table 16.8 Composition of Colostrum

	Water %	Fat %	Total protein %	Lactose %	Vitamin A (μg/kg)
Colostrum	68	15	13.6	3.1	–
Normal milk	73	9.55	9.59	7.54	1.8

16.1.7 Alterations of milk composition

Milk composition can be altered both before and after the milking. If the change occurs inside the udder it is mostly due to a disease or treatment of the disease by antibiotics or other type of medication. Feeding can alter the normal composition, however, these changes are seldom extreme, but within normal intervals. Season can affect the normal milk composition, although these changes are mostly due to differences in feeding during different seasons.

(a) Stage of lactation and milk yield

The fat percentage varies with stage of lactation and with milk yield. A study on Nili-Ravi buffaloes in Pakistan showed that the fat percentage increased steadily from 5.5% in the first month of lactation to 7.5% in the 10th month of lactation.

There is a negative correlation between lactation yield and percentage of total solids, fat and protein. However, the total amount of solids, fat and protein is higher in a high yielding buffalo than in a low yielding one.

(b) Feedstuff

A rule of thumb is that roughage increases fat content in milk, whereas concentrate depresses it. This depends on the differences in VFA production in the rumen from the different carbohydrate sources. Digestion of fiber results in a higher proportion of acetic acid and thereby more milk fat. Digestion of concentrate on the other hand, results in a higher proportion of propionic acid which is unfavorable for milk fat synthesis. If too much concentrate is given, fat depression might occur.

Higher energy diets seem to give better coagulation properties of the milk. Long chain fatty acids increases when the energy concentration in feed is low.

Glucosinolates in *Brassica* spp. are hydrolyzed by the ruminal microbes into thiocyanates, iso-thiocyanates and some other products. Thiocyanate is then excreted in the milk. High feeding levels with *Brassica* spp. may therefore lead to unsatisfactory levels of thiocyanate in the milk. Thiocyanate may cause thyroid enlargement in animals as well as human ingesting it. A common feed stuff of *Brassica* spp. is mustard fodder and mustard oil cake. Even 15 days after withdrawal of mustard feed, circulatory high levels of thiocyanate exists and is secreted in milk.

(c) Disease and medication

Mastitis changes the milk composition dramatically. The alterations can some times be used as detection of the disease. If antibiotics are used to cure for example mastitis, these will be excreted in the milk. Controlling of external parasites with *e.g.* diazinon affects milk yield as well as composition. The chemical is detected in the milk up to 48 hr after dermal application.

16.2 Physico-chemical properties of buffalo milk

The compositional differences between buffaloes and cows milk are reflected in their physico-chemical properties, too. Although physico-chemical properties of buffalo milk has not been extensively studied, the available literature provides sufficient background information related to these aspects. (Table 16.9) gives average values for various physico-chemical properties of buffalo and cows milk.

Table 16.9 Physico-chemical Properties of Buffalo and Cow Milk

Property	Buffalo milk	Cow milk
Acidity, % lactic acid	0.13	0.15
Buffer value at pH 5.1	0.0417	0.0359
Curd tension, g	32–85	28–54
Density at 20 °C	1.0310	1.0287
Electrical conductivity, mmhos	6.69 ± 0.223	6.615 ± 0.271
Fat globules size, nm	5.01	3.85
Fat globules, no. in millions per mm^3	3.2	2.96
Fluorescence under UV light	Greenish yellow	Pale bluish
Freezing point, °C	-0.552 to -0.558	- 0.522
Heat capacity, Cal/g/°C at 20 °C	0.852 ± 0.017	0.933 – 0.954
Oxidation reduction potential, Eh	+ 0.31 V	+ 0.258 V
pH at 20 °C	6.74	6.60
Phosphatase activity (units/100z)	28	82
Reflective index	1.3448	1.3338
Surface tension	55.4 (49.51–50.70)	55.9
Thermal conductivity, Kcal / h m °C	0.5689 ± 0.00734	0.460
Thermal expansion	4.106×10^{-4}	–
Viscosity, centipoises	2.04	1.86

Compiled from different sources.

16.3 Buffalo milk products

Buffalo milk is used for preparing the same products as those made from cow milk such as yoghurt, sweets, ice-cream and various types of cheese. Soft cheese made from buffalo milk in the Philippines and Iraq are delicacies, and mozzarella (the Italian pizza cheese made from buffalo milk in southern Italy) is a product of world fame and a gourmet's delight. In India, good quality processed cheese is manufactured from buffalo milk. Ice-cream made from buffalo milk, with its rich cream, tastes better than any other available in the market.

Tropical peoples developed many milk products peculiar to a local situation and climate. Several of these products have been made for generations and a few are important today in the dairy processing industry of particular areas. Milk products common in world markets are also made in tropical dairy plants, primarily for domestic use. Attention is mainly directed in this section to the more important indigenous milk products.

In warm climates coagulation of milk might occur relatively soon because acid quickly develops in milk by bacterial action if the milk is not cooled. Early man observed that coagulated milk kept a little longer, was not usually harmful and had a flavour which he thought pleasant or which he could soon learn to enjoy. The nature of the spontaneous fermentations of milk varies from time to time and from place to place. Curd forms so universally if raw milk is held unprotected for a few hours at temperatures, characteristic of warm climates or hot seasons, that it was surely one of the first products of milk known to man.

Coagulation of milk by action of acid may be said to take place in three stages. In the first stage small concentrations of acid develop, but they produce no appreciable visual affect. The milk remains liquid, but acquires an acid flavor and odor and a lower heat stability which may be indicated by coagulation on heating. In the second stage, as the concentration of acid increases, the stability of the milk is so adversely affected that at ordinary temperature a distinct visual change may be observed as the milk coagulates. This is indicated by the milk changing to a gel like or semi solid condition. Sometimes very suddenly the third stage will follow, either on heating the coagulated milk or on further development of the concentration of acid. A distinct shrinking and solidifying of the curd takes place as it separates from the whey– a nearly clear, watery liquid. Certain enzymes, produced by bacteria or obtained from other sources, will also coagulate milk to form a curd.

Man gradually learned to influence the natural fermentations of milk by changing and controlling the conditions under which milk was handled until it coagulated. This contributed ultimately to the wide use of fermented milk and its products as human food, possibly even before the Phoenician era (Kosikowski, 1966). Great numbers of these products are now known and relished in the tropics (Bhattacharya and Srinivasan, 1967; Davies, 1939a; Mahanta, 1964b). They include the full unbroken curd; the mixed curd, usually broken finely enough to give a thick liquid product like cultured or artificial buttermilk; the drained or semi-dried curd; the why; and many forms and modifications of each.

16.3.1 Curd

Curd may be from whole milk or skim milk. In a tropical home or village the milk is usually boiled, and allowed to cool to atmospheric, *i.e.* to incubation, temperature. It is then inoculated and incubated until the desired fermentations are complete (Davies, 1939a). Curd made one day is usually used as an inoculum or starter for making curd the next day. In addition to acid fermentation that gives a clean, sharp, acid flavour, other fermentations are often encouraged to assure other flavour characteristics according to local conditions and preferences. A fine texture and firm body which does not whey off easily are also desired in the curd.

The unbroken or solid curd is greatly enjoyed plain, or it may be flavoured, usually with sugar, salt, fruit juices, or jam. It may also be coloured. A refreshing drink or beverage is made from the mixed or broken curd (Warner, 1951). It is mixed with an approximately equal volume of water and flavoured to taste, or coloured (lassi). Essences and sometimes spices are used for flavouring solid curd in addition to the items mentioned above. Curd is a common ingredient of numerous fine dishes or preparations included in the diet in many tropical homes (Bhanumurthi and Trehab, 1970; Rangappa, 1965; Wrner, 1951).

The practise of mixing fruit products with curd has spread in recent years to temperate countries. The dairy industry in several of these countries has begun to manufacture and distribute large quantities of fruit flavoured yogurt (Gillies, 1971).

In western Asia and in Africa, nomadic people who move their animals over large areas to provide them with adequate grazing, and people who regularly travel with caravans across large expanses of desert, continue by tradition to depend upon milk from their animals as one of this main foods. Some of this milk is used fresh, but sour milk, fermented milk, curd and cheese are commonly made from the remainder. Some curds are dried and saved for lean seasons when they are reconstituted and eaten. Ghee is also made and used. Small amounts of this may be sold in the local markets near a route being followed. Cow, sheep, goat and camel milks are generally used; mare milk is utilized in some regions of Asia.

16.3.2 Cheese

Products of the third stage of acid and/or enzyme coagulation are usually called cheese. Several hundred kinds and varieties of cheese are known. Most cheese does not ordinarily keep for long periods of time in areas or seasons of high atmospheric temperatures without special facilities. For this reason cheese is not so common an article of diet in the tropics as in the temperate parts of the world. There are, however, several kinds of cheese made and enjoyed in hot countries. Domiati and Kariesh cheese are well known in Egypt (Hamdy, 1970; Imail and Elnahat, 1964; Kosikowski, 1966). In the Indian subcontinent several soft, high moisture cheeses such as Surati, Dacca and Bandal are made. The latter two may be smoked (Davies, 1939a; Warner, 1951). Indigenous cheeses named Requeijao and Minas (Assis and Frensel, 1966) are made in Brazil.

16.3.2.1 *Surati cheese*

Surati cheese is made by standardizing milk to 6% fat and heating it to 77.5 °C (172 °F) for about 20 seconds; the milk is then cooled to 35 °C (95 °C) and approximately 14 g (0.5 oz) of an active acid starter, preferably *Streptococcus lactis*, is added for each 45 kg (100 lb) of milk, followed by sufficient rennet solution to give a firm coagulum in about one hour. By use of a spoon, thin spoon-sized pieces of the curd are dipped into a small basket. During dipping each piece is placed flat, one over another, to form layers. Salt is added to the curd throughout the dipping, at a rate equal to about 2% of the weight of the milk. During draining the cheese is turned several times by first inverting an empty basket over that containing the curd and turning the two together. The whey is collected. After the prints are well formed and firm they are immersed in the whey where they are allowed to ripen at room temperature for up to a day and a half, according to the flavour and body desired. This process gives about a 40% yield of cheese from standardized milk (Warner, 1951).

16.3.2.2 *Cheddar cheese*

While many varieties of cheeses are made using buffalo milk, Cheddar cheese has greater production in India and elsewhere. Several reviews and scientific papers published so far highlight the problems associated with use of buffalo milk for manufacturing various dairy products, including cheeses. Patel *et al*. (1994) have critically reviewed the work on cheddar cheese from buffalo milk.

Unlike cow milk, buffalo milk is regarded as a poor raw material for manufacture of hard varieties of cheese like Cheddar, owing to the quantitative and qualitative differences. Use of technique(s) standardized for cow milk does not yield typical Cheddar cheese and several difficulties listed in (Table 16.10) are encountered and as a result standardized procedures are developed for making good quality Cheddar cheese from buffalo milk.

Table 16.10 Problems Associated with Buffalo Milk for Cheddar Cheese

Property	Effect	Probable reason
Acid development	Slow	High buffering capacity
Curd tension	High	High Casein and Ca content and large size casein micelle
Rennet action		
Primary	Slow	Difference in size of casein micelle, less ê casein on micelle
Secondary	Fast	Higher amount of Ca
Fat losses in whey	More	Larger size fat globules
Moisture retention in curd and cheese	Low, higher evaporative loss in ripening	Larger size and lower voluminosity of casein micelles, higher amount of minerals
Body and texture	Hard, dry,	Larger size casein micelles, higher mineral short, crumbly makeup of casein micelles, low moisture
Flavour development	Slow	Low moisture, slow proteolysis, slow lipolysis

16.3.2.3 *Mozzarella cheese*

Mozzarella is a well known variety of Italian cheese having popularity world over. In Italy, name Mozzarella is used for the cheese prepared only from buffalo milk. The term Mozzarella di bufala in Italy has a legal protection as a product made strictly from buffalo milk. Mozzarella is popular as a pizza topping in the United States and Europe. Buffalo milk is preferentially used for making mozzarella cheese.

Methods of Mozzarella cheese making

Manufacture of buffalo milk mozzarella cheese has been standardized employing two different approaches. The conventional approach or "starter culture method" involves fermentation of milk by starter cultures, rennet coagulation, separation of curd, stretching and brining of the product. The other procedure referred as "direct acidification technique" involves addition of acids instead of starter culture before renneting.

Starter Culture method for Mozzarella cheese

In this process, buffalo milk standardized to casein: fat ratio of 0.7:1 is pasteurized (72 °C/no holding) and inoculated with 2% starter culture of *Streptococcus thermophilus* and *Lactobacillus bulgaricus* (1:1). The milk is incubated at 37 °C for 40–45 minutes, until an acidity of 0.01 to 0.02% lactic acid develops.

Rennet is then added at 37 °C, and the milk is allowed to set for about 30–45 minutes. The curd is cut and cooked with the whey at 40 °C for about 2 hr and 30 min until an acidity of about 0.4% lactic acid is reached. After draining of the whey, 2.5–3.0% sodium chloride is added, and the curd immersed in boiling water for 4–5 minutes. The curd is then plasticized, manually or mechanically at 85–90 °C and shaped into balls or rectangular blocks. The product is immersed in pasteurized cold water at 4–5 °C for 2 hr, and finally packaged in polyethylene bags or other suitable packages and stored at 5–8 °C (Upadhyay *et al.,* 1986).

Ghosh and Singh (1991) also standardized the starter culture method for production of buffalo milk mozzarella cheese. As per the process pasteurized milk containing 4% fat is added with 1:1 culture of *Streptococcus thermophilus* and *Lactobacillus bulgaricus* and coagulated with microbial rennet (Meito rennet). The curd pieces are heated to 41 °C in the whey. One-third of the whey is drained at this stage, and cooking is continued till a titratable acidity of 0.40–0.43% lactic acid is reached. The curd is plasticized in hot water at 80–83 °C and moulded into pats or blocks. The cheese is cooled in chilled water, and transferred to chilled brine. The cheese is dipped in brine till it attained a salt content of about 1.75%. The product was drained and finally packaged.

Direct acidification technique

Good quality Mozzarella cheese from buffalo milk can be prepared by this technique. In this process fermentation of milk by starter culture is done away by addition of known quantities of acid to get specific desired pH at cutting. Addition of 1.6–3.5 ml of HCl or 2–4 ml acetic acid per liter of buffalo milk gives the desired pH at 6–8 °C. Rennet is added @ 0.9–1.0 g calf rennet or 0.4–0.5 g meito rennet per 100 liters of milk. After addition of rennet and thorough mixing the temperature of the mixture is raised and maintained at 35 °C. When the curd sets, it is cut and stirred for about 20 minutes in whey at 35 °C. It is further dipped in whey for 30 minutes, which improves the melting and stretching properties of the cheese. The whey is drained off and 3.0 % common salt is added. The curd is plasticized at about 90 °C and shaped into blocks. The product is immersed in water at 4–5 °C for 2 hr and packaged.

In another direct acidification process standardized by Upadhyay *et al.* (1986), buffalo milk was acidified at 20 °C with lactic acid to a pH of 5.0. It was renneted at the same temperature for 30–35 minutes, cut and cooked at 37 °C for 30 minutes to obtain an acceptable quality mozzarella. The product had a superior meltability and lower fat leakage than mozzarella made by the conventional process.

16.3.3 Paneer

Paneer was originally found in the area that today encompassess Iran, India and Pakistan by nomads. It is the persian word for "cheese." The nomads carried it with then in camel skin bags. It is a high protein food generally substituted for meat in India and Mughlai and Arab recipes. Paneer is a unripened fresh and delicate cheese made from milk in Indian subcontinent. It is rich in nutrients but not heavy on calories, involves somewhat delicate making procedure.

Traditionally milk is boiled and either fresh lemon juice or a spoon full of citric acid is added to the milk while it is still hot. When the milk gets fully curdled, the liquid is drained out by pouring it on to a clean fine muslin cloth. Once all the liquid has been drained, it is laid flat and some weight has to be kept on it for some time. It becomes solid if kept for an hour or two. This can be used in various interesting recipes. Paneer needs to be refrigerated and used within 3–4 days. It develops a sour and stale taste if kept for long. Some of the most common recipes made with paneer are palak paneer, shahi paneer, paneer masala and matar paneer.

Some nutrition facts about paneer:

1. Paneer is a good source of calcium, which in later years, helps prevent osteoporosis.
2. Paneer in moderation, is associated with lower body weight as well as reduced risk of developing insulin resistance syndrome.
3. Paneer is a good source of protein and it reduces cancer risk.
4. According to Ayurveda, paneer is a health food and it can prevent stomach disorders.
5. It is supposed to help in lower back and joint pain.

16.3.4 Butter and ghee

Acid coagulated whole milk, particularly buffalo milk with its high fat content, may be churned to produce types of butter and buttermilk (Davies, 1939a; Mahanta, 1964a). An Indian name for this butter, *i.e.* '*deshi*, country or cooking butter' indicates that it is usually made by a village method and is intended for cooking purposes. The buttermilk obtained by this churning is generally used in the home as a food beverage. It contains much of the lactose, protein and ash, and some of the fat of the whole milk from which it was made (Warner, 1951). *Deshi* buttermilk is an excellent food. Throughout its preparation, therefore, it should be protected from undesirable bacterial contamination to assure its safety for food use.

Deshi butter is commonly used in homes in tropical Africa and Asia for making ghee. The butter is heated in an open pan to evaporate its moisture. The non-fat milk solids then settle to the bottom and the butterfat or ghee can be decanted off practically free of any of the other constituents. The heating should continue initially until active boiling ceases, indicating that the free moisture has been separated from the fat by evaporation. The boiling will take place at approximately 100 °C (212 °F). Heating should continue thereafter with considerable caution. During the second part of the heat in process only very slow boiling should be permitted and no more than slight foaming should occur if the control is adequate. The temperature of the butterfat during this time should not exceed 118 °C to 123 °C (244 ° to 253 °F). This higher range of temperature, compared to that during earlier heating, results from the absence of free moisture. Great quantities of heat are needed to change water into steam. This consumes much of the heat during the first stage of heating. Therefore, during the second stage of the heating the temperature is free to rise still higher, for the butterfat is itself then receiving almost all the heat. Skilled ghee makers exercise greatest possible care at this time to prevent excessive heating. Sufficient heating of the product is indicated by the non fat solids drying out to a light amber colour. Even slight over heating at this stage will give a burnt flavour and a dark colour to the ghee, both undesirable in fine quality market ghee.

Ghee may also be made form unsalted and uncoloured factory butter, or directly from cream. These two methods are more common in organized dairy operations. Culturing the cream for making the butter or for heating directly is said to improve ghee flavour.

During the churning of the butter, a certain type of 'separation' of fat and the non fat solids takes place. The physical forces that hold the fat in globule form, and dispersed throughout the serum of the curd or the cream, break down during churning to give a separation of the fat as it gathers in the form of butter granules. The serum no longer holds the fat and the two fractions become more distinct or separate. This change in the product reduces the heat required for making ghee from butter as compared to that from cream. Churning cream to produce factory butter gives greater degree of this change than does churning curd to produce deshi butter. Fuel requirements and the duration of heating of cream in making ghee, therefore, may be significantly greater than those for *deshi* butter, and those in turn somewhat greater than for factory butter.

For ghee making, butter made by churning cream has not only the advantage of this 'separation' of the fat and non-fat solids, but also has a high fat content and comparatively low non-fat solids and moisture contents. Such butter, if not for a sale as such, may contain over 80% of fat, 1 to $1^1/_2$% of milk solids-not-fat, and

16 to 20% of moisture. Cream may contain only up to 50 to 60% of butter fat, often somewhat less, but 5 to 7% of non-fat solids and as much as 60% or more of moisture.

The first stage of heating of cream is prolonged by the relatively large amounts of moisture and non-fat solids present. Furthermore, the undisturbed physical condition of the fat and other solids of cream complicate this part of the *ghee*-making process. Cream may boil up and froth and sputter during heating, making special care essential to prevent these troubles and the fat losses that may cause. Cream requires longer heating at this time because of its high moisture content.

The second stage of heating of cream may also require more time and fuel, and more caution, than that of factory butter, because of the higher content of non-fat solids. Not only is longer heating required in order adequately to dry out these extra solids, but they are often present in larger and thicker particles, from which it is more difficult for the moisture to escape than it is from smaller or thinner curd particles.

Experimentally, cream has been washed as a means of reducing the foaming and sputtering during earlier heating and the danger of over heating during later heating. It was hoped that this practise might reduce fuel requirements. Washing involves adding sufficient water to the cream to restore the fat content to approximately that of whole milk, then recuperating this mixture. The added water dilutes the solids in the non-fat fraction of the cream and most of these are then removed on reseparation. The addition of water and reseparation may be repeated a second time. Washed cream is not so likely to froth and foam during heating and contains very little non-fat solids. Fat losses during heating may be noticeably reduced, and the final heating more easily controlled.

An advantage of heating cream directly for ghee-making is that it avoids the extra processing and possible fat losses on churning. Washing cream, however, involves extra processing and fat losses during reseparation.

Induced stratification (De and Srinivasan, 1958; Ray, 1955) may conserve fuel during the first stage of heating when making ghee. This procedure is most effective with factory butter. The butter or cream is heated in a pan or kettle having a drain cock or valve in its bottom. This product is held at about 80 °C (176 °F) until it separates into two layers, the fat above and the water and most of the non-fat solids below. This may take 30 to 40 minutes, depending on the size of the batch. As much as possible, the water and non-fat solids layer is then drained off from beneath the fat. Heating then proceeds in the usual way.

Ghee made for use in the village home and not for sale might contain 2 or 3% moisture, possible more; *deshi* butter in such cases may have 20 to 30% moisture (Davies, 1939a). Regulations in India require that for market purposes ghee shall contain not more than 0.3% moisture; *deshi* butter if for sale as such, should contain at least 76% of butterfat.

Products comparable to ghee and recently developed in temperate countries such as anhydrous or dry butterfat (Gunnis, 1967) and butter oil (Jenness and Patton, 1969) are not often made as yet in most tropical countries.

Butterfat deteriorates like any other organic material. However, it usually does so more slowly than other solid constituents of milk, except possibly some parts of the ash fraction, butterfat changes chemically and physically if exposed to air, sunlight and/or certain metals from which dairy utensils and equipment are sometimes made, especially copper and iron. Some of these changes cannot be arrested or reversed after they have once started; most of them occur slowly. If ghee is made and packed so as to protect the butterfat throughout from such contacts, and also from bacterial contamination, it can be expected to keep several weeks without refrigeration. Because of its comparatively longer keeping quality, ghee is normally moved from a home or village to a market only once each week or so. Both its keeping quality and its high value per unit of weight permit its economic transport over comparatively long distances.

Factory butter for market is made in tropical countries wherever the cream supply is well organized in an intensive milk producing area. The method used in manufacturing this product is familiar throughout the butter industry elsewhere, *i.e.* by batch churning mechanically separated cream. Continuous butter making procedures have not yet been adopted widely by the dairy industry in the tropics.

Butter is packaged for sale in 'prints' which are generally rectangular and which may be wrapped with parchment paper or in a parchment and foil laminated wrap with the foil outside. Wrapped may sometimes be placed in a waxed or plastic coated cardboard carton. In many tropical areas butter is tinned. Butter requires refrigeration so as to preserve it and prevent it from melting. If it is not tinned it may flow out of the package if it melts. Refrigerated transport vehicles and retail sales cabinets are becoming more common in many cities in tropical areas and so fine quality packet butter is consequently more widely available than it was a few years ago. In the shops of small towns and villages and at times in larger cities, packet butter may be held immersed in water on the sales counter until it is sold. This practise usually keeps the butter within the package, but contributes to rapid bacterial growth both within the butter and on its surface.

If made and packaged under condition which are highly aseptic, tinned butter made from pasteurized cream may keep several weeks without chemical or bacterial deterioration. A physical defect, however, frequently appears in tinned butter as a result of the high temperatures to which it maybe exposed in storage, in transit, in the market or in the home. Butter is likely to melt at 38 °C (100 °F). Should butter melt, it separates into fat and water fractions or layers inside the tin. Butter which has physically broken down in this way will not return automatically to its original physical condition, nor can the water readily be remixed into the fat as it was in the original butter. This reduces the market quality of tinned butter.

16.3.5 Khoa and Chhena

Buffalo milk owing to its high solids concentration, is extensively used for making various indigenous dairy products. Two products, made mainly in Asia, *i.e.* khoa and chhanna are primarily for use in sweets (Rastogi *et al.*, 1966).

Khoa is a partially heat desiccated milk. It is made by rapidly boiling milk in an open pan until the volume of milk is reduced 75 or 80%. The milk must be stirred vigorously and constantly scraped from the bottom of the pan during boiling so that no scorching or charring of the milk solids occurs. On cooling, this product is often as hard as cheddar cheese and has a crystalline texture. Quick cooling gives a comparatively fine texture by producing smaller lactose crystals throughout the body of the khoa; slow cooing gives a coarse texture. A fine texture is preferred. If made from buffalo milk, khoa has a grayish white colour; if made from Zebu milk, its colour may be slightly yellowish. Considerable developments have taken place with respect to various aspects of this product, including its mechanized large scale continuous production.

Chhena or chhanna is made by bringing a small quality of cow milk quickly to a boil, up to 1.8–2 kg seems most suitable (Davies, 1939a). Buffalo milk is usually not used for making chhena. As the milk boils, or immediately after lifting it off the fire or stove while boiling, sufficient sour whey or lime juice and occasionally a strong citric-acid solution is mixed into coagulate the milk. Complete coagulation is desired, but an excessive acid flavour will result from even a slight surplus of the coagulant. The chhena or curd is then recovered by draining off the whey through a cloth. Often the curd is pressed several hours to reduce its moisture content still further before it is used. Persons of Bengal origin particularly enjoy sweets made from chhena, such as the *rasgolla* (also called *rasogolla*), *sandesh* and many other such delicacies.

Standard methods of the manufacture of several products indigenous to tropical areas are not widely adopted either because they are not required under public health regulations or because competition makes them uneconomic. Indigenous tropical dairy products may not be of such uniform composition as are dairy products in the world's markets. Davies (1939a) indicated that in chhena the fat may vary from 25 to 35%, protein from 15 to 20%, lactose from 2.0 to 2.5%, and ash from 0.3 to 0.4%. Standard chhena, however, should contain not less than 15% of fat and not more than 70% of moisture. Khoa may have 25 to 30% protein and up to 50% moisture (Rastogi *et al.*, 1966). As it is made by evaporating water from whole milk, the ratio of one solid to another will be similar in khoa to that between the same two solids in the original milk.

16.3.6 Condensed milk

Condensed milk is made by concentrating milk by removing much of its water by evaporation. Whole milk or skim-milk is used, whole milk being usually standardized before evaporation to assure compliance with legal requirements. Sugar may or may not be added before condensing. Condensed milk was first made by Appart in France; its commercial production was started in 1856 by Borden in the United States (Hunziker, 1949). Brazil began the manufacture of this milk product in 1913. One of the oldest condenseries continuing to operate at present in the tropics was started in Jamaica in 1944. A small condensery was also operating at about this time in Lahore, Pakistan.

India now has twenty plants producing more than 12000 tons sweetened condensed milk a year. The country requires sweetened condensed milk to contain not less than 9% milk fat, 31% total milk solids, and 40% cane sugar; composition limits are also prescribed for condensed milk-unsweetened or evaporated milk, condensed skimmed milk unsweetened or evaporated skimmed milk and condensed skimmed milk-sweetened.

Sweetened condensed milk is usually retailed in tins containing 350 to 400 g (12 to 14 oz) of product. Unsweetened condensed milk may be tinned and then heat sterilized to assure its keeping quality. The high sugar content contributes to the keeping quality of sweetened condensed milk; the added sugar would caramelize, producing a serious defect in this product if it were heat sterilized.

16.3.7 Dried milk

Dried or powdered milk has become increasingly familiar in many tropical countries. In many of these areas it is now made, usually by one of two processes-roller or spray drying.

When made by the roll or roller process, milk is first partially concentrated, then dried in a thin film on the surface of a metal roller that is heated from the inside by steam. Several methods are used to ensure that the thickness of the milk film on the roller surface is as uniform as possible; this is important. Commonly two rollers are used, arranged so that between them they form a trough into which the milk flows and from which it is picked up by the roller surfaces as they move upwards and outwards while turning. The temperature at which the roller operates evaporates the water from the milk film, leaving on each roller a sheet of dried milk solids that is scraped off by means of a special knife.

Irregularities in the thickness of the milk film and other factors may give rise to improper heating of some portions of this milk film. Some browning often results from overheating, given irregularly coloured powder. The intense heating denatures some of the milk protein, reducing the solubility of this type of powder.

Partial concentration of milk prior to spray drying is practiced for economic reasons and product quality. This milk is sprayed or atomized directly into a blast of hot dry air which forcibly carries the milk droplets across a drying chamber or cyclone while removing water by evaporation. This reduces each tiny droplet of milk to a particle of powder. Heat application in this method of drying is comparatively uniform. It is less variable in its intensity, and consequently somewhat less denaturing of the milk protein takes place. Overheating is not so liable to occur. These factors contribute to the relatively higher solubility of spray powder which is without the colour defects of roller dried powder resulting from browning or charring of some of the milk solids. Spray dried milk powder is more often used for human food purposes.

Technologies have been developed for manufacture of various kinds of milk powders including infant milk foods and such other products using buffalo milk.

Imports of non-fat dried milk solids, or skim-milk powder, have been increasing in many developing areas throughout the past 20 years or so. In the last decade the local production of this product has become important in several tropical areas. India now has thirteen plants licensed to produce up to 30480 t (29970 tons) of milk powder a year. In 1970 eight of these plants were in production and others were under construction. Among these, the plant with the largest licensed capacity, 7200 t (7080 tons) a year, is the cooperative plant at Anand, India. Brazil has had national inspection of powdered milk production since 1961. The growing popularity of toned, double toned, standardized, filled and recombined milks in several countries increases the demand for the domestic production of skim milk powder. The cost of importing milk powder often makes its use too expensive, utilizing foreign exchange that is not always readily available to a developing economy.

16.3.8 Ice-cream

Ice-cream is greatly increasing in popularity in hot countries. A few manufacturers are equipped generally with a batch freezer to produce large volumes of this fine milk product. The hand operated freezer, using ice and salt for cooling and freezing the ice-cream, is still common where small quantities are made. In several areas the sale of soft ice-cream is increasing particularly rapidly. The special shapes and sizes of hard ice cream on sticks are very common, as are ice cream cones. Ice cream is also frequently sold in waxed paper or plastic cups. In certain areas local laws and ordinances effectively define the composition of ice cream.

A frozen product resembling ice-ream is made in the homes and villages of India and Pakistan (Davies, 1939a; Warner, 1951). *Kulafi* mix contains milk, sugar, and flour of either wheat, pulses or both; it is always flavoured and usually coloured. It is frozen in small covered containers, holding about 60 g (2 oz) in a cold ice-salt brine. Dry ice might be used for this. Traditionally the containers were made of tinned or galvanized iron, but waxed paper or plastic may now be used. The freezing is without agitation, giving a coarse texture that varies with the composition, particularly with the solids content. *Kulafi* is not made commercially on a large scale, but it is commonly served in the home as a dessert at a tea or dinner for small groups. Minhas *et al*., (2002) studied the flow behavior characteristics of ice-cream mix made with buffalo milk and various stabilizers. The viscosity of the mixes having various stabilizers (optimum levels) was found to be in descending order: sodium alginate, gelatin, karaya, guar gum, acacia, ghatti and control.

CHAPTER 17

UTILIZATION OF CARCASS BY-PRODUCTS

There is a prevalent misconception that full use of animal offal can be achieved only with costly equipment and under meat packaging conditions. In many countries this obsolete approach creates the paradox that while certain raw materials, such as blood, bones or intestines, are completely wasted, costly blood meal, bones or intestines, or sausage casings are imported.

Even hides and skins often receive extremely careless treatment because it is not appreciated that first class leather can be made from the local material, using simple methods and equipment, if it were only properly handled and prepared. Too often it is assumed that quality hides can only be achieved in hide cellars fitted with temperature and humidity controllers and using chemically pure salt.

It is economically unimaginative to consider an animal only as a source of meat and hide; it can yield an extensive range of by-products of benefit to the individual and the whole country. The aim should always be to make the fullest use of the whole animal. No matter how small the amount of each by-product, the aggregate on a countrywide basis will certainly add substantially towards the economy of the territory.

While it is true that large scale utilization of slaughterhouse offal requires costly and elaborate machinery, by-products of high quality can be obtained with locally made equipment and at very low cost.

17.1 Hides and Skins

Good leather can only be made from good hides and skins. With the increased cost of buildings and labour, the competition from plastic materials and the appearance on the market of properly flayed and cured hides and skins from countries with an established meat packaging industry or an efficient hide improvement service, there is today no chance of obtaining an attractive price for hides and skins unless they are as free from visible or invisible damage as possible.

The various processes to which hides and skins are subjected in the tannery, are a combination of physical and chemical action which tends to emphasize every hole, gouge mark, pox mark or even incipient putrefaction. This is contrary to the common belief that tanning covers blemishes so that the raw material has little bearing on the finished leather.

17.1.1 Pre-slaughter care

It is not sufficient to start the care of the hide or skin from the moment the beast appears at the slaughterhouse, as irreparable damage can be done on live animals prior to slaughter. Indiscriminate branding, either for identification or curative purposes, applied in places which produce the best raw material, will invariably reduce the value of the whole hide. The ornamental branding carried out by certain indigenous people renders the hide practically worthless.

A brand should be small and placed on an inferior part of the hide or skin, such as the ear, cheek, or the foreleg below the shoulder or below the stifle of the hind leg. On humped beasts, branding on the top of the hump does not damage the hide, as during the process of tanning the hide is split in two so that the brand appears on the edge of the finished leather.

Scratches from barbed wire and thornbushes, or horn gores, yoke calluses, wounds, whip lashes and other mechanical injuries to the hide or skin during the life of the beast, will have a bearing on the quality of the leather. Parasitic diseases like ringworm, streptothricosis, mange, or other diseases such as cowpox, hyperkeratosis, sweating sickness, etc., injure the grain. Warble-fly damage, resulting in holes on the back, which is the best part of the hide, is a serious menace in countries infested with this fly. Even tick bites leave spots on the finished leather.

17.1.2 Avoiding damage in the slaughterhouse

Most of the damage, however, occurs either through careless handling or lack of knowledge. It may be inflicted during killing and flaying or appear in the form of bruises when the animal has been beaten prior to slaughter or cast forcibly on a hard floor. Hauling the carcass over a rough surface damages the grain, and results in so called ‘dragged’ or ‘rubbed’ grain, especially obvious where the bones are prominent. Bad bleeding, resulting in the small cutaneous vessels being gorged with blood, may lead to putrefaction by encouraging bacterial growth and results in a raised grain ('veiny' leather).

Cuts and gouge marks caused either by the use of a pointed knife or the insecure position of the carcase, delays between slaughter and flaying, or the unnecessary use of a knife where the hide or the skin can be fisted or pulled off, all contribute greatly towards loss of value.

Bad shape, caused through improper ripping lines, is another fault. The tanner requires a square hide, which can only be obtained by removing it according to standard cuts. To obtain the best results the following basic incisions should be made in large animals:

1. One long, straight incision, down the midline, from the chin to the anus (cuts reaching only to the udder or scrotum are not to be recommended as the shape of the hide is affected; two unnecessary flaps are left which have to be trimmed off, thus affecting the shape and size of the hide)
2. Two circular incisions on the forelegs round the knees;
3. Two similar incisions round the hocks;
4. Two straight cuts on the inside of the forelegs from the knees to the fore end of the breast bone; and
5. Two straight cuts on the hind legs from the back of each hock joint to a point midway between the anus and the scrotum or udder.

For sheep and goats the ripping cuts are the same, but it is much better to remove the skin in cased form. This method has been spread by the Arabs and can be highly recommended as it causes less scores and cuts than any other way. The whole skin can be drawn off like a glove once the proper initial cuts have been made.

In certain countries where whole skins were used as rafts or as containers for liquids, a method of inflation, *i.e.* injecting air between the skin and the carcass, was developed. When the skin is pummeled the compressed air tears the subcutaneous tissue and the skin can then be removed with the minimum use of the knife. This method should not be discouraged; on the contrary, it is strongly recommended, as skins produced in this manner are of the highest quality.

17.1.3 Treatment after flaying

Damage after flaying can be avoided once the basic principles of hide handling are understood. These are as follows:

1. Contamination in the form of blood, ingesta or manure may lead to rapid decomposition;
2. The delay between flaying and subsequent treatment should be reduced to a minimum to prevent bacterial action;
3. Rapid arrest of bacterial action is essential and can be achieved by depriving the micro-organisms of moisture, either by suspension drying or by salting.

These principles are achieved by the following:

1. Washing

After flaying, while the blood is still liquid, the hide or skin should be washed, using a scrubbing-brush, preferably under running water. Soaking in drums or troughs filled with thick, dun-coloured water, as is often seen in tropical countries, has an adverse effect, as the hide is further infected with bacteria rather than cleansed. The washed hides should be drained by hanging them over a pole or wire, and fleshed immediately.

2. Fleshing

This should be done so that all non-leather-forming substances, such as meat and fat, are removed, either by knife or scraper. Over-flaying leads to loss of corium, *i.e.* the leather-forming substance, and is as dangerous as leaving meat or fat on the hide.

3. Trimming

All irregular flaps at the edges and corners should be removed so that the hide or skin is as 'square' as possible. The switch should be cut off 15 cm (6") below the root.

4. Preserving

When hides and skins cannot be passed absolutely fresh to the tanner, some method of preservation will generally be essential. There is a choice of three methods:

(a) Wet salting, where hides are built up with alternate layers of salt and cured until the salt has absorbed so much moisture that putrefaction cannot commence. This process takes approximately 3 weeks.

Salting is done by piling alternate layers of salt and hides, usually on a slatted platform, until a stack 1–2 to 1–5 m (4 to 5 ft) high has been built. After about 10 days the pile must be overhauled and a new stack built in which the top hides from the original stack form the bottom of the new. If a slatted platform is not used and the brine retained, there is a definite danger of spoilage. 'Red heat', which is a discoloration of the hide through bacterial action, can be avoided by the use of certain chemicals, such as sodium silicofluoride or 'Santobrite'.

Wet salting should be employed only where the throughput is large and a cellar with a temperature not exceeding 15–16 °C (60 °F) through out the year can be built. Whenever the relative humidity exceeds 90% or the temperature reaches 30° to 42.2 °C (86 ° to 108 ° F) deterioration of the corium, *i.e.* the leather forming substance, begins.

So called 'pickling' of hides, *i.e.* immersing the hides in a brine filled drum for many days before collection, should be discontinued, as this results in loss of hide substance, hair slip and putrefaction. This method is a distorted version of the frigorifico method of brining, *i.e.* immersing the hides overnight in a saturated salt solution prior to wet salting.

(b) Dry salting, which is a combination of methods (a) and (c). The hides are salted for a short period to protect them from damage during the most critical time, *i.e.* the first few days after they have been removed from the animal, and the rest of the moisture is then removed by air drying.

(c) In suspension drying, the hides suspended in frames or from one or three wires in such a way that there is free circulation of air from all sides.

Suspension drying is the cheapest method in the tropics: it requires little control and can be carried out with simple equipment; the maximum reduction in weight is obtained; the hide will keep almost indefinitely, provided it is protected from beetle infestation; and it facilitates the assessment of weight and quality.

17.1.4 Suspension drying equipment

Frames can easily be constructed from any locally available timber and they should measure 2–7 × 3–1 m (9 × 10 ft) for larger beasts and 2–7 × 2–4 m (9 × 8 ft) for smaller animals. Large cattle frames can be subdivided into four sheep or goat frames by inserting cross-pieces. The distance between the frames is customarily 23 cm (9").

In order to ensure the fullest air circulation, care must be taken so that the hides do not touch each other and that the hump, if present, is properly stretched out with a stick.

Drying in a loop made from green branches, a very simple and efficient method, is recommended in cases where a producer sells a beast only occasionally, where straight timber is costly or unavailable, and where the nomadic activities of a people precludes the carrying of heavy frames. Thick *galvanized* wire may be used in place of green branches.

Good results are obtained by drying hides and skins over galvanized wires instead of in frames, provided they are hung over one or preferably three wires in such a way that they do not touch each other at any point and that care is taken to stretch all small infolds by means of a stick or twigs.

In places where, there may be depreciation from theft, vermin or rain, hide-drying sheds should be provided.

The hides or skins are laced into the frames using string, bark or preferably hide strips, from perforations around the edges. Wire should never be used, as not being elastic it tears the hole out, or rusts, leading to stains during tanning. Small pieces of wire left in the skin may cause serious damage to hundreds of skins if they are tanned in drums.

The time taken to dry naturally depends on the weather and the thickness of the hide or skin. Too slow drying, experienced in high-rainfall areas, often leads to mould growth and hair slip, while over drying may cause cracks. It is essential to remove the hide or skin from the frame before it is completely dry, otherwise it will crack when folded. If hides and skin are orientated east-west to offset the heat of the midday sun they may be left uncovered, but in countries with excessive insolation it is preferable to protect them by providing overhead shade.

17.1.5 Treatment after drying

Hides and skins should be folded only once lengthwise down the line of the backbone. During transit they must be protected from wetting, rubbing, contamination by oil, paint or grease and from binding wire.

If they are to be stored they must be kept dry by proper roofing and by being placed on a slatted platform raised not less than 15 cm (6") from the floor. They must also be protected from vermin. Today the hide beetle *(Dermestes maculatus)* can be completely controlled by the use of such insecticides as Gammexane or DDT powder if the floors, walls and rafters of the store are dusted as well as the hides and skins.

Good results can be obtained by applying fungicides mixed with insecticides to achieve protection both from moulds and beetles. A proved formula for such a powder is as follows:

BHC (13% gamma isomer)	6%
Boracic acid	4%
'Santobrite' (Monsanto)	2%
Kaolin, or other filler	88%

This powder will contain 0–75% of BHC gamma isomer. Periodic aeration by exposure to the sun and light will help to check mould growth.

Another widely used method to control hide beetle is to dip the hides in a solution containing 0–2% arsenious oxide. Alternatively, this liquid may be sprayed over both sides using a pump. The hides and skins must be very carefully dried after spraying, otherwise moulds will develop and hair slip may begin.

This method of protecting hides and skins from beetles must not be confused with the soaking of ground-dried goods for a prolonged period in a strong solution of arsenic, with subsequent stretching in frames. This fraudulent practice has only one aim, namely to make an inferior hide look better than it really is. The damage caused by this procedure is enormous, as once the resoaked hides are mixed with suspension-dried ones, the tanner accepts the consignment bona fide as all suspension-dried, and then during the process of soaking or liming, the ground-dried hide reveals serious damage to the grain or hide substance, or falls to pieces and has to be rejected. Naturally the tanner reduces the price for all goods coming from countries where this malpractice is prevalent, thus penalizing all producers for his loss. Resoaking of hides should be prohibited by legislation.

17.1.6 Tools and equipment

Despite their simplicity, the tools and equipment required for the proper preparation of hides and skins are of the greatest importance. Proper flaying knives with a convex edge and curved, blunt point are essential. Their edge should always be kept keen. Separate knives should be used for butchering and flaying operations.

A hoist, which allows proper bleeding, is of the greatest importance in hide production. The use of skinning cradles to protect the hide from dirt and damage by the cement floor is another important item. The provision of ample clean water, frames of the correct size and proper transport and storage facilities all contribute towards the production of quality goods.

17.1.7 By-products

The by-products of hides and skins are trimmings, such as pieces of shank, lips, ears and masks, which can be dried or cured in the same way as hides and sold for gelatine or glue production. In addition tail switches, which should be cut off approximately 15 cm (6 inch) from the base of the tail, can be dried and sold for local brush manufacture. For export the tail hair should be cut as near the skin as possible, washed in cold water then immersed for several hours in warm water to which 2% washing soda has been added, washed again in running water and dried.

17.1.8 AGMARK standards for buffalo meat

Hides grading and marking rules

1. Short title and application

(a) These rules may be called the hides grading and marking rules, 1937

(b) They shall apply to hides produced in India

2. Grade designation

Grade designations to indicate the quality of hides (kids, buffaloes and calves) are set out in column 1 of schedule-1.

3. Definition of quality

The quality indicated by such grade designation is set out against such designations, in columns 2 to 11 of schedule 1.

4. Grade designation marks

The grade designation mark shall be indicated by 2, 3 or 4 holes of $^1/_4$ inch diameter, representing A, B or C grades respectively.

5. Method of marking

The holes shall be punched on each hide at the root of the tail at a distance of $^1/_4$ inch from each other and shall be arranged as shown in schedule II-A lead seal bearing the word AGMARK on one side and a letter representing one of the weight groups, namely H, M, L or C as the case may be, on the other, shall be

Table 17.1 Grade Designations and Definition of Quality of Hides (Kids, Buffaloes and Calves) Produced in India

Definition of Quality

Special Characteristics Respecting State or Condition

Grade designation	Constitutional	Branding iron marks	Hair side pox, scab, sores and skin disease marks	Marks due to external injuries	Scratches
1	2	3	4	5	6
A	Hides from young or middle aged animal with no rib marks or signs of emaciation.	Not allowed	No pox, score or marks of skin diseases allowed but one small scab and a small sore may be allowed if not on the butt.	Not allowed	One small superficial scratch not affecting the grain allowed if within 4" of the outer edge of the neck or belly portion only.
B	Slight rib marks allowed	Not more than one brand mark allowed which may only be on the belly neck or cheek	No pox or skin disease marks allowed. Two small scabs or sores allowed if not on the butt In the case of kids and buffalo hides these may be on the butt provided they are within 2" of the line of the spine.	Permitted on belly $	Superficial scratches not affecting the grain allowed.
C**	Rib marks freely allowed	At least, one half of the hide shall be free from brand marks	At least half the hide shall be unaffected	Allowed if on neck belly or below hip joint $.	Reasonably free from deep scratches.

General characteristics					
Flesh side		State or condition	Trimmed weight prior to curing		
Flayer's marks and cuts	Warbles and warble holes*		Weight groups	Limits of weight buffalo hides	Kids
7	8	9	10	11	12
No deep cuts allowed. Not more than 4 shallow cuts other than on the butt permitted $	One open and not more than 3 blind warbles allowed	The hides shall be reasonably clean and free from flesh, fat and extraneous matter. Horns ears feet (from above the fetlocus) and tail (beyond 4 inched) shall be removed			
Shallow cuts allowed and one deep cut allowed if not on the butt	Not more than 5 open and 8 blind warbles allowed	-do-	Heavy (H)	60 lb and over	28 lb and over
Shallow cuts allowed a reasonable number of deep cuts or holes permitted if not on the batt	Not more than 10 open and a reasonable number or blind warbles allowed	-do-	Medium (M) Light (L) Calf (C)	40 to 60 lb 25 to 40 lb under 25 lb	18 to 28 lb 8 to 18 lb Under 8 lb

* Warble holes are holes or spots which can be opened with a blunt skewer.

** But hides weighing over 12 lb. shall not be graded higher than C. Young bull hides upto 12 lb. may be placed in B grades but not in A.

$ In case of heavy and medium weight buffaloes only "One sore or cut mark on the hair of flesh side within two inches of the spine" may be permitted.

secured to each hide by a piece of twine tag which should be passed through all the holes. The place of packing shall also be indicated on the lead seal in code letters.

17.2 Bones

Bones amount to approximately 15% of the weight of the dressed carcase, depending on the age and condition of the beast. In extremely fat beasts it may beonly 12%, while in emaciated cattle it can be up to 30%. Fresh bones yield an appreciable amount of fat, which should always be recovered either by boiling the bones in open kettles or under pressure.

The bones are obtained from a number of sources, *viz.* rendered long bones tallow manufacturing units; bones like seapula, ribs, skull etc. from retail stall; and bones from naturally dead animals etc. Bones are stored in the open, exposed to the natural vagaries like heat, cold, rain, and wind. Later, these are despatched to bone-crushing factories which, in turn, pay about Rs 700/tonne.

Bones are used for several purposes:

1. For the production of stock feeds, such as meat and bonemeal when processed together with other offal.
2. As steamed bone meal, which is produced from defatted bones digested under pressure. The inorganic matter of bones consists of slightly more than 32% calcium and a little more than one-half of this amount of phosphorus. Both are invaluable supplements for feeding to livestock.
3. For gelatine and glue production, as ossein, the organic substance forming the bone structure, amounts to 33 to 36% of the total.

A fact of immense importance is that bones, before use either as a fertilizer or as mineral supplement for stock feed, should be properly sterilized.Otherwise they may be a source of many dangerous diseases, such as anthrax and botulism.

Bones can be sterilized merely by burning. To fire the bones they should be piled on a metal frame, for example old motorcar springs or a chassis.

Burning deprives the bones of all organic matter, leaving the mineral components in a friable form, which can easily be pulverized. This calcined bone powder is equal in feeding value to the best steamed bone-meal, although it looks like charcoal. The yield of calcined bones from freshly slaughtered beasts is approximately one-third of the green weight of bone, while from bones which have been exposed to climatic conditions for a long time (so called 'desert' bones) the yield is approximately two-thirds.

While prolonged boiling in open kettles can be used to sterilize bones, it is safer and more economic to use retorts where pressure can be applied. When moist heat is used, 15 minutes under 1–4 kg per cm (201b per inch) pressure is sufficient; with dry heat, the temperature must be 154 °C (310 °F) for not less than 3 hr.

Every precaution must be taken to prevent sterilized bones coming in contact with unsterilized material, and the strictest division observed between processed and unprocessed material.

For use as soil dressing or stock feed, the sterilized bones should be converted to a fine flour.

It can be seen, therefore, that the basic equipment of every bone plant, either stationary or mobile, should be:

1. A boiler to provide the necessary steam pressure;
2. A retort or digester, single or double jacketed, where the bones can be subjected to pressure or dry heat; and
3. A disintegrator where the sterilized bones are converted to a powder.

Bones intended for gelatine manufacture can be piled in large heaps, exposed to the elements and then crushed with a stone-crusher. The loss in weight due to exposure saves freight.

17.3 Blood

The way in which blood is often mishandled is typical of the manner in which valuable protein is continually wasted. At the same time, there are considerable difficulties in its disposal. It clogs drains, attracts vermin and may spread disease. On the other hand, the blood wasted from every 454 kg beast slaughtered deprives the country of a potential 2–7 kg of blood meal for livestock feed.

The use of fresh blood for stock feed is limited by two factors:

1. It has a very poor keeping quality and unless fed immediately decomposes rapidly under tropical conditions; and
2. If fed unsterilized, it may transmit disease.

The blood from healthy animals is collected hygienically, defibrinated immediately before clotting sets in and centrifuged to separate out the serum. The resulting product, a thick, viscous, dark fluid containing mostly red blood cells, is eventually used as an important ingredient in haematinic formulations. Before the introduction of pharmaceutical use, it was converted into blood-meal either by sun drying or kettle-heating procedures.

The rights over collection of blood is auctioned annually by the slaughterhouse.

17.3.1 Methods of preparing blood meal

1. Absorption

Blood is mixed with any of the locally available cereal products, such as bran, pollards, maize or cassava flour, and then spread out for drying on mats or on trays heated from below. In this way the low-protein vegetable matter is enriched with first class animal protein. This process may be repeated several times, each treatment increasing the protein content of the meal.

2. Treatment with lime

The addition of approximately one per cent of unslaked lime to fresh blood causes it to coagulate in the form of a black, rubber like mass. If slaked lime is used, three times as much will be needed. The advantages of this method are that the keeping quality is greatly enhanced. The coagulated blood can be carried away in baskets, boxes or bags, and there is no discharge of serum. Lime treated blood may be fed fresh or sundried. Drying further improves its keeping quality.

As previously mentioned, the feeding of unsterilized blood is not recommended. Sterilization of blood, treated either by absorption or with lime, can be achieved easily by boiling it together with other food, such as potatoes or cassava, immediately before feeding, or by heating the dried blood on galvanized iron trays over a fire.

3. Coagulation, pressing and drying

Where larger quantities of blood are available, a different system is recommended. The blood should first be cooked for 15 to 20 minutes to make it coagulate; this expresses nearly half the water. Further reduction of the moisture can be achieved by putting the coagulated blood in a hessian bag and hanging it over a pole or pressing it between wooden planks weighted with stones. The mass is then sundried or dried over artificial heat.

17.3.2 Commercial by-products plant

The brief description given here for the production of meat meal and blood meal is applicable only to places where offal is available in small quantities. Where the raw material, *i.e.* blood, bones, condemns, fleshings, trimmings, and other inedible offal, amounts to approximately one ton per day, it is economically and hygienically preferable to install a standard by-products plant. That amount of offal may be derived from one slaughterhouse or from many dispersed slaughterhouses.

A by-products plant can be operated economically, not only in the vicinity of a slaughterhouse, producing stock feed from slaughterhouse offal, but also where, for instance, a large amount of surplus unproductive stock is available which is a burden to man and to the land. It must be borne in mind that most of the livestock with which the sellers are willing to part are animals unable to walk any great distance to an abattoir without excessive shrinkage or mortality.

Where the throughput does not warrant the establishment of a stationary plant, a mobile processing unit can be used, which by virtue of its mobility can progress either with the movement of the livestock or into unexploited areas.

The establishment of a by-products plant to utilize undersized, diseased and unproductive stock can on the one hand lead to a dramatic improvement in animal husbandry in the area, and on the other hand supply badly needed stock feed to other areas where intensive production can then be established.

The basic equipment for a by-products plant is:

1. A boiler to provide the necessary steam;
2. A drier (called a milter or cooker) where the products are first sterilized under pressure and then dried;
3. Equipment to remove excess fat from the dried material that may be a press, an extractor, an expeller or a solvent extraction plant; and
4. A mill to reduce the dried, sterilized defatted material to a powder.

17.4 Casings

Conversion of the intestines into casing is a highly remunerative operation despite its simplicity.

Casings are the sub-mucosal coat of gastro-intestinal tract of food animals us natural containers for meat products. Four varieties of casings are prepared from buffalos (i) weasand, (ii) dry casings, (iii) fat-ends, and (iv) bladders from oesophagus, small intestine up to rectal end of gut and urinary bladder respectively. These raw materials are procun casings procured by traders at a nominal price. The technique of process is almost similar for all casings and includes steps like flushing out the contents, turning (as in bladders), fermenting and sliming (not applicable in weasand), inflating with air, sundering pressing, measuring and storage. Exporters periodically purchase the casings from different by-products processing centres in the country.

'Stripping', which follows, is a term applied to the squeezing out of the contents by pulling the guts with one hand and passing them between the first and second fingers of the other hand. Removal of the fat is called 'fatting'. This process can be performed on sheep and goat intestines by a plain wooden or bone scraper, while beef and pig casings require a pair of scissors or a knife. In order to loosen the different layers, the intestines are now subjected to fermentation. Depending on the temperature of the water and climatic conditions, this can be achieved within a few hr.

The process of 'sliming' follows until the casing is completely clean and transparent. Sliming means the removal of the mucous lining by means of a sliming knife, which is a piece of wood or bone resembling a knife. By pressing the sliming stick along the intestines the slime is gradually removed, leaving a thin transparent membrane deprived of fat, meat and dirt. This is the true casing.

During processing, care must be taken to avoid damage which might result in one or several tears in the casing. The ideal is a complete, undamaged length of 24.4 to 30.5 m (80 to 100 ft) of sheep or 19.8 to 22.9 m (65 to 75 ft) of goat intestine.

The completely clean casings are now ready for curing. A choice of two methods is available:

1. Rubbing the casings with fine salt and leaving them to drain and cure for approximately 2 weeks. After this period they are shaken out and coiled in neat bundles of equal length.
2. Brining in a saturated salt solution. A wooden cask is gradually filled with casings which displace the brine. When filled, the cask is ready for shipment. If oil and patrol drums are used they should be lined with a suitable plastic film.

Cattle and pig casings are similarly prepared, but the market demand in the tropics is for sheep and goat.

17.5 Tallow

The rendered buffalo fat is tallow. It is mainly used for soap manufacture. However, the reprocessed and bleached tallow is used even in confectioneries. Soft tissues like caul fat, mesenteric fat, offal trimmings, and hard materials like bones are subjected to prolonged heating for 8–12 hr in huge metallic. Semi-spherical vessels mounted permanently on fire-pits. The clear fat, as it melts, transferred to another vessel, filtered through a muslin cloth and stored in $15^1/_2$ kg tins with residual oil, upon overnight simmering, separates out to the top and is removed.

17.6 Hooves and Horns

There is a firm market for hooves and horns provided they are properly handled. Neither hooves nor horns should be mixed with bones intended for stock feed, but should be treated separately, if only for the reason that the price realized for hoof-and-horn meal is much higher than that for one-meal, and its presence in bone meal makes the mixture very unpalatable.

The horn core, or horn pith, is a bony support for the horn and contains a high percentage of gelatine. It should therefore be separated from the horn proper immediately after the horn has been cut off. This can be done by immersion in hot water and then knocking with a hammer. After this, the horns are stacked for drying in ventilated sheds, before being despatched to gelatine or glue factories.

An appreciable quantity of horns is exported for button, comb and handle manufacture. The price depends on the method of treatment, the length, hardness and colour. Exposure to the sun will cause warping and cracking and thus render horns useless. Washing in warm water and drying in a shed is the only treatment they require.

Hoof and horn meal is a rich, nitrogenous fertilizer which is obtained by digesting the hooves and horns from cattle, sheep and goats under pressure for approximately 10 hr.

17.6.1 Hooves into fertilizers

The hooves are separated from feet by immersing the hoof portion in boiling vat for about 30 min and by striking it with the blunt end of an axe. These are stored in gunny bags and sold to cottage industries for manufacture of 'hoof meal' which is considered as a very good fertilizer far potato crops.

17.6.2 Horns into decorative

Buffalo horns vary widely in size, shape, curvature and colour depending upon the breed, sex and age. After sun-drying for about 4 days, the inner horn pith is separated from the outer horn shell by tactfully striking on a heavy metallic object. Horn shells, in good condition and free from damage on the outer surface, are stored and sold to cottage industries for manufacture into combs and a number of household decorative materials.

17.7 Miscellaneous Products

17.7.1 Rumen mucosa into musical drums

The rumen mucous membrane, after peeling of muscular coat, is developed into drum covers for musical instruments. The processing steps are as follows: soaking the raw materials in lime water for about 15 min, scraping papillomatous surfaces, cutting into circular pieces of about 30 and 35 cm diameter size drying in pairs (one being placed over the other with papillated sides up) and packing in bundles of 100 pieces of same size. Usually about 8–9 pieces of 30 cm or 5–6 pieces of 35 cm size can be prepared from one rumen mucosa.

17.7.2 Intestinal serosa into musical strings

The serous covering of the small intestine is stripped manually; the operation is made easy because it has already been cut longitudinally along the mesenteric line during the detachment of mesenteric fat by knife. It is folded twice along its length to make it into 4 fold thick and is cut into standard lengths of 1.8 m each. Each piece is twined by spinning for about 2 min on the twisting rod of a hand operated wheel, sun dried for 1–2 days, polished with emery paper, and sold in the market.

Instead of multilayer strings, as described above, single layer strings of about 8 m length can also be manufactured for end use as sport guts and knitting materials for bamboo articles. Processed single layer strings are sold in the market.

17.7.3 Bile for pharmaceuticals

The yield of bile varies considerably from animal to animal (40–400 ml). Raw bile can be stored up to 2 months without processing. The processing consists essentially of concentrating the liquid bile into a thick, viscous paste by prolonged heating for 30–36 hr in shallow and concave metallic vessels. The pharmaceutical firms purchase the bile concentrate for further processing into bile salts, cholic acid etc.

17.7.4 Feet bones into ornaments

The distal extremities of long bones (metacarpal and metatarsal) of feet are cut transversely and the shafts are subjected to wet rendering. The rendered long bones are sold to cottage industries for manufacture into ornaments.

17.7.5 Tail hairs into brushes

Tail-tips, comprising last $^1/_3$ portion of the tail and tail switch, are washed thoroughly in water and hung from a rope for a day; then the hairs are pulled out manually, packed into bundles and sold for manufacture of brushes. One kg of tail hairs can be collected from about 60–80 animals.

17.7.6 Trims into wood finishing materials

By-products like tail-tips, hide trimmings, feet ligaments and penis are sun-dried for a long period for eventual use as "polishing, filling and finishing materials" for furniture and other wood works.

17.7.7 Gut contents into manure

The excreta and ingesta of slaughtered animals, along with the floor sweepings of layerage and slaughter hall, are dumped for ultimate use as manure.

17.7.8 Remote utilities of by-products

The butchers have a regular habit of searching into the reticular meshes for fortunes like gold rings, coins etc. which might have been ingested inadvertently by the animals; they also collect the amphistomes from infested rumen mucosal folds for feeding to poultry. Such remote utilities are ample proofs for the dictum "all parts of the slaughtered animal, other than its last bleat, are harnessed for utilitarian purposes".

CHAPTER 18

ECONOMICS OF BUFFALO PRODUCTION

Buffalo has become the main milch animal in this country contributing 55% of the total milk production followed by 42% by cows and 3% by goats. Buffaloes can utilize low-quality feeds, more efficiently than cattle. Buffaloes also have unique advantages over cattle; they are less susceptible to ticks and other ecto-parasites they can better withstand wet conditions underfoot and are therefore more suitable in many areas. They are more docile (children/old people can manage them relatively easily).

18.1 Unique Features of Buffalo Contributions

(a) Richer milk

Buffalo milk contains higher total solids (protein, fat, minerals) of 18–23% as compared to 13–16% in cow milk. This confers advantage to the production of cheese and some other dairy products.

(b) Leaner meat

Buffalo meat is tasty and lean. It contains lower saturated fat than beef and pork, and hence is considered a meat of good dietary value.

(c) Efficient converter of low quality feed

Buffaloes can utilize less digestible feeds (*eg*. rice straw, maize stovers, sugar-cane wastes, etc.) better than cattle to grow. This makes buffaloes easy to maintain using locally available roughage and crop residues.

(d) Best draught power for wet environments

Buffaloes are superior to other draught animals in wet or waterlogged conditions, such as in muddy paddy fields. They can also be used for cart haulage, carrying heavier loads than cattle.

(e) Enrich soil fertility

Buffaloes improve soil structure and fertility while treading paddy fields. Each year, an adult buffalo produces 4 to 6 tonnes of wet manure plus additional urine as bio-fertilizer to the land. This reduces or eliminates the need for chemical fertilizers as well as provides essential soil humus which chemicals cannot provide.

(f) Secure socio-economic status of farmers

Buffaloes are often used as cash savings, can be sold when need arises. Thus, the animals ensure the farmers' socio-economic security.

18.2 Economics of Buffalo Rearing

Dairying is an important source of subsidiary income particularly to small/marginal farmers and agricultural labourers. Apart from milk, the manure from animals provides a good source of organic matter for improving soil fertility and crop yields. The gober gas from the dung is used as fuel for domestic purposes as also for running engines for drawing water from wells. The surplus fodder and agricultural by-products are gainfully utilized for feeding the animals. Buffalo males are good source of meat apart from draught power. Since agriculture is mostly seasonal, there is a possibility of finding employment throughout the year for many persons through dairy farming. Thus, dairy provides employment throughout the year. The main beneficiaries of dairy programmes are small/marginal farmers and landless labourers. A farmer can earn a gross surplus of about Rs12000 per year from a unit consisting of 2 milking buffaloes. The capital investment required for purchase of 2 buffaloes is Rs18223. Even after paying a sum of Rs 4294 per annum towards repayment of the loan and interest, the farmer can earn a net surplus of Rs 6000–9000 approximately per year. Even more profits can be earned depending upon the breed of animal, managerial skills and marketing potential.

Specialization is another means to increase the productivity of mixed farming systems, as long as the complementary economic and environmental benefits and linkages can be maintained (LEAD, 2004). *e.g*, in India, landless ownership of cattle is becoming increasingly common. An agreement may be made between a crop producer and cow owner that crop by-products will be made available to the animal. The cow may then be fed on the farm, and its manure returned to the fields. Thus, the farmer can focus on his crop production, and the livestock owner can focus on the health and nutrition of the animal and consequent milk production. The fact that the land owner does not also own the animal does not invalidate the nutrient transfer benefits of the system. Thus mixed farming systems can transcend ownership boundaries of land and livestock. On a larger scale, with increasing industrial livestock systems there may be opportunities to continue producing in a mixed farming context by integrating industrial, land-detached units and arable farms in a 'regional mixed farm' (Blackburn, 2004).

It is evident from the study of Ramarao *et al.* (2005) who have published a detailed study on integrated mixed farming system. They have shown that the net income from crops (rice wheat system) was Rs 7843 only when livestock income was added the net income rose to Rs 33076 on an area of 1.5 acres in the rain fed farming.

The results of this study have indicated that integration of various enterprises on 1.5 acre size of land holding were financially viable (Table18.1).

Table 18.1. Income and Expenditure of Different Mixed Farming Modules for Marginal Holder (Mean ± SE)

S.No	Treatment	Expenditure (Rs)	Gross Income (Rs)	Net Income (Rs)	Cost return ratio	Employment days
T_1	Crop (1.5 acre)	$12396^a \pm 83.37$	$20239^a \pm 63.42$	$7843^a \pm 93.29$	$1.63^a \pm 0.01$	$165^a \pm 1.55$
T_2	Crop + 2 bullocks + 1 cow	$18920^b \pm 81.74$	$33104^b \pm 52.90$	$14184^b \pm 93.25$	$1.75^b \pm 0.007$	$273^b \pm 1.15$
T_3	Crop + 2 bullocks + 1 buffalo	$19188^c \pm 85.27$	$37449^c \pm 54.47$	$18260^c \pm 92.76$	$1.95^c \pm 0.008$	$273^b \pm 1.02$
T_4	Crop + 2 bullocks + 1 cow + 1 buffalo	$21341^d \pm 82.60$	$42803^d \pm 51.26$	$21462^d \pm 92.04$	$2.00^d \pm 0.007$	$291^c \pm 0.99$
T_5	Crop + 2 bullocks + 1 cow + 1 buffalo + 10 goats	$23294^e \pm 80.56$	$52695^e \pm 51.56$	$29400^e \pm 89.96$	$2.26^e \pm 0.007$	$308^d \pm 0.98$
T_6	Crop + 2 bullocks + 1 cow + 1 buffalo + 10 goats + 10 poultry + 10 ducks	$24899^f \pm 79.40$	$57975^f \pm 54.69$	$33076^f \pm 90.84$	$2.23^f \pm 0.007$	$316^e \pm 0.90$

Values with different superscript in the same column differ from each other significantly ($P<0.01$).
Source: *Livestock Research for Rural Development* **18** (7) 2006 -Ramarao *et al.* (2005).

In spite of the clear demonstration of high dividends from livestock enterprises. The entire focus of the government projects is in improving crop based agriculture with small and marginal farmers, being while they continue to be below the poverty line and in perpetual penury.

The present need is for implementing the livestock entrepreneurship model developed in 2002 using livestock, showing income per entrepreneur, raising in multiples of 3 crossbred cows, 10 buffaloes, goats, sheep, poultry individually and coupled with each other so that marginal farmer can come out of the poverty line with the help of the present programmes if these are opened up by change in the policy frame work.

The basic model envisages a Central Livestock Farm with high producing crossbred/buffaloe dairy cows as a backbone for the bull production farm comprising of a semen and embryo bank, feed plant, diagnostic centre, a services and goods company, a milk processing plant, a milk and milk products sale unit and an R and D facility. This would support Satellite Livestock unit, located in an area 50–100 villages around central complex from where the necessary goods and services will be made available to the farmers in the operational area. Each farmer who possess cattle or buffaloes will be converted into a unit of 3 crossbred cow with adequate land for fodder production. Similar units of 6 crossbred cows and multiples thereof starting upto and 20 crossbred cows which could go upto 500 crossbred cows with or without adequate land for fodder, in such case green fodder was to be purchased. Similar units could be generated for indigenous cattle and buffaloes, a dairy unit of 6 to 10 buffaloes and multiples thereof. The model has been tested in two locations in district Sonepat of Haryana one under complete commercial conditions led by HBN dairy and the other by Ayurvet Research Foundation in 50 villages around their feed plant at Chidana.

Financial assistance is currently available from the banks and NABARD need to broad based their project funding wherein fund availability could be given to financially viable unit as detailed above. As an example for an investment (capital as well as working cost) of Rs 1.44 lakhs one could establish a unit of 3 crossbred cows to generate an annual income of Rs 1.08 lakhs in a year. The cash flow statement of 3 crossbred (Table 18.2) and 6 crossbred cows (Table 18.3) are given below as an example based on data accumulated from 2003 to 2009.

The green fodder when grown at farmer's land, the profitability will increase by Rs 675 per month when he will operate 2 acres of land with hired labour.

Table18.2 Cash Flow Statement for a Unit of 3 Crossbred Cows

Item description	I year	II year	III year	IV year	V year	VI year
Gross income	27719	28219	55551	37663	46607	37663
Loan	67000	53600	40200	26800	13400	–
Outstanding interest	6700	5360	4020	2680	1340	–
Loan instalment	13400	13400	13400	13400	13400	–
Net surplus	7169	9459	38131	31583	31867	37663
Net income/ month	635	788	3178	1800	2656	3138

Table 18.3 Cash Flow Statement for a Unit of 6 Crossbred Cows

Item description	I year	II year	III year	IV year	V year	VI year
Gross income	73798	75298	134362	104830	104830	104830
Loan	126000	100800	75600	50400	25200	–
Outstanding interest	12600	10080	7560	5040	2520	–
Loan installment	25200	25200	25200	25200	25200	–
Net Surplus	35998	40018	101602	74590	77110	104830
Net income/ month	3000	3335	8466	6215	6426	8735

The green fodder when grown at farmer's land, the profitability will increase by Rs 1140 per month when he will operate 2 acres of land with hired labour.

If a major thrust is placed on animal husbandry which is the only viable option in developing a safety net for small and marginal farmers, landless labours and women in particular, if animal husbandry based entrepreneurship are made the corner stone of agricultural development it will provide around the year employment and food for the self and the country.

A model scheme involves creation of a livestock complex comprisisng of a model dairy production unit, bull farm with semen and embryo freezing, feed plant, milk processing plant, a slaughter facility from where a services and good company can operate in 50 km radius to convert small and marginal farmers in dairy enterpress.

It is proposed to conduct an asset inventory of small and marginal farmers, medium and large farms so that 100–200 livestock entrepreneurship, which are financially viable, can be created in an area of 50 kms around the livestock plant complex and slaughterhouse. These units may have the following structure. Their viability analysis is given at the end of this section.

1. Dairy unit for 6 buffaloes with adequate land for fodder production/or not.
2. Dairy unit for 10 buffaloes with adequate land for fodder production/or not.
3. Dairy unit for 20 buffaloes with adequate land for fodder production/or not.
4. Dairy unit for 50 buffaloes with adequate land for fodder production/or not.
5. Dairy unit for 500 buffaloes with adequate land for fodder production/or not.

18.2.1 Financial assistance presently available from banks/NABARD for dairy farming

NABARD is an apex institution for all matters relating to policy, planning and operation in the field of agricultural credit. It serves as an apex refinancing agency for the institutions providing investment and production credit. It promotes development through formulation and appraisal of projects through a well organized Technical Services Department at the Head Office and Technical Cells at each of the Regional Offices.

Loan from banks with re-finance facility from NABARD is available for starting dairy farming. For obtaining bank loan, the farmers should apply to the nearest branch of a commercial or co-operative Bank in their area in the prescribed application form which is available in the branches of financing banks. The Technical Officer attached to or the Manager of the bank can help/give guidance to the farmers in preparing the project report to obtain bank loan.

For dairy schemes with very large outlays, detailed reports will have to be prepared. The items of finance would include capital asset items such as purchase of milch animals, construction of sheds, purchase of equipments etc. The feeding cost during the initial period of one/two months is capitalized and given as term loan. Facilities such as cost of land development, fencing, digging of well, commissioning of diesel engine/pumpset, electricity connections, essential servants' quarters, godown, transport vehicle, milk processing facilities etc. can be considered for loan. Cost of land is not considered for loan. However, if land is purchased for setting up a dairy farm, its cost can be treated as party's margin up to 10% of the total cost of project.

18.2.2 Scheme formulation for bank loan

A Scheme can be prepared by a beneficiary after consulting local technical persons of State animal husbandry department, DRDA, SLPP, dairy co-operative society/ union/federation/commercial dairy farmers etc. If possible, the beneficiaries should also visit progressive dairy farmers and government/military/agricultural university dairy farm in the vicinity and discuss the profitability of dairy farming. A good practical training and experience in dairy farming will be highly desirable. The dairy co-operative societies established in the villages as a result of efforts by the Dairy Development Department of State Government and National Dairy Development Board would provide all supporting facilities particularly marketing of fluid milk. Nearness of dairy farm to such a society, veterinary aid centre, artificial insemination centre should be ensured. There is a good demand for milk, if the dairy farm is located near urban centre.

The scheme should include information on land, livestock markets, availability of water, feeds, fodders, veterinary aid, breeding facilities, marketing aspects, training facilities, experience of the farmer and the type of assistance available from State Government, dairy society/union/federation.

The scheme should also include information on the number and types of animals to be purchased, their breeds, production performance, cost and other relevant input and output costs with their description. Based on this, the total cost of the project, margin money to be provided by the beneficiary, requirement of bank loan, estimated annual expenditure, income, profit and loss statement, repayment period, etc. can be worked out and shown in the Project report.

18.2.3 Scrutiny of schemes by banks

The scheme so formulated should be submitted to the nearest branch of bank. The bank's officers can assist in preparation of the scheme for filling in the prescribed application form. The bank will then examine the scheme for its technical feasibility and economic viability.

(a) Technical feasibility would briefly include:

1. Nearness of the selected area to veterinary, breeding and milk collection centre and the financing bank's branch.
2. Availability of good quality animals in nearby livestock market.
3. Availability of training facilities.

4. Availability of good grazing ground/lands.
5. Green/dry fodder, concentrate feed, medicines etc.

(b) Economic viability this would briefly include:

1. Input cost for feeds and fodders, veterinary aid, breeding of animals, insurance, labour and other overheads.
2. Output costs *i.e.* sale price of milk, manure, gunny bags, male/female calves, other miscellaneous items etc.
3. Income-expenditure statement and annual gross surplus.
4. Cash flow analysis.
5. Repayment schedule (*i.e.* repayment of principal loan amount and interest).

Other documents such as loan application forms, security aspects, margin money requirements etc. are also examined. A field visit to the scheme area is undertaken for conducting a techno-economic feasibility study for appraisal of the scheme. Model economics for a 6-buffaloe unit to 500- buffaloe unit are given in the following pages.

18.2.4 Sanction of bank loan and its disbursement

After ensuring technical feasibility and economic viability, the scheme is sanctioned by the bank. The loan is disbursed in kind in 2 to 3 stages against creation of specific assets such as construction of sheds, purchase of equipments and machinery, purchase of animals and recurring cost on purchase of feeds/fodders for the initial period of one/two months. The end use of the fund is verified and constant follow-up is done by the bank.

18.2.5 Lending terms in general

Unit Cost

Each Regional Office (RO) of NABARD has constituted a State Level Unit Cost Committee under the Chairmanship of RO-in-charges and with the members from developmental agencies, commercial banks and cooperative banks to review the unit cost of various investments once in six months. The same is circulated among the banks for their guidance. These costs are only indicative in nature and banks are free to finance any amount depending upon the availability of assets.

Margin money

NABARD had defined farmers into three different categories and where subsidy is not available the minimum down payment as shown below is collected from the beneficiaries.

Category of farmer	Level of pre-development return to resources	Beneficiary's contribution
Small farmers	Upto Rs 11000	5%
Medium farmers	Rs 11001–19250	10%
Large farmers	Above Rs 19251	15%

Interest rate

As per the RBI guidelines the present rate of interest to the ultimate beneficiary financed by various agencies are as under:

Loan Amount	CB's and RRB's	SLDB/SCB
Upto and inclusive of Rs 25000	12%	As determined by SCB/SLDB subject to minimum 12%
Over Rs 25000 and up to Rs 2.0 lakh	13.5%	-do-
Over Rs 2.0 lakh	As determined by the banks	-do-

Security

Security will be as per NABARD/RBI guidelines issued from time to time.

Repayment period of loan

Repayment period depends upon the gross surplus in the scheme. The loans will be repaid in suitable monthly/quarterly installments usually within a period of about 5 years. In case of commercial schemes it may be extended up to 6–7 years depending on cash flow analysis.

Insurance

The animals may be insured annually or on long term master policy, whichever is applicable. The present rate of insurance premium for scheme and non scheme animals are 2.25% and 4.0% respectively.

18.2.6 Model of dairy unit for buffaloes

The model economics of buffalo dairy unit of 6, 10, 20, 50 and 500 buffaloes with adequate land for fodder production is provided in Tables 19.4 to 19.8 respectively.

Table 18.4 Dairy Unit for 6 Buffaloes with Adequate Land for Fodder Production

Item description	6 Buffaloes amount in (Rs)
A. Capital Investment	
Cost of buffaloes 8 litres/day @ Rs 14000/-	84000.00
Buildings	
Covered area for buffaloes 10 sq.ft/buffalo @ Rs 150/ sq.ft.	9000.00
Open area for buffaloes 20 sq.ft/buffalo @ Rs 25/ sq.ft.	3000.00
Cost of calf-shed	1500.00
Cost of small tank and hand pump	3000.00
Cost of equipment (buckets, can, measuring jar etc.)	1000.00
Miscellaneous	500.00
Total Capital Investment	102500.00
B. Working Cost (Variable Costs)	
Feed and fodder	
Green fodder @ 25 kg/day/buffalo @ Rs 50/q when purchased	27375.00
@ Rs25/q when fodder is grown	(13687)*
Treated Straw @ 4 kg/day/buffalo (@ Rs 1.00/kg)	8760.00
Concentrates @ 3.5 kg/day/buffalo(@ Rs 7.50 per kg)	7487.00
Labour cost @ Rs 1500/month	18000.00
Mineral mixture, salt etc.	500.00
Veterinary aid @ Rs 1000/buffalo/annum	6200.00
Technology aid @ Rs 1500 for confirm pregnancy	9000.00
Total Variable Cost	127122.00
C. Working cost/fixed costs	
Insurance on cost of buffaloes @ Rs 2.25%	1890.00
Depreciation @ 10% on (a) cost of buffaloes, (b) cost of equipment and (c) cost of shed	10250.00
Total fixed costs	12140.00
D. Income	
By sale of milk @ Rs 14.00 per litre	201600.00
By sale of manure 6 ton per buffalo @ Rs 300/ton	10800.00
Total Income	212400.00
E. Bank Loan Required	
Farmer's Equity @ 25% of capital investment	25625.00
Working capital needed for a period of 3 months	31780.00
Amount required as loan (capital investment Farmer's Equity + 3 months working capital/unit)	108655.00
Interest rate (% per annum)	10%
Total	108655.00

Table 18.4 (*Contd.*)

Item description						6 Buffaloes amount in (Rs)
F. Birth and death register						
Item description	**I year**	**IIyear**	**III year**	**IV year**	**V year**	**VI year**
Female buffalo at hoof	06	–	–	–	–	–
Births						
Male	–	03	03	03	03	03
Female	06	03	03	03	03	03
Heifers	–	–	–	–	–	–
Pregnant heifers	–	–	06	03	03	03
She buffalo	–	–	–	–	–	–
Disposal						
Male	–	03	03	03	03	03
Female	–	–	–	–	–	–
Heifers	–	–	–	–	–	–
Pregnant heifers	–	–	06	03	03	03
Income by selling of pregnant heifer and male calf	–	1500	67500	34500	34500	34500
Income by selling of manure of pregnant heifers	–	–	7200	3600	3600	3600
Income year wise		1500	74700	38100	38100	38100
Cost in raising a calf up to pregnant heifer is assume to be 30% of variable cost per buffalo (30% = 6356)	–	–	38136	19068	19068	19068
Extra total income (year wise)	–	1500	36564	19032	19032	19032

G. Profitability

Gross income = income – (fixed costs + variable cost)	3138.00
Repayment of Loan ($^1/_5$ annually) of Rs108655/-	1730.00
Employment (man months) when fodder is purchased	24.00
Employment (man months) when fodder is grown	48.00
Time to pay back loan	5 years

Note: Figures in the brackets indicate when fodder is grown by the farmer

H. Cash Flow Statement for a Unit of 6 Buffaloes

Item description	I year	II year	III year	IV year	V year	VI year
Gross income	73138	74638	109702	92170	92170	92170
Loan outstanding	108655	86925	65195	43465	21735	–
Interest	10865	8692	6519	4346	2173	–
Loan instalment	21730	21730	21730	21730	21735	–
Net surplus	40543	44216	81453	66094	68262	92170
Net income/month	3378	3685	6787	5507	5688	7680

Table 18.5 Dairy Unit for 10 Buffaloes with Adequate Land for Fodder Production

Item description	10 Buffaloes amount in (Rs)
A. Capital investment	
Cost of buffaloes 8 litres/day @ Rs 14000/-	140000.00
Buildings	
Covered area for buffaloes 10 sq.ft/buffalo @ Rs 150/ sq.ft.	45000.00
Open area for buffaloes 20 sq.ft/buffalo @ Rs 25/ sq.ft.	5000.00
Cost of calf-shed	2500.00
Cost of small tank and hand pump	3000.00
Cost of equipment (buckets, can, measuring jar etc.)	1500.00
Miscellaneous	500.00
Total Capital investment	197500.00
B. Working cost (variable costs)	
Feed and fodder	
Green fodder @ 25 kg/day/buffalo @Rs 50/q when purchased	45625.00
@ Rs 25/q when fodder is grown	(22812)*
Treated straw @ 4 kg/day/buffalo (@ Rs 1.00/kg)	14600.00
Concentrates @ 3.5 kg/day/buffalo (@ Rs 7.50 per kg)	95812.00
Labour cost @ Rs 1500/month	18000.00
Mineral mixture, salt etc.	800.00
Veterinary aid @ Rs 1000 /buffalo/annum	10000.00
Technology aid @ Rs 1500 for confirm pregnancy	15000.00
Total Variable Cost	199837.00
C. Working cost: fixed costs	
Insurance on cost of buffaloes @ Rs 2.25%	3,150.00
Depreciation @ 10% on (a) cost of buffaloes, (b) cost of equipment and (c) cost of shed	16750.00
Total Fixed costs	19900.00
D. Income	
By sale of milk @ Rs 14.00 per litre	336000.00
By sale of manure 6 ton per buffalo @ Rs 300/ton	18000.00
Total Income	354000.00
E. Bank Loan Required	
Farmer's equity @ 25% of capital investment	41875.00
Interest rate (% per annum)	10%
Total	175000.00

Table 18.5 (*Contd.*)

Item description					10 Buffaloes amount in (Rs)	
F. Birth and death register						
Item description	**I year**	**IIyear**	**III year**	**IV year**	**V year**	**VI year**
Female buffalo at hoof	10	–	–	–	–	–
Births						
Male	–	5	5	5	5	5
Female	10	5	5	5	5	5
Heifers	–	–	–	–	–	–
Pregnant heifers	–	–	10	5	5	5
She buffalo	–	–	–	–	–	–
Disposal	–	–	–	–	–	–
Male	–	5	5	5	5	5
Female	–	–	–	–	–	–
Heifers	–	–	–	–	–	–
Pregnant heifers	–	–	10	5	5	5
Income by selling of pregnant heifer and male calf	–	2500	112500	57500	57500	57500
Income by selling of manure, of pregnant heifers (4 tons upto pregnancy @ Rs 300/ton)	–	–	12000	6000	6000	6000
Cost in raising a calf up to pregnant heifer is assume to be 30% of variable cost per buffalo (30% = 5995)	–	–	59950	29975	29975	29975
Net surplus income (year wise)	–	2500	64549	33525	33525	33525

G. Profitability

Gross income = income – (fixed costs + variable cost)	167768.00
Repayment of Loan ($^1/_5$ annually) of Rs198000/-	39600.00
Employment (man months) when fodder is purchased	24
Employment (man months) when fodder is grown	60
Time to pay back loan	5 years

Note: Figures in the brackets indicate when fodder is grown by the farmer

H. Cash flow statement for a unit of 10 buffaloes

Item description	**I year**	**IIyear**	**III year**	**IV year**	**V year**	**VI year**
Gross income	134263	136763	198812	167788	167788	167788
Loan outstanding	175000	140000	105000	70000	35000	–
Interest	17500	14000	10500	7000	3500	–
Loan instalment	35000	35000	35000	35000	35000	–
Net surplus	81763	87763	153312	125788	129288	167788
Net income/month	6813	7313	12776	10482	10774	13982

Table 18.6 Dairy Unit for 20 Buffaloes with Adequate Land for Fodder Production

Item description	20 Buffaloes amount in (Rs)
A. Capital investment	
Cost of buffaloes 8 litres/day @ Rs 14000/-	280000.00
Buildings	
Covered area for buffaloes 10 sq.ft/buffalo @ Rs 150/ sq.ft.	90000.00
Open area for buffaloes 20 sq.ft/buffalo @Rs 25/ sq.ft.	10000.00
Cost of calf-shed	5000.00
Cost of small tank and hand pump	3000.00
Cost of equipment (buckets, can, measuring jar etc.)	3000.00
Miscellaneous	1600.00
Total Capital Investment	392600.00
B. Working cost (variable costs)	
Feed and fodder	
Green fodder @ 25 kg/day/buffalo @ Rs 50/q when purchased	91250.00
@ Rs 25/q when fodder is grown	(45625)*
Treated Straw.@ 4 kg/day/buffalo (@ Rs 1.00/kg)	29200.00
Concentrates @ 3.5 kg/day/buffalo(@ Rs 7.50 per kg)	191624.00
Labour cost @ Rs 1500/month	36000.00
Mineral mixture, salt etc.	1600.00
Veterinary aid @ Rs 1000 /buffalo/annum	20000.00
Technology aid @ Rs 1500 for confirm pregnancy	30000.00
Total Variable Cost	399674.00
	(354049)*
C. Working cost fixed costs	
Insurance on cost of buffaloes @ Rs2.25%	6300.00
Depreciation @ 10% on (a) cost of buffaloes,	33260.00
(b) cost of equipment and (c) cost of shed	
Total fixed costs	39560.00
D. Income	
By sale of milk @ Rs14.00 per litre	672000.00
By sale of manure 6 ton per buffalo @ Rs 300/ton	36000.00
Total income	708000.00
E. Bank loan required	
Farmer's equity @ 25% of capital investment	98150.00
Working capital needed for a period of 3 months	99918.00
Bank loan needed = 75% of capital investment +	394368.00
3 months working capital / unit	
Interest rate (% per annum)	10%
Total	394000.00

Table 18.6 (*Contd.*)

Item description					20 Buffaloes amount in (Rs)

F. Birth and death register

Item description	I year	IIyear	III year	IV year	V year	VI year
Female buffalo at hoof	20	–	–	–	–	–
Births						
Male	–	10	10	10	10	10
Female	20	10	10	10	10	10
Heifers	–	–	–	–	–	–
Pregnant Heifers	–	–	20	10	10	10
She Buffalo	–	–	–	–	–	–
Disposal						
Male	–	10	10	10	10	10
Female	–	–	–	–	–	–
Heifers	–	–	–	–	–	–
Pregnant Heifers	–	–	20	10	10	10
Income surplus						
Income by selling of pregnant heifer and male calf	–	5000	225000	115000	115000	155000
Income by selling of manure of pregnant heifers	–	–	24000	12000	12000	12000
Cost in raising a calf up to pregnant heifer is assume to be 30% of variable cost per buffalo (30% = 5995)	–	–	119900	59950	59950	59950
Income surplus						
Net surplus income (year wise)	–	5000	81100	43050	43050	43050

G. Profitability

Gross income = income – (fixed costs + variable cost)	311816.00
Repayment of Loan ($^1/_5$ annually)	69800.00
Employment (man months) when fodder is purchased	36.00
Employment (man months) when fodder is grown	84.00
Time to pay back loan	5 years

Note: Figures in the brackets indicate when fodder is grown by the farmer

H. Cash flow statement for a unit of 20 buffaloes

Item description	I year	IIyear	III year	IV year	V year	VI year
Gross Income	268766	273766	349866	311816	311816	311816
Loan outstanding	349000	279200	209400	139600	69800	–
Interest	34900	27920	20940	13360	6980	–
Loan instalment	69800	69800	69800	69800	69800	–
Net Surplus	164066	176046	259126	228056	235036	311816
Net income/month	13672	14670	21593	12004	19586	25984

Table 18.7 Dairy Unit for 50 Buffaloes with Adequate Land for Fodder Production

Item description	50 buffaloes amount in (Rs)
A. Capital investment	
Cost of buffaloes 8 litres/day @ Rs 14000/-	700000.00
Buildings	
Covered area for buffaloes (10 sq.ft / buffalo @ Rs 150 / sq.ft.)	75000.00
Open area for buffaloes (20 sq.ft / buffalo @ Rs 25 / sq.ft.)	25000.00
Cost of calf-shed	12500.00
Cost of small tank and hand pump	7500.00
Cost of equipment (buckets, can, measuring jar etc.)	6000.00
Miscellaneous	4000.00
Total capital investment	830000.00
B. Working cost (variable costs)	
Feed and fodder	
Green fodder @ 25 kg/day/buffalo @ Rs 50/q when purchased	228125.00
@ Rs 25/q when fodder is grown	(114062)*
Treated Straw @ 4 kg/day/buffalo (@ Rs 1.00/kg)	73000.00
Concentrates @ 3.5 kg/day/buffalo(@ Rs 7.50 per kg)	479062.00
Labour cost @ Rs 1500/month	90000.00
Mineral mixture, salt etc.	4000.00
Veterinary aid @ Rs 1000 /buffalo/annum	50000.00
Technology aid @ Rs 1500 for confirm pregnancy	75000.00
Total variable cost	999187.00
C. Working cost fixed costs	
Insurance on cost of buffaloes @ Rs 2.25%	15750.00
Depreciation @ 10% on (a) cost of buffaloes, (b) cost of equipment and (c) cost of shed	83000.00
Total Fixed costs	98750.00
D. Income	
By sale of milk @ Rs 14.00 per litre	1680000.00
By sale of manure 6 ton per buffalo @ Rs 300/ton	90000.00
Total income	1770000.00
E. Bank loan required	
Farmer's equity @ 25% of capital investment	245500.00
Working capital needed for a period of 3 months	249796.00
Bank loan needed = 75% of capital investment + 3 months working capital/unit	984796.00
Interest Rate (% per annum)	10%
Total	872000.00

Table 18.7 (*Contd.*)

Item description	50 Buffaloes amount in (Rs)

F. Birth and death register

Item description	I year	II year	III year	IV year	V year	VI year
Female buffalo at hoof	50	–	–	–	–	–
Births						
Male	–	25	25	25	25	25
Female	50	25	25	25	25	25
Heifers	–	–	–	–	–	–
Pregnant Heifers	–	–	50	25	25	25
She Buffalo	–	–	–	–	–	–
Disposal						
Male	–	25	25	25	25	25
Female	–	–	–	–	–	–
Heifers						
Pregnant Heifers	–	–	50	25	25	25
Income surplus						
Income by selling of pregnant heifer and male calf	–	12500	562500	287500	287500	287500
Income by selling of manure of pregnant heifers	–	–	60000	30000	30000	30000
Cost in raising a calf up to pregnant heifer is assume to be 30% of variable cost per buffalo (30% = 5995)	–	–	299750	149875	149875	149875
Net surplus income (year wise)	–	12500	322750	167625	167625	167625

G. Profitability

Gross income = income – (fixed costs + variable cost)	839688.00
Repayment of Loan (1/5 annually)	174400.00
Employment (man months) when fodder is purchased	120
Employment (man months) when fodder is grown	300
Time to pay back loan	5 years

Note: Figures in the brackets indicate when fodder is grown by the farmer

H. Cash flow statement for a unit of 50 buffaloes

Item description	I year	II year	III year	IV year	V year	VI year
Gross income	672063	684563	994813	839688	839688	839688
Loan outstanding	872000	697600	523200	348800	174400	–
Interest	87200	69760	52320	34880	17440	–
Loan instalment	174400	174400	174400	174400	174400	–
Net surplus	410463	440403	768093	630408	647848	839688
Net income/month	34205	36700	64007	52534	53987	69974

Table 18.8 Dairy Unit for 500 Buffaloes with Adequate Land for Fodder Production

Item Description	500 Buffaloes Amount in (Rs)
A. Capital Investment	
Cost of buffaloes 8 litres/day @ Rs 14000/-	7000000.00
Buildings	
Covered area for buffaloes (10 sq.ft/buffalo @Rs 150/ sq.ft.)	2250000.00
Open area for buffaloes (20 sq.ft/buffalo @Rs 25/ sq.ft.)	250000.00
calf pen	
250000.00	
Store	
100000.00	
Cost of small tank and hand pump	75000.00
Cost of equipment (buckets, can, measuring jar etc.)	50000.00
Miscellaneous	40000.00
Total capital investment	9890000.00
B. Working cost (variable costs)	
Feed and fodder	
Green fodder @ 25 kg/day/buffalo @ Rs 50/q when purchased	2281250.00
@ Rs25/q when fodder is grown	(1140625)*
Treated Straw @ 4 kg/day/buffalo (@ Rs 1.00/kg)	730000.00
Concentrates @ 3.5 kg/day/buffalo(@Rs 7.00 per kg)	4471250.00
Labour cost @ Rs 1500/month	630000.00
Mineral mixture, salt etc.	40000.00
Veterinary aid @ Rs 1000 /buffalo/annum	500000.00
Technology aid @ Rs 1500 for confirm pregnancy	750000.00
Total variable cost	9402500.00
C. Working cost fixed costs	
Insurance on cost of buffaloes @ Rs 2.25%	157500.00
Depreciation @ 10% on (a) cost of buffaloes, (b) cost of equipment and (c) cost of shed	839000.00
Total fixed costs	996500.00
D. Income	
By sale of milk @ Rs 14.00 per litre	16800000.00
By sale of manure 6 ton per buffalo @ Rs300/ton	900000.00
Total income	
17700000.00	
E. Bank loan required	
Farmer's equity @ 25% of capital investment	2472500.00
Working capital needed for a period of 3 months	2350625.00
Amount needed as loan = 75% of capital investment + 3 months working capital/unit	8643125.00
Interest rate (% per annum)	10%
Total	8643000.00

Table 18.8 (*Contd.*)

Item Description					500 Buffaloes Amount in (Rs)
F. Birth and death register					

Item description	I year	II year	III year	IV year	V year	VI year
Female buffalo at hoof	500	–	–	–	–	–
Births						
Male	–	250	250	250	250	250
Female	500	250	250	250	250	250
Heifers	–	–	–	–	–	–
Pregnant heifers	–	–	500	250	250	250
She buffalo	–	–	–	–	–	–
Disposal						
Male	–	250	250	250	250	250
Female	–	–	–	–	–	–
Heifers	–	–	–	–	–	–
Pregnant Heifers	–	–	500	250	250	250
Income surplus						
Income by selling of pregnant heifer and male calf	–	125000	5625000	2875000	2875000	2875000
Income by selling of manure of pregnant heifers	–	–	600000	300000	300000	300000
Cost in raising a calf up to pregnant heifer is assume to be 30% of variable cost per she buffalo (30% = 5641)	–	–	2820500	1410250	1410250	1410250
Net surplus income (year wise)	–	125000	3404500	1764750	1639750	1764750

G. Profitability

Gross income = income – (fixed costs + variable cost)	7301000.00
I year = gross profit + net surplus income	7301000.00
II year	7426000.00
III year	10705500.00
IV year	9065750.00
V year	8940750.00
VI year	9065750.00
Repayment of loan ($^1/_5$ annually) of Rs 86m 43m 125	1953600.00
Employment (man months) when fodder is purchased	1000.00
Employment (man months) when fodder is grown	2500.00
Time to pay back loan	5 years

Note: Figures in the brackets indicate when fodder is grown by the farmer

Table 18.8 (*Contd.*)

Item Description						500 Buffaloes Amount in (Rs)
H. Cash flow statement for a unit of 50 buffaloes						
Item description	**I year**	**II year**	**III year**	**IV year**	**V year**	**VI year**
Gross income	672063	684563	994813	839688	839688	839688
Loan outstanding	872000	697600	523200	348800	174400	–
Interest	87200	69760	52320	34880	17440	–
Loan instalment	174400	174400	174400	174400	174400	–
Net surplus	410463	440403	768093	630408	647848	839688
Net income/month	34205	36700	64007	52534	53987	69974

CHAPTER 19

MARKETING OF BUFFALO MEAT

Meat animals, fresh meats, processed meats and animal casings are the four commodities exported by Indian meat industry. It may be categorically stated that the export meat trade is dominated by buffaloes. Agricultural and Processed Food Export Development Authority (APEDA), Ministry of Commerce, Government of India is the promotional and regulating authority for export activities in meat industry.

19.1 Current Practices in Buffalo Meat Industry

19.1.1 Trading of slaughter buffaloes

The primary producers are both at countryside and urban areas. The stocks in countryside comprise adult females, condemned due to senility, infertility, mastitis, metritis etc; work buffaloes, discarded for reasons of senility, yoke gall, lameness etc., and male calves, which are either surplus or poor in growth. But, stocks from commercial dairies of cities and peri-urban areas consist only of adult females of fairly young age and well built conformation. This is because there is a constant drain of she-buffaloes from breeding tracts to urban milking centres, where after lactation, they are sold for slaughter at very low prices. The owners do not have the facilities or money to maintain them during long dry periods. Male calves are grossly neglected at birth and starved to death slowly. Even female calves are not attended to properly. The slaughtered buffaloes reach the hands of meat traders and exporters for the production of meat to meet the domestic and export requirements.

The various trading channels from farm to slaughtering are depicted in (Fig 19.1}. The weekly livestock markets are the primary focal points in trading.

Livestock merchants are able to supply animals from upcountry to far-flung terminal markets situated in close proximity to major slaughterhouses.

Fig.19.1 Various trading channels from farm to slaughtering

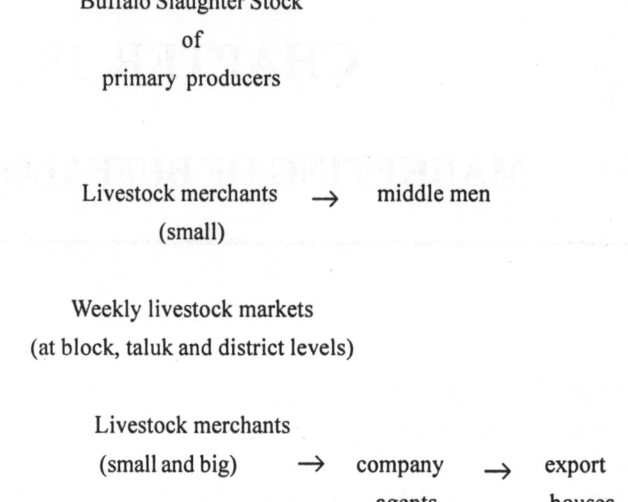

The agents, appointed by export houses, usually trade in group purchases. A group consists of a fixed number of animals, all of similar age, weight and finish. The prevailing prices for well finished females, medium finished females, poor females, medium finished work male and medium to poor male calves are approximately Rs 900, Rs 700, Rs 500, Rs 600 and Rs 180 respectively. A general account of livestock marketing in India is given by Bailur (1975). However, so far, no authenticated survey has been carried out to unravel the very many intricacies such as demand and supply patterns, sources of supply, pricing, seasonal variations, effect of natural calamities, trade financing, stake of primary producers etc.

19.1.2 Marketing of meat and edible offals

Hide, carcass and offal (head, entrails and feet) are the three major produces traded at wholesale level. The wholesale price, in respect of each item, is determined through bargain between meat traders and buyers.

19.2 Slaughter Technology and Meat Hygiene

Slaughter procedures

Out of the 3000 slaughterhouses in the country, some are meant exclusively for slaughter of buffaloes. Halal method is uniformly followed in all abattoirs. Line slaughter is practised at Mumbai (Deonar), Aurangabad (Allana) and Usgaon in Goa. Floor slaughter is practised in others. The animals are cast and bled on the floor. Casting may be regular (in rows) or irregular (haphazard). Floor slaughter may be either complete or partial; in the latter, the animals after partial flaying and evisceration are hoisted on to a rail or a bamboo. Regular, complete floor slaughter is practised at Bareilly; irregular, complete floor slaughter is followed in Delhi and Hyderabad; and regular, partial floor slaughter is practised in Kolkata, Madras and many abattoirs in Kerala. Kondaiah and Panda (1987) suggested a semi-modern abattoir design which consists essentially of multi-hoist points with mechanical lifting for improvement of floor slaughtering under Indian conditions.

The traditional slaughter art deserves merit for production of hygienic meat; the hide serves as a clean-mat for the carcass; the evisceration operations are most scrupulous and scientific, *i.e.* two fingers are held just below the knife opening, at, the time of making mid-ventral incision, to guard against cutting open the viscera, the rectal contents are pushed back before making anal cut and a knot is made out of the terminal portion to prevent contamination with excreta; the blood in the visceral cavity is removed completely by repeated sponging with a cloth. But, it is a pity that such an excellent workmanship is displayed only in respect of good carcasses. Further, post-dressing contamination is of an alarming magnitude because the cleaning facilities are either absent or awesomely low ; the guts are opened close to the carcass ; no immediate removal of wastes is practised. The butchers' tools are inadequate and in sanitary and clothing are very dirty.

Meat inspection

A methodical and orderly ante-and postmortem inspection is carried out only in a very few slaughterhouses as in Mumbai, Aurangabad and Usgaon. In other places, the job, by and large, is restricted to routine passing of the animals and stamping of the carcasses, more often by abattoir attendants than by veterinarians. However, butchers voluntarily trim out the grossly infected portions from carcasses and offal.

19.3 Transport of Carcasses

The carcasses, in the form of quarters and primal cuts, and edible offal are transported from slaughterhouses to retail shops by a variety of means like cycles, cycle-rickshaws, auto-rickshaws, animal-drawn carts, and four-wheelers. The carts are manually lifted by workers from the wholesale marketing yard to vehicles parked outside. Four wheelers are fully covered containers. In other modes of transport, the material is covered with gunnies or cloth or plastic sheets and fastened with ropes to protect it from crows and dogs en route.

19.4 Meat Handling at Retail Shops

Retail meat and offal traders possess enormous skills in converting a visibly unhygienic material into an appealing product. The outer surface of carcasses and offals, by the time it reaches the retail shops, is heavily contaminated with dung, dust, dirt, hair, mud etc., but the retailers are able to meticulously knife out all such foreign materials and give a beautiful shape by spreading out the superficial fascia and fat. This practise, no doubt, is camouflage hygiene; but it infuses a sense of satisfaction on the unwary customers, who are not fully aware or experienced with the appalling conditions through which the meat has passed.

19.5 Wholesale Trading of Meat and Offal

The offal are the first to be disposed; in fact the dressing operations are suspended up to settlement of sale for offal. A part of rumen, cauls fat (which, in most cases, is inflated by blowing air through mouth) and small intestine are pulled out through mid-ventral incision for display to the prospective retail offal traders. The degree of finish, reflected by the amount of fat on the displayed portions, influences the price. A poor animal may fetch a very low price where as a well finished animal may fetch a very high price. Carcass is the second commodity to be sold. It is divided into halves by a circular cut between 9th and 10th ribs. Seldom carcasses are sold in the slaughter hall; these are brought to the wholesale meat marketing yard, just outside the hall, and split into 4 quarters (stage I cuts). The halves and quarters are displayed in unique styles for attracting the retail meat traders. If the demand is insufficient, they are further split into primal and sub-primal cuts (stage II and stage III cuts) and disposed. Each cut has a colloquial name. Variations with reference to carcass cutting methodologies, primal cuts, their nomenclature etc., are noticed in different regions of the country. Hides are the last to be disposed.

19.5.1 Retail trading of meat

The quarters, primal and sub-primal cuts, undercuts etc. are displayed at retail stalls after a meticulous face lifting. Splitting into smaller cuts are helpful to the retailers for better display, customer attraction, easy handling and efficient deboning.

Many customers are familiar with primal cuts and demand meat from selective locations. Barring brisket, chops and shanks, meat is usually sold without bones. It is remarkable that retail price of meat has remained unchanged over a long period. There is no differential pricing based on consumer preference and carcass quality. This is, indeed, an unhealthy situation because the wholesale price fluctuates highly depending upon carcass grade and demand and supply factors.

19.5.2 Retail trading of offal

There are exclusive retail shops for offal. Of course, offals are also sold at retail meat stalls; but, here, the sale is restricted to organ meats and for fulfilling the specific demands of regular customers.

The art of separating the offal into its constituent portions is a specialized job and can be appreciated better by actually seeing the techniques. Head is split into check musculature, ears, tongue, and bones. Pluck is separated into lungs, heart, diaphragm meat, and oesophagus meat. The muscular layer of rumen is pulled off as rumen meat. Reticulum, omasum and abomasum are emptied and washed thoroughly in water. The cauls fat is set by hanging from a rope after spreading its surface using wooden sticks. Offals are sold substantially at lower prices than carcass meat. They mostly cater to the needs of people from lower strata of the society. There is a strong belief that eating buffalo lungs increases milk secretion in lactating mothers and lungs are invariably purchased for this purpose.

19.6 Meat Export Trend

India's agro export products trend over the last 10 years and the share of meat products within this is:

Figures in Rs Crores

	Agro products export (APEDA)	Meat export	As % of agro products
1995–96	7915	614	8
1996–97	7723	690	9
1997–98	7271	792	11
1998–99	9682	770	8
1999–00	7365	797	11
2000–01	9213	1453	16
2001–02	10169	1177	12
2002–03	13828	1345	10
2003–04	14184	1647	12

19.7 World Bovine Meat Scenario

Opportunity for India to grow its international buffalo meat trade very rapidly is indeed encouraging. Historically, there have been 4 major bovine meat producing clusters:

- North America (USA and Canada)
- South America (including Brazil, Argentina and Uruguay)
- Oceanic countries (Australia and New Zealand)
- Europe (British Isles and continental Europe).

FAO's estimate of world bovine meat production in 2003 is 62 million tonnes carcass weight equivalent (CWE). World statistics of bovine meat is always denoted as Carcass Weight Equivalent. Carcass Weight Equivalent is roughly 1.66 times boneless meat weight, the form in which India exports. The four bovine meat producing clusters in 2003 produced 35 million tonnes. Due to the mad cow disease or bovine spongiform encephalopathy (BSE) crisis of the 90s, Europe as a major force in bovine meat export has declined substantially.

Further as a result of EU's decision to reduce agro subsidy gradually and eliminate it over the next 5–10 years, bovine meat prices in Europe have already risen and with a strengthening Euro, are just not viable for exports. Infact with declining production, for the first time in 2003, became a net importer of bovine meat. The total production of EU in 2003 was 7.3 million tonnes (with net imports of 73000 tonnes). Net exports in 1993 was 832000 tonnes. It is expected that in next 5 years the biggest new market for bovine meat will be Europe with expected annual imports towards the end of this decade of over a million ton.

North America with one incidence each of BSE in USA and Canada in 2003 and a further 2 in Canada later, lost most export markets, importantly Japan, but after political intervention at the highest level, conditional exports have resumed recently. It is not expected that North America will come out of the crisis, with more incidences of dormant BSE likely to surface and investments in the sector not increasing. However USA and Canada continue to be very large bovine meat consumers, majority of which is produced domestically. The total production and consumption of USA and Canada in 2003 is 13.3 million and 13.5 million tonne respectively.

That leaves two of the original four supply clusters, *viz*. South America and Oceanic countries. Both have abundant grasslands and its a major boon to their bovine meat industry.

Bovine meat production in south America, particularly Brazil has expanded substantially to fill the world bovine meat supply gap. The total production in south America in 2003 is 12.7 million tonnes (in 1993, 9.7 tonnes) out of which Brazil accounted for 7.2 million tonnes which included 1.2 million tonnes export. But there are serious apprehensions that Brazilian bovine meat supply will not be able to keep pace with export demand, under their current bovine meat production structure. Further, due to their improving economies, domestic demand in the south American countries is growing even more rapidly, leaving exportable volumes lower, whilst firming up prices.

Similarly, though to a lesser extent, the Oceanic countries are also growing (total production in 2003 was 1.8 million tonnes v/s 1.6 million tonnes in 1993). Draughts in recent years affected production in Australia but is likely to rise in next few years. Interestingly, the following table shows per capita consumption is very high in countries where production is high. Countries in the Middle east, west Asia, north Africa and south east Asia have a comparatively low per capita consumption but expected to grow substantially in the next few years resulting from economic growth and changing life styles and eating habits. This is good for India's future exports since they are our major markets.

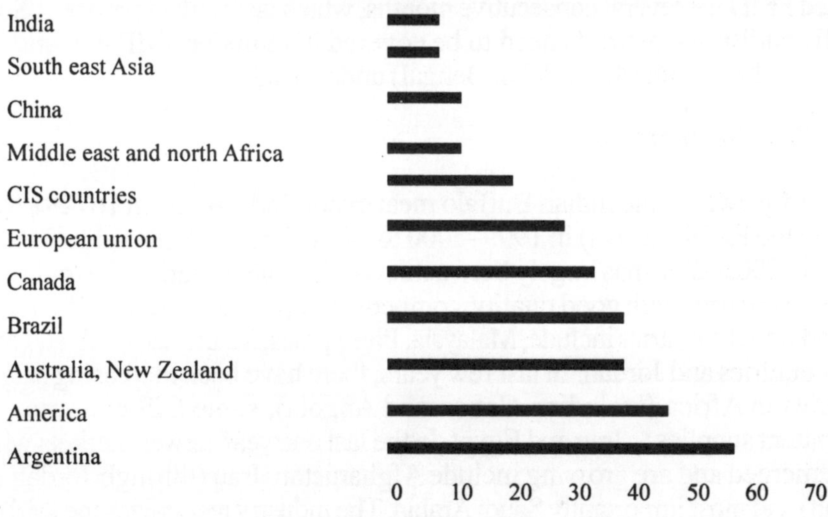

2003 per capita consumption (kg cwe).

Central America and the Caribbean, Russia and CIS, Africa, Middle East, and east Asia constitute the chief bovine meat deficit regions. Future aggregate increases in global bovine meat demand is expected primarily from South East Asia, Russia and CIS countries, Middle East, North and west Africa and EU.

19.7.1 Livestock health

Livestock health status plays a major role in world meat trade. The livestock health situation in India is definitely improving. India has always been free from the dreaded mad cow disease (BSE) and has been free from rinderpest since 1995. There has not been a single incidence of contagious bovine pleuro pneumonia (CBPP) in India during the previous 12 years. Foot-and-mouth disease remains the only issue of concern, though better controlled.

Importantly, a major programme has been initiated by the Central Government since August 2003 through the FMD-Control Programme (FMD-CP) covering 54 districts across the country. With the Indian Veterinary fraternity's successful accomplishment of eradication of rinderpest, a more difficult and dreaded disease, it will not be wishful thinking to believe given the right resources, substantial and relevant areas in India could become FMD free, duly recognized by the OIE, in the next 5–6 years.

Uttar Pradesh and Maharashtra, the prominent States where buffalo meat is processed for export, have covered most of the balance districts in their states under Assistance to States for Control of Animal Diseases (ASCAD) where the Centre's contribution is 75%. Already Uttar Pradesh and Maharashtra have not reported FMD for several consecutive months, which is excellent progress. Some specific additional districts need to be covered both under FMD-CP and in 2 states (Andhra Pradhesh and West Bengal) under ASCAD.

19.7.2 Existing markets

The good growth of the Indian Buffalo meat export industry from 167291 MTS (FOB value Rs 797 crores) in 1999–2000 to 343817 MTS (FOB value Rs 1647 crores) in 2003–2004 has largely been achieved through nurturing the traditional markets/countries with good quality, competitively priced, frozen Buffalo meat. The traditional countries include, Malaysia, Philippines, Mauritius, UAE and other GCC countries and Jordan. In last few years, there have been increasing sales to countries in Africa (including Gabon and Angola), some CIS countries and intermittent supplies to Iran and Egypt. In the last one year, newer markets which have emerged and are growing include Afghanistan, Iraq (through Jordan and Kuwait) and most importantly Saudi Arabia. The industry recognizes the lead role of the Ministry of Commerce, WANA Division, in pursuing opening of the Saudi Arabian market for both, sheep/goat meat and buffalo meat.

19.7.3 New markets

Some of the countries which are major importers of bovine livestock/meat and could be targeted for supply of frozen buffalo meat from India are shown, along with our best estimates of their equivalent bovine meat imports. Some of them like Egypt and Iran have previously intermittently sourced buffalo meat from India.

19.8 Government's Crucial Role

In India, the growth of the Buffalo meat export industry (since pioneered by Allanasons Ltd in 1969) has been largely undertaken by the exporters. However, the support of APEDA and the Central and State Animal Husbandry Departments must be singularly recognized. Recently, government support has been forthcoming. Positive government support/intervention, is the industry's expectation, particularly since the meat export industry touches rural development at the very basic level of the poor, marginal, landless farmers whose primary source of livelihood and existence comes from small livestock holdings.

Table 19.1 Import of Meat by Different Countries

Country	Current imports of potential markets
Egypt	100000
Iran	30000
Algeria	50000
Syria	30000
Indonesia	50000
Thailand	30000
Philippines (retail segment)	80000
Russia	100000
Ukraine	50000
Kazakhastan	25000
Krygystan	25000
Tajkistan	25000
	595000

This is in line with and supports the progressive, pro-farmer, rural development oriented policies of the Government of India. One of the most important virtues of Indian buffalo meat, we must emphasis, is its competitive price. It is a real value for money product and is also its Unique Selling Proposition (USP). For continued success, competitive price must be maintained. A number of issues from time to time affect the economics of the meat export industry detrimentally and are immediately taken up with the authorities concerned. Issues which persist and currently affect the industry, many of which were projected at the first meeting of the Monitoring Committee on Agriculture held on 27th December 2004, are listed below.

- Inclusion of meat as an eligible item in the "Vishesh Krishi Upaj Yojana" (Special Agricultural Produce Scheme)
- Restoration of APEDA Financial Assistance for upgradation of export oriented abattoirs/processing plants as was applicable during 1997–2002
- Inclusion of buffalo meat under APEDA's Transport Assistance Scheme for new markets in Africa/CIS where freight cost from India for refrigerated containers is much higher than from competing countries
- Restoration of DEPB rates for frozen buffalo meat
- Exemption from Service Tax on transportation of meat products processed for exports. This is presently applicable only for fruits, vegetables, eggs or milk even for domestic consumption
- Ministry of Food Processing Industry's grants as per their published schemes (currently being denied)

Most of these points pertain to steps which would help Indian buffalo meat retain its competitive price. Meat export industry does not desire any special or out of the way benefits/subsidies. What it seeks is legitimate application (and non-denial) of existing schemes/policies to the meat export sector, wherever it is logical.

CHAPTER 20

LAWS, RULES AND REGULATIONS (JURISPRUDENCE) AS APPLICABLE TO LIVESTOCK-BUFFALOES

By

M.C. Sharma, Umesh Dimri and P.N. Bhat

Veterinary Jurisprudence (legal veterinary medicine) is defined as the part of veterinary medicine including the use of principles and know-how of veterinary medicine for purposes of law (civil/criminal) and deals with rights of the animal owners and harm caused to the animals (Block, 2003; Clutton, 1999; Anderson, 2003). A veterinarian must be very sound in his knowledge about Veterinary Science in relation to law, as he is the only evidence on who depends the action to the taken against an involved person.

A veterinarian may appear in a court as a witness or complainant. He must be well acquainted with the acts affecting his privileges and obligations. He must also be well versed about criminal courts. The code of criminal procedure, Act X, of 1872, in sections 4, defined a criminal court as every Judge or Magistrate or body of Judges or Magistrates enquiring into or trying any criminal case or engaged in any judicial proceeding.

The courts for criminal dealings are of three types, namely, high courts, session's courts and courts of Magistrates. The main classes of Magistrates are the Judicial Magistrates and the Executive Magistrates. The Judicial Magistrates include the Chief Judicial Magistrate. Additional Chief Judicial Magistrates, Judicial Magistrates of the first or second class, Sub Divisional Judicial Magistrates, Special Judicial

Magistrates, Chief Metropolitan Magistrates and the Special Metropolitan Magistrates. The Executive Magistrates include the District Magistrates, Additional District Magistrates and the Special Executive Magistrates.

The Supreme Court in Delhi is the highest judicial tribunal within the territory of India and a law declared by the Supreme Court is binding on all courts within India. The high court is at the peak tribunal at the State level. The session's courts can only try cases committed by a Magistrate. They may pass any sentence under the jurisdiction of law but a death sentence by courts needs a high court confirmation before it is carried out. The sentences authorized by law include death, life imprisonment, imprisonment with solitary confinement and reformatory detention.

A first, second and third class Magistrates may pass maximum imprisonments up to two years, 6 months and one month respectively. The first, second and third class Magistrates may pass a maximum fine of one thousand rupees, two hundred and rupees fifty respectively. Only a Magistrate can take cognizance of an offence.

A subpoena (or summon) is a document to compel the presence of a witness in a court of law under a penalty. Once served, the witness must do so without a fail or a justifiable excuse. A fail without a justifiable excuse may make him vulnerable to a fine or imprisonment (Sharma, 1981).

Once called into a witness-box, a witness, before giving evidence has to take an oath which includes a solemn affirmation substituted by law for an oath, and any declaration required or authorized by law to be made before a public person or to be used for the purpose of proof, whether in a court of justice or not. In oath one has to state touching the matter that the evidence which he shall put forth shall be the truth, the whole truth and nothing but the truth.

Under the prosecution process, when a buffalo owner seeks redress of his grievances in a court of law for criminal injury to his buffalo, he lodges a report at the nearest police station. The veterinarian examines the buffalo at the spot/hospital and submits his report to the police. This pin points the cause and manner of death in criminal cases. The evidence is defined in the Indian Evidence act (1872) as all statements which the court permits or requires from a witness related to facts under enquiry (oral evidence) or all documents in this regards (documentary evidence). The documentary evidence includes Veterinary Certificates and vetero-legal reports (Sharma, 2003).

The evidence recording involves Examination-in-chief, cross-examination, re-examination and questions put by the judge. Examination-in-chief is the first examination of a witness by the calling party. The cross-examination tests the value of the evidence and is held by counsel for the accused who tries to elicit facts. The council, conducting the examination-in-chief, has the right of reexamination of the witness to explain misunderstandings, if any, occurring during cross-examination. The judge, to clarify matters, may put any question (s), at any time during the process.

Witnesses are of four kinds namely, common, expert, skilled and hostile. A common witness testifies the facts as he observed. An expert witness, based on his professional competence, deduces opinions or inferences from the facts observed by him or by others (Lall, 1971).

To give evidence a veterinary witness should adhere to the following points: before going to a court consider the possible questions with their answers which one may be asked; one should speak clearly, slowly and audibly; use simple plain language avoiding technical terms; speak minimum possible; do not lose temper, appear cool and dignified and avoid quotations from books. In case if a veterinarian has to stand against another Veterinarian, then it is possible to disagree without being disgraceful. The Veterinary witness must volunteer statements. Under section 126 of the Indian Evidence Act, a veterinary witness should, on no account, volunteer professional secrets but should divulge them under protest to show his moral duty, when pressed by the court to do so.

Postmortem examination and the law

The post-mortem examination aims at identifying, if not known and at ascertaining the time and cause of death.

The rules for post-mortem examination are as follows: (i) under law a post mortem examination should be done by written order from police or District Magistrate, (ii) Prior to examination, the police report should be read thoroughly, (iii) It should be done in day light and be thorough and complete, (iv) Look for all details on the body surface, and one must have a knowledge of the normal anatomy and physiology (Dabas *et al.*, 2007).

The post-mortem examination must include an external examination followed by internal examination.

The main portions to be examined are the lungs, heart, brain, digestive tract, parenchymatous organs, genital tract, skin. Blood stains also be examined in veterolegal cases (Broom, 2006).

Death in relation to law

Death includes both somatic and molecular deaths. Somatic death is complete cessation of the vital functions of brain, heart and lungs where as molecular death is that of tissues and cells. accompanied by cooling of the body, reducing its temperature to an equilibrium.

Death is due to three primary reasons, *viz.* coma, syncope and asphyxia. Coma insensibility involving central portion of brain stem and is mainly due to compression of the brain due to injury/disease or toxins (from outside/inside the body). There is stupor followed by unconsciousness, There is relaxation of sphincters, insensibility of pupils to light, body temperature is normal/subnormal, pulse is slow and breathing is irregular. On post-mortem, skull bone or brain injuries may be present with blood effusion into the cranial cavity, brain with its membranes are congested.

Syncope is death due to stoppage of heart's action and may be due to anaemia, shock, asthenia or exhausting disease. There is pallor of mucous membranes, dim vision, dilated pupils, restlessness, gasping respiration, marked fall in blood pressure. The heart is contracted on postmortem and its chambers empty in anaemia while in asthenia they contain blood.

Asphyxia causes death when there in stoppage of respiratory functions before stoppage of heart action. It may be due to mechanical causes; deficiency of atmospheric oxygen, exhaustion of respiratory muscles; wounds due to non entrance of blood in lungs as in embolism. The main symptoms include dyspnoea, convulsions and exhaustion. In dyspnoea, the carbon dioxide accumulation in blood stimulates the respiratory centre in the medulla, with respiration becoming deep, hurried and labored. In convulsions, the expiratory muscles show spasmodic movements, spreading to all muscles of the body. In exhaustion, there is paralysis of the respiratory centre when pupils are dilated with fall in blood pressure.

The concept of death w.r.t. law states that the cessation of the vital functions depends upon tissue anoxia, brought about by any of the following four ways: (i) Defective oxygenation (ii) Reduced oxygen-carrying capacity, (iii) Histo-toxic anoxia and (iv) Inefficient blood circulation through tissues. The tissue anoxia due to any of these causes produces circulatory failure leading to death.

Sudden death may occur from natural or unnatural causes. The possibility of death due to disease and injury together is also to be kept in mind. The natural causes producing sudden death are mainly the diseases of the cardio-vascular system (enteric fever, pericarditis diphtheria, hypothyroidism, right ventricular failure).

The signs of death are cessation of circulation and respiration; changes in eye and skin; cooling of the body; cadaveric lividity and changes; putrefaction; adipocere and mummification. Cadaveric changes in muscles include primary relaxation or flaccidity, cadaveric rigidity or rigor mortis and secondary relaxation. Rigor mortis is simulated by heat stiffening, cold stiffening and cadaveric spasm.

From a legal point of view it is very important that a veterinary jurist rightly assesses the time since death. The points to be considered here are warmth or cooling of the body, absence or presence of cadaveric hypostasis, rigor mortis and putrefaction. These conditions may vary between cadavers and thus only an approximate time of death can be ascertained.

Suffocation is the form of death resulting from exclusion of air from the lungs by means other than compression of the neck. It may be due to either by forceful closure of the mouth and nostrils or choking of the air passages from within and inhalation of irrespirable gases.

Death is usually due to asphyxia, but may be due to shock. On an average, the death occurs from ten to fifteen minutes after complete withdrawal of air from lungs, although cases with death occurring almost instantaneously after blockage of the windpipe by a foreign body are reported. Recovery may occur if treated properly within five minutes.

Post mortem appearances may be external or internal. External means death due to suffocation or asphyxia. Whereas in internal, a foreign body may be found in the mouth, throat, larynx or trachea when suffocation is caused by a foreign body impaction in the air passage.

Drowning is the death in which air is prevented from entering the lungs by submersion of respiratory orifices in a fluid. During drowning a body alternatively rises and sinks thrice, until all air has been expelled from the lungs and its place is taken up by water. Only after this, does the animal become insensible and sinks to the bottom and dies. The mode of death may be asphyxia, shock, concussion, apoplexy or exhaustion.

Starvation or inanition is deprivation of a regular and constant food supply necessary to maintain the body nutrition. In absence of food, an acute feeling of hunger lasts for first thirty to forty-eight hours, after four or five days of this, the absorption of subcutaneous fat begins. On post mortem, externally the body is debilitated, eyeballs are sunken, tongue is dry, the subcutaneous fat is very less or absent. Internally, the lungs are pale and collapsed, and exude very little blood when cut, intestines do not contain any digested material and are contracted. The liver, spleen, kidneys and pancreas are small and shrunken. Urinary bladder contains no urine and the gall bladder is enlarged in size. To confirm the death due to starvation it must be differentiated from death due to wasting disease like tuberculosis.

Heat, lighting and electrical injuries

Injuries due to flame, radiant heat, lightening, electricity, x-ray and chemical are classified as burns under law. Scalds are injuries due to moist heat e.g. steam. Burns are more serious than scalds.

Ordinary scalds are not very serious, causing in general, hyperaemia and vesiculation. Scalds from oils or other sticky substances, boiling at a higher temperature than water, closely resemble burns.

X-rays cause burns with dermatitis and hair shedding. Prolonged exposure may result into ulcers. Longer exposure to sun may cause burns with eczematous dermatitis. Burns due to chemicals do not have a red line of demarcation and vesicles. Burns are classified as follows:

(i) First degree burn is due to instantaneous contact with the burning agent and there is no scar formation.
(ii) Second degree burn is due to prolonged contact with the agent and as only superficial epithelial layers are damaged so no scar is formed.
(iii) Third degree burn has a scar formed after healing maintaining the integrity of the part.
(iv) Fourth degree burn has destruction of the whole skin and the scar left is of dense fibrous tissue.
(v) Fifth degree burn has effect on deeper muscles and forms scarring and deformity.
(vi) Sixth degree burn has either death or charring of whole limb with inflammation of adjacent tissues and organs.

The severity of burns and scalds depends on the degree of heat, exposure period, extent of surface and site. The death here is because of shock, suffocation, accidents, inflammation of visceral organs, hypoproteinaemia. Death may occur mostly in the first week or in cases with pus formation it may occur after one to one and a half month. On post mortem, externally, in burns due to radiant heat, skin is whitened; in burn by flame, the skin is blackened; in prolonged burns by molten metal, there is roasting and charring of the body parts; in burns from kerosene oil, the characteristic odour is there. Internally, the brain and its meninges are congested with inflammation of the pleurae.

The ante-mortem and post-mortem burns are mainly differentiated on the basis of line of redness, vesication and separative processes. Regarding finding the time of burn, immediate to burning is redness, within two to three hours is vesication, in two to three days is pus, sloughs separate within a fortnight. Later to this, the granulation covers the burn surface. A cicatrix or deformity appears after several weeks or months.

Lightning occurs during thunderstorm and near tall objects, so animal keeping should be avoided near tall objects during thunder storm. After being affected by lightening, there is syncope or concussion and death is due to nervous paralysis or from effects of burns after some days. Reddish brown arborescent markings indicating the paths taken by the branching discharge are seen on the surface of the skin. To protect the buffalo, oxygen is provided along with cardiac stimulants. On post mortem, rigor mortis appears soon after death and passes off quickly.

Deaths due to shock may occur due to electricity as well. The points of touch of electrical item get burned. The treatment involves removal of the electrical source, providing artificial respiration, cardiac and respiratory stimulants. On post mortem, lesions are present at the contact point of the electrical source and the body with congestion of the viscera. Mechanical injuries include bruises/contusions, abrasions and wounds. Bruises or contusions are produced by a blunt weapon blow. The age of bruise is adjudged by colour changes which its ecchymosis undergoes during absorption. The colour is red in the beginning, blue during the first three days, greenish on fifth to sixth day, and yellow from seventh to twelfth day and after fourteenth or fifteenth day, the skin regains its normal colour. In an antemortem bruise there is swelling and colour changes which are absent in a postmortem bruise.

Abrasions are injuries with loss of superficial epithelium of the skin. The antemortem abrasion has bleeding while the post-mortem abrasions do not have bleeding.

A wound is forcible breach in the continuity of soft tissues of body like skin, mucous membrane and cornea. The main types of wounds are incised, punctured, lacerated and firearm wounds. An incised wound is created by a sharp cutting object and is always broader than the edge of the weapon creating it with its superficial extent being greater than its depth. Hemorrhage is stronger in an incised wound. In punctured wounds, the object while passing through the tissue pierces an body cavity like thorax or abdomen. The aperture of this wound in skin is usually a little smaller in length than the breadth of the weapon used. A lacerated wound is a tear or split produced by blows with blunt objects and missiles where hemorrhage is not extensive. Firearm wounds are caused by firearms like shot guns and rifles, where larger bullets cause greater damage to the natural organs than small ones. A bullet with a high velocity produces a clean, circular, punched-out slit.

Points to be considered by a veterinarian while examining a wound, nature of injury, kind of weapon and age of injury. Death form wounds is mainly due to hemorrhages, vital organ injury or shock. Cardinal features of an ante-mortem wound are hemorrhage, retraction of the wound edges and inflammation.

Fractures during life have blood effusion, muscular laceration, pouring of lymph and callus formation but these features are absent in a post-mortem fracture.

Defaults in the sale of livestock and their products

The frauds in livestock sale are punishable under section 420, Indian Penal Code (I.P.C.). Before sale, a buffalo should be examined by history, general examination (including circulatory, respiratory, digestive, nervous and reproductive systems), agglutination test for brucellosis, tuberculin test for tuberculosis, mallein test for glanders in equines, examination of body excretions and secretions for presence of any pathogenic organism. Animals used for food purposes should be considered with ethics (Mephan, 1996).

Alteration is a fraud in livestock sale where its description is altered and is done by castration, tail and mane clipping, docking, use of hair dye and bishoping. Bishoping is to make the animal look younger.

The adulteration of milk is done by reduction of fat, addition of preservatives and accidental adulteration. Reduction of fat is done by either adding water, skimming milk or by both together. The addition of water is detected by determining the specific gravity and total solids in milk or by the presence of nitrates in milk as nitrates are present in water but not in pure milk. The skimming of milk is detected by determining the fat percentage in milk by Gerber's method as skimmed milk has high specific gravity, low fat content and high solids not fat content. Under skimming and watering, fat is removed and desired water is added to make specific gravity of such milk equal to that of normal milk and is detectable by finding fat content and total solids in such milk. Solids in milk are determined by gravimetric method. Mathematically total solids percentage (TS) and solids not fat content of milk are calculated from the following formula:

$$a)\ Babcock's\ formula\ T.S = \frac{G}{4} + 1.2F$$

$$b)\ Fleischmann's\ formula\ T.S = \frac{G}{4} + 1.2F + 0.25$$

$$c)\ Richmond's\ formula\ T.S = \frac{G}{4} + \frac{1}{2}F + 0.14$$

Where, G = Corrected lactometer reading, and F = percentage of fat.

Under skimming and watering, the milk becomes less viscous and to increase its viscosity, thickening agents like starch, gelatin, cane sugar etc. are added which can be detected by using iodine solution (5%), acid mercuric nitrate solution and hydrochloric acid respectively. The addition of colouring agents in milk is detected by annatto test and coal-tar dye.

Preservations are added to retard decomposition of milk and include boric acid, formalin, salicylic acid, benzoic acid, hydrogen peroxide and sodium bicarbonate.

Accidental adulteration of milk occurs by mistake when animal urine, dung, hairs etc. get into the milk. The quality of milk can be detected by the methylene blue reduction test where milk is graded as good, fair, poor or very poor.

The addition of heated milk in fresh milk is detected by storch's test. The adulteration of ghee is mainly done by adding vegetable oils, hydrogenated oils or animal fat. Phytosteryl acetate test detects the adulteration of vegetable oils in ghee. Sesame oil is necessarily added to all vegetable ghee in our country and thus the presence of sesame oil in marked pure ghee is a sure indication of adulteration with vegetable ghee which can be detected by boudouin test.

Adulteration of meat includes mixing meat from different animals or mixing inferior meat with the superior one e.g. horse flesh in place of beef, goat's meat for mutton, cat for rabbit. Such adulteration can be detected by physical, chemical or serological examinations. Mutton is duck red with ammonical odour; white, hard and firm fat; slightly red bone marrow; rich fat between muscles and firm and dense muscle fibre density. Chevon is paler than mutton, with very little fat in muscles and has a buck odour. Pork is whitish grey with soft consistency; white, granulated fat intermixed with muscles, pink-red bone marrow and urine like odour. Dog's meat has slightly inter mixed white fat with muscles and a dark red colour. Beef is highly red in colour and its fat acquires a firm consistency after chilling.

Horse meat has a dark red colour becoming blackish on exposure to air, the fat is golden to dark yellow and a greasy bone marrow. In poultry meat, muscular tissues are firm with fine fibres and has fat intermixed with muscles. Fish meat is white, with finely distributed fat in muscles and a higher water content of meat. The methods of chemical examination include the glycogen test, test for fat of animals (iodine value) and examination of refractive index. The serological examination aims up to differentiate meats of different species. It includes the precipitation test, agar gel diffusion test, chromatographic test, electrophoretic test.

The code of Federal Regulation, USA, 1970 has indicated the following rules for meat inspection: Ante-mortem inspection includes an examination of all buffaloes to be slaughtered on the day of slaughter, within the slaughter house and any buffaloes with a problem requiring condemnation of a part should be marked 'suspect', condemn buffalo with high temperature, likely to die or dead, signs of rabies, scrapie, tetanus, arthrax. A careful post mortem examination of all buffaloes and their parts including head, tail, tongue, thymus and all viscera should be made just after slaughter. Fit carcass should be marked 'inspected and passed'. Condemn parts/carcasses which are unhealthy or suspected for anthrax, tuberculosis, black leg and in case of actinobacillosis or actino mycosis the affected parts should be removed.

Determination of age of buffaloes

From the legal view point, a Veterinarian may be required to assess the age of a buffalo which is indicated by the teeth appearance, rings in the horns, external appearance and bones. Age is also indicated by time of appearance and extent of wear of temporary and permanent teeth. Temporary teeth are smaller and whiter than permanent teeth. For age assessment, knowledge of deciduous and permanent dental formulae, time of eruption and difference in appearance between deciduous and permanent teeth is needed.

In buffaloes, the incisors are absent in upper jaws and are replaced by dental pads. The canines are absent in cattle. The deciduous dental formulae of cattle is 2 (Di 0/4 Dc 0/0 Dp 3/3) = 20 and the permanent formulae is 2 (I 0/4 C 0/0 P 3/3 M 3/3) = 32. Regarding eruption, Di1 and Di2 are before birth, Di3 is birth to 1 week, Di4 is birth to 2 weeks, Dp1, Dp2 and Dp3 are birth to 3 weeks. With respect to the permanent teeth eruption, I_1 is 1a to 2 years, I_2 is 2 to 2a years, I_3 is 3 years, I_4 is 3a to 4 years, P_1 is 2 to 2a years, P_2 is 1a to 2a years, P_3 is 2a to 3 years, M_1 is 5 to 6 months, M_2 is 1 to 1a years and M_3 is 2 to 2a years.

The centres of ossification and the progress of ossification in unifying bones give an approximate idea of an animal's age. Thus, radiographs help in estimating an animal's age.The horn rings may also indicate a buffalo's age. In buffalo, the first ring appears at 3 years of age and then one ring is added each year. So age is 2 + no. of rings.

Euthanasia

Legally Veterinarians are also required, at times, to sacrifice the buffalo suffering with certain incurable diseases for which euthanasia remains the only solution.

Euthanasia is the production of quite painless death in a buffaloes for humane reasons. The method should produce painless death, be reliable, minimize undesirable psychological stress, economical, minimum environmental impact and safe for humans (Edney, 1989).

Euthanasia can be achieved by any of the following agents:

1. Inhalant agents these include:

a) Inhalant anaesthetics chloroform, ether, halothane and methoxyflurane
b) Carbon monoxide
c) Carbon dioxide
d) Hydrogen cyanide gas

2. Non-inhalant agents these include:

a) Barbiturates pentobarbital sodium
b) Chloral hydrate
c) Strychnine
d) Hydrocyanic acid
e) Magnesium sulphate
f) Curariform drugs

3. Physical methods Electrocution, shooting, before euthanizing a buffalo, consent should be taken from its owner.

The Prevention of Cruelty to Animals Act, 1960

The act extends to the whole of India, except the state of Jammu and Kashmir. Under this act, unless stated otherwise: (a) "animal" means any living creature other than a human being (Dawkins,1990); (b) "captive animal", means any animal not being a domestic animal in captivity or confinement, whether permanent or temporary, or which is subjected to any appliance or contrivance for the purpose of hindering or preventing its escape from captivity or confinement or which is pinioned or which is or appears to be maimed: (c) "domestic animal" means any animal which is tamed or which has been or is being sufficiently tamed to serve some purpose for the use of man or which, although it neither has been nor is intended to be so tamed, is or has become in fact wholly or partly tamed (Duncan, 2006); (d) "local authority", means a municipal committee, district board or other authority for the time being invested by law with the control and administration of any matters within a specified local area; (e) "owner" used with reference to an animal, includes not only the owner but also any other person for the time being in possession or custody of the animal, whether with or without the consent of the owner (Dawkins, 2001); (f) "Phooka" or "doomdev" includes any process of introducing air or any substance into the female organ of a milch animal with the object of drawing off from the animal any secretion of milk; (g) "prescribed" means prescribed by Rules made under this Act; (h) "street" includes any way, road, lane, square, court, alley, passage or open space, whether a thorough fare or not to which the public have access.

It shall be the duty of every person having the care or charge of any animal to take all reasonable measures to ensure the well being of such animal and to prevent the infliction upon such animal of unnecessary pain or suffering (Benson, 2004; Blecha, 2000). Animals should be transported with due care (FAWC, 1991).

If any person (a) beats, kicks, over-rides, over drives, over loads, tortures or otherwise treats any animal so as to subject it to unnecessary pain or suffering or causes or being the owner permits, any animal to be so treated or (b) employs any animal unfit to be so employed (for age or disease etc.) or permits such unfit animal to be employed or (c) willfully and unreasonably administers or attempts to cause any injurious drug or injurious substance to be taken by any animal or (d) causes any unnecessary pain or suffering or (e) puts an animal at a place which hinders reasonable opportunity for movement or (f) keeps an animal unreasonably chained for an unreasonable time (EFSA Animal Health and Welfare Panel, 2006) or (g) being an owner, fails to supply sufficient food, drink or shelter or (h) without reasonable cause, abandons an animal that it will suffer pain by starvation or thirst or (i) wilfully permits his animal with contagious or infectious disease to go into street or permits him to die in street or (j) has an animal suffering pain by mutilation, starvation, thirst, overcrowding or other ill treatment or (k) with a view to provide entertainment causes him to become an object if prey for any other animal or incites any animal to bite or pray any other animal or (l) promotes any shooting match or competition where animals are released from captivity for the purpose of such shooting : he should be punishable (in the case of a first offence, with fine which shall not be less than ten rupees but which may extend to fifty rupees, and in the case of a second or subsequent offence committed within three years of the previous offence, with fine which shall not be less than twenty five rupees but which may extend, to one hundred rupees or with imprisonment for a term which may extend, to three months, or with both (Balm, 1999).

If any person is involved in 'phooka', he shall be punishable with fine which may extend to one thousand rupees, or with imprisonment for a term which may extend to two years, or with both, and the animal on which the operation was performed shall be forfeited to the government. Animal emotions must be given due weightage (Mendl and Paul, 2004; Dawkins, 1993; Webster, 2005).

Where the owner of an animal is convicted of an offence, it shall be lawful for the court, if the court is satisfied that it would be cruel to keep the animal alive, to direct that the animal be destroyed and to assign the animal to any suitable person for that purpose; and the person to whom such animal is so assigned shall, as soon as possible, destroy such animal or cause such animal to be destroyed in his presence without unnecessary suffering, and any reasonable expense incurred in destroying the animal may be ordered by the court to be recovered from the owner as if it were a fine. Sick animals need to be properly treated and taken care of (Campbell *et al.,* 2006; Collier *et al.,* 2006; Corke and Broom, 1999; Bartussek, 1999; Brambell, 1965; Sharma, 1987; Singer, 1990).

Animal welfare board of India

This was established for the promotion of animal welfare generally and for the purpose of protecting animals from being subjected to unnecessary pain or suffering (Aubert, 1999; Bertenshaw and Rowlinson, 2001). All necessary measures have to be taken to avoid putting any animal to unnecessary pains (Eicher *et al.,* 2006; Sandoe, 2003; Webster, 1994). The board shall be a body corporate having perpetual succession and a common seal with power, subject to the provisions of Act 26 of 1982, to acquire, hold and dispose of property and may by its name sue and be sued. The Board shall consist of the following persons:

(a) The Inspector General of Forests, Government of India, Ex-officio,
(b) The Animal Husbandry Commissioner to the Government of India, ex-officio,
(c) Two persons to represent respectively the Ministers of Central Government dealing with Home Affairs and Education, to be appointed by the Central Government,
(d) One person to represent the Indian Board for Wild life, to be appointed by the Central Government,
(e) Three persons who, in the opinion of the Central Government, are or have been actively engaged in animal welfare work and are well-known humanitarians, to be nominated by the central government,

(f) One person to represent the association of veterinary practitioners,
(g) Two persons to represent practitioners of modern and indigenous systems of medicine,
(h) One person to represent Municipal Corporation,
(i) One person to represent three organizations actively interested in animal welfare,
(j) One person to represent three societies dealing with the prevention of cruelty to animals,
(k) Three central Government nominees,
(l) six members of Parliament, four to be elected by the House of the people, (Lok Sabha) and two by the Council of States (Rajya Sabha).

The term for which the Board may be reconstituted under Section 5A shall be three years from the date of the reconstitution and the Chairman and other members of the Board as so reconstituted shall hold office till the expiry of the term for which the Board has been so reconstituted. The funds of the Board shall consist of grant made to it from time to time by the Government and of contributions, donations, subscriptions, bequests, gifts and the like made to it by any local authority or by any other person. Animal sufferings need to be proper studied (Gregory, 1999; Gregory, 2004; Gregory, 2007).

The functions of the Board shall be:

(a) To keep the law in force and advice the Government for timely amendments.
(b) To advise the Central Government on making of rules related to animal welfare.
(c) To advise on improvements in design of vehicle to lessen burden on draught animals.
(d) To take step for betterment in animal houses, watering and veterinary assistance.
(e) To advise on matters of slaughter houses to make them more humane.
(f) To take steps to remove unwanted animals.
(g) To encourse by finance or otherwise, formation of rescue homes for animals.
(h) To cooperative with associations involved in reaching animal pain and suffering.
(i) To give advice or financial help to animal welfare organizations.
(j) To advise for improvement in health services.
(k) To educate people for humane treatment of animals.
(l) To advise the Government on any matter related to animal welfare.

Care has to be given to welfare of all animals including farm and wild animals (FAWC, 1993; FAWC, 2001).

Experimentation of buffaloes

Under act 26 of 1982, S.12, nothing contained in this act shall render unlawful the performance of experiments on buffaloes for the purpose of advancement by new discovery of physiological knowledge or of knowledge which will be useful for saving or for prolonging life or alleviating suffering or for combating any disease, whether of human beings, animals or plants. The Central Government, on advice of the animal welfare board, may by notification in the official gazette, constitute a committee consisting of such number of officials and non-officials, as it may think fit to appoint thereto. The funds of this committee related to experimentation of animals shall consist of grants made to it from time to time by the Government and of contributions, donations, subscriptions, bequests, gifts and the like made to it by any person. It shall be the duty of the committee to take all such measures as may be necessary to ensure that buffaloes are not subjected to unnecessary pain or suffering before, during or after the performance of experiments on them, and for that purpose it may, by notification in the Gazette of India and subject to the condition of previous publication, make such rules as it may think fit in relation to the conduct of such experiments.

If the committee is satisfied, on the report of any officer or other person made to it as a result of any inspection under section 18 or otherwise, that the rules made by it under section 17 are not being complied with by any person or institution carrying on experiments on buffaloes, the committee may, after giving an opportunity to the person or institution of being heard in the matter, by order, prohibit the person or institution from carrying on any such experiments either for a specified period on indefinitely, or may allow the person or institution to carry on such experiments subject to such special conditions as the committee may think fit to impose. If any person

(a) Contravenes any order made by the committee or

(b) Commits a breach of any condition imposed by the committee; he shall be punishable with fine which may extend to two hundred rupees, and when the contravention or breach of condition has taken place in any institution the person in charge of the institution shall be deemed to be guilty of the offence and shall be punishable accordingly.

Performing buffaloes

Under this 'exhibit' means exhibit or any entertainment to which the public are admitted through sale of tickets and `train' means train for the purpose of any such exhibition and the expressions `exhibitor' and `trainer' have respectively the corresponding meanings.

No person shall exhibit or train (a) any performing buffalo unless he is registered in accordance with specified provisions; (b) as a performing buffalo, any buffalo which the Central Government may, by notification, in the official gazette, specify as a buffalo which shall not be exhibited or trained as performing buffalo.

The Prevention of Cruelty to Draught and Pack Animals Rules, 1965

In exercise of the powers conferred by sub-section (2) of section 38 of the Prevention of Cruelty to Animals Act, 1960 (59 of 1960), the Central Government made the Prevention of Cruelty to Draught and Pack Animals Rules, 1965.

The maximum prescribed loads for draught animals are

Animal	Type of vehicle	Maximum permissible load
1. Small bullock or	two wheeled	
small buffalo	a) with ball bearings	1000 kg
	b) with pneumatic tyres	750 kg
	c) not with pneumatic tyres	500 kg
2. Medium bullock or	two wheeled	
medium buffalo	a) with ball bearings	1400 kg
	b) with pneumatic tyres	1050 kg
	c) not with pneumatic tyres	700 kg
3. Large bullock or	two wheeled	
large buffalo	a) with ball bearing	1800 kg
	b) with pneumatic tyres	1350 kg
	c) not with pneumatic tyres	900 kg
4. Horse or mule	two wheeled	
	a) with pneumatic tyres	750 kg
	b) not with pneumatic tyres	500 kg
5. Pony	two wheeled	
	a) with pneumatic tyres	600 kg
	b) not with pneumatic tyres	400 kg
6. Camel	two wheeled	1000 kg

No person shall cause any animal specified in column 1 of the table below to carry any load in excess of the weight specified in the corresponding entry in column 2 thereof

	1	2
1.	Small bullocks or buffalo	100 kg
2.	Medium bullock or buffalo	150 kg
3.	Large bullock or buffalo	175 kg
4.	Pony	70 kg
5.	Mule	200 kg
6.	Donkey	50 kg
7.	Camel	250 kg

General conditions for use of draught and pack animals:

No person shall use or cause any animal to be used for drawing any vehicle or carrying any load:

(i) For more than nine hours in a day in the aggregate
(ii) For more than five hours continuously without a break for rest for the animal
(iii) In any area where the temperature exceeds 37 °C (99 °F) during the period between 12.00 noon and 3.00 pm

The Prevention of Cruelty to Buffaloes (licensing of farriers) Rules, 1965

Under this no person shall, after the commencement of these rules, begin to carry on the business of a farrier, and no person carrying on the business of farrier at the commencement of these rules, shall, after the expiration of three months, from such commencement, continue to carry on such business, except under a license.

Any person fulfilling the following criteria is entitled to apply for license

a) Has completed the age of eighteen years
b) Has undergone any such training in the business of shoeing buffalo as may be approved by the licensing authority; or
c) Has been carrying on the business of a farrier for not less than two years before the commencement of these rules.

Transportation of Buffaloes, Rules, 1978

Transport of buffalo

It must be accompanied with a valid health certificate indicating fitness of the animals for transport and theirselves being free from any contagious or infectious disease and in the absence of this certificate, the carrier shall not accept the consignment; the average space provided per cattle in Railway wagon vehicle shall not be less than two square meters; hungry and thirsty cattle should not transported; cattle in advanced pregnancy shall not be mixed with young cattle to avoid stampede during transportation.

Prevention of Cruelty to Animals (application of fines) Rules, 1978

Fines levied and realized under this act shall, subject to any deductions relating to the cost of collection, be made over by the State Government to the Board as soon as may be after due appropriation by law (by State Legislature) in this behalf. Fines made to the Board shall be used for buffalo welfare activities. Fines realized in one State shall be used within that state only.

Prevention of Cruelty (Capture of Animal's Rules, 1972)

No buffalo shall be captured for the purpose of sale, export or for any other purpose except by sack and look method.

Acts related to buffaloes

The Livestock Importation Act, 1898 (Act No IX of 1898)

As modified up to 1st September, 1952, this act is to make better provision for the regulation of the importation of livestock. In this Act, unless there is anything repugnant in the subject or context : (a) the expression "infectious or contagious disorders" includes tick-pest, scabies and other disease or disorder which may be specified by the Central Government by notification in the official Gazette, and (b) "livestock" includes buffaloes, horses, camels, sheep, and any other animal which may be specified by the Central Government notification in the official Gazette, regulate, restrict or prohibit, in such manner and to such extent as it may think fit, the bringing or taking, by sea or land, into the territories to which this Act extends or any specified place therein, of any livestock which may be liable to be affected by infectious or contagious disorders, and of any fodder, dung, stable litter, clothing, harness or fittings pertaining to livestock or that may have been in contact there with. No suit, prosecution or other legal proceeding shall lie against any person for anything in good faith done or intended to be done under this Act.

The Poisons Act, 1919 (Act No. XII of 1919)

As modified up to 1st January, 1960, it is an act to consolidate and amend the law regulating the importation, possession and sale of poisons. The Central Government may, by notification in the official Gazette, prohibit, except under and in accordance with the conditions of a license, the importation into India across any customs frontier defined by the Central Government of any specified poison, and may by rule regulate the gront of licenses.

The State Government may by rule regulate the possession of any specified poison in any local area in which the use of such poison for thepurpose of committing murder or mischief by poisoning buffalo appears to it to be ofsuch frequent occurrence as to render restrictions on the possession thereof esirable.

The District Magistrate, the Sub-Divisional Commissioner of Police, may issue a warrant for the search of any place in which he has reason to believe or to suspect that any poison is possessed or sold in contravention of this Act or any rule there under, or that any poison liable to confiscation under this Act is kept or concealed.

The person to whom the warrant is directed may enter and search the place in accordance there with, and the provisions of the code of criminal procedure, 1898, relating, to search-warrants, shall as far as may be, deem to apply to the execution of the warrant.

The Dangerous Drugs Act, 1930 (Act No. II of 1930)

This Act aims to centralize and vest in the Central Government the control over some operations relating to dangerous drugs and to increase and provide uniform penalties for offences relating to such operations "Dangerous drug" includes coca leaf, hemp and opium and all manufactured drugs.

No one shall (a) cultivate any coca plant, or gather any of its portions (b) manufacture or possess prepared opium unless it is prepared from opium lawfully possessed for the consumption of the person so possessing it or (c) import into India, export from India, transship or sell prepared opium. Provided that these shall not apply to the cultivation of any coca plant or to the gathering of any portion thereof on behalf of the Government. The State Government may make rules restricting and regulating the manufacture and possession of prepared opium from opium which is lawfully possessed under clause (b) above. No one shall engage in or control any trade whereby a dangerous drug in obtained outside India, and supplied to any person outside India-save in accordance with the conditions of a license granted by the discretion of the State Government.

Whoever: (a) cultivates any coca plant or gathers any portion of a coca plant (b) manufactures or possesses prepared opium other than as permitted or (c) imports, exports, transships or sells prepared opium, shall be punishable with imprisonment which may extend to three years, or with fine or both.

The Collector or any officer authorized by the state Government in this regards, or a presidency Magistrate or a Magistrate of the first class, or a Magistrate of second class specially, empowered by the State Government in this behalf, may issue a warrant for the arrest of any person when he has reason to believe to have committed an offence punishable or for the search, whether by day or by night, of any building, vessel or place in which he has reason to believe any dangerous drug in respect of which on offence punishable has been committed is kept or concealed.

The Drugs and Cosmetics Act, 1940 (Act No. XXIII of 1940)

This act aims at regulating the import, manufacture, distribution and sale of drugs and cosmetics. The provisions of this Act shall be in addition to, and not in derogation of, the dangerous drugs in Act, 1930 and any other law for the time being in force.

In this Act, unless there is anything repugnant in the subject or context:

(a) Ayurvedic (including siddha) or unani drug includes all medicines intended for internal or external use for or in the diagnosis, treatment, mitigation or prevention of disease in human beings, mentioned in, and processed and manufactured exclusively in accordance with the foumulae described in the authoritative books of ayurvedic (including siddha)andunanisystems of medicine, specified in the First Schedule. "Cosmetic" means any article intended to be rubbed, poured, sprinkled or sprayed on or introduced into, or otherwise applied to the human body or any of its parts for cleansing, beautifying, promoting attractiveness, or altering the appearance, and includes any article intended for use as a component of cosmetic, but does not include soap.

A drug shall not be deemed to consist, in whole or in part, of any decomposed subtance only by reason of the fact that such decomposed substance is the result of any natural decomposition of the drug within the period, if any, specified on the label of the drug within which the drug is to be used. Provided that such decomposition is not due to any negligence on the part of the manufacturer of the drug or the importer or the dealer thereof and that it does not render the drug injurious to health.

From such date as may be fixed by the Sate Government by notification in the official Gazette in this behalf, no person shall himself or by any other person on his behalf sell or stock or exhibit for sale, or distribute, any Ayurvedic or Unani drug other than that manufactured by a manufacturer.

The Drugs and Cosmetics Rules, 1945 (under the Drugs and Cosmetics Act, 1940)

Schedule F (1), Part-I vaccines

(a) Provisions applicable to the production of bacterial vaccines

A competent expert in bacteriology with sufficient experience in manufacture and standardization of biological products shall be in charge of the establishment responsible for the production of bacterial vaccine and he shall be assisted by a staff adequate for carrying out the tests required during the preparation and standardization of the vaccines.

Cultures used in the preparation of vaccine before being manipulated into a vaccine, should be thoroughly tested for identity by the generally accepted test applicable to the particular micro-organisms.

The vaccines shall be tested for sterility, purity, safety and potency before use. Some common vaccines are Anthrax spore vaccine (living), Black quarter vaccine, Brucella abortus (Strain 19) vaccine (living), Salmonella abortus equi vaccine, Streptococcus equi vaccine, (b) provisions applicable to the production of viral vaccines. The establishment in which viral vaccines are prepared must be under the direction and control of an expert in virology or an expert in bacteriology with specialised training in virology and sufficient experience in the production of viral vaccines, and he shall be assisted by a staff adequate for carrying out the tests required during the preparation and standardisation of the vaccines. Some viral vaccines are Fowl Pox vaccine, Pigeon Pox virus (living); Ranikhet Disease vaccine (living), Ranikhet disesae vaccine F strain (living); sheet pox vaccine (living); rabies vaccine (inactivated); Rabies vaccine (living); Rinder pest lapinised vaccine (living).

The Indian Penal Code

Section 44. Deals with "injury" which denotes any illegal harm caused to any animal, in body, mind, reputation or property.

Section 47. States that an "animal" denotes any living being, other than a human being.

Section 51. Indicates that an "oath" includes a solemn affirmation substituted by law for an oath, and any declaration required or authorised by law to be made before a pubic servant or to be used for the purpose of proof, whether in a Court of Justice or not.

Section 52. Nothing is said to be done or believed in "good faith" which is done or believed without due care and attention.

Section 53. Under the provisions of this Code, the offenders are liable for punishments like death, transportation, imprisonment, forfeiture of property and fine.

Section 80. It indicates that anything is not an offence, which is done by accident or misfortune, and without any criminal intention or knowledge in the doing of a lawful act in a lawful manner by lawful means and with proper care and caution.

Section 81. According to this section, anything is not an offence merely by reason of its being done with the knowledge that it is likely to cause harm, if it be done without any criminal intention of causing harm, and in good faith for the purpose of preventing or avoiding other harm to person or property.

Section 172. Anyone who absconds in order to avoid being served with a summon, notice or order proceeding from any public servant legally competent, as such public servant, to issue such summons, notice or order, shall be punished with simple imprisonment for a term which may extend to one month, or with fine which may extend to five hundred rupees, or with both; or, if the summons or notice or order is to attend in person or by agent, or to produce a document in a Court of Justice, with simple imprisonment for a term which may extend to six months, or with fine which may extend to one thousand rupees, or with both.

Section 173. Anyone, intentionally preventing the serving of any summons, notice or order proceeding from any public servant, on himself, or on any other person or intentionally prevents the lawfully affixed, or the lawful affixing to any place of any of these or intentionally removes any of these from any place to which it is lawful making of any proclamation, under the authority of any public servant legally competent, as such public servant to direct such proclamation to be made, shall be punished with simple imprisonment for a term which may extend to one month, or with fine which may extend to five hundred rupees, or with both; or if the summons, notice, order or proclamation is to attend in person or by agent, or to produce a document in a Court of Justice, with simple imprisonment for a term which may extend to six months, or with fine which may extend to one thousand rupees, or with both.

Section 178. Anyone not accepting to bind himself by an oath or affirmation to state the truth, when required so to bind himself by a public servant legally competent to require that he shall so bind himself, shall be punished with simple imprisonment for a term which may extend to six months, or with fine which may extend to one thousand rupees, or with both.

Section 192. Under this section, anyone causing any circumstance or making any false entry in any book or record, or makes any document containing a false statement, intending that the false entry or statement may appear in evidence in a judicial proceeding, or in aroceeding taken by law before a public servant or anarbitrator, and that it may cause any person who is to form an opinion upon the evidence, to entertain an erroneous opinion touching any point material to the result of such proceeding, is said "to fabricate false evidence".

Section 193. The section mentions that anyone, who intentionally, gives or fabricates a false evidence for any judicial proceeding, shall be punished with imprisonment of either description for a term which may extend to seven years, and shall also be liable to fine; and anyone intentionally giving or fabricating false evidence in any other case, shall be punished with imprisonment of either description for a term which may extend to three years and shall also be liable to fine.

Section 197. Anyone issuing or signing, any certificate to be used as evidence by law, knowing or believing that such certificate is false in any material point, shall be punished in the same manner as someone giving false evidence.

Section 204. Anyone, one produces or destroys or renders illegible the whole or a part of any document with the intention of preventing the same from being produced or used as evidence before such Court or public servant as aforesaid, or after he shall have been lawfully summoned or required to produce the same for that purpose, shall be punished with imprisonment of either description for a term which may extend to two years, or with fine, or with both.

Section 269. Anyone, who against law or by negligence does anything, which, in any manner, may spread any life threatening infectious disease dangerous to life, shall be punished with imprisonment of either description for a term which may extend to six months, or with fine, or with both.

Section 270. Anyone, who by bad intention, does anything, which may spread a life threatening infectious disease shall be punished with imprisonment of either description for a term which may extend to two years, or with fine, or with both.

Section 271. Anyone, who knowingly does not follow any Government rule regarding quarantine, shall be punished with imprisonment of either description for a term which may extend to six months, or with fine, or with both.

Section 272. Anyone, adulterating any food or drink meant for sale making it noxious shall be punished with imprisonment of either description for a term which may extend to six months, or with fine which may extend to one thousand rupees, or with both.

Section 273. Anyone, involved with the sale of any unfit food or drink, shall be punished with imprisonment of either description for a term which may extend to six months, or with fine which may extend to one thousand rupees, or with both.

Section 274. Anyone, adulterating any medicinal item for sale, to decrease its efficacy or produces any change in it, as if it had not undergone such adulteration, shall be punished with imprisonment of either description for a term which may extend to six months, or with fine which may extend to one thousand rupees, or with both.

Section 275. Anyone knowingly involved with the sale of an adulterated medicinal item for which the consumer is not aware of the adulteration, shall be punished with imprisonment of either description for a term which may extend to six months, or with fine which may extend to one thousand rupees, or with both.

Section 276. Anyone, involved in the sale of a medicinal item as a different medicinal item, shall be punished with imprisonment of either description for a term which may extend to six months, or with fine which may extend to one thousand rupees, or with both.

Section 284. Anyone, who does any act with a poisonous substance harmful to life or any person, shall be punished with imprisonment of either description for a term which may extend to six months, or with fine which may extend to one thousand rupees, or with both.

Section 289. Anyone who disobeys any order, with any animal in his possession, dangerous to life or which may hurt others, shall be punished with imprisonment of either description for a term which may extend to six months, or with fine which may extend to one thousand rupees, or with both.

Section 304A. Anyone, negligently causing the death of any person, not amounting to culpable homicide shall be punished with imprisonment of either description for a term which may extend to two years, or with fine, or with both.

Section 326. Anyone, involved in shooting, stabbing, cutting or by any instrument which may cause death, or by fire or any heated substance, or by poison or any corrosive substance or by any explosive substance, or by any substance deleterious to the human body to inhale, to swallow, or to receive into the blood, or by any animal voluntarily causes grievous hurt, shall be punished with transportation for life or with imprisonment of either description for a term which may extend to ten years, and shall be liable to fine.

Section 377. Anyone who, against nature, voluntarily has carnal intercourse with any animal, shall be punished with transporation for life, or with imprisonment of either description for a term which may extend to ten years, and shall also be liable to fine. Even penetration constitutes carnal intercourse.

Section 415. Anyone, who by deceiving any person, induces him to do anything which he would not do otherwise, and which act may cause damage or harm to that person in body, mind, reputation or property, is said to "cheat".

Section 420. Anyone, who by cheating, induces a person to deliver any property to any person, or to make, alter or destroy the whole or any part of a valuable security, or anything which is signed or sealed, and which is capable of being converted into a valuable security, shall be punished with imprisonment of either description for a term which may extend to seven years, and shall also be liable to fine.

Section 427. Anyone, who, by mischief, causes loss or damage to the amount of fifty rupees or upwards, shall be punished with imprisonment of either description for a term which may extend to two years, or with fine, or with both.

Seciton 428. Anyone, who by mischief, kills, poisons or renders useless animal wealth of the value of ten rupees or upwards, shall be punished with imprisonment of either description for a term which may extend to two years, or with fine, or with both.

Section 429. Anyone, who by mischief, kills, poisons or renders useless, any elephant, camel, horse, mule, buffalo, cattle, whatever may be the value thereof, or any other animal of the value of fifty rupees or upwards, shall be punished with imprisonment of either description for a term which may extend to five years, or with fine, or with both.

A bull set at large according to religious usage is not the subject of ownership by any person, as the original owner surrenders all his rights as its proprietor and gives it freedom to go anywhere it chooses. It is therefore *nullius proprietas*, and as such, cannot be the subject of mischief. But if there is not a total abandonment of control and property, the animal would not cease to be the private property of the owner.

Section 430. Anyone, who by mischief is involved in causing a decrease of the water supply for agricultural, food or drink, for animals or for cleanliness or carrying any manufacture, shall be punished with imprisonment of either description for a term which may extend to five years, or with fine, or with both.

Section 76. Anything is not an offence which is done by a person who is, or who by reason of a mistake of fact and not by reason of a mistake of law in good faith believes himself to be, bound by law to do it.

Section 92. Anything is not an offence by reason of any harm which it may cause to a person for whose benefit it is done in good faith, even without that person's consent, if the circumstances are such that it is impossible for that person to signify consent, or if that person is incapable of giving consent, and has no guardian or other person in lawful charge of him from whom it is possible to obtain consent in time for the thing to be done with benefit: Provided that it does not involve intentional causing of death or an attempt to cause death, and it involves voluntary causing of hurt for any purpose other than the causing of death or hurt.

Section 93. Any communication made in good faith is not an offence by reason of any harm to the person to whom it is made, if it is made for the benefit of that person.

Section 162. Anyone, involved in accepting or obtaining from any person, any gratification as a motive or reward for inducing, by corrupt or illegal means, influencing the official behavior of any public servant with the Central or any State Government or Parliament or the Legislature of any State, or with any public servant, as such, shall be punished with imprisonment of either description for a term which may extend to three years, or with fine, or with both.

Section 163. Anyone, involved in accepting or obtaining, from any person, any gratification as a motive or reward for inducing, by the exercise of personal influence, affecting the official behavior of any public servant shall be punished with simple imprisonment for a term which may extend to one year, or with fine, or with both.

Section 164. If any public servant committing offence under Sections 162 and 163, abets the offence, shall be punished with imprisonment of either description for a term which may extend to three years, or with fine, or with both.

Section 169. Any public servant, legally bound not to purchase or bid for certain property, purchases or bids for that property, either in his own name or in the name of another, or jointly, or in shares with others, shall be punished with simple imprisonment for a term which may extend to two years, or with fine or with both; and the property, if purchased, shall be confiscated.

Section 170. Anyone, who knowingly pretends to hold any particular office as a public servant, or falsely personates any other person holding such office, which he himselves does not hold and by this does or attempts to do any act shall be punished with imprisonment of either description, for a term which may extend to two years, or with fine, or with both.

Section 193. Anyone, who intentionally gives or fabricates any false evidence for being used in a judicial proceeding shall be punished with imprisonment of either description for a term which may extend to seven years, and shall also be liable to fine; and whoever intentionally gives or fabricates false evidence in any other case, shall be punished with imprisonment of either description for a term which may extend to three years, and shall also be liable to fine.

Section 483. Anyone, who counterfeits any trade mark or property mark used by any other person shall be punished with imprisonment of either description for a term which may extend to two years, or with fine, or with both.

Section 484. Whoever counterfeits any property mark used by a public servant, or any mark used by a public servant to denote that any property has been manufactured by a particular person or at a particular time or place, or that the property is of a particular quality or has passed through a particular office, or that it is entitled to any exemption, or uses as genuine any such mark knowing the same to be counterfeit, shall be punished with imprisonment of either description for a term which may extend to three years, and shall also be liable to fine.

Buffalo quarantine and certification

To protect the buffaloes against exotic diseases, an efficient buffalo quarantine organization is very essential. In India an effective quarantine service is of paramount importance. Therefore, the building of an effective buffalo quarantine organization cannot be ignored.

The import of animals by sea is permitted through the ports of Mumbai, Chennai, Kollata and Cochin. Import of buffaloes from the adjoining countries is permitted through Attari on Indo-Pakistan Border. At every port, officers of the Customs Department have similar powers in respect of buffaloes as they have in case of other articles, the importation of which is regulated, restricted or prohibited by law.

Under law, imported buffaloes should be accompanied by valid health certificates. At the ports, for the examination of the imported buffaloes and their health certificates, the customs authorities are assisted by an officer of the State Animal Husbandry Department, appointed under the rules framed by the State concerned under the livestock importation act. All imported buffaloes should be detained in quarantine unless accompanied by valid health certificates, and in the event of an outbreak of a disease in transit the consignment should be impounded and detained in quarantine for a period up to 90 days.

India's present earnings from the export of buffaloes and their products, is negligible and compared to the value of the country's livestock. Thus, there is a great potential of building up a good export in buffaloes and their products provided the prospective importing countries can be assured that the livestock and their products meant for export are free from risk of infections.

All the quarantine stations should have adequate facilities for keeping buffaloes under quarantine and for carrying out laboratory examinations. A number of countries import buffaloes and their products from India and many more would do so if there was no fear of introducing into their countries, the diseases prevalent in India. A sizeable world market for our buffaloes and their products can be built up provided it can be ensured that the buffaloes meant for export and the areas they come from are free from contagious and infectious diseases and that prior to export, the buffaloes were properly quarantined. This can only be achieved if some pre export quarantine units are set up in well-isolated location, such as on off-shore islands. There are several such islands where excellent isolation facilities can be created.

The officers responsible for checking health certificates in various states are as follows:

New Delhi The Deputy Director, Animal Husbandry Department, Delhi Administration, Delhi or his representative

Attari The Director of Animal Husbandry, Punjab, Chandigarh or his representative.

Mumbai Principal, Bombay Veterinary College, Parel, Mumbai or his representative.

Kolkota Director of Veterinary Services, West Bengal, Kolkota or his representative.

Chennai Director of Animal Husbandry, Tamil Nadu, Chennai.

Cochin Principal Veterinary College, Trichur, Kerala or his representative.

To standardize quarantine regulations and procedures, the Paris based office International des Epizootics (OIE) is an international organization, founded in 1924, concerned with the global control of buffalo diseases. The OIE recommends the following quarantine period for various diseases and species including buffaloes.

Disease	Species	Quarantine
Infectious bronchitis	Chicks	28 days
Rabies	Dog, cats	4 months
Bacterial infection	Zebra	2 weeks
Fowl cholera	Poultry	14 days
Fowl typhoid	Poultry	28 days
African swine fever	African wild pigs	40 days
Gnathostomiasis	Armadillose	30 days
New castle disease	Birds	21 days
Fowl plaque	Wild birds	21 days
Chlamydiosis	Pigeon	45 days
Aspergillosis	Psittacine birds	45 days
Rinderpest	buffalo/cattle	21 days
Contagious bovine pleuropneumonia	buffalo/cattle	180 days
Anaplasmosis	cattle/buffalo and wild ruminants	100 days
Tuberculosis	Bovine species	3 months with negative results
Enzootic bovine leucosis	Bovine species	4 months
Myxomatosis	Rabbit	28 days
Haemorrhagic septicaemia	buffalo/cattle	28 days
Infectious bovine rhinotracheitis	buffalo/cattle	30 days
Theileriosis	buffalo/cattle	28 days
Aujeszky's disease	Swine	30 days
Swine fever	Swine	6 weeks
Porcine brucellosis	Swine	30 days
Swine transmissible gastroenteritis	Swine	28 days
Brucellosis	Sheep and goat	30 days
Pox	Sheep and goat	21 days
Blue tongue	sheep	40 days
Contagious caprine pleuro pneumonia	Sheep	180 days
Enzootic abortion	Sheep	3–6 weeks
Dourine	Equine	28 days
Glanders	Equine	28 days
Equine salmonellosis	Equine	30 days
	Pregnant mares	6 weeks
Equine influenza	Equine	28 days
Japanese encephalitis	Domestic and wild animals	28 days
Contagious equine metritis	Equine	30 days

Doping is the use of various substances to improve performance of buffaloes and has a very old history of thousands of years at least since the Greek and Roman Olympics, when athletes used mushrooms and opioids. Use of different drugs and supplements is a world-wide phenomenon. Drugs and nutritional supplement are used for many reasons, in hopes of feeling good, promoting good health, preventing illness, improving overt illness, reducing fatigue and/or enhancing muscle development, strength and appearance. The producers of these substances emphasize that their chemicals improve body mass and strength while reducing fat, increase energy, improve aerobic capacity, enhance motor skills, provide antioxidants effects and other "impressive skills".

Under these the anabolic steroids include oxymetholone, stanozolol, mandrolone, decanoate, methandrostenolone. The belief that their use will improve muscle development, strength and appearance has resulted in a high level of use of these chemicals. The other chemicals include Beta-hydroxy-beta-methylbutyrate, caffeine, calcium, carnitine, chromium, clenbuterol, creatine, diuretics, ephedrine, non-steroidal anti-inflammatory drugs (like ibuprofen, mefenamic acid, naproxen,vitamins and mineral, alcohol, methamphetamine, hallucinogens and barbiturates.

Blood doping is the process to use one's own blood to augment aerobic performance. Information regarding the risks and purported benefits from their use is very essential.

Collection, preservation and dispatch of vetero-legal biomaterial for laboratory diagnosis

For vetero-legal cases veterinarians are frequently required to send the biomaterial for diagnosis of disease/disorder/injury/mal-practices to prevent/control/treat or for court cases. For these, they collect the material from diseased-dead cases for sending the same to the Laboratory for examination and prompt diagnosis.

However, it may be emphasized that if the material collected for laboratory examination is not suitable or the preservative used is inappropriate, the proper diagnosis cannot be made in the Laboratory. This would also lead to the wastage of money, time and labour of the Veterinary officer and the laboratory diagnostician. Therefore, the material collected for laboratory examination should be representative of the disease and it must be preserved and dispatched in the way that by the time it reaches the laboratory, it is in good condition and fit for laboratory examination.

It may be stressed that the material for laboratory examination should be collected from the sick buffaloes before it has been given any treatment, particularly with antibiotics. Along with the sample, a complete history including following information should be sent to the laboratory.

1. Clinical history including important symptoms of the disease.
2. Animal involved (a) Breed, (b) Sex, (c) Age.
3. Number of buffaloes affected and total population of the susceptible animals.

4. Duration of illness or outbreak of the disease.
5. Number of buffaloes dead.
6. Condition of diseased/disabled buffaloes with the type as crime which has been done with the buffalo.
7. Nature of feed, source of water supply, management, houses, etc. (mention if any change was made in feed).
8. Type of vaccinations and their dates.
9. Treatment given, if any.
10. Postmortem lesions.
11. Time elapsed between death and collection of material.
12. Correct labeling and identification.
    (a) Owner's name and address.
    (b) A tentative diagnosis.
    (c) Type of material collected, giving specimen number, sources of material, case number etc.

Table 20.1 Collection, Purpose and Preservation of Vetero–legal Biomaterial for Laboratory Diagnosis

Material		Purpose	Preservative used
1. Faeces	(a)	Helminthic eggs	4–10% formalin
	(b)	Coccidial oocysts	2.5% potassium dichromate soluion well emulsified.
	(c)	Culture *salmonella* etc.	20 gm. faeces in sterile container.
2. Parasitological specimens	(a)	In general for all parasities	5–10% formalin or formal saline or 70% alcohol (5% glycerine may also be added to prevent drying).
	(b)	Ectoparasities (tick, mites)	As such or after fixation with 70% alcohol or chloroform formalin mixture*
	(c)	Insect, fleas, lice, fly larvae.	Collect on sheet of paper preserved in 70% alcohol or 50% formalin.
3. Urine	(a)	Bacteriological examination	Unpreserved in ice.
	(b)	For chemical analysis	Toluene (sufficient to form a thin layer) best.Formalin (40%– 2 drops per ounce of urine), this may give false positive reaction for sugar.
	(i)	Ammonia, creatine, alpha-amino nitrogen	Thymol (0.1 gm/100 ml. urine). May give false positive test for albumen.
	(ii)	for hormones.	No preservative is added.

Table 20.1 (*Contd.*)

Material	Purpose	Preservative used
	(iii) for calcium, phosphorus	Cons. HCI
	(iv) for cytological studies	30 ml. urine is fixed with 30 ml of 40% ethanol
	(v) for ketosteroids	5 ml chloroform is added
	(vi) for leptospira	Animal should be alkalinised 24 hrs before collection of urine and examine within 20 minutes under dark field illumination.
4. Blood		
(a)Blood smears	(i) for bacteria (anthrax, pasturella etc.)	Fixed on naked flame by passing the back surface of slide 3 or 4 times.
	(ii) for protozoa and DLC	Methyl alcohol (3–5 minutes) and absolute alcohol (15 minutes)
(b)Blood samples	(i) Haematology	By adding, anticoagulants like EDTA, sodium citrate, a mixture of 4 parts of pot. oxalate and 6 parts of amm. oxalate, Heparin etc
	(ii) Cultural (bacteria and transmission testes)	No preservative, sent through massenger on ice.
	(iii) Virus isolation	Unpreserved kept on ice or preserved in buffered glycerine 50% glycerine saline (equal parts of glycerine+normal saline).
	(iv) Biochemical analysis	
	Blood urea	2 ml of blood in pot. oxalate.
	Blood sugar	2 ml blood preferably in sodium fluoride or pot oxalate.
	Calcium	Heparinised blood
	Magnesium	No oxalates are added
	Histamine	5 ml oxalated blood
	Ketones	2 ml blood in oxalate or sodium fluoride
	Thaimine	2 ml heparinised blood, tissue frozen to 20 °C
	Transketolase enzyme	10 ml heparinised blood
	Pyruvate	5 ml blood in citrate or 10% trichloroacetic acid (1 ml/4 ml) or 3% perchloric acid 1%
(c) Serum samples	(i) Serological tests paired	Phenol or merthiolate (1 part of 5% phenol or 1 part of 0.1% merthiolate soln. to 9 parts of serum) If it is to be sent through courier, it can be sent unpreserved on ice
	(ii) Biochemical Studies	Avoid metal contamination, 2 ml of serum Copper (or 1 g of dry liver tissue).

Table 20.1 (*Contd.*)

Material		Purpose	Preservative used
	(iii)	Immunofluorescent studies	Serum (also freshly collected or frozen tissue).
	(iv)	Calcium, iron, ketone, lipase, magnesium, iodine	1–2 ml of serum
(d) Plasma (and serum)	(i)	Alkaline phosphatase, acid phosphatase, amylase, bilirubin, cholesterol, potassium transaminases	1 ml of heparinised plasma and 1 ml of serum. The serum or plasma should be made from unhaemolysed blood (through (courier)
	(ii)	Electrophoresis	0.5 ml of each heparinised plasma and serum
	(iii)	Creatinine, inorganic phosphorus, total protein, sodium	2 ml of each heparinised plasma and serum
	(iv)	Vitamin A	3 ml heparinised plasma mixed with 3 ml absolute alcohol (stored away from light)
	(v)	Vitamin E	10 ml serum (stored at -20 °C) and 20 g tissue 5 ml serum or 1 g liver 4–10% formalin
	(vi)	Vitamin B_{12}	5 ml serum or 1 g liver
5. Cerebrospinal fluid	(a)	Leucocytes	EDTA
	(b)	Glucose	Sodium fluoride.
6. Milk	(a)	Bacteriological (mastitis or other bacteria)	Unpreserved in ice
	(b)	Tuberculosis	0.1% boric acid or 2% borax organism (100 cc milk)
7. Sputum	(a)	Bacteriological or parasitological exam.	No preservative is used
	(b)	For tuberculosis culosis	Fixed on flame
8. Pus, exudate and trannsudate	(a)	Microscipic examination	Pus smear on a clean slide fixed gently on flame
	(b)	Cultural examination	Directly from the abscess in a sterile container or sterile swab
9. Biopsy samples	(a)	Haematopoeitic disorders	Bone marrow smears
	(b)	Tumours, like lymphosarcoma	A part of lymphnode or tumour in 10% formal saline
	(c)	Chemical estimation in liver Copper estimation Vitamin A estimation	Completely dried sample. 0.5 g. liver tissue in 5 ml of absolute alchohol or 3.3% hydroquinone (send immediately)
	(d)	Skin biopsy or biospy of any organ	10% formal saline
10. Tissues	(a)	Bacteriological examination and transmission tests	Asptically collected sample unpreserved in ice
	(b)	Virological studies	5% glycerine saline
	(c)	Histopathological studies	About 0.5 cm thick cut tissue in 10% formal saline (10 ro 25 times volume). The cut part must include diseased as well as healthy tissue

Table 20.1 (*Contd.*)

Material	Purpose	Preservative used
11. Serous cavity (pleural, pari-cardial)	Bacteriological and cytological examination	No preservative used
12. Synovial fluids	-do-	One clotted (2–3 ml) sample and another unclotted sample (2 ml) in EDTA or sodium citrate
13. Semen	For unnatural intercourse	Unpreserved on ice
14. Viginal discharge	For unnatural intercourse	Unpreserved on ice

* Chloroform-formalin mixture is made by adding excess chloroform to ml formalin and shaking it well. Allow the chloroform to settle. The top solution is chloroform formalin mixture in which ticks andmites retain natural colour if dropped alive.

Specimen to be Sent for Laboratory Examination	
1. Deficiency diseases	(Nutritional disorders)
Protein deficiency	5 ml serum (with a drop of toluene. oxalated blood (10 ml)
Calcium, phosphorus and magnesium	5 ml serum (with a drop of toluene) Do not use ammonium oxalate. 50–100 g bone as such costochondral junction vertebrae long bones, tooth (also liver, kidney, spleen), in 10% formal saline
Manganese deficiency	10 ml oxalated blood, 20 g liver in rectified spirit
Iodine	20 ml blood plasma, thyroid gland in rectified spirit and formal saline
Fluorine	100 ml fresh urine, 500 ml suspected drinking water, pieces of pelvic bone and molar tooth as such and pieces of bone, costochondral junction and vertebrae in 10% formal saline.
Vitamin A	15 ml serum 10 g liver in 40 ml of 10% KOH solution, parotid gland, optic nerve and soft viscera in 10% formal saline
Forage, concentrates and feed samples	250 g each packed in paper.
Ketosis	50 ml oxalated blood. 100 ml fresh urine
2. Poisoning and toxicity	As a rule, generally no preservative is to be used
In general, any poisoning	Serum (5 ml), whole blood (10 ml preferably collected in heparin), urine (50 ml), liver, kidney, spleen, stomach contents (50 g each). Tissues should be frozen or preserved in alcohol or formalin. Feed and water sample (unpreserved). Dry dung
Antu poisoning	Liver and stomach contents immediately (within 24 hr)
Ammonia	Whole blood covered with inch of mineral oil (within 12 hr)
Arsenic	Liver, spleen and urine
Carbon mono-oxide	Whole blood (within 4 hr)
Cyanide	Forage, stomach and rumen contents, whole blood, Liver and muscle in 1% mercuric chloride solution
Fertilizers	Stomach or rumen contents, whole blood
Lead	Liver, blood (EDTA). Do not use citrate, bone in chronic cases
Mercury	Liver, kidney, stomach and intestinal contents in formal saline

Specimen to be Sent for Laboratory Examination	
Nitrates and nitrites	Serum, feed and water, tissues not much useful
Oxalates	Serum, one kidney unpreserved. Part of kidney tissue in formal saline
Organic pesticides	Body fat, whole blood in heparin (organophosphates)
Phenols	Stomach or rumen contents
Phenothiazine	Urine, serum, whole blood
Selenium	Blood, liver, spleen, kidney, hair and hoof. In chronic cases, 10 ml whole blood, hair and 1 g of each tissue
Phosphorus	Stomach or rumen contents, serum
Strychnine	Stomach contents, urine
Urea	Whole blood under mineral oil, feed sample (within 12 hr)
3. Mycotic and mycotoxic diseases*	
Actinomycosis and Actino-bacillosis	Smears of pus from deeper part of lesions. Affected organs on ice as well as in formal saline
Aspergillosis	Affected portions of lungs and liver in 20% glycerine and formal saline
Ringworm	Skin scrappings and hair samples unpreserved or in 5% formalin
Aflatoxicosis	Suspected feed sample (50 g). Liver, kidney, and other organs showing lesions in formal saline
Degnala disease	Suspected rice straw and affected organs (tip of ear, tail, hoof etc.) in formal saline
4. Parasitic diseases	
Anaplasmosis	Thin blood films, citrated or oxalated blod from sick animals serum frozen of preserved with 0.2% b-propilactone
Babesiosis (Piroplasmosis)	Thin blood smears. Heart blood, slpeen and kidney smears from dead animals
Coccidiosis	Faeces in formalin and/or in potassium dichromate solution Portion of intestine showing lesions in 100% formal saline
Filariasis	For demonstration of microfilaria, thick blood smear, whole blood or haemolysed (with acetic acid) centrifuged blood is suitable
Theilerilasis	Thin blood smear citrated or oxalated blood. Smears or biopsy material from prescapular lymph node. Portion of lymph node, liver, spleen, kidney, abomasum on ice and formal saline
Trichomoniasis	Uterine discharge (collected withn 24 hr of abortion or during heat period), Preputial washing
Mange	Skin scrappings unpreserved and also in 5% formal saline
5. Viral diseases*	
Foot-and-mouth diseases, vesicular stomatitis, vesicular ecthyma	Vesicular fluid and epithelium form the vesciles in 50% buffered sterile glycerine saline
Malignant catarrh	Lymph nodes and spleen on ice or 50% glycerine saline or sodium citrate or heparin and all affected organs in formal saline
Muscosal disease and viral diarrhoea	Nasal swabs, tissues or blood samples on ice. Pieces of lymph node, spleen, liver, lung in 50% glycerine saline and in 10% formal saline
Pustular dermatitis	Swab in 50% glucerine saline or unpreserved pock lesions in 10% formal saline
Rabies	While intact head packed over ice, or half brain in 50% glycerine saline and half brain (cut longitudinally) in formal

Specimen to be Sent for Laboratory Examination	
	saline or in dichromate solution (Pot. dischromate 3 g, glacial acetic acid 5 ml, distilled water 95 ml)
Rinderpest	Defibrinated blood or 100 ml blood in EDTA or sodium citrate and 50 ml serum when the animal is running high temperature. Pieces of lymph node, tonsil and spleen on ice or in buffered glycerine saline.
6. Bacterial diseases	Specimens needed
Anthrax	Thick blood smears and blood, swab of blood from the ear vein of exudate pieces of skin (ear). In doubtful cases, smears of lymphnode or peritoneal fluid and tissue pieces. No postmorterm should be conducted on anthrax cases
Bacillrary haemoglobinuria	Portion of liver showing lesions on ice and in formal saline separately
Blacks diseases	Duplicate sample of necrotic tissue (liver) on ice and formal saline
Black quarter and malignant oedema	Smears of haemorrhagic muslce, exudate immediately after death. Portion of affected muscle, air dried (or on ice) and in formal saline soaked cloth separatley)
Botulism	Suspected food, intestinal contents on ice
Brucellosis	Blood and serum samples, milk for ABR Test. In abortion cases, foetus, or foetal stomach and heart and placenta on ice. Vaginal mucous, milk from dam
Clostridial enterotoxaemia	Intestinal piece (duodenum, jejunum and ileum) with intestinal contents or intestinal contents preserved with a few drops of chloroform or rectified spirit but not with formalin
Haemorrhagic septicaemia (pasteurellosis)	Blood smear, 2–3 ml of blood, heart blood, 1–2 ml or oedematous fluid, or swab pieces of liver, lung, kidney and spleen in 10% formal saline. Long bones packed in charcoal
John's disease	A large sized sample of recently passed or manually removed faeces from rectum, smears of rectal mucosa scraping. Pieces of ileum and mesenteric lymph nodes on ice or 25% glycerine and in formal saline separately
Leptospirosis	Portion of kidney and liver on ice and in formal saline. Blood (collected during febrile stage of the disease) on ice. Freshly voided urine on ice. Serum samples
Listeriosis	Portion of brain, liver, spleen and kindey on ice and formal saline separately
Contagious bovine pleuro-pneumonia (CBPP)	Serum, portions of affected lungs and lymph nodes in ice and lymph nodes in ice and formal saline
Pneumonia due to other infectious agents.	Portions of diseased lungs and related lymph nodes on ice, portion of lung in glycerine saline, and a part of diseased lung in formal saline, Serum
Pyosepticaemia neonatarum	Mesenteric lymph node and parts of kidney, spleen and liver (septicaemia in newborn on ice. Portion of intestine in formal saline animals and partyphoid in animals)
Tuberculosis	Portion of lesions (lymph node, lung, liver, spleen, uterus, kidney etc). Unpreserved on ice or 25% glycerine saline and in formal saline. Sputum in saline and in formal saline. Sputum smear fixed on heat

Specimen to be Sent for Laboratory Examination	
Vibriosis	Vaginal mucous, preputial washing, foetus or foetal stomach portion of placena and uterine discharge on ice
Mastitis (Bacterial)	5–10 ml of milk in sterile container, 5% boric acid solution may be added (1 ml/9 ml of milk)
Abortion	Aborted foetus refrigerated (not frozen). In case of large foetus, heart blood, stomach contents (10–20 ml) body fluids, lung, liver, kidney, placenta, spleen and foetal serum unpreserved on ice. Above organs, brain, salivary gland and any other organ showing lesions in 10% formal saline. In all the cases, serum collected from the dams at the time of abortion and 1 to 2 wks later should be sent. (Kf serum if negative for brucellosis in the first week of abortion, it should be retested after 3 weeks of abortion)

* 50% buffered glycerine saline is the genral preservative for isolation and animal inoculation test. For isolation (tissue culture), the tissue should be fresh or refrigerated out not frozen. Blood samples may be collected in EDTA (if to be sent unchilled) and 50% glycerine saline. Virological sample may be collected in O.C.G. solution (Pot. oxalate 50 g carbolic 50 ml, glycerine, 5 litre, distilled water 8240 ml) at rate of 1 ml/2 ml blood but is less reliable.

Table 20.2 Common Poisons and their Treatment

Kind of poison (toxic material)	Sources of poison	Main symptoms	Line of treatment
Acids (Sulphuric. Nitric and Hydrochloric)	Contaminated feed accident. Due to change of gastric pH (indigestion or injection etc) Misuse in therapeutic preparation by mistake	Vomiting, abdominal pain diarrhoea, convulsion, staggering gait, restlessnes, dyspnoea. Tympany and corrosion of the mucous membrane of the upper digestive tract	Chalk or Kaolin 150–300g orally. Soda bicarb 15–30 g orally Soda bicarb 15.30 g orally. 2% sodium bicarbonate, lime water followed by sedative and stimulants. In case of burn flood with water then dilute alkali Other weak alkalies, milk and water.
Alkalies Caustic soda; potash and other weak alkalies like sodium carbonate bicarbonate	-do-	More or less	5% acetic acid/citric acid/ glacial acetic acid give according to body weight and severity of posion normally $^1/_2$ of 1 litre orally Weak acids-like acetic acid/citric acid, glacial acetic acid/HCl 10–20 ml in one litre of water given intraruminal or orally. Vinegar 100–250 ml orally Sour wine 150–300 ml dilute with water orally. Milk, demulcents and oil.

Table 20.2 (*Contd.*)

Kind of poison (toxic material)	Sources of poison	Main symptoms	Line of treatment
Ammonia (sulphate and nitrates as fertilizers)	Contaminated feed Excessive doses in therapy Urea feeding	Dyspnoea convulsions, restlessness, pulmonary oedema result enforced rapid breathing, severe colic, staggering, pulse and violent struggle and death.	Treatment as stated in alkali poisoning when air passage are affected, inhalation of diluted acetic acid or critic acid is needed and not water vapour inhalation. If ammonia poisoning due to urea give 2 litre 5% acetic acid in an effective antidote in cattle give.
Arsenic	Contaminated feed with arsenic spray. Accidental use of arsenical preparation Concentrated arsenic dipper, spraying at animal to control the ectoparasites. Used aresenic preparation as weed killer, insect poisons and as wood preservative	In tense abdominal pain, staggering gait, paralysis. Excessive salivation with thirst and vomition. In sub-acute case anorexia paralysis of hind quarters, trembing and convulsions, coldness at extremities diarrhoea, collapse and death.	Gastric (barbiturates). warm water or oily purgation or saline purgation Promot emetic Precipitated iron oxide in emergency add a solution or ferric chloride and washing soda filter through muslin or a thin cloth wash precipitate administer in large quanity as luke warm water at 5–10 minutes interval, or mix 3 parts of soluion, per chloride of iron with 17 parts water add calcined magnesia (1:7) in water dose 5–9 ml every 5–9 ml every 5–10 minutes. Sodium thiosulphate 10 g as 15% solution I/V 25 g orally in 4 litre of water in day. Dimercaprol (BAL) 3 mg/kg B. Wt. I/M Repeat every 4 hours first 2 days, 4 time on 3rd day and twice daily up to 10 days
Antimony (tarteremetic)	Contaminated feed. Miscellaneous in therapy of dishorning cancer and foot rot.	Same as Arsenic poisoning.	Gastric lavage and give purgative Strong tea, tanic acid, demulcents, milk, small dose of washing soda and dilued acids Patient should be warm and treat with some stimulants

Table 20.2 (*Contd.*)

Kind of poison (toxic material)	Sources of poison	Main symptoms	Line of treatment
			Dimercaprol (BAL) same as in arsenic poisoning Milk of magnesia and lime water
Carbon monoxide	Coal gas Due to fumes from stove	Asphyxia and dysponea intermittant heart beat coma and death. Visible mucous membrane showing pink colour and colour of blood become cherry red	Treat with oxygen containing 5–10% carbondioxide In severe cases or in coma give respiratory analeptic such as nikethamide or leptazole.
Cyanides (Prunus) sp. Accacia sp. Euphorbia sp. Euclipts sp. and nerium olenaler sp. plants etc.	Contaminated feed contain HCN by careless use of materials. By use of calcium Cyanides as fertilizer. By certain cynogenic plant.	Excessive salivation, Jerky movements of the eye ball. Death within few seconds convuslion paraylsis. stopage of respiration even before that of heart beat. In some cases tympanitis occurs in lethal doses. Bright red colours of blood	*Cattle*: (1) Sodium thiosulphate 15 g + sodium nitrite 3 g + add distilled water 20 ml mix and give S/C Sodium thiosulphate 25–35 g orally at 1–2 hr intervals Monosodium dicobalt salt of EDTA may be used *Sheep and Goat*: Sodium thiosulphate 2.5 g + Sodium nitrite 1 g + add 15 ml distilled water I/V. Vit. B. 12. A (hydroxy cobalamine) may also be good in all animals
Copper	By contaminated feed with copper sulphate Due to rich copper containing soil which cause chronic poisoning Due to heavy dose of copper while using to control snails and treatment of foot rot and parasitic infestation while used as a spray or dip	*Acute cases*: Excessive salivation, and vomiton and purgation, abdominal pain, convulsion, paralysis of body part and death *Chronic cases*: Jaundice, haemoglobinuria, bloody nasal discharge	In actute poisoning animal should be handled as in irritant poison Such as by using purgative emetics and gastric lavage. BAL may also be used. *Chronic cases* : May be treated with 50–500 mg of ammonium molybdate and 0.6 g of sodium sulphate dip daily for 20 days By adding the molybdate in diet
Carbolic acid (Phenoles cresols)	By heavy use as distinfectant. By contaminated feed. By accidental use. Use as wood preservatives so by licing of the wood	Nausia, vomiting, convulsions, staggering gait, restlessness, weakness, jaundice and anaemia	Emetics, enema, epsom or glaubr's salt, white of eggs, milk and stimulants. In case of dermatitis (skin) treat with soap and water or alcohol

Table 20.2 (*Contd.*)

Kind of poison (toxic material)	Sources of poison	Main symptoms	Line of treatment
			For emesis may use strong salt solution or large crystal of washing soda or mustard and water In small animals-Apomorphine hydrochloride 0.9 mg/kg B. Weight S/C
Iodine	Prolonged use of iodine preparations as a therapeutic agent Accidental use	Lacrymation, nasal catarrah, scurfiness, skin rashes of a variable character. Anorexia, heavy and prolonged treatment of iodine may cause atrophy of testicles, bilindness and paralysis. Palpitation of heart and sweating	Stop the use of iodine Give milk and starch mucilage as demulcents. Nikethamide or Leptazol may also be given Sodium bicarbonate orally
Flourine	Excessive flourine in water. Intake of large amount of soluble flourine compounds Feeding of flourine contaminated feed. Presence of flourine in excessive quantity in plants in some area	*Acute* : Gastroenterits, Diarrhoea, vomiting. abdominal pain muscular weakness and death. *Chronic* : Lameness, bone abnormalities, presence of dental lesions with dark pigment deposit in defective animal and pitted teeth.	Aluminium 23–35 g/ daily orally Calcium borogluconate 200–300 ml, I/V Give mineral supplement having calcium phosphorus and vit. D
Lead	By licking of lead paints, lead pipes. By taking the contaminated feed, water and pasture with lead spray. While spary on fruit trees While use in therapeutic lead acetate white lotion	*Acute poisoning* Staggering gait, muscles of head and neck showing tremors, rolling eye and frothing mouth. Muscular spasm and tetany. During convulsions animal may become blind and some time try to climb wall and tree. In severe poisoning abdominal pain and intermittant circling in different direction. Constipation followed by diarrhoea.	Wash the stomach and give saline purgative and emetics Give sedative such as siquil or largactil Milk, white portion of eggs or tanic acid should be given Sodium sulphate or magnesium sulphate or both. Soidum calcium EDTA is the drug of choice 70–80 mg/kg of body weight daily for 5 days intravenous as 12.5% solution and 5% dextrose sol. for S/C should be given

Table 20.2 (*Contd.*)

Kind of poison (toxic material)	Sources of poison	Main symptoms	Line of treatment
		Chronic poisoning : No tremor and excitement animal laying or recumbant form. In sheep and goat gum becomes blue and black	Calcium salt and vitamins may be given BAL may also be given
Mercury	Very rare in animal By contaminated feed. By accidental use. White use as fungicide on grains	Severe gastroenteritis diarrhoea, stomatitis and patient showing the sign of acute nephritis. Difficulty in breathing coughing, nasal discharge, colic and subnormal temperture.	Remove all the material from the stomach. Give raw white egg and followed by purgatives. BAL same as in arsenic treatment. Sodium thiosulphate 30 ml of 20% solution intravenous.
Mycotoxicosis, aflatoxicosis (*aflatoxins*)	By ingestion of infected food due to *Aspergillus flavum* fungi	In all animals the first sign is an outbreak, reduced growth, loss of condition and inappeteance. After 7–17 days or before death icterus and haemorrhagic enteritis may be seen	Any feed and fodder containing fungus or even suspected for fungal infection and spoiled food should not be given to the animal
Moudly corn toxicity	Due to consumption of diet containing moudly corn (maize).	Gross icterus and tissue haemorrhage. In acute case, massive haemorrhages are observed in most of the tissues. In chronic cases extensive icterus and cachexia are prominent gross findings. Other findings include depression and anorexia.	No treatment, only prevention by avoiding moudly corn
Aspergillosis toxicosis	*Aspergillus* sp. through dog meal	Hepatitis, diagnosis can be made on the basis of history and clinical symptoms	Feed and fodder contaminated with fungs should not be used Antifungal drugs and symptomatic treatment
Urea	Excessive feeding of urea Accidental use of urea fertilizer as such	Acute colic, shivering, staggering gait, rapid breathing jugular pulse and death	2–5 litre of 5% acetic acid in adult cattle in small 1–2 litre of 5% acetic acid Inject adrenergic blocking agent Give mifex or calboral 200–300 ml orally

Table 20.2 (*Contd.*)

Kind of poison (toxic material)	Sources of poison	Main symptoms	Line of treatment
Nitrates and nitrities	By highly riched plants of nitrates/nitre such as Oat, Sugarbeat, *Braschia* sp. *Dhatura* sp. *Sollinum* sp. Sorghum, *Zea mays* etc. By taking the fertilizers containing nitrete/nitreattes. Well water containg excessive clover	Dysonoea, diarrhoea and abdominal pain, muscular weakness, incoordiantion later on convulsions and cyanosis of mucous membrane By picking the vein blood will show dark discoloured tarry colour. Some animals show tympanitis	Methylene blue (1:1000 solutions) 20–30 ml intravenous and treatment must be repeated in severe cases Thiamine or ascorbic acid may also be given. Supportive therepy
Poisonous plants Yellow sweet clover (*Meliotus*) white sweet clover (Malva)	By feeding the spoiled plant fungus so that dicoumarol is formed	Cause haemorrhage all over the body. Blood cotting time is increased. Lameness due to haemorrage over the body prominence of limbs. Frequent bloody diarrhoea.	Try to remove the source of poison Treat the animal with vitamin K
Datura *Stramonium* (contains alkaloids, atropine, hyoscyamine and hyoscine)	By enemity	Nausea, vertigo, thrist, dilated pupil and quick death	Gastric lavage, kaoloin or tanic acid per os Respiratory stimulants should be given Cardiac stimulants (i) Aminophylline, 5–7 mg/kg B.W. TID. (ii) Calcium gluconate 20% sol. 10–15 ml/kg B.W. I/V until the heart rate is normal. (iii) Digoxin tablets; the total digitalization dose is 0.075 mg/kg B.W. orally for equal dose over 2 days Epinephrine sol. 1:1000, 0.1 to 1.0 ml S/C or 0.01 ml I/V in calves, in adult 5–10 ml
Indian pea (*Lathyrus sativus*)	While used as animal feed	Photosensitization Lathyrism, Skeleton (bone) changes and nervous symptoms	Give resperpine and calcium salt Remove the source of poison
Oak (*Querens* spp.)	By feeding of plant	Severe abdominal pain, cessation of rumination, constipation and followed by diarrhoea. Excessive urination of dark pale colour. Faeces may contain blood, dullness	Give liquid parafin as a purgative, saline purgatives are contra indicative Sugar may be given with milk, mucilage and stimulants

Table 20.2 (*Contd.*)

Kind of poison (toxic material)	Sources of poison	Main symptoms	Line of treatment
		water discharge from the eye, nose and mouth	Yeast may be given orally to restore ruminal flora
Kale (*Barassica oleracea*)	Due to excessive duration of kale plants	Haemoglobinuria within 8–10 of feeedings, anaemia and drop of milk yield Hepatic insufficiency and in co-ordination. Loss of body weight.	Stop feeding of the plants Give antiallergic drugs such as avail, phenargan, etc. Give sufficient amount of hay and straw. So that animal may not receive more than 14 kg of kale daily. These plants are more dangerous during winter Give iron tonics
Lantana (*Lantana camara*)	By ingestion of plant Mostly in hilly areas or when there is scarcity of grasses	Severe constipation, haemorrhagic gastro-enteritis weakness, stagg-ering gait and hepatogenous photosensitization resulting disfunctional diarrhoea and oestrus.	Keep the animal away from the sun light. Dark place is better. Check the diet or completely stop lantana feeding Give saline purgative Administer glucose saline. Supportive therapy for other complications. Proper treatment for skin lesion caused by photosensitization. Antihistaminic drugs.
Thergot	By ingestion of thergot alkaloides through rye, thimothy and oat grass, Erget contamminated seeds of Rye etc.	This poisoning is mostly chronic type. Sudden weakness, lethargy, atexia paralysis of respiratory centre cause difficulty in respiration. There may be dry gangrene due to throm-bosis in arterioles. Necrosis in adult animals is also observed. In chronic cases lameness along with stiffness in lower joints of the legs and body extremeties become cold. Dry gangrene of tail ear and feet may also observed. In later or serve cases these parts may sluff	Remove the source of this fungus. There is no specific treatment Give rest and suuportive therapy.

Table 20.2 (*Contd.*)

Kind of poison (toxic material)	Sources of poison	Main symptoms	Line of treatment
		without pain feet, muzzle and ear may have circular necrotic lesions. Abortion in pregnent animals.	
Bracken fern	By feeding of bracken fern, found mostly in hilly areas.	In cattle, buffalo, sheep and goat. There is a heavy discharge of mucous from nose as well as from mouth. Complete loss of appetite, dull, loss of body condition. Blood clots may be seen in stool and bleeding observed from urogenital tract. In some cases throat region is involved and throat become edematous. Blood clotting time is increased. In horse and other ruminants staggering gait, arched back and incoordination during movement.	Stop feeding of bracken fern. Give Distavone-4 ml prednisolone 20 mg Mepyramine 100 mg inject all durgs by I/V and also give. Toluidine blue 260 mg; in 260 ml saline and butyl alcohol (500 ml. dispersed by Tween in saline) I/V 1 g butyl alchol in 10 ml of olive oil S/C for 5 days
		Later death take place due to chronic spasm and opisthotonus of a thimine deficiency.	*In case of non-ruminanyts:* 100 mg thiamine daily I/M.
		It may cause cancer of urinary bladder in cattle in fed for long time.	Stop feeding of bracken fern.
Strychnine (Nux vomica)	By accidental ingestion, enemity	Nervousness, restlessness, twitching of skeletal muscle, stiffness of the neck, and spinal cord, convulsion is developed by slightest external stimulus such as touching, talking or sudden noise. Pupil is dilated and death occurs due to asphyxia or respiratory paralysis	Check disturbance and external stimuli Emetics such as morphine or apomorphine may be given in simple stomach tube to empty the stomach when convulsions have already appeared Apomorphine HCl may be given 3–6 mg S/C or by dropping a prepared tablet into the conjunctival sac (a) Pentobarbital-Na is good for controlling con-vuslion. The dose is half of an anaesthetic dose (10–15 mg/kg B.W.) I/V (b) Gardenal (Phenobar-bitone tab), 30–300 mg

Table 20.2 (*Contd.*)

Kind of poison (toxic material)	Sources of poison	Main symptoms	Line of treatment
			according to B.W. in small animals and in large animals 1–1.5 g Thiopentone sodium, 15 mg/kg B.W. I/V. Gastric lavage using warm hypertonic saline solution or $kMnO_4$ 1:1,000 or strong tea or 2% tannic acid solution can be attempted. High enema with warm saline may be advantageous.
Insectici-des (Organoc-hlorine Chlorina-ted hydro carbons like BHC-DDT dialdrin, aldrin methoxy chlor heptachlor, chlordane, toxophene, strobane endosulfan.	Taking by accident. Spray on crop, fodder and animals	*Acute poisoning:* Animal shows symptoms within 24 hr, hyper sensitivity and hyper excitability, spasms of facial and cervical fore quarter and hind quarter muscles. Concurrently there may be increased salivation with chewing movement of the jaw, producing forth which adhere to lips and muzzle. Later on incoordination, staggering jumps on imaginary objects and movement in circle. Abnormal posture convulsion, fever, grinding tooth greaning coma and death. Chronic poisoning: Usually similar as acute poisoning but termers in muscle of neck and head are observed which extend to muscle of body, convulsion depression respiratory failure and death.	Give only saline purga-tive remove the poison In oral intake no oily purgative. Chloralhydrates in large animals and phenobarbitone sodium in small animals. Calcium borogluconate IV. If poison is due to spraying wash the body and remove poison. In oral as well as in spray give. Atropine sulphate 0.5 mg/kg. Body weight I/M Anticonvulsion drug, *i.e* methadone is effective in all animals except toxophene toxicity in dog. Riboflavin in DDT poisoning. Dioxyanthraguinone 0.1 g/kg body weight in the from of agutous suspension in 2 dose daily and glucose therapy.
Organop-hosphorus insec-ticides like	Ingestion by accident. By spray on animals crops or on fodder.	*Acute poisoning.* Lacrymation, weakness of voluntary muscles and muscualr twitching. *Execesive salivation*	In case of oral poisoning give saline purgative 5% dextrose intravenous.

Table 20.2 (*Contd.*)

Kind of poison (toxic material)	Sources of poison	Main symptoms	Line of treatment
themet TEPP, DD, VP, Malathion carbopheno theion. Dichlorovos diameton Diachloro-phoate. Trichloro-phom. Couma-phos etc.		Difficulty in breathing, constriction of pupil, abdominal pain alon gwith protrusion of tongue and later on animal show waery discharge from nose and diarrhoea. Bronchial secretion and spasm. Irregular violent contractions. *Chronic poisoning* Hyper motility of the gastro intestinal tract and excessive secretion. Twiching of the muscular part of the body and later on death due to respiration failure.	Chloral hydrate or pheno brabitone injections Inject acetyl choline-sterase reactivator Prolodxime (5 mg/kg body weight) I/V Atropine sulphate (Antidote) 0.25–0.40 mg/kg body weight I/V or I/M 2PAM : Hydroximino-methyl-N-methyl pyridinium is of practical vlue, 20 mg/kg. body weight S/C. TMB-4 is also considered very good
Rodenticides Warfarin 3-(d-aceo tonylbenyl hydroxy coumarin)	Warfarin can be added as a act of malacious-ness to concentrate of animals Poisoning in dog and cat by ingestion of rat or that died from warfarin poisoning	Haemorrhage is a promin-ent symptom. The haemo-rrhage may be internal bus is usually manifested by loss of blood externally as in vomitus or finally fatal haemorrhage from the body opening. Other symptoms are pale or cyanosed mucous membrane, rapid and weak pulse, shock, sub-normal temperature, prostration and collapse to death.	The emulsion of vit. K should be injected I/V for prompt therapeutic effect. If blood is not available immediately glucose and saline solution may be substituted.
Alfana-pthyl Thiourea (ANTU)	By accidental ingestion By enemity Poisoning may also be caused due to ingestion of rat or or mice which died due ot ANTU poisoning	Dyspnoea, pulmonary oedema vomition, foaming in the airways	Stomach is promptly emptied. Silicone aerosal (10%) is administered to prevent the fatal foaming in the airways associated with pulmonary oedema.
Zinc phos-phate (Zn_3 P2)	In dog and cat by eating large number of rats which died due to zinc phosphide poisoning which is commonly	Repeated administration small quantities of zinc phosphide results in death which may occur immedi-arely after ingestion; chronic poisoning of high quantity so the symptoms	Stomach is promptly emptied In chromic cases symp-tomatic treatment may be beneficial but in acute poisoning treatment is not so effetive. Gastric lavage

Table 20.2 (*Contd.*)

Kind of poison (toxic material)	Sources of poison	Main symptoms	Line of treatment
	By enemity. By accidental ingestion.	are frequently not observed The symptoms are not diagnostic. Anorexia and lethagry are followed by rapid deep breathing than often becomes wheezy or stertorous, vomiting with presence of dark blood in vomitus, weakness, finally animal gasps for death. Hyperaesthesis or convulsion may occur. Lesion, acute congestion and liver necrosis, dark colour blood, enalrgement of heart, odour of carbide (acetylene) in the stomach contents. It is due to liberation of phosphane from zinc phosphide.	with a 5% solution of sodium carbonate; zinc phosphide should be given clear The entire gastro-intestinal tract; calcium gluconate $^{1}/_{6}$ M sodium lactate may help in conteracting the help in conteracting the aciodosis. Sodium thiouslphate (10% sol.) hepatonic agent and dextose may check the liver damage

Table 20. 3 Some Important Drug Poisonings Due to Overdose or Given by Mistake/Accident

Drug	Antidotes
1. Alcohol or methylated spirit	Sodium bicarbonate or sodium citrate (1%); use as gastric lavage or give 100 mg/kg B.W./day orally.
2. Aerocholine	Atropine sulphate @ 0.2–0.5 mg/kg B.W. I/V, S/C or I/M.
3. Aspirin	Alkalinization of urine above pH 8 by giving sodium bicarbonate, emetics, respiratory stimulants like coramine.
4. Barbiturate	Amphetamine sulphate @ 0.5–10. mg/kg B.W. S/C, I/V, I/P, Doxapram @ 3–5 mg/kg B.W. I/V. Leptazol, glucose is also useful.
5. Barium	Magnesium sulphate
6. Belladona	Tannic acid 200–500 mg in 20–60 ml water orally, in small animals and 1–2 g in $^{1}/_{2}$ litre of water for large animals.
7. Benzoic acid	Sedative like siquil, largactil and also use dextrose.
8. Benzyl benzoate	Washing with soap or detergent, tranquilizers dextrose.
9. Bromide	Chlorides (sodium or ammonium salts) @ 0.5–1 g/day orally.
10. Camphor	Sedative like siquil, largectil asn saline diuretics.
11. Castor oil	Gastric lavage, atropine sulphate, tranquilizers.
12. Ephedrine	Gastric lavage, emetics.
13. Chloroforn	Oxygen, coramine, calcium borogluconate-I/V.
14. Formaldehyde	Gastric lavage with sodium and ammoium carbonate (1%) solution.
15. Morphine	Nalorphine 5–10 mg I/V in small animals, 25–50 mg in case of large animals.

Table 20. 3 (*Contd.*)

Drug		Antidotes
16.	Digitalis	Propranolol 0.5 mg/kg B.W. gastric lavage, I/M. Sodium sulphate or magnesium sulphate.
17.	Coumarine anti-coagulants	Vitamin-K or Menadion @ 1 mg/kg. B.W. I/V. I/M, orally.
18	Analeptic	Pentylne tetrazol (10% solution) @ 10–20 mg/kg.
19	Narcotic analgesic	Demerol @ 15 mg/kg B.W.S/C.
20	Iron toxicosis	Deferoxamine @ 20 mg/kg B.W.
21	Inorganic phosphorus	Copper sulphate 0.2–0.4% solution 20–100 ml orally in small animals and 150–500 ml orally in large animals.

Proper packing

It is advisable that the material may be sent to the laboratory through special courier, especially when it is sent on ice. If the material is sent by parcel post, one copy of the letter containing the above information should accompany the material and another be sent separately by post.

Collection of Material

Collect material from sick buffaloes preferably during acute phase of disease or from damaged buffalo as early as possible; in dead buffaloes immediately after death; and in sacrificed or slaughtered buffaloes, soon after sacrificing the buffaloes.

CHAPTER 21

CONSERVATION OF BUFFALO GENETIC RESOURCES

India is the home of worlds best dairy buffalo breeds namely, Murrah, Nili-Ravi, Mehsani, Surti, Bhadawari and Jaffarabadi. Other breeds are Nagpuri, Pandharpuri, Sambalpuri and Gangatiri etc; Murrah has been used as an improver breed not only in India but also all over the world.

Buffaloes are well known to convert poor quality roughages into milk and meat efficiently. Buffaloes in India, pose a large number of problems, especially related to delayed sexual maturity and long inter calving period (primarily arising from seasonal calving, which results their non-exhibition of behavioral symptoms during summer), high calf mortality–partly due to neglect and partly due to higher susceptibility to neonatal diseases as the antibodies are not transferable from dam to fetus through placental membranes and the mortality incidence is maximum in the age group of up to 30–40 days.

21.1 Buffalo Genetic Resource in India

Little information is available about the performance of Indian buffaloes; moreover, it is mostly restricted on Murrah, Nili-Ravi and Surti and on the basis of data from different breeds, strains of institutional farms and utility *vis-à-vis* production system in their home tracts and under farmers management conditions. The situation is further complicated by the fact that there exists no breed societies or breeds registration/improvement societies to register animals of specific breeds, maintain herd books and ensure the purity of the breeds. There is no controlled breeding in the breeding tract of any of the breeds. Rather uncontrolled breeding is the common order under widely prevalent extensive grazing situations throughout the country.

In majority of the cases, the true productive potential of individual breeds in their breeding tracts has not been adequately documented. This has affected the detailed description of the breeds and also their genetic potential. There is thus an urgent need for differentiating the real breed differences by conducting systemic scientific studies. There is an urgent requirement to uniformly describe all the Indian buffalo breeds by utilizing common breed descriptions, by studying their native environment, management practices, qualitative and quantitative aspects of morphological, physiological and functional traits, blood groups and biochemical polymorphisms, cytogenetic parameters, DNA analyses, utility and demographical and geographical distributions. This will lead to the identification of the types of genes and gene combinations available in different breeds and will also assist in formulating breeding policies and selection of animals for conservation, propagation and improvement programmes.

21.2 Conservation of Buffalo Genetic Resources

The country's buffalo genetic resources need to be used judiciously. The rich biological diversity of this species is progressively being eroded due to unplanned breeding. Except in few organized farms which maintain small herds of pure breed, there is almost unrestricted interbreeding among different breeds and there is a marked decline in the availability of unique animals conforming to the attributes of defined breeds, particularly in their native breeding tracts. There has been a non-judicious utilization of buffalo genetic resources in the country. The males are only partially utilized in the form of bulls and bullocks. There is always a scarcity of breeding bulls of superior genetic merit. Above all, the high producing milch buffaloes from the breeding tract, representing the best germplams, are taken to metropolitan cities in large numbers for milk production (NDRI, 2006). After completion of lactation, these buffaloes are slaughtered, causing a serious erosion of elite germplasm.

Though the livestock census in India is conducted specie wise and not breed-wise, it is extremely difficult to determine the exact number of animals of a particular breed. Preliminary surveys conducted by the National Bureau of Animal Genetic Resources in the breeding tracts of various breeds of buffaloes indicate that the buffalo breeds which need urgent attention for conservation due to their vulnerable status are Bhadawari and Toda. Bhadawari buffaloes are famous for the very high content of fat (10–14%) in their milk. But due to their low milk yield (600–1000 liters) these buffaloes have been used for upgrading with Murrah and consequently the number of pure Bhadawari buffaloes is restricted to a few thousands only. If immediate attention is not paid, the combination of genes imparting high fat percent may be lost. Similarly, the population of Toda buffaloes of Nilgiri hills of Tamilnadu is limited to 10–15 thousand. These animals have been reared traditionally by the

Toda tribes due to their suitability in the hilly regions of their home tract and their superior meat production potential. The number of Nili-Ravi buffaloes in its breeding tract in Punjab is also decreasing at a faster rate due to the preference for Murrah or Murrah graded buffaloes. Yet, a large population of this breed is available in neighboring areas of Pakistan.

21.3 Conservation Approach

Broadly, there are two means of conservation *i.e. in situ* and *ex situ*. Conserving the live animals that exist in nature is *in situ* conservation. The animals are maintained in their original habitats under native conditions with no interference in their mode of management, feeding and other conditions. The main problem of *in situ* conservation is inbreeding and genetic drift typical of small populations. The *ex situ* conservation is to be used when the endangered population is dismally low in numbers, as this process has its own innate problems. It may suffer from spread of disease, or neglect during periods of institutional weakness, besides being costly in long term preservations and losing the relatedness of current genotype with environment when one of these is preserved for long time (Singh *et al.*, 2004).

21.4 Conservation Strategies

Ex-situ conservation

Generally sperm, oocytes, embryos, DNA and embryonic stem cell are conserved. It is possible now to store a wide variety of living cells for long periods of time. The techniques can be used for the conservation of endangered breeds as follows:

Sperms and oocytes

Deep freezing is suitable for most of the species of domestic animals.

Embryos

Cryopreservation of embryos of cows, buffaloes, sheep, goats and horses has successfully been done to produce offspring. This is a better tool for conservation as all the genetic information is stored in one diploid zygote.

Storage of DNA

Cryogenic storage of DNA is another method of preservation of genetic material.

Cloning of somatic cells

Cloning offers the advantage of producing series of exact replica/copy of the concerned animals.

Embryonic stem cells

Embryonic stem cells are derived from culture of inner cell mass of a young blastocyst. These embryonic cells are potent and have potential to develop into viable embryos.

In-situ conservation

Explicit efforts to select males from superior dams under farm conditions and making wider use of the selected best bulls and also preserving their semen are necessary. The process has been initiated for some of the breeds by NBAGR.

Data bank strategy

Maintenance of a database containing all relevant breeds, population census and ecological data is essential for designing and implementing conservation strategies. Several agencies are engaged in generation and dissemination of data/information on Animal Genetic Resources. A useful body of knowledge has already been generated/gathered at NBAGR and at other locations.

Gene bank strategy

Semen from indigenous breeds has been cryopreserved for use in the future. Ideally sufficient doses should be stored at least at two locations remote from each other. The preserved material should be periodically evaluated and put into use.

DNA bank strategy

Genetic material can be preserved in the form of DNA fragments under cryogenic conditions. This has the advantage over storage of live cells as it is economical, occupies less space and there is no spread of diseases. Within and across different countries the storage of DNA has been made feasible.

Somatic cell strategy

With the advent of Dolly sheep, somatic cell technology has received a great fillip. In future it may be possible to produce a live animal from stored somatic cells.

This possibility is very important since the protocols for collecting somatic cell samples are less demanding and inexpensive than for collection of spermatozoa and embryos.

21.5 Reproductive Bio-techniques for Genetic Improvement of Buffaloes

Landmarks of bio-technical developments during recent past have been:

- Cryopreservation of semen and artificial insemination.
- Implementation of embryo transfer and genome analysis by means of molecular genetic methods.
- Cloning of genes and generation of functional gene constructs.
- Production of transgenic animals.
- Cloning of animals employing embryonic, fetal and even adult cells.

The techniques developed have been used for achieving various goals in animal production such as efficient utilization of germ cell production in males and females (AI and ET as standard procedures, control of estrus, ovulation and parturition, collection of semen and oocytes and their micromanipulation), world wide exchange of genetic material through use of frozen semen and embryos, implementation of hygienic concepts using AI and ET (*in vitro* maturation, fertilization and culture aloof mass production of embryos for practical and scientific purposes as well as for the exploitation of oocyte cells from donor), improvement of genetic progress though integrated breeding programmes based on biotechnological procedures (MOET), sexing, cloning and *in vitro* production of embryos and producing improved or completely new products from farm animals (gene transfers for pharmaceutical protein synthesis, improving animal health and introduction of specific traits), estimating breeding values and genetic progress accurately and eradicating genetic disorders and genetically determined diseases (genome analysis, gene mapping).

Uses of animal biotechnology have shown its impact on genetic selection, animal health, production and efficiency, animal nutrition and reproduction, and niche markets for specialized animal products (Heap and Spencer, 1998). Studies of livestock genomes using molecular genetic have helped in gene mapping and their resolution. Maps are being based on micro-satellite loci covering majority of genome (*e.g* the European Pig-map, Bov-map, and Chick-map). Diagnostic procedures are also being developed to aid genetic selection *e.g.* screening for the variants of a fatty acid binding protein in pigs for positive selection and gene associated with stress related deaths in pigs for negative selection. Selection for disease resistance is another long-term aim in livestock and poultry. Further for an important application of molecular genetics is in the preservation of biodiversity and to overcome the continued threat for the erosion of domestic animal diversity. Advances in animal health are notable in the diagnostic use of monoclonal antibodies to monitor fertility and infertility and to characterize a wide range of bacterial and viral diseases including FMD, trypanosomiasis, brucellosis, rinderpest and bovine Herpes virus. New molecular vaccines under investigation include those for *Babesia*, *Boophilus microplus*, rinderpest, trypanosomiasis and helminthes, infestation. The vaccines may eventually include DNA based vaccines. The biotechniques for improving production and efficiency include the use of recombinant growth hormone and related growth factors which provide the earnest animal equivalent to enhanced nitrogen fixation improving volume of milk secretion and output of protein and an efficiency gain. Introduction of growth hormone gene into germ line by oocytes microinjection stimulates growth rate. Alternatively use a DNA construct coding for growth hormone releasing hormone (GHRH) in suitable vector, prospects its using in food animals seems remote.

Specialized reproductive technologies continue to provide new opportunities for consistent improvement and safety of products from economically valuable food animals and safeguarding rare species. This has been achieved by dissemination of elite stock through artificial insemination and it is expected that the livestock industry will be dominated with progeny tested bulls screened for genetic markers of economic traits. Sex-selected semen, embryo sexing and cloning are expected to produce animals of preferred genotype and the gender. Though the results of these biotechnologies seem promising but prospects that they will have an early impact on food production are regarded by many as overoptimistic.

Product diversification and adding value to the animal products using gene transfer techniques in animals have been well demonstrated in the past decade. Selective transfer of genes both within and between species has been skillfully used to modify the milk of dairy animals and potential market value of some proteins is substantial. Over 50 different proteins have been studied, (i) to enhance nutritive properties of milk, (ii) to reduce putative milk allergens, and (iii) to produce substantial concentrations of high value pharmaceutical proteins in milk such as factor VIII, factor IX, protein C and al-anti-trypsin. The horizon of diversification for specialist stock-keepers has been expanded further with the transgenic modification of animal organs for xeno-transplantation (Heap and Moor, 1995).

These examples illustrate the extraordinary range of possibilities that exists for design, transfer and targeting of gene constructs tailored in precise ways to meet specific requirements (Mercier and Villote, 1997). The biotechnological procedures applied so far in buffaloes are for increasing breeding efficiency and for preservation of genetic resources. For improvement of product quality or for new production strategies and animals products have to be explored and employed according to needs, suitability and their public perception by addressing ethical principles. Except the technique of artificial insemination, the application of biotechnology for genetic improvement of buffaloes through enhanced reproductive efficiency is a recent development. The new techniques include the techniques of oestrus control (Rao, 1979; Diaz *et al.*, 1994), multiple ovulation and embryo transfer (Drost *et al.*, 1983), IVM and IVF (Totey *et al.,* 1996; Madan *et al.*, 1994), others including production of transgenic and bio farming (Bhat, 2007).

In spite of various attempts for effective utilization of these techniques they are yet to become commercially viable. During last decades there has been a tremendous progress in application of biotechnology for improved reproduction and production and multiplication of genetically superior or transformed animals for the benefit of animals, man and environment. The techniques have also been used in buffaloes but not to the extent that applied in cattle, pigs or chicken. Most important priorities for future research in buffaloes would include:

- Standardization of embryo transfer techniques for application at farmer's herds
- Development of an efficient cryo preservation of oocytes
- Development of defined in vitro culture systems for buffalo embryo to comply with hygienic requirements of international movement of embryos
- Understanding the mechanisms of 'large offspring syndrome' following transfer of embryos derived from *in vitro* production or nuclear transfer and subsequent elimination of this phenomenon
- Improvements in separation of X and Y spermatozoa
- Understanding the reprogramming of the donor nucleus and improvements in nuclear transfer
- Development of efficient gene targeting in buffalo.

REFERENCE

Abdul-Samad, Awaz, K.B. and Samad, A. (1997). A rapid spot (MAUM) test based on detection of trypsin inhibitors in milk to diagnose bovine subclinical mastitis. *International J. Anim. Sci.* **12**: 97–101.

Acone, P., Gnarino, C. and Izzi, R. (1970). An outbreak of enterotoxaemia. Cl. perfringens type D in buffaloes. *Atli. Soc. Ital. Sci. Vet.* **23**: 1011–1014.

Acres, S.D., Saunders, J.R. and Radostits, M. (1977). Acute undifferentiated neonatal diarrhoea of beef calves. The prevalence of enterotoxoxigenic E. coli, Reo-like (Rota) virus and other enteropathogens in cow calf herds. *Can. Vet. J.* **18**: 113–121.

Adlakha, S.C. and Sharma, S.N. (1992). Infectious diseases. In: Buffalo Production. World Animal Science C6. Tulloh, N.M. and Holmes, J.H.G. (editors), Elsevier, Amsterdam, pp. 271–303.

Afjal, H., Ilahi, A. and Sarwar, M.M.(1966). Geographical distribution of black quarter in West Pakistan and some observations on therapeutic trials in the disease. Bull. Off. Int. Epizoot. 65: 825–832.

Agarwal, V.P. (1974). Studies on the growth rate and carcass quality of buffalo calves as influenced by different plance of nutrition. Ph.D. Thesis. Agra University, Agra, India.

Agarwala, O.P. (1956). Annual genetic gain due to selection in a buffalo herd. *Indian J. Dairy Sci.* **9**: 105–108.

Agnihotri, M.K. (1988). Studies on keeping quality an despoiled pattern of buffalo meat. Ph.D. thesis. Indian Veterinary Research Institute, Izatnagar.

Agrawal, G.S. and Batra, H.V. (1999). Comparison of an inhibition enzyme linked immunosorbent assay with other serological tests for detection of antibodies to Brucella. *Indian Vet. J.* **76**: 10–12.

Agrawal, M.C. and Dutt, S.C. (1977). Chemotherapy of Stephanofilaria zaheeri Singh, 1958, infection in dairy buffaloes. *Indian Vet. J.* **54**: 77–79.

Agrawal, M.C. and Dutt, S.C. (1977). A note on the prevalence of Stephanofilaria zaheeri Singh, 1958, infection in buffaloes of different age groups. *Indian J. Anim. Sci.* **48**: 232–234.

Agrawal, M.C., Vegad, J.L. and Dutt, S.C. (1977). Pathology of naturally occurring Stephanofilaria zaheeri Singh, 1958, infection in buffaloes. *Indian J. Anim. Sci.* **48**: 261–265.

Ahmad, Z. (1961). The role of Stephanofilaria zaheeri in causing ear sore of buffaloes. *Indian Vet. J.* **38**: 257–262.

Ahmed, I.A. and Tantawy, A.O. (1959). Breeding efficiency of Egyptian cows and buffaloes. *Emp. J. Exp. Agric.* **27, 17**. (fide Anim. Breed. Abstr. 27, 741).

Ahmed, T.M. and Abd-El-Aal, A. (1997). Brucellosis in normally slaughtered cattle and buffaloes. *Assiut Vet. Med. J.* **36**: 97–102.

Ahmed, T.M. and Abd-El-Aal, A.A. (1996). Investigations on brucellosis in a large dairy buffalo herd. *Assiut Vet. Med. J.* **35**: 105–113.

Ahuja, A.K., Makkar,G.S. and Kakkar, V.K. (1986). Studies on the nitrogen metabolism in the buffalo rumen by feeding uromin lick. SARAS *J. Livst. and Poult.*, **2**: 85–89.

Ahuja, A.K., Sethi, R.P., Makkar, G.S. and Langar, P.N. (1983). Bioconversion of poultry dropping through fungal fermentation. *J. Anim. Sci.*, **53**: 445–447.

Ainsworth, G.C. and Austwick, P.K.C. (1973). Fungal Diseases of Animals. fied. C.A.B. Farnham Royal, Slough, U.K., p. 216.

Akhtar, M.Z. and Anjum, A.D. (1997). Haematological and some biochemical values in indigestion in buffaloes. *Buffalo J.* **13**: 187–193.

Alexiev, A. (1989). Genetic aspects of embryo transfer in buffalo breeding. Proceedings II World Buffalo Congress, New Delhi, 12–16 Dec. 1989, Invited papers and special lectures Vol. II, part I, p 330.

Alexiev, A. (1998). The Water Buffalo, St. Kliment Ohridski University Press, Sofia.

Aliev, E.A. and Sultanova, F.G. (1980). Use in buffaloes of the live vaccine Brucella abortus strain 82 Veterinary, Moscow 3: 50–52.

Alim, K.A. (1953). Studies on the Egyptian buffaloes. I. Selection for milk yield. *Can. J. Agric. Sci.*, **15**: 1814.

Alim, K.A. (1957). Environmental and hereditary effects on calving intervals in milking buffalo in Egypt. *Emp. J. Exp. Agric.* **25**: 229–36.

Alim, K.A. (1967). Repeatability of milk yield and length of lactation of the milking buffalo in Egypt. *Emp. J. Exp. Agric.* **44**: 159–63.

Alim, K.A. (1978). The productive performance of Egyptian buffalo in a dairy herd. World Rev. Anim. Prod. **14**: 57–64.

Alim, K.A. and Ahmed, I. A. (1954). Month of calving, age at first calving and calving intervals of the buffaloes in a dairy herd in Egypt. *Emp. J. Exp. Agric.* **22**: 37–41.

Alim, K.A. and Taher, A. (1979). The performance of friesian and buffalo calves. Wld. Rey. Anim. Prod. 15 : 71–79.

Amano, T. (1978). Coat colour variation and blood protein polymorphism of water buffaloes in Philippines. Rep. Soc. Res. Native Livestock. **8** 40–50.

Amble, V.N., Gopalan, R., Malhotra, J.C. and Mehrotra, P.C. (1970). Some vital statistics and genetic parameters of Indian buffaloes at military dairy farms. *Indian J. Anim. Sci.* **40** : 377–88.

Amble, V.N., Krishnan, K.S. and Srivastava, J.S. (1958). Statistical studies in breeding data on Indian herds of dairy cattle. *Indian J. Vet. Sci.* **28** 33–82.

Amer, A.A., Raghib, M.F., Gazia, N., Salim, H.A., El-Nawawi, F. and Abdel-Monim, M. (1982). The effect of dietary vitamin A deficiency on the clinical condition, blood constituents and internal organs of adult buffaloes. *Assiut. Vet. Med. J.* **9**: 137–142.

Anantwar, L.G. and Singh, B. (1998). Therapeutic evaluation of triamcinolone acetamide in ketotic buffaloes. *Indian Vet. J.* **75**: 559–560.

Anderson, N.G., Zurbrigg, K. (2003). Auditing cow comfort video behind barn doors. Advances in Dairy Technology 15: 97–14.

Anjaneyulu, A.S.R., Senger, S.S., Lakshmanan, V. and Joshi, D.C. (1985). Meat quality of male buffalo calves maintained on different levels of protein. *Buffalo Bulletin.* **4 (3)**: 45–47.

Anon. (1977). Handbook of Animal Husbandry. I.C.A.R., New Delhi.

Arora, G.S., Chauhan, P.P.S., Agrawal, R.D. and Ahluwalia, S.S. (1975). Observations on the seasonal dynamics of the lazed eye worms in buffaloes. *Indian J. Anim. Sci.* **45**: 953–957.

Arora, S.P., Bajpai, L.D. and Muley, A.R. (1962). Influence of dry period on milk yield and effects of season on lactation yield and lactation days of Murrah buffaloes. *J. Vet. Anim. Husb. Res., Mhow.* **5**, 51.

Arya , J.S., and Madan, M.L. (2001). Post partum reproductive cyclicity based on ovarian steroids in suckled and weaned buffaloes. *Buffalo Journal.* **17 (3)***:* 361–369.

Arya, Y.V. and Desai, R.N. (1969). Growth rate and its relationship with weight and age at first calving in buffaloes maintained in military farms. *Indian vet. J.* **46**: 61–68.

Aschaffenburg, R., Sen, A. and Thompson, M.P. (1968): The caseins of buffalo milk Comp. Biochem. Physiol., 27, 621–623.

Asghar, A.A. (1987). In 4th Annual Report. Studies required for development of progeny tested bulls to improve buffalo milk production in Punjab. PK-ARS-123 Project. Livestock Prod. Inst. Bahadurnagar, Okara.

Ashfaq, M. (1961). Buffalo production in Pakistan. *Pakistan J. Anim. Sci.* **1 (1)**: 4–12.

Ashfaq, M. and Mason, I.L. (1954). Environmental and genetical effects on milk yield in Pakistani buffalo. *Emp. J. exp. Agric.* **22**: 161–75.

Ashok kumar (1971). Personal Communication.

Ashour, G. and Shafie, M.M. (1992). Water balance in riverine buffaloes. I. Mobilization of body fluids under heat and dehydration stresses. Proceedings of the Joint Symposium of ESAP, EAAP, FAU, ICAMAS and OIE on prospects of buffalo production in Mediterranean and the Middle East. Pudoc Scientific Publisher, Wageningen, 1993, pp. 194–98.

Asif, M. and Akhtar, S. (1997). Prevalence of rinderpest disease virus antigen in slaughtered buffaloes at Landhi cattle colony, Karachi. *Buffalo J.* **13**: 203–207.

Askar, A.A., Ragab, M.T. and Ghazy, M.S. (1973). Repeatability and heredity of some characteristics in Egyptian buffalo. *Ind. J. Dairy Sci.* **6**: 61–65.

Asker, A.A. and El-Itriby. A.A. (1958). Frequency of using bulls for service and the distribution of calving in the Egyptian buffaloes. *Alex. J. Agric. Res.* **6**: 25–38

Asker, A.A., Bedeir, L.H. and El-Itriby, A.A. (1967). The inheritance and relationship between some dairy characters in the Egyptian buffaloes. *J. Anim. Prod. UAR* **5**: 119–30.

Asker, A.A., Bedeir, L.H. El-Itriby, A.A., Ahmed, I.A. and Khishin, S.S. (1971). Longevity in Egyptian buffaloes. *J. Anim. Prod. UAR* **9**: 173–79.

Asker, A.A., Ragab, M.T. and Ghazy, M.S. (1953). Repeatability and heritability of some characters in Egyptian buffaloes. *Indian J. Dairy Sci.* **6**: 61–65.

Asker, A.A., Ragab, M.T. and Hilmy, S.A. (1955). Genetic improvement in milk yield in two herds of cattle and buffaloes in Egypt. *Indian J. Dairy Sci.* **8**: 39–46.

Assis, R.J. and Frensel, O. (1966). Dairy Production and industry in Brazil. Proc. *17th Int. Dairy Cong. EF. pp.* 53–66.

Aubert, A. (1999). Sickness and behaviour in animals: A motivational perspective. Neuroscience and Biobehavioural Review, **23**: 10291–036.

Awad, Y.L., Abrahim, M.S. and Georgy, M.E. (1979). Susceptibility of buffalo calves to hypomagnesemia. *J. Egyptian Vet. Med. Assoc.* **39**: 51–55.

Award, F.I., El Molla, A., Fayed, A., Abd El Halim, M. and Refai, M. (1982). Studies on mycotic mastitis in Egypt. *J. Egyptian Vet. Med. Assoc.* **40**: 35–41.

Award, F.I., Farrag, I., Shawkat, M.E. and Abeid, M.H. (1979). Studies on enterotoxaemia in young buffalo calves. *Egyptian J. Vet. Sci.* **14**: 25–29.

Azimov, I. H. (1963). Trikhofitiya buivlovi Zebu. Veterinary, Moscow. 40: 30.

Babu, R.N., Venkataramanujam, V. and Shanmugam, A.M. (2002). Effect of modified atmosphere pacakging on fresh buffalo meat quality. *Buffalo Journal,* **18 (2)**:171–182.

Bachhil,V.N.(1985) Studies on the occurrence of selected food poisionin bacteria and and the sanitary indices in buffalo meats and slaughter house, Ph.D. thesis, IVRI, Izatnagar, Rohilkhand University, Bareilly (U.P).

Badran, A.E.(1985). Genetic and environmental effects on mastitis disease in Egyptian cow and buffaloes. *Indian J. Dairy Sci.* **38**: 230–234.

Badreldin, A.I. and Ghany, M.A. (1952) Adaptive mechanisms of buffaloes to ambient temperature Nature (Lond.), 170: 457–8.

Bagherwal, R.K. and Shukla, P.C. (1996). Use of pefloxacin in the treatment of bovine mastitis. *Indian J. Vet. Med.* **16**: 16–17.

Bailur, D.M. (1975). Livestock markets and marketing or livestock. Ind. Fd. Packer. July-Aug. 56–59.

Bain, R.V.S. (1957). The problem of haemorrhagic septicemia in cattle. *Ceylon Vet. J.* **5**: 2–7.

Bain, R.V.S.(1963). Haemorrhagic septicemia. FAO Agriculture studies, no. 62: F.A.O., Rome.

Bain, R.V.S., De Alwis, M.C.L., Carter, G.R. and Gupta, B.K.(1982). Haemorrhagic septicemia. F.A.O. *Anim. Hlth. Prod. Paper* 33. F.A.O., Rome.

Bakshi, M.P.S. and Wadhwa,M. (2004). Effect of herbal feed additives on the nutrient utilization in buffalo calves. *Bubalus Bubalis*. 10(1): 65–70.

Bakshi, M.P.S., Gupta, V.K. and Langar, P.N. (1987). Effect of moisture level on the chemical composition and nutritive value of fermented straw. Biological Wastes. 21: 343.

Bala, A.K. and Sidhu, N.S.(1981). Studies on disease resistance vis-à-vis susceptibility in cattle and buffaloes. II. Genetic group difference in the incidence of tuberculosis. *Indian J. Anim. Hlth.* **20**: 25–29.

Balakrishnan, C.R., Goswami, S.L. and Yadav, B.R. (1980). Studies on chromosomes of farm animals and their application in animal breeding. Annual Report, pp. 72–73, NDRI, Karnal.

Balani, A.S. and and Barnabas, J.(1965) Polypepetide chains of buffalo haemoglobines. Nature(Lond.), 205: 809–815.

Balasubramanium, H.K., Krishnamoorthy,U., Mallikarjunappa, S. and Rai,A.V., (1979). Nutrient intake and digetibility of Surti heifers on feeding maize straw imprignated with urea and molasses. Proce., Sym. On protein and NPN utilization in ruminants, NDRI, Karnal, May.

Balm, H.M. (1999). Stress physiology in animals. Sheffield, UK: Academic Press.

Bandyopadhyay, S.K. , Roy, D.J. and Banerjee, A.K. (1974). The effect ofglycerol level and equilibration time onthe freezability of buffalo spermatozoa in egg yolk citrate and reconstituted skim milk extender. *Indian J. of Ani,. Health*. **13** (**1**): 155–160

Baradat, R. (1949). The livestock of Cambodia. Rev. Elev. Med. Vet. Pays Trop. N.S. 3, 29–37. (fide Anim. Breed. Abstr. 18–249: 1950).

Barman, K. and Mohini, M. (2002). Nutrient utlization for growth andcost of feeding in bufflao calves as influence by rumension supplementation. *Buffalo Journal*. **18(1)**:71–81.

Bartussek, H. (1999). A review on the animal needs index (ANI) for the assessment of animals' well-being in the housing systems for Austrian proprietary products and legislation. Livestock production science 61: 179–192.

Baruah, K.K. (1982). Effect of different planes of nutrition on growth, nutrient utilization and carcass characteristics of desi male buffalo claves. Ph.D. thesis IVRI, Izatnagar, Rohilkhand University, Bareilly (U.P).

Baruah, K.K., Ranjhan, S.K., and Pathak, N.N. (1983). Nutrient requirement for grwoth of Desi male buffalo calves. V World Congress, Tokyo, August 14–19, 111.

Basavaiah, P. (1970). Genetic studies on blood groups, haemoglobin and transferrin polymorphism in Indian water buffaloes. M.Sc. Dissertation, Punjab University, Chandigarh, Punjab.

Basavaiah, P. (1978). Annual Report, All-India Co-ordinated Research Project on Buffaloes, Dharwar.

Basavaiah, P., Siddappa, L., Kulkarni, V.S. and Mukundkrishna, A. (1983) Study on genetic parameters of the first lactation in Surti buffaloes, Dharwad. VI Workshop on "AICRP on Buffalo". Report of Dharwad Centre, pp. 33–35.

Basirov, E.B. (1964). The biology of reproduction and artificial insemination of buffaloes (in Russian). Proc. 5th Int. Congr. Anim. Reprod. Al (Torento), 6., 6, pp. 4–10. (fide Anim. Breed. Abstr., 33–60: 1965.)

Basu, S.B. and Ghai, A.S. (1978). Studies on milk production in Murrah buffaloes. *Indian J. Anim. Sci.* **48**: 593–96.

Basu, S.B. and Rao, M.K.(1979). Growth pattern in Murrah buffalo calves. *Indian vet.* J. **56**: 570–74.

Basu, S.B. and Sarma, P.A. (1982). Cross breeding in buffaloes. A. Reviewof Agro-animal Sciences and Health **6**: 465–66.

Basu, S.B., and Ghai, A.S. (1978). Studies on milk production in Murrah buffaloes. *Indian J. Anim. Sci.* **48**:593–96.

Basu, S.B., Bhosrekar, M., Goswami, S.L. and Sarma, P.A. (1978). Sourcesof variance affecting reproductive performance of Murrah buffaloes. *Indian J. Dairy Sci.* **3**: 294–96.

Basu, S.B., Tomar, O.S. and Ghai, A.S. (1984). Genetic studies on sexual maturity traits of buffalo heifers. *Indian J. Anim. Sci.* **54**: 14–18.

Batra, H.V., Chand, P., Sadana, J.R. and Mukherjee, R.(1995). Coagglutination and counter-immunoelectrophoresis for the detection of brucella antigens in fetal stomach contents of aborted cattle and buffaloes. *Indian Vet. J.* **72**: 319–323.

Belorkar, R.M., Khire, D., Kadu, M.S. and Kaikini, A.S. (1977). Studies on the effect of service period on lactational yield, lactational period and dry period in Nagpuri (Berari) buffaloes. *Indian vet. J.* **54**: 384–88.

Benson, G.J. (2004). Pain in farm animals: Nature, recognition and management. In: G.J. Benson and B.E. Rollins (eds), The well-being in farm animals: Challenges and solutions (pp.61–84). Oxford: Blackwell.

Bertenshaw, C., Rowlinson, P. (2001). The influence of positive human- animal interaction during rearing on the welfare and subsequent production of the dairy heifer. Proc. of the Bri. Soc. of Ani. Sci. **17**.

Bhadekar, A.R., Khot, J.B., Sherikar, A.T., Jayarao, B.M., Sherikar, A.A. and Pillai, S.R. (1987). Microbial flora of buffalo meat cuts and effect of chilling and freezing on them. In advances in meat research. Pp. 174–181. Publihsed by department of Food Hygiene and Public Health, Bombay Veterinary College and Red and Blue Cross Publishers, Bombay.

Bhadula, S.K and Desai, R.N. (1972). Genetic Studies on production efficiency of early calvers in Murrah buffaloes. *Indian J. Anim. Prod.* **3**: 66–73.

Bhadula, S.K. (1966). Genetic studies on production traits of buffaloes maintained at military farms in southern plateau. M.V.Sc. Thesis, G.B.Pant University of Agriculture and Technology, Pantnagar, UP.

Bhadula, S.K. and Desai, R.N. (1973). Genetic studies on breeding efficiency of buffaloes on military farms. *Indian Vet J.* **50**: 777–84.

Bhalaru, S.S. and Dhillon, J.S. (1978). Estimatesof phenotypic and genetic parameters of some measures of efficiency of milk production in buffaloes. *J. Agric. Sci. Camb.* **91**: 181–83.

Bhalaru, S.S. and Dhillon, J.S. (1981). First lactation milk yield versus some measures of efficiency of milk production as the selection criterion for buffaloes. *Indian J. Anim. Sci.* **51**: 153–56.

Bhalla, R.C., Sengar, D.P.S. and Soni, B.K. (1967). Study on the birth weight of Murrah buffalo and Sahiwal calves and factors affecting them. *Indian J. Dairy Sci.* **20**: 139–41.

Bhandari, N., Chauhan, R.A.S., and Mathew, A. (1982). Note on the effect of rate of freezing on freezability of buffalo spermatozoa. *Indian J. of Anim. Sci.* **52** (12): 1237–1238.

Bhanumurthi, J.L. and Trehan, K.S. (1970). Preservation and bottling of buttermilk. *Ind.Dairyman.* **22**: 275–278.

Bhardwaj, R.M.(1991). Clinical management of incisive papillitis induced nasal reflux of water in buffaloes *Indian Vet. J.* **68**: 981–982.

Bhardwaj, U. (1998). Therapeutic efficacy of herbal anti-tympanitic agent for recurrent tympany in buffaloes. *Indian Vet. Med. J.* **22**: 327–328.

Bhat P.N., and Taneja V.K. (1987). Genetics and breeding of river buffalo. Research bulletin No. 5, Indian Veterinary Research Institute, Izatnagar, U.P.

Bhat, P.N. (1975). Exploitation of dairy animals for meat production. Proceedings of the symposium on introduction, improvement and evaluation of germplasm for milk production, NDRI, Karnal, Haryana, India pp 1–8.

Bhat, P.N. (1983). Constraints in genetic improvement for livestock production in tropics. Proc. IV International Congress of Sabrao. pp 51–65.

Bhat, P.N. (1977). Personal communication.

Bhat, P.N. (1991). Personal communication.

Bhat, P.N. (1994). Biodiversity-Issue problems and approach in conservation of Animal Genetic Resources. *Indian J. Anim. Genet. and Breed.* **5** (**1**): 30–40.

Bhat, P.N. (2007). Personal communication.

Bhat, P.N. and Patro, B.N. (1978). Effect of various non-genentic factors on milk yield and lactation length in Indian buffaloes. *Indian J. Dairy Sci.* **31**: 321–25.

Bhat, P.N., Kumar, R. and Koul, G.L. (1982). Measures of persistency of milk yield in Murrah buffaoes. *Indian J. Anim. Sci.* **52**: 621–27.

Bhat, P.N., Kumar, R. and Raheja, K.L. (1983). Breeding behaviour of Murrah buffaloes. Proc. IV International Congress of SABRAO, pp. 91–97.

Bhat,P.P., Mishra,B.P. and Bhat, P.N., (1990). Polymorphism of mitochondrial DNA (mtDNA) in cattle and buffaloes, *Bichem. Genet.* **28**: 311–318.

Bhatia, K.C., Gupta, R.K.P., Ram, G.C. and Veno, Y.(1983). Preliminary studies on rice straw incriminated with the causation of deg nala disease. International mycotoxin conference, Cairo, Egypt, March 1983, pp. 19–24.

Bhattacharya, D.C. and Srivastava, M.R.(1967). New Varieties of *dahi*(fermented whole milk).*Ind.Dairyman.*B **19**: 35–38.

Bhattacharya, S.D., Ray Chaudhary, A.K. and Sinha. A. (1963). Inherited (a and b) lactoglobin polymorphism in Indian Zebu cattle. Comparison of zebu and buffalo a-lactalbumin. Nature 197: 797–99.

Bhavasar, B.K., Patel, K.S., Kerur, V.K. and Kodagali, S.B. (1986). Seminal characters, freezability and fertility in Mehsana and Murrah buffalo bulls. *Indian J. of Anim. Repro.* **7** (**1**): 7–14.

Bhikane, A.U., Anantwar, L.G. and Bhokre, A.P. (1999). Epidemiology, clinicopathology and treatment of haemoglobinuria in buffaloes in and around Udgir. In Compendium: National Symposium on Sustainable Development in Animal Health Care Measures- Vision for the Future. Department of Medicine, Orrisa Veterinary College, Bhubaneswar, p. 17.

Bhosrekar, M.R. (1993). Mannual on buffaloes, pp. 1–49. BAIF Development Research Foundation Pune, India.

Bhullar, M.S. (1974). Factors affecting the accuracy of progeny testing in buffaloes. Ph.D. Thesis, Punjab Agricultural University, Ludhiana.

Biswal, G. (1948). Coccidiosis in buffalo calves. *Indian Vet. J.* **25**: 36–38.

Biswas, T.K., Bhattacharya, T.K., Narayan, A.D., Badola, S., Pushpendra Kumar and Arjava Sharma (2003). Growth hormone gene polymorphism and its effect on birth weight in cattle and buffalo.*Asian Australasian J. Anim.Sci.* **16** (**4**): **494–497.**

Blackburn, H. (2004). Livestock production, the environment and mixed systems, Agriculture and Natural Resource Department, World Bank. Accessed Feb.

Blecha, F. (2000). Immune responses to stress.In: G.P. Moberg and J.A. Mench (eds). The biology of animal stress: Basic principles and implications for animal welfare (pp. 111–121). Wallingford, UK: CAB International.

Bloch, B. and Lal, S.M. (1975). A study of the ultra-structure of the buffalopox virus. Acta. Path. Microbiol. (Scand). 83: 191–200.

Block, G. (2003). The moral reasoning of believers in animal rights. Society and Animals **11** (2): 1–10.

Bodhipaksha, P., Chantaraprateep, P., Lohachit, C., Kunawongkrit, A. and Virakul, P., (1980). The horns and the buffalo. Veterinary yearbook. Jointly published by the veterinary faculties of Chulalongkorn and Kasetsart universities, Bangkok, Thailand (in Thai).

Bongso, T.A., Hassan, M.D. and Nordin,W (1984). Relationship of scrotal circumference and testicular volume to age and body weight in the swmap (*Babalus bubalis). Theriogenology,* 22 (2):127–134.

Bongso, T.A., Hilmi, M. and Basrur, P.K. (1983). Testicular cells in hybrid water buffaloes (Babalus bubalis). *Res. Vet. Sci.* **35**: 253–258.

Borghese, A. (2005). Personal information.

Bosgiraud, C., Menudier, A., Champagnol, M.P., Hangard-Vidaud, N., Lamachere, M. and Nicolas, J.A. (1991). Epidemiological study of distribution of Listeria species and Listeria monocytogenes serotypes in human and animal diseases and in food. *Rev. Med. Vet.* **142**: 463–468.

Brambell, F.W.R. (1965). Report of the technical committee to enquire into the welfare of livestock kept under intensive conditions. HMSO, London.

Britten,R.J. and Davidson,E.H. (1969).Gene Regulation for higher cells. A theory., **165:**349–357.

Brock, J.C.(1989). A natural history of domesticated mammals, Cambridge University Press (British Museum) Cambridge, UK., pp. 140–142.

Brody, S., Ragasdale, A.D. and Turner, C.W. (1923). The rate of decline of milk secretion with the advance of the period of lactation. *J. Physiol*. **5**: 441–44.

Broom, D.M. (2006). Behaviour and welfare in relation to pathology. App. Ani. Beh. Sci. 97: 73–83.

Brown, W.M. (1980). Polymorphism in mitochondrial DNA of humans revealed by restriction endonuclease analysis. *Proc. Natl. Acad. Sci.* **77**: 3605–3609.

BSTID, Board on Science and Technology for International Development.(1981). Report of an Ad hoc Panel of the Advisory Committee on Technology Innovation, Commission on International Relations.

Burridge, M.J. and Odeke, G.M.(1973). Thelaria lawrencei infection in the Indian water buffalo, Bubalus bubalis. Exptl. Parasitol. 34: 257–261.

Byomi, A., Herbst, W. and Baljer, G.(1996). Some viral agents associated with neonatal calf diarrhoea. *Assiut Vet. Med. J.* **35**: 96–104.

Cady, R.A., Shah, S.K., Schermerhorn, E.C. and McDowell, R.E. (1983). Factors affecting performance of Nili-Ravi buffaloes in Pakistan. *J. Dairy Sci.* **66**: 578–86.

Caldwell, J. and Cumberland, P.A. (1978). Cattle lymphocyte antigens. *Trnsplant. Proc.* **10**: 889–892.

Callow, L.L., Parker, R.J., Rodwell, B.J. and Ottley, M.L. (1976). Piroplasmosis in buffaloes and its serological diagnosis based on homology between buffalo and bovine immunoglobulins. Australian Vet. J. 52: 40–41.

Campbell, J.B., Boxler, D.J., Adams, D.C., Applegarth, A.F. (2006).Efficacy of several insecticide ear tags for control of horn flies (diptera: Muscidae) on Nebaska beef cattle. *J. of the Kansan Entomol. Soc.* **79**: 113–118.

Castillo, L.S., Ranjhan, S.K., Roxas, D.B.,Lapitan, R.M. and Pascual, F.Sd. (1985) Maintenance requirement for digestible energy and protein of mature carabao. FAO/UNDP report.

Chabra, Aruna and Arora, S.P (1979). The relative influence of feeding liquid protein supplement on the growth rate of cow and buffalo calves. Proc. Symposium on Protein and NPN utilization in Ruminants, Karnal (May)

Chaisrisongkarm, W. (1973). Some diseases found in slaughterhouse. *J. Thai. Vet. Med. Assoc.* **24**: 5–25.

Chakrabarti, A., Chatterji, A., Chakrabarti, K. and Sen Gupta, D.N.(1981). Human scabies from contact with water buffaloes infested with Sarcoptes scabiei var. bubalis. *Annals Trop. Med. Hyg.* **75**: 353–357.

Chakrabarty, S. (1977). Studies on chromosomal polymorphism in cattle and buffalo. Ph.D. Thesis, Agra University, Agra.

Chamnanpood, P., Cleland, P.C., Baldock, F.C., Gleeson, I.J., Copland, J.W., Gleeson, L.J. and Chamnanpood, C. (1994). Epidemiological investigations of foot and mouth disease outbreaks in northern Thai villages. Diagnosis and epidemiology of foot and mouth disease in Southeast asia: Proceedings of an international workshop. Pp. 33–37.

Chand, P., Behra, G.D. and Chand, P. (1995). Factors influencing occurrence of mastitis genetic and environmental factors. *Indian J. Dairy Sci.* **48**: 271–273.

Chander, R., Chawla, S.P. and Kanatt, S.R. (2003). The role of irradiation on microbiological safety and shelf- life extension of non-sterile and sterile convenience meat products sotred at ambient temperatures.

Chandra, R., Garg, S.K., Rana, U.V.S. and Rao, V.D.P. (1987). Pox infection of buffaloes. Farm Anim. 2: 57–69.

Chantalakhana, C. (1978). Performance of swamp, riverine and crossbred buffaloes in Southeast Asia. Proceedings of the FAO/SIDA/GOI/Seminar on Reproduction and Artificial Insemination of Buffaloes, NDRI, Karnal and FAO Animal Production and Health Paper., 13: 129–42.

Chantalakhana,C. (1999). Research priorities for small holder dairying. In *Small holder dairying in the tropics.* (Eds.L. Falvey and C.Chantalakhana). Chapter 23.p.403. International Livestock Research Institute, Kenya.

Chantaraprateep, P., Kabayashi, G., Variceal, P., Kunavongkrit, A., Techakumphu, M., Prateep, P. and Dusitsin, N. (1989a). Success in embryo transfer in Thai Swamp buffalo. *Buffalo Bulletin* **8 (1)**, 4–5.

Chaturvedi, M.L., Singh, U.B. and Ranjhan, S.K. (1973). Effect of feeding chopped and ground wheat straw on the utilization of nutrients and VFA production in growing cow calves and buffalo calvaes. *Indian J. Anim. Sci.*, **43**: 382.

Chaudhary, I.A. and Shaw, A.O. (1965). Production traits of Pakistani buffaloes. Pakist. J. Sci. 17, 252–58. (fide Anim. Breed. Abstr., 35: 3381).

Chaudhry, M.A., Malik, S.Z., Ahmad, N. and Zia-ur-Rahman. (1994). In vitro and in vivo evaluation of different antibiotics for the treatment of mastitis in buffaloes. Pakistan Vet. J. 14: 258–261.

Chauhan, P.P.S. and Pande, B.P. (1972). Onchocercal lesions in aorta and ligamentum nuchae of buffaloes and bullocks- a histological study. *Indian J. Anim. Sci.* **42**: 809–813.

Chauhan, P.P.S., Bhatia, B.B. and Pande, B.P.(1973). Incidence of gastro-intestinal nematodes in buffalo and cow calves at State Livestock Farms in Uttar Pradesh. *Indian J. Anim. Sci.* **43**: 216–219.

Chauhan, P.P.S., Bhatia, B.B., Arora, G.S. and Agrawal, R.D. (1978). Viability of Neoascaris vitulorum (Goeze, 1782) eggs under different physical and chemical conditions. *Indian Vet. J.* **55**: 1011–1013.

Chauhan, R.A.S., Bhandari, N. and Mathew, A. (1981). Method for successfully deep freezing of buffalo semen. Livestock advisor 6: 13–14.

Chauhan, T.R, Sharma, N.D., Dahiya, S.S., Punia, B.S., Hooda, O.K. and Lall, D. (2000). Effect of plane of nutrition on nutrient utilization in pregnant buffalo heifers. *Buffalo Jouranal* **16 (1)**:47–52.

Chauhan, T.R. (1999). Growth production and nutrient utlization by buffalo calves offered ammoniated wheat straw and alfalfa hay ad lib supplemented with available cereal energy sources. J. *Bubalus Bubalis*. **5 (3)**: 67–72.

Chavananikul, V. (1994). Cytogenetic aspects of crossbreeding for improvement of buffalo In: Long-Term Genetic Improvement of the Buffalo (Ed). Bunyavejchewin, P., Chantalakhana, C. and Sangdid, S. Proceedings of the First Asian Buffalo Congress (ABA) held in Khon Kaen, Thailand during 17–21 January, 1994.

Chet Ram, Khanna, N.D. and Prabhu, S.S. (1964). Studies on bovine blood groups. 1. Buffalo blood antigenic factors detected through cattle blood group reagents. *Indian J. vet Sci.* **34** 84–88.

Chhabbra, M.B., Sharma, S.M., Tikaram, S.M. and Chawla, S.K.(1978). An outbreak of bovine trypanosomiasis. Haryana Vet. **2**: 116–120.

Chhabra, R.C., Baxi, K.K. and Chawla, S.K. (1972). Theileriasis in exotic cattle and a buffalo in Punjab. *Indian Vet. J.* **49**: 321–325.

Chhikara, B.S., Balaine, D.S., Chaudhary, S.R. and Chopra, S.C. (1978). Performance levels and culling patterns in Murrah buffaloes. *Indian J. Dairy Sci.* 31: 292–93.

Chopra, A.K., Kakkar, V.K., Gill, R.S. and Kaushal, J.R. (1974). Preparation of Uromol urea molasses complex and rates of breakdown in vitro. *Indian J. Anim. Sci.* **4**: 970–72.

Chou, C.S. (1986). A study on the freesing method of buffalo pelleted semen.*Buffalo Bulletin,* **5 (2)**: 30–33.

Chowdhury, M.S. and Barhat, N.K. (1979). Factors affecting measures of milk production efficiency in buffaloes. *Indian J. Anim. Sci.* **49** :6–9.

Clutton- Brock, J. (1999). A natural history of domesticated mammals (p.284). London: Natural History Museum.

Cockrill, W. Ross (1966). A key animal for a hungry world. New Scientist. 20: 37–2.

Cockrill, W. Ross (1967). The water buffalo Scient. Am. **217**: 118–25.

Cockrill, W. Ross (1968). The draught buffalo (Bubalus bubalis). The Veterinarian (Oxford). **5:** 265–72.

Cockrill, W. Ross (1970). The water buffalo. *Sci. J.* (Lond.), **6**: 34–40.

Cockrill, W.R. (1974). The husbandry and health of the domestic buffalo, FAO, Rome. FAO (2003) www.FAO.org/DAD-IS.

Collier, R.J., Dahl, G.E., VanBaale, M.J. (2006). Major advances associated with environmental effects on dairy cattle. *J. of Dairy Sci.* **89**: 1244–1253.

Corke, M.I., Broom, D.M. (1999). The behaviour of sheep with sheep scab, *Psoroptes ovis* infestation. Vet. Parasitol. 83: 291–300.

Crews, S. Barth,R., Hood, L., Perhn, J. and Calame, K. (1982).Mouse c-myc oncogene is located on chromosome 15 and translocated to chromosome 12 in plasmacytomas. *Science*.**218:**1319–1321

Dabas, Y.P.S., Verma, M.C. and Kumar, M. (1993). Use of long acting oxytetracycline in management of foot and mouth disease cases. *Indian J. Vet. Med.* **13**: 36–37.

Dabas, Y.P.S., Saxena, O.P., Dabas, R. (2007). Veterinary Jurisprudence and Post-mortem, 3rd edn. International Book Distributing Co., Lucknow, India.

Dabash, A.K. (1976). Macro-element deficiency at Minya. I. Pregnancy hypomagnesemia in buffalo cows. *J. Egyptian Vet. Med. Assoc.* **35**: 1–6.

Dabbas, A.H.A. and Sharma, G.L. (1981). An epizootic of cutaneous anthrax in donkeys, horses and cattle in southern Iraq. *Indian Vet. J.* **58**: 1–9.

Dahiya, B.S. and Kapur, M.P. (1984). Bacteriological examination and therapeutic trials with cloxacillin sodium in clinical cases of mastitis in buffaloes. *Indian J. Vet. Med.* **4**:1–4.

Dandapat, P. and Verma, R.(1998). Deg nala disease: An overview. Intas Polivet 1: 27–34.

Das, K.M., Tripathy, S.B. and Misra, S.K. (1975). Studies on Stephanofilaria zaheeri Singh, 1958, infection in buffaloes. *Indian J. Anim. Sci.* **45**: 949–952.

Dash, S.K, Patro,B.N and Rao,P.K (2005).Genetic analysis of Kalahandi buffaloes of Orissa.*Indian Vet. J.* 2005; **82 (2)**: 158–160.

Dave, B.K. (1971) First lactation curve of the Indian water buffalo. *JNKVV Res. J.* **5**: 93–98.

Davies, W.L. (1939a) Indian Indigenous mMilk Products. *Thacker Spink:* Calcutta.

Dawkins, H.J.S., Johnson, R.B. and Spencer, T.L. (1990). Rapid detection of Pasturella multocida organisms responsible for haemorrhagic septicemia using enzyme-linked immunosorbent assay. Res. Vet. Sci. 49: 261–267.

Dawkins, M.S. (1990). From an animal's point of view: motivation, fitness and animal welfare. Behavioural and Brain Sci. 13: 1–61.

Dawkins, M.S. (1993). Through our eyes only? The Search for Animal Consciousness. Freeman, Oxford.

Dawkins, M.S. (2001). How can we recognize and assess good welfare? In: Coping with Challenge: Welfare in Animals Including Humans (ed. D.M. Broom), pp. 63–76. Dahlem University Press, Berlin.

Dayal, J.S., Krishna, N., Rao, E.R. and Reddy, T.J. (2002). Evaluation of urea-NH3 treated palm press fibre by in vitro, in sacco and in vivo techniques. Indian J. of anim. Nutrition. 19 (4):293–300.

Dayal,S., Bhattacharya,T.K., Vohra,V., Kumar,P. Sharma, A (2005). Genetic polymorphism of alpha-lactalbumin gene in riverine buffalo. *DNA Sequence J.***16(3):** 173–179.

De Alwis, M.C.L.(1981). Mortality among cattle and buffaloes in Sri Lanka due to haemorrhagic septicemia. Trop. Anim. Hlth. Prod. 13: 195–202.

De Alwis, M.C.L., Jayasekera, M.U. and Balasundram, P.(1975). Pneumonic pasturellosis in buffalo calves associated with Pasteurella multocida, serotype 6B. *Ceylon Vet. J.* **23**: 58–60.

De Jesus, Z. (1962). Report on worm parasites of domesticated mammals in Thailand. U.P. Veterinarian 7: 17–19.

De Los Santos, E.B. and Momongan, V.G. (1987). Comparative evaluation of the work ability of Philippine carabao and its crosses with Murrah (F_1) and Nili-Ravi (F_1) in terms of physiological responses. M.S. Thesis, University of the Philippines at Los Banos, College, Laguna, Philippines.

De S. and Srinivasan, M.R. (1958). Production and Merketing of Ghee. Part 1. Production. Part 2. Marketing. *Ind. Dairyman.* **10:**156–160,165, 216, 217, 220.

Deb, R.N. and Kadu, M.S. (1977). Some observations on production characters in Nagpuri buffaloes. *Indian J. Anim. Hlth.* **16**: 35–38.

Desai, R.N. and Kumar, D. (1964). Methods of fitting constants and weighted squares of means to study the effect of season of claving on production traits in Murrah buffaloes. *Indian Vet. J.* **41**:275–80.

Deshpande, A.P., Anantwar, L.G., Digraskar, S.U. and Deshpande, A.R. (1993). Clinico-pathological and biochemical alterations incalf scour. *Indian Vet. J.* **70**: 679–680.

Dhakal, I.P., Kapur, M.P. and Bhardwaj, R.M. (1991). Diagnosis of subclinical mastitis in buffaloes using somatic and viable cell counts. *Indian J. Dairy Sci.* **44**: 585–586.

Dhanda, M.R.(1977). Diseases caused by bacteria and fungi. In: Handbook of Animal Husbandry, I.C.A.R., New Delhi, p. 788.

Dhanda, M.R., Das, M.S., Lall, J.M. and Seth, R.N.(1956). Immunological studies on Pasturella septica. I. Trials of adjuvant vaccine. *Indian J. Vet. Sci.* **26**: 273–284.

Dhanda, O.P. (2004). Developments in water buffalo in Asia and Oceania. Proc. of the 7th World Buffalo Congress, Manila, Philippines, 20–23 Oct 17–28.

Dhillon, K.S. (1973). Preliminary observations on the deg nala disease in buffaloes. *Indian Vet. J.* **50**: 482–484.

Dhillon, K.S., Singh, T.J., Sodhi, S.S., Sandhu, H.S., Dwivedi, P.N., Singh, J. and Gill, B.S. (1995). Milk bacteriology: pre and post-trisodium citrate mastitis treatment in buffaloes. *Indian J. Anim. Sci.* **65**: 9–11.

Dhinsa, H.S. (1963). Inheritance of some economic characters in Murrah buffaloes. Indian Vet. J. 40:352–61.

Diaz, J.S., Fritsch,M. and Rodrigues, J.L.(1994). Prefixed A.I. in water buffaloes with synchronized estrus using protagandin. Proc. 4th World Buffalo Congress Sau. Paulo, Brazil, 27–30 June, **Vol. 3**: 588–590.

Dickerson, G.E. (1973). Inbreeding and heterosis in animals. Proc. Anim. Breed. and Genet. Symp. in Honor of J.L. Lush. ASAS and ADSA, Champaign, Illinois.

Dissamarn, R. and Thirapat, K. (1965). Studies on the carrier stage of anaplasmosis in Thai cattle and buffalo by capillary tube agglutination test. *J. Thai. Vet. Med. Assoc.* **16**: 82–86.

Drost, M., Wright, J.M., Cripe, W.S. and Richter, A.R. (1983). Embryo transfer in water buffalo (*Bubalus bubalis*). Theriogenology. 20: 579.

Dubal, D.B., Waghmare, A.V., Samad, A. and Jagadish, S.(1999). Tiamutin residue in buffalo and cow milk after intra mammary infusion. *Indian Vet. J.* **76**: 105–107.

Dubey, J.P., Speer, C.A. and Fayer, R. (1989). Sarcocystosis of animals and man. CRC Press Inc., Boca Raton, Florida, p. 86: 91 and 137.

Dumbell, K. and Richardson, M. (1993). Virological investigation of specimens from buffaloes affected by buffalopox in Maharashtra state, India between 1985–1987. Archives Virol. 128: 257–267.

Duncan, I.J.H. (2006). The changing concept of animal sentience. App. Ani. Behaviour Sci. 100: 11–19.

Dundar, B. and Ozer, E. (1996). Sarcocystis species and their development in buffaloes. Etlik Vet. Microbiyoloji Dergisi 8: 58–69.

Dutt, M., Singh, S.P. and Desai, R.N. (1965). Significance of age at first calving and 305 days first lactation yield in relation to lifetime production, longevity and productive life in Murrah buffaloes. *Indian vet. J.* **42**: 28–39.

Dutt, M., Singh, S.P. and Desai, R.N. (1965a). Significance of age at first calving and 305 days first lactation hield in relation to lifetime production, longevity and productive life in Murrah buffaloes. *Indian vet. J.* **42**: 28–39.

Dutt, M.K. and Bhattacharya, P. (1952) Chromosomes of the Indian water buffalo. Nature (Lond.), 170: 1129–30.

Dutt, T. (1991). Genetic analysis of early and life time performance traits in Murrah buffaloes. Ph.D. Thesis, Indian Veterinary Research Institute (Deemed University), Izatnagar, U.P., India.

Dutt, T. and Taneja, V.K. (1994): Phenotypic and genetic parameters of age at first calving, first lactation yield and some measures of efficiency of production in Murrah buffaloes. *Indian J. Anim. Sci.* **64**: 966–968.

Dutta, P.K., Sarma, G. and Das, S.K. (1983). Foot and mouth disease in Indian buffaloes. Vet. Rec. 113: 134.

Dwivedi, J.N. (1966). 'Supernumirary teats in buffaloes'. M.V.Sc Thesis, Agra University, Agra.

Dwivedi, S.K., Mallick, K.P. and Malhotra, M.N.(1979). Babesiosis clinical cases in Indian water buffaloes. *Indian Vet. J.* **56**: 333–335.

Dwivedi, S.K., Shivnani, G.A. and Joshi, H.C. (1971). Clinical and biochemical studies in Lantana poisoning in ruminants. *Indian J. Anim. Sci.* **41**: 948–953.

Dzafraov, S. (1968). Buffalo breeding, a valuable source of butter and meat production. Mol. Miasn. Zivotn. 3: 16–19.

Edney, A. (1989). Killing with kindness. Veterinary Record 124: 320–322.

Edwards, J.T. (1927). Rinderpest: some properties of the virus and further indications for its employment in the serum simultaneous method of protective inoculation. Trans. 7th Congress Eastern Assoc. *Trop. Med.* **3**: 699–706.

EFSA Animal Health and Welfare Panel (2006). Scientific report on the risk of poor welfare in intensive calf farming systems. An update on the report of the veterinary committee on the welfare of calves. *The EFSA Journal* **366**: 1–36.

Eicher, S.D., Cheng, H.W., Sorrells, A.D., Schutz, M.M. (2006). Short communication: Behavioural and physiological indicators of sensitivity or chronic pain following tail docking. *J. of Dairy Sci.* **89**: 3047–3051.

El-Arian, M.N. (1986). Genetic and nongenetic factors affecting components of reproductive efficiency and their relationship with milk production in buffaloes. Ph. D. Thesis. Rohilkhand University, Bareilly.

El-Gindy, H., Farrag, H.F. and Abou El-Asm, L. (1964). A study of the treatment of mastitis in buffaloes and cows in Egypt. Vet. Med. Small Anim. Clin. 59: 380–2. (*fide Vet. Bull.* **34**: 451 1964.)

Elimwadi, D.S., Kupfer, V. and Flukiger, A. (1974). Deep freezing of buffalo semen with egg yolk citrate diluent XIX International Dairy Congress, New Delhi (India) IE:59.

El-Kirabi, F. (1995). Buffalo population and production in Egypt. *Buffalo Newsletter* **3**: 8–9.

Ellis, T.M. and Masters, A.M. (1997). Use of magnetic particles to improve the diagnosis of sheep-associated malignant catarrhal fever by polymerase chain reaction. *Aust. Vet. J.* **75**: 520–521.

El-Refai, A.H., El-Abdin, Y.Z. and Hamsa, S.M. (1979). Control of ringworm in buffalo calves with fulcin. *J. Egyptian Vet. Med. Assoc.* **37**: 111–123.

El-Said, W.A.G., Salem, S.E., El-Garhy, M.M.M., El-Rashidy, A.A. and Tawfik, M.S. (1996). Immunopotentiated K99 vaccine for buffalo dams to protect neonates against E. coli enterotoxicosis. *Vet. Med. J. Giza.* **44**: 749–759.

Elsawaj, S.A., Goher, M.E. and Schmidt, K. (1964). The inter-relationship of milk yield, dry period, calving interval and nutrient requirements in the buffalo. *Vet. Med. J., Cairo.* **10**: 101–11.

El-Shabiny, L.M. (1994). Enzyme linked immunosorbent assay for the diagnosis of mycoplasmal mastitis in cows and buffaloes. *Vet. Med. J. Giza.* **42**: 51–53.

El-Sheikh, A.S. (1967). The reproductive performance of the buffalo in the UAR. *Indian J. Dairy Sci.* **20**: 89–95.

El-Sherif, M.M.T., Shawkat, E. and Fayed, A.A. (1981). The efficacy of different antimycotics in the treatment of trichophytosis in buffalo calves. *Vet. Med. Rev.* **2**: 186–189.

Epstein, H. (1969). Domestic animals of China. Farnham Royal, Bucks, Common wealth Agricultural Bureax. p. 166.

FAO (2003). www.FAO.org/DAD-IS

FAO Statistical Database (2004). website (http:www.fao.org/) Food and agricultural organization of the United Nations, Rome.

FAO (1994). FAO Year Book, *Vol.* **48** FAO, Rome

FAO (2006). FAOSTAT.

FAO (1977). Water Buffalo. Food and Agriculture Organization of the United Nations. Animal production and health series No. 4, FAO, Rome, pp. 283.

FAO 1997. Water Buffalo. Food and Agriculture Organization of the United Nations. Animal Production and Health Series No. 4, FAO, Rome, pp. 283.

Fahimuddin, M. (1975). Domestic Water Buffalo. Oxford and IBH Publishing Co., Calcutta.

Fahmy, S.K. (1972). Genetic and environmental relationship among some traitsof Egyptian buffaloes.Indian J. Dairy Sci. 25: 80–87.

Fahmy, S.K., Shaheen, M.A. and Youssef, F.G. (1975). Some aspectsof age and milk production in Egyptian buffaloes. Agric. Res. Rev., Cairo 57:1–15.

Farber, J.M. and Peterkin, P.I.(1991). Listeria monocytogenes, a food borne pathogen. Microbiol. Rev. 55: 476–511.

Farid, A., Ibrahim, M.S. and Refai, M.(1976). Studied on colibacillosis in calves in Egypt. *Zbl. Vet. Med.* **23**: 38–43.

Farm Animal Welfare Council (FAWC) (1991). Report on the European commission proposals on the transport of animals. DEFRA, London.

Farm Animal Welfare Council (FAWC) (1993). Second report on priorities for research and development in farm animal welfare. DEFRA,London.

Farm Animal Welfare Council (FAWC), (2001). Interim report on the animal welfare implications of farm assurance scheme. DEFRA, London.

Faruque, O. (2000). Buffalo breeding in Bangladesh. Proc. Workshop on Animal Recording and Management Strategies for Buffaloes. ICAR Technical Series. 4: 51–54.

Fernando, S.T., Gunawardena, V.K., Weilgama, D.T., Tennakoon, B. and Wickramatilaka, G.N.(1981). Life cycle and migratory behaviour of Neoascaris vitulorum in buffaloes and some observations on the immunological response of the hosts. Proceedings 2nd Research Coordinated Meeting of the Use of Nuclear Techniques to Improve Domestic Buffalo Production in Asia, Chulalongkorn University, Bangkok, Thailand, pp. 75–80.

Fischer, H. (1974). Cytogenetic observations on crossbred between swamp and Murrah buffaloes. Zuchthyg. Fortpft Stor. Besam. Haustiere. 9: 105–10.

Fischer, H. and Richards, B.G, (1965). A polled water buffalo. Z. Tierzuchtg. Zuchtungsbiol., 81: 355–357.

Galal, E.S.E. and Fahmy, M.N. (1969). Birth weight and gestation length in the Egyptian buffalo and factors influecning them. Trop. Agric. Trin. 4:329–36.

Ganguli, N.C. (1981). Buffalo as a candidate for milk production. Federation Internationale De Laiterie–International Dairy Federation Bulletin 137.

Gard, G.P., Cybinski, D.H. and Zakrgewski, H.(1984). The isolation of a fourth bovine ephemeral fever group virus. *Aust. Vet. J.* **61**: 332.

Garillo, E.P., Salas, C.G., Neric, S.P., Qunes, Q.D. and Ranjhan, S.K. (1987). Studies on the draftability of swamp buffaloes and its crosses with Murrah (F_1) under wet and dry land tillage operation in the Philippines. *Indian J. Anim. Sci.* 57: 575–580.

Gautam, O.P., Bansal, S.R. and Dey-Hazra, A. (1976). Field trials with fenbendazole against Neoascaris vitulorum in buffalo-calves. *Indian Vet. J.* **53**: 965–966.

Gautam, O.P., Malik, K.S., Nagpal, M.C. and Sharma, R.M. (1972). Phosphorus deficiency haemoglobinuria in buffaloes in India. *Haryana Agric. Univ. J. Res.* **2**: 4–8.

Gautam, O.P., Sharma, R.D. and Singh, B. (1970). Anaplasmosis-II. Clinical cases of anaplasmosis in cattle, buffaloes and sheep. *Indian Vet. J.* **47**: 1012–1019.

George, M. (1981). Studies on the relation between red cell potassium and sodium levels and red cell surface antigensin Murrah buffaloes. *Annual Report, pp.* **70** NDRI, Karnal.

George, M. and Balakrishnan, C.R. (1980). Genetic studies on pottassium concentration in the erythrocytes of cattle and buffalo. *Annual Reprot. pp.* **71** NDRI, Karnal.

Georges, M., Lequarre, A.S., Hanset, R. and Vassart, G. (1987). Genetic variation of the bovine thyroglobulin gene studied at the DNA level. Anim. Genet. 18: 41–50.

Ghanem, Y.S., Abul Fadal, A., Zalur,A. and Soliman, F.A. (1955). Environmental causes of variation in the length of gestation of buffaloes. *Indian J. vet. Sci.* **25**: 301–306.

Ghildyal, N., Garg, S.K. and Chandra, R. (1986). Comparison of immunodiagnostic tests in buffalo pox. *Indian Vet. Med. J.* **10**: 68–72.

Ghosh, T.K., Arora, R.R., Sehgal, C.L., Ray, S.N. and Wattal, B.L.(1977). An investigation of buffalo pox outbreak in animals and human beings in Dhulia District (Maharashtra State). 2. Epidemiological studies. *J. Com. Dis.* **9**: 93–101.

Ghosh,B.C. and Singh, S. (1991). A new method for vegetarian Mozzarella cheese from buffalo milk. *Indian J. of Dairy Sci.* **44**:140–142.

Gill, B.S., Singh, B. and Gupta, P.P. (1977). Pulmonary aspergillosis in Indian buffaloes. Mykosen **20**: 65–70.

Gill, G.S. and Dev, D.S. (1970). Studies on some economic traits in a herdof Murrah bufaloes. Indian *J. Anim. Prod.* **1**: 168–75.

Gillies, A.J. (1971). Flvoured yoghurt for the home. *Queensland Agric. J.* **97**:139–144.

Girard, H.C., Choonpicharna, P., Smitinondana, P., Charutamra, U., Supavilai, P., Punyaoopapatt, S. and Longswasdi, P. (1959). Buffalo foot and mouth viruses are adopted viruses. Bull. Off. Int. Epizoot. **51**: 395–402.

Giri, K.V. and Pillai, N.C., (1956). Multiple haemoglobins in the blood of animals, Nature (Lond.), 178: 1057

Gogoi, P.K., Das, D., Goswami, R.N., Nahardeka, N. and Das, G.C. (2002). Studies on age at first calving in Murrah and Surti buffaloes maintained in Assam. *Indian Vet. J.* **79** (**8**): 854–855.

Gogoi, P.K., Johar, K.S. and Singh, A. (1985). Factors affecting age at first calving in Murrah buffaloes. *Indian vet. med. J.* **62**: 1058–62.

Gokhale, D.R. (1958). Glucose sodium bicarbonate and sulphamezathine buffer as a diluent of buffalo semen. *Indian vet. J.* **35**: 573–81.

Gokhale, S.B. (1974). Inheritance of part lactations andtheir use in selection among Murrah buffaloes. Ph.D, Thesis, Agra University, Agra.

Gokhale, S.B. and Nagaracenkar, R. (1980). The selection indices based on part lactation of Murrah buffaloes. *Indian J. Hered.* **12**: 57–66.

Golem, S.B. (1943). Serological examination for brucellosis of man and domestic animals in Turkey (in Turkish). Turk. Z. Hyg. Exp. Biol., 3, 115–16. (fide, *Vet. Bull. 16 No.* 1712).

Gomes, I., Ramalho, A.K. and Auge De Vello, P. (1997). Infectivity assays of FMD virus contact transmission between cattle and buffalo in the early stages of infection. *Vet. Rec.* **140**: 43–47.

Gopal Dass and Sadana, D.K. (2000). Factors affecting some economic traits in Murrah buffaloes. *Indian J. of Ani.Res.* **34**: (**1**): 43–45.

Gopal Dass and Sharma, R.C. (2000). Effect of season and period of calving on predicted milk yields in Murrah buffaloes. *Indian J. of Ani.Res* **34** (**1**): 68–70.

Gorbelik, V.I. (1935). The breeding of buffaloes (in Russian). Trud. Azerbaidzhan. Stanc. Zhivotn., 4, 5–26. (fide Anim. Breed. Abstr. 4 162–3: 1936.)

Goswami, S.B and Kumar, S. (1968). Influence of season and off-season of calving and average daily milk yield of the Indian water buffaloes on their fertility. *Indian J. Dairy Sci.* **21**: 27–36.

Goswami, S.B. and Nair, A.P. (1964). Effect of some climatological factors on reproduction in buffaloes. *Indian J. vet. Sci.* **18**: 104–108

Goswami, S.L., Balakrishnan, C.R.and Yadav, B.R. (1980). Studies on heterochromatin in livestock chromosomes. Annual Report, pp. 68–69, NDRI, Karnal.

Grant, R.J., Van Soest, P.J., McDowell, R.E. and Perez Jr., C.B. (1974). *J. Anim. Sci.* **39**: 423.

Griffiths, R.B. (1974). Parasites and parasitic diseases. In: The Husbandry and Health of the Domestic Buffalo. Cockrill, W.R. (editor), FAO, Rome, pp. 236–275.

Gregory, N. (2004). Physiology and behaviour of animal suffering. Oxford: Blackwell.

Gregory, N.G. (1998). Animal welfare and meat science. Wallingford, UK: CAB International.

Gregory, N.G. (2007). Animal welfare and meat production, 2nd edn. Cromwell, Trowbridge.

Gudding, R. and Naess, B. (1986). Vaccination of cattle against ringworm caused by Trichophyton verrucosum. *Am. J. Vet. Res.* **47**: 2415–2417.

Gudding, R., Naess, B. and Aamodt, O.(1991). Immunization against ringworm in cattle. *Vet. Rec.* **128**: 84–85.

Guha, D.K. and Chatterjee, A. (1999). Studies on detection of antibodies after rinderpest vaccination by enzyme–linked immunosorbent assay and virus neutralization test. *Indian Vet. J.* **76**: 189–190.

Gunnis, L.F. (1967). Techniques for the production of anhydrous milk fat. Report Commission IV (Dairy Technique). *Int. Dairy Fed.Bull.* **(6)** 54 pp.

Gupta, B.D. (1988). Genetic and nongenetic factors affecting components of reproductive efficiency and their relationship with milk production in buffaloes. Ph.D. Thesis, Rohilkhand University, Bareilly, Uttar Pradesh, India.

Gupta, B.K. and Malik, N.S. (1990). Preliminary Studies on a new block-lick of subabul leaves. *Indian J. Anim. Sci.* **61**: 113–116.

Gupta, G.C., Joshi, B.P. and Rai, P. (1977). A note on some biochemical studies in haemorrhagic septicemia in buffaloes (*Bubalus bubalis*). *Indian J. Anim. Sci.* **45**: 503–504.

Gupta, L.R., Hariom, Yadav, R.S., Grewal, G.S., (2004). Effect of different housing systems on the performance of Murrah buffalo heifers during winter. *Indian J. Anim.Sci.* **74 (12)**: 1239–1240

Gupta, P. and Ray-Chaudhuri, S.P. (1978). Robertsonian changes in the chromosomes of Indian Murrah buffalo. Nucleus 21: 80–97.

Gupta, P.K., Singh, R.P. and Singh, I.P.(1970). A study of dermatomycoses (ringworm) in domestic animals and fowls. *Indian J. Anim. Hlth.* **9**: 85–89.

Gupta, S.C., Sharma, M.C. and Gupta, O.P. (1983). Effect of Banminth-II (morantel tartrate 4%) against nematode infection in suckling buffalo calves. *Indian Vet. Med. J.* **7**: 102–106.

Gupta, S.K.(1997). Serological evidence of Brucella and leptospiral infection in cases of abortion in bovines. *Indian J. Anim. Reprod.* **18**: 75–76.

Gupta, V.K., Bakshi, M.P.S. and Langar, P.N. (1984). Use of natural inoculum for enrichment of wheat straw-urea mixture by fermentation. *Indian J. Anim. Nutr.* **1**: 41–44.

Gupta. H.P. and and Saxena, V.B. (2000). Deep freesing of buffalo spermatozoa in tris and citrate dilutors. *Indian J. of Anim. Res.* **34 (2)**: 139–141.

Gurnani, M., Nagarcenkar, R., Gupta, S.K. and Singh, A. (1976). Seasonality of calving and its influence on economic characters of Murrah buffaloes. *Indian J. Dairy Sci.* 29: 173–78.

Gurung, B.S. and Johar, K.S. (1982). Studies on factors affecting first lactation milk yield in Murrah buffaloes. *Indian vet. J.* **59**: 521–26.

Hadi, M.A. (1965). A preliminary study of productive and reproductive characters of Marathwada buffaloes of Maharashtra State. *Indian vet. J.* 42: 692–99.

Hafez, E.S.E. (1952). The buffalo-a review. *Indian J. vet. Sci.* **22**: 257–63.

Hafez, E.S.E. (1953). Conception rate and periodicity in the buffalo. *Emp. J. exp. agric.* **21**: 15–21.

Hafez, E.S.E. (1954). Oestrus and some related phenomena in the buffalo. *J. agric. Sci.* (*Camb.*) 44: 165–72.

Hafez, E.S.E. (1955). Puberty in the buffalo cow. *J. Agric. Sci.* (Camb.) **46**: 137–42.

Hajela, S.K. and Sharma, S.K. (1978). A note on foot and mouth disease atypical in buffalo calves. *Indian Vet. Med. J.* **2**: 103–104.

Hamdy, A.(1970). The manufacture of Domiati cheese using a microbial rennet. *Proc. 18th Int. Dairy Congr.,* ***IE*** 350.

Handa, B.N. (1938). Buffalo breeding in the Punjab. Punjab (Lahore) Civil Veterinary Department. *Vet. Bull.* **128** 6.

Harisha, M. and Azmi, T.I. (1989). Heterosis estimates for growth rates in crossbred water buffaloes. In: Seminar on Buffalo Genotypes for Small Farms in Asia.

Hazarika, R.A., Deka, D.K., Phukan, S.C. and Saikia, P.K. (1995). Sarcoptic mange in buffalo calves and treatment with Pestoban. *J. Vet. Parasitol.* **9**: 143–145.

Heap, R.B. and Moor, R.M. (1995). Reprodcutive techniques in farm animals : ethical issues. In issues in Agricultural Bioetchis. Ed by Mepham,T.B., Tucker,G.A. and Wiseman, I. 247–368. Nottingham University Press.

Heap, R.B. and Spencer, C.G. W. (1998). Animal Biotechnolgy:convergence of science, law and ethics. In Proc. Special Symposium and Plenary Session, 8th WACP, Seoul, Korea, 100–114.

Hein, W.R. and Tomasovik, A.A.(1981). An abattoir survey of tuberculosis in feral buffaloes. *Aust. Vet. J.* **57**: 543–547.

Hoda, I., Ahmad, S. and Srinivasan, P.K.(2002).Effect of microwave oven processing, hot air oven cooking, curing and polyphosphate treatment on physioc-chemical, sensory and textural characteristics of buffalo meat products. *J. Food Sci. and Tech. Mysore.* B39(3):240–245.

Hoffmann, D., Soeripto, S., Sobironingsih, S., Campbell, R.S.F. and Clarke, B.C. (1984). The clinico-pathology of malignant catarrhal fever syndrome in the Indonesian swamp buffaloes (*Bubalus bubalis*). *Aust. Vet. J.* **61**: 108–112.

Hosney, G.A., Omrah, R.M.A., Shehab, G.G., Aly, N.M. and Karim, I.A. (1996). Pathological, immunopathological and virological findings in a field outbreak of mixed infection of bovine diarrhoea and FMD in cattle and buffaloes. In: Proceedings of fourth scientific congress- Vet. Med. and Human Health, Cairo. 44: 349–369.

Hua, L.C. (1957). The use of Malayan swamp buffaloes and Murrah/Malayan crossbreds for milk production in Taiping. *J. Malay vet. med. Ass.* **1**: 141–43.

Hultnaes, C.A. (1982). Deep freeze preservation of water buffalo semen. World animal review 42: 45–46.

Hunziker, O.F. (1949). Condensed milk and milk powder (7th edn*)*. Hunziker: La Grange, Illinosis.

Huong, L.T.T., Dubey, J.P., Nikkila, T. and Uggla, A. (1997). Sarcocystis buffalonis N. sp. (Protozoa: Sarcocystiae) from the water buffalo (*Bubalus bubalis*) in *Vietnam. J. Parasitol.* **83**: 471–474.

Hussain, K.A., Saleem, A.N. and Fatoohi, F.A.M. (1996). Prevalence of brucellosis in buffaloes, cattle and sheep in Mosul region. *Iraqi J. Vet. Sci.* **7**: 233–239.

Hussan, F.M., Khan, M.A., Zahid, T., Akhtar, M.N. and Khan, A.G.(1994). Epidemiological investigations and therapeutic trials against deg nala disease. Proceedings of 8th International Congress on Animal Hygiene, St. Paul, USA, September 12–16 1994.

Hutchison, C.A., Newbold, J.E., Potter, S.S. and Edgell, M.H. (1974). Maternal inheritance of mammalian mitochondrial DNA. Nature (Lond.), 251: 536–538.

Iannelli, D., Bettini, T.M. and Oreste U. (1973). Water buffalo blood groups. IV. Antigenic determinant located beneath the surface of red cells. *The J. of Immunology* **111**: 278–280.

Iannuzzi, L., Di Meo, G.P., Perucatti, A., F. Liotola, Incarnato, D., Di Palo, R., Peretti, V., Campanile, G. and Zicarelli, L.(2005). Free martinism in river buffalo: clinical and cytogenetic observations. Cytogenet. Genone Res. 108: 355–358.

ICAR, 1939. A brief survey of some of the important breeds of cattle in India. 2nd Prize winners at the First All-India Cattle Show, Delhi. Misc. Bull. No.24, Imperial Council of Agricultural Research, New Delhi.

ICAR, 1941. A brief survey of some of the important breedsof cattle in India. 3rd Prize winners at the second and third All-India Cattle Shows, Delhi. Misc. Bull. No. 46, 30p., Imperial Council of Agricultural Research, New Delhi.

Ichhponani, J.S. and Sidhu, G.S. (1965a). Production of VFA when cattle and buffalo were fed on green berseem and wheat straw. *Indian J. vet. Sci.* **34**: 331.

Ichhponani, J.S. and Sidhu, G.S. (1965b). Comparative concentration of rumen VFA in cattle and buffalo. *J. Anim. Sci.* 24: 291.

Ichhponani, J.S. and Sidhu, G.S. (1966). Relative performance of zebu cattle and the buffalo on urea and non-urea rations. *Indian J. of Dairy sci.* **19**: 33–38.

Ichhponani, J.S., Gill, R.S., Makkar, G.S. and Ranjhan, S.K. (1977). Work done on buffalo nutrition in India - A Review. *Indian J. Anim. Sci.* 30: 177.

Ichhponani, J.S., Makkar, G.S. and Sidhu, G.S. (1971b). Effect of different levels of urea nitrogen on in vitro digestion of cellulose by rumen micro-organisms from zebu cattle and water buffaloes. *Indian vet. J.* **48**: 356.

Ichhponani, J.S., Makkar, G.S. and Sidhu, G.S. (1969a). Effect of different carbohydrates on 'in vitro' digestion of cellulose and utilization of urea nitrogen by rumen micro-organisms from zebu cattle and buffalo. *Indian J. Anim. Sci.* **39**: 27.

Ichhponani, J.S., Makkar, G.S. and Sidhu, G.S. (1969b). Production of VFA in vitro by rumen micro-organisms obtained from rumen of zebu cattle and water buffalo. *Indian J. Anim. Sci.* **39**: 201.

Ichhponani, J.S., Makkar, G.S. and Sidhu, G.S. (1971a). Cellulose digestion in water buffalo and zebu cattle. *Indian vet. J.* **48**: 267.

Ichhponani, J.S., Makkar, G.S. and Sidhu, G.S. (1972). The significance of molar proportions of acetic and propionic acids in the rumen on the growth rate of buffalo and zebu. *Indian vet. J.* **49**: 995.

Ichhponani, J.S., Makkar, G.S., Moxon, A.L. and Sidhu, G.S. (1962). Cellulose digestion in water buffalo and zebu cattle. *J. Anim. Sci.* **21**: 1001.

Ichhponani, J.S and Sidhu, G.S. (1965). Relative performance of Zebu cattle and the buffalo on urea and non-urea rations. *Indian J. Dairy Sci.* **18**: 33–38.

Ilahi, A., Afzal, H. Sarwar, M.M. and Ayaz, M. (1966). Coccidiodomycosis in buffalo in West Pakistan. *Bull. of Int. Epizoot.* **66**: 907–914.

Imail, A.A. and and Elnahat,A. (1964). The variation between buffalo's and cow's milk in 'Kariesh' cheese making. *Alex.J agric.Res.* **12**: 195–203.

Iovane, G, Abdi Farah, A., Tempesta, M., Palomba, E., Buonavoglia, D. and Pagnini, P.(1995). Isolation and characterization of rotavirus strains from buffaloes with diarrhoea. Acta Medica Veterinaria. 41: 69–73.

Iqbaluddin, Singh, B.P. and Dutt, M. (1970). Association of part yield of first lactations and age at first calving with complete lactation yield in Murrah buffaloes. *Indian J. Anim. Hlth.* **9**: 163–69.

Irfan, M.(1971). The clinical picture and pathology of deg nala disease in buffaloes and cattle in West Pakistan. Vet. Rec. **88**: 422–424.

Ishaq, S.M. (1956). Oestrus and allied problems in buffaloes. Agric. Pakist. 7: 361–5.

Ismail, A.A. and Amer, A.A. (1978). Efficacy of Dursban, diazinon, lindane and DDT for treatment of mange in camels and buffaloes. *Assuit. Vet. Med. J.* **3**: 199–207.

Jafery, M.S. and Rizvi, A.R.(1975). Aetiology of mastitis in Nili Ravi buffaloes of Pakistan. Acta Tropica. 32: 75–78.

Jain, A. (1982). 'Genetic analysis of some economic traits in Murrah buffaloes'. M.Sc. Thesis, Kurukshetra University, Kurukshetra, Haryana.

Jain, A. and Taneja, V.K. (1982). Effects of genetic and non-genetic factors on reproduction and production traits in Murrah buffaloes. *Asian J. Dairy Res.* **1**: 123–29.

Jain, A. and Taneja, V.K. (1984). Prediction of lifetime production in buffaloes. *Indian J. Anim. Sci.* **54**: 104–105.

Jain, J.P. and Malhotra, J.C. (1971). Comparative study of different methods of indexing sires. *J. Anim. Sci.* **41**: 1101–1108.

Jain, T.B., Pal, R.N. and Mahadevan, V. (1970). *Indian Vety. J.* **47**: 26.

Jainudeen, M.R. (1989) Embryo transfer technology I buffalo. A review: *Paper presented to joint FAO/IAFA/ACIAR Research Co-ordination and Planning Meeting on Buffalo Productivity,* held at Queensland, Australia. Feb.20–24.

Jand, S.K. and Dhillon, S.S.(1975). Mastitis caused by fungi. *Indian Vet. J.* **52**: 125–128.

Jani, V.R, Vashi, B.G, Derasri, H.J. and Kodagali, S.B. (1986). Fertility following heterospermic insemination in Surti buffaloes. *Indian J. Anim. Reproduciton*, **7(1)**: 134–136.

Jarez, J.A., Pinto, A.A., Arrunda, M.V.M., Koseki, I., Abuhah, T.G and Rodrigues, M.A.La. (1979). Foot and mouth disease in buffaloes (Bubalus bubalis). Detection of antibodies to infection associated antigen and isolation of virus. Arquivos do Institute Biologic, Brazil. 46: 111–116.

Jat, R.P., Gupta, L.R, Yadav, B.L. (2005). Effect of roof modification in loose house on intake and utilization of nutrients in buffalo calves during ainy season. *Indian J. of Dairy Sci.* **58(1)**: 54–57.

Jawarkar, K.V. and Johar, K.S. (1974). A genetic study of milk yield of Murrah buffaloes. *JNKVV Res. J.* **8**: 98–101.

Jenness, R. and Patton, S. (1969). Principles of Dairy Chemistry.

Johari, D.C. and Bhat, P.N. (1978). Selection indexes based on growth, reproduction and production traits of Indian buffaloes. *Indian J. Anim. Sci.* **49**: 79–84.

Johari, D.C. and Bhat, P.N. (1979). Effect of genetic and non-genetic factors on production traits in buffaloes. *Indian J. Anim. Sci.* **49(12)**: 984–991.

Johari, D.C. and Bhat, P.N. (1979a). Effect of genetic and non-genetic factors on reproductive traits in Indian buffaloes. *Indian J. Anim. Sci.* **49**: 1–6.

Johari, D.C. and Bhat, P.N. (1979b). Effect of genetic and non-genetic factors on body weight in buffaloes. *Indian J. Anim. Sci.* **49**: 597–603.

Johari, D.C. and Bhat, P.N. (1979b). Effect of genetic and non-genetic factors on reproductive traits in buffaloes. *Indian J. Anim. Sci.* **49**: 984–91.

John, B. and Miklos, G.L.C. (1979). Functional aspects of satellite DNA and hetrochromatin. Int.Rev.Cytol. **58**: 1–114.

Joseph, P.G. (1979). Haemorrhagic septicemia in Peninsular Malaysia. Kajian Veterinar. 11: 65–79.

Joshi, D.C., Khan, M.Y., Lakshmanan, V., Prasad, M.C. and Sharma, G.C. (1984). Effect of using unground rich husk as a roughage substitute on the performacne and carcass characteristics of bufffalo calves. *The Ind. J. Nutr. Dietet.* **21:** 19–26.

Joshi, H.C., Kumar, M., Saxena, M.J. and Chhabra, M.B. (1996). Herbal gel for the control of subclinical mastitis. *Indian J. Dairy Sci.* **49**: 631–634.

Joshi, H.C., Sharma, M.C. and Kumar, M. (1999). Advances in Veterinary Medicine. In: Handbook of Animal Husbandry. I.C.A.R., New Delhi.

Joshi, H.D. and Shrestha, H.K. (1995). Study on the prevalence of clinical mastitis in cattle and buffaloes under different systems in the western hills of Nepal. Working Paper Lumle Regional Agri. Res. Centre, No. 95/64, p. 22.

Joshi, H.D., Joshi, B.R. and Shrestha, H.K. (1997). Investigation on clinical mastitis in cattle and buffaloes in western hills of Nepal. Working Paper Lumle Agri. Res. Centre, No. 97–11, p. 11.

Joshi, R.C., Chaudhary, P.G. and Bansal, R.P. (1977). Occurrence of cutaneous eruptions in rinderpest outbreak among bovine. *Indian Vet. J.* **54**: 871–873.

Joshi, S.C., Tomar, S.P.S. and Desai, R.N. (1969). Relative importance of maternal and environmental influences on pregnancy in buffaloes on military farms in the North. *Indian J. Dairy Sci.* **21**: 37.

Juneja, R.K. and Choudhary, R.P. (1971). Albumin polymorphism in some breeds of Indian cattle and water buffaloes. *J. Anim. Morphol. Physiol.* **18**: 186–181.

Juyal, P.D., Sahai, B.N., Srivastava, P.S. and Sinha, S.R.P.(1982). Heavy sarcocystosis in the ocular musculature of cattle and buffaloes. Vet. Res. Communication **5**: 337–342.

Juyal, P.D., Singla, L.D., Dhami, D.S., Walia, P.S. and Ahuja, S.P.(1998). Diagnosis and immunobiochemical basis of Trypanosoma evansi infection in cattle and buffaloes. *Indian J. Anim. Prod.* **30**: 152.

Kadu, M.S., Khire, D.W., Belorkar, R.M. and Kaikini, A.S. (1978). Studies on reproductive efficiency in Nagpuri (Berari) buffaloes. *Indian vet. J.* **55**: 94–98.

Kaikini, A.S. and Paragaonkar, D.R. (1969). Nagpuri buffalo. *Indian Dairyman* **21**: 47–48.

Kakkar, V.K., Ahuja, A.K. and Malik, N.S. (1983). Uromol impregnated wheat straw for buffaloes. *Indian J. Anim. Sci.* **53**: 650.

Kakkar, V.K., Malik, N.S. and Makkar, G.S. (1991). Protein requirement of male buffalo calves. *Indian J. Anim. Nutr.* **8**: 191–194.

Kalra, D.S., Bhatia, K.C., Gautam, O.P. and Chauhan, H.V.S.(1973). Chronic ergot poisoning like syndrome in buffaloes and cattle in Haryana and Punjab states. *Indian Vet. J.* **50**: 484–486.

Kalsi, J.S. and Dhillon, J.S. (1982) Performance of buffaloes in the first three lactations. *Indian J. Dairy Sci.* **35**: 218–19.

Kalsi, J.S. and Dhillon, J.S. (1984). Estimates of phenotypic and genetic parameters of some measures of life-time production in buffaloes. *Indian J. Dairy Sci.* **37**: 269–71.

Kamel, Y.Y., Ahmad, M.A. and Ismail, A.A. (1977). Dermatophytoses in animals, birds and man. I. Animals as potential reservoir of dermatophytes to man. *Assiut Vet. Med. J.* **4**: 149–159.

Kanameda, M., Ekgatat, M., Pachimasiri, T., Wongkashmenchit, S., Sirivea, C., Kongkrong, C., Apiwatanakorn, B., Naronwanichagan, W., Shoya, S. and Boontarat, B. (1997). The pathology of bovine tuberculosis in swamp buffaloes (Bubalus bubalis). *Buffalo J.* **13**: 351–362.

Kanaujia, A.S. (1978) Genetic and economic investigations in buffaloes. Thesis Abstr., Haryana Agric. Univ. **4 (88)**: 281–283.

Kanaujia, A.S. and Singh, D. (1975). Factors affecting some production traits in Indian buffaloes. *Indian J. Dairy Sci.* **28**: 57–62.

Kanaujia, A.S., Balaine, D.S. and Rathi, S.S. (1974). Factors affecting some economic traits of reproduction in Indian buffaloes. *Indian J. Dairy Sci.* **27**: 264–70.

Kanaujia, A.S., Balaine, D.S. and Rathi, S.S. (1975). Effect of some environmental and physiological factors on reproduction and production traits in Indian buffaloes. Haryana agric. *Univ. J. Res.* **5**: 179–87.

Kannan, A., Sihag, S., Yadav, K.R. and Khirwar, S.S. (2000). Effect of replacing mustard cake with decoiled sunflower cake on intake and digestibility of nutrients in lacating Murrah buffaloes. *Indian J. of Anim. Nutrition.* **17(3)**: 217–221

Kapoor, S., Singh, A. and Pandey, R.(1996). A dot immunobinding assay for detection of rotavirus in faeces of neonatal bovine calves. *Indian J. Virol.* **12**: 119–121.

Kapur, M.P. and Singh, R.P.(1978). Studies on clinical cases of mastitis in cows, buffaloes and goats in Haryana State. *Indian Vet. J.* **55**: 803–806.

Karaivanov, C., Vlakhov, K., Petrov, M., Kachava, D., Alexiev, A. and Polikhronov, A. (1987). Superovulation in buffaloes with FSH (in Bulgaria). *Vet. Med. Sci.* **24**: 57.

Karam Shah, S., Ahmad, M. and Shah,I.H. (1979). Effect of cooling during summer months on milk production of Nili-Ravi buffaloes. *J. anim. Sci. Pakistan* **1**: 35–40.

Karthikeyan, M.K. Iyue,M., Kandazamy, N. and Panneerselvam,S.(2002). Characteristics and performance of Toda buffaloes of the Nilgiris,India:I. Habitat, morphology and morphometry. *Buffalo Journal.* **18 (3)**: 303–313.

Kasarilikar, V.R., Kumar, S.P., Udupa, K.G., Shivaprakash, B.V., Patil, N.A. and Mutnal, A.B. (1999). Comparative efficacy of levamisole, piperazine and albendazole against ascariasis in buffalo calves. *Indian Vet. J.* **76**: 150–152.

Kasiraj, R., Reddy, N.S.R., Rao, M.M., Misra, A.K. and Pant, H.C. (2001) Embryo transfer in the buffalo under field conditions. *Bubalus Bubalis* 7(**4**): 75–82.

Kassem, M.H. and Soliman, K.N. (1966). Diseases of the buffaloes. In: International Encyclopedia of Veterinary Medicine. Sir Thomas Dalling (editor), W. Green and Sons, Edinburgh, pp. 527–533.

Katra, D.S. and Dhanda, M.R. (1964). Incidence of mastitis in cows and buffaloes in North-west India. *Vet. Rec.* **76**: 219–22.

Kaura, Y.K. and Sharma, V.K.(1981). Some ecological aspects of Samonella infection in young buffalo calves. *Indian J. Anim. Sci.* **51**: 415–418.

Kaushal, J.R., Kochar, A.S. and Chopra, A.K. (1972). Utilization of residual sugars of factory bagasse as energy source of urea-supplemented feeding in ruminants. *Indian J. Anim. Sci.* **42**: 399.

Kearl, L.C. (1982). Nutrient requirements of ruminants in the developing countries. International Feedstuffs Institute, Utah Agriculture Experiment Station, Utah State University, Logan, UT. USA.

Kendall, S.B.(1953). Report to the Government of Pakistan on fascioliasis. Rome, FAO/ETAP Report No. 83.

Kendall, S.B.(1954). Fascioliasis in Pakistan. Annals Trop. Med. Paraitol. 48: 307–313.

Kesava Rao, V. (1988). Effect of Procesing and storage on buffalo meat lipids. Ph.d. thesis. Indian Veterianry research Instiute, Izatnagar.

Keshava Rao, V. and Kowale, B.N. (1986). In Advances in meat research, Khot, J.B.,Sherikar, A.T., Jaya Rao, B.M. and Pillai, S.R.(Eds.), Bombay, India: Red and Blue Cross Publishers.

Keskar, D.V., Venkataraman, R., Srinivasan, S.R. and Dhanapalan, P. (1995). Hepatic dysfunction in neonatal buffalo calves and its treatment. *Indian Vet. J.* **72**: 1170–1173.

Khalacheva, M. and Sherkov, S.H. (1981). Outbreak of leptospirosis among buffaloes in Bulgaria. Veterinarna Sbirka. **79**: 33–36.

Khan, A. and Anjum, A.D. (1994). Liver paramphistomiasis in a buffalo. *Buffalo J.* **10**: 185–188.

Khan, A. and Khan, M.Z. (1996). Neonatal calf mortality in Pakistan. I. Prevalence and factors influencing mortality in buffalo and cow neonates. *Buffalo J.* **12**: 219–229.

Khan, A. and Khan, M.Z. (1997). Bacteria isolated from natural cases of buffalo and bovine neonatal calf diarrhoea, pneumonia and pneumoenteritis. Veterinarski-Archiv. 67: 161–167.

Khan, H.Z. (1964). Listeriosis in Pakistan. Bull. Off. Int. Epizoot. 61: 429–436.

Khan, M.A. (1986) 'Genetic analysis of a purebred herd of Nilli-Ravi buffalo'. Ph.D. Thesis, Department of Animal Breeding and Genetics, University of Agriculture, Faisalabad, Pakistan.

Khan, M.A. and Ahmed, M.D. (1991). *Indian J. Anim. and Pl. Sci.* **1(4)** 203–206.

Khan, M.Z., Mohammad, G. and Sabri, M.A.(1981). Deg nala like disease in buffalo calf. *Pakistan Vet. J.* **1**: 122–123.

Khanna, B.M., Kapur, M.P. and Gupta, S.L.(1995). Clinical trial of Tiamutin R for bovine mastitis treatment. *Indian J. Vet. Res.* **4**: 33–35.

Khanna, N.D. (1969). Studies on transferrin polymorphism in Indian water-buffaloes. *Indian J. Anim. Sci.* **39**: 21–26.

Khanna, N.D. (1973). Biochemical polymorphism and blood groups in Indian buffaloes (Bubalus bubalis). Dr. Met. Vet. Thesis pp. 154.

Khanna, N.D. and Braend, M. (1968). Haemoglobin and albumin polymorphisms in Indian water buffaloes. Acta Vet. Scand. 9: 316–27.

Khanna, N.D. and Tandon, S.N. (1978). Red cell carbonic anhydrase polymorphism in Indian buffaloes. *Indian vet. J.* **57**: 863–64.

Khanna, N.D., Chet Ram, Majumdar, N.K. and Tandon, K.N. (1968). Studies in bovine blood groups. 5. Breed differences in the concurrence of blood antigenic factors in Indian water buffaloes. *J. Anim. Morph. Physiol.* **15**: 46–52.

Kharole, M.U., Gupta, P.P., Singh, B., Mandal, P.C. and Hothi, D.S. (1976). Phycomycotic gastritis in buffalocalves (Bubalus bubalis) Vet. Path. **14**: 409–416.

Knox, M.R., Kennedy, P.M., Hennessy, D.R., Steel, J.W. and Le Jambre, L.F.(1994). Comparative pharmacokinetics of fenbendazole in buffalo and cattle. Vet. Res. Communications **18**: 209–216.

Kohli, M.L. and Malik, D.D. (1960). Effect of service period on total milk production and lactation length in Murrah buffaloes. *Indian J. Dairy Sci.* **13**: 105–11.

Kondaiah, N. and Panda, B. (1987). A semi modern abattoir concept for hygienic meat production. *Indian Fd. Industry* **6: 58–61.**

Kondaiah, N., Anjaneyulu, A.S.R., Kesava Rao, V. and Sharma, N. (1986a). Effect of different handling conditions on quality of minced buffalo meat. *Indian J. Anim. Sci.* **56:** 677–679.

Kondaiah, N., Anjaneyulu, A.S.R., Kesava Rao, V., Sharma, N. and Joshi, H.B. (1985). Effect of salt and phosphate on the quality of buffalo and goat meats. Meat Science **15**: 183–192.

Kondaiah, N., Lakshmanan, V., Anjaneyulu, A.S.R. and Sharma, N. (1983). Evaluation of aged and spent buffaloes for meat. *Indian J. Anim. Sci.* **53:** 1208–1212.

Kondaiah, N., Lakshmanan, V., Kesav Rao, V, and Sharma, N. (1981). Studies on type and conformation of slaughter buffaloes and carcass iin India. *Indian J. Anim. Sci.* **15: 107–111.**

Kondaiah, N., Lakshmanan, V., Sharma, N. and Kesava Rao, V.(1982). Meat quality traits in buffaloes. *Indina J. Anim. Sci. 52: 889–893.*

Kondaiah, N., Rao, V.K., Anjaneyulu, A.S.R. and Sharma, N. (1986b). Evaluation of variety meats form buffaloes for some quality parameters. *J. of Fd. Sci. and Technol.* **23: 113–115.**

Kosikowski, F. (1966) Cheese and Fermented Milk Food. Kosikowski: Ithaca, NY.

Krishnan, K.R. and Nagarcenkar, R. (1979) Studies on the influence of body weight at slaughter in male buffalo calves on the production and quality of meat. Annual Report, pp. 186, NDRI, Karnal.

Kulkarni, S.G., Khot, J.B., Sherikar, A.T., Jayrao, B.M., Pillai, S.R. and Sherikar, A.A. (1987). Microbial profiles of frozen buffalo meat cuts. In: advances in Meat Research. Pp. 182–190.

Kulshreshtha, R.C., Singh, J. and Chandiramani, N.K. (1980). A study on prevalence of tuberculosis and Johne's disease in cattle and buffaloes in Haryana State. Haryana Vet. 19: 139–141.

Kumar, A., Sharma, S.D. and Joshi, B.P. (1999). Studies on the epidemiology of bubaline foreign body syndrome. In Compendium: National Symposium on Sustainable Development in Animal Health Care Measures– Vision for the Future. Department of Medicine, Orrisa Veterinary College, Bhubaneswar, p. 8.

Kumar, B. (1984). Index method of sire evaluation in buffaloes. Ph. D. Thesis, Kurukshetra University, Kurukshetra, Haryana, India.

Kumar, B., Joshi, H.C. and Kumar, M.(1986). Biochemical changes in buffaloes naturally infected with microfilariae of Setaria cervi. *Indian J. Anim. Sci.* **56**: 1009–1012.

Kumar, G.S. and Suryanarayana, C.(1995). Clinico-biochemical and therapeutic studies on mange in buffalo calves. *Indian Vet. J.* **72**: 77–79.

Kumar, M. and Sharma, S.P. (1994). Haematological and biochemical studies in setariasis in buffaloes. *J. Appl. Anim. Res.* **5**: 41–46.

Kumar, M. (1985). Recent advances in the diagnosis of filariasis. In: Proceedings of Summer Institute on Recent Concepts in Preventive Veterinary Medicine and Planned Livestock Health Programme. G.B. Pant University of Agri. and Tech. Pantnagar, pp. 83–89.

Kumar, M., Maru, A. and Pachauri, S. P.(1983). Role of serum enzymes in evaluating flukicidal action of disophenol in Fasciola-affected buffaloes. *Br. Vet. J.* **139**: 262–264

Kumar, M., Joshi, H.C. and Garg, S.K. (1993). Prevalence of Setaria cervi in buffaloes of Kumaon region. *Indian J. Vet. Res.* **2**: 1–6.

Kumar, M., Joshi, H.C. and Varshney, K.C.(1991). Pathological changes in experimental Setaria cervi infection in rabbits. *Indian J. Anim. Sci.* **61**: 924–932.

Kumar, M., Pathak, K.M.L. and Pachauri, S.P.(1981). Clinico-haematological studies in scouring buffalo calves. *Indian Vet. Med. J.* **5**: 121–124.

Kumar, M., Pathak, K.M.L. and Pachauri, S.P.(1982). Clinico-pathological studies on naturally occurring bovine fascioliasis in India. *Br. Vet. J.* **138**: 241–246.

Kumar, P. (1971) Ph.D. Thesis, Agra University, Agra.

Kumar, R and Dass, R.S.(2005). Effect of niacin supplementation on rumen metabolites in Murrah buffaloes (Bubalus bubalis). *Asian Australia J. of Anim. Sci.* **18(1)**:38–41.

Kumar, R. and Bhat, P.N. (1978). Effect of some non-genetic sources of variation on milk yield in Indian buffaloes. *Indian J. Anim. Sci.* **48**: 639–42.

Kumar, R. and Bhat, P.N. (1978). Effect of non-genetic factors on lactation length in Indian buffaloes. *Indian J. Anim. Sci.* **48**: 559–62.

Kumar, R. and Bhat, P.N. (1979). Lactation curve in Indian buffaloes. *Indian J. Dairy Sci.* **32**: 156–60.

Kumar, R., Bhat, P.N, Dwivedi, I.S.D. and Garg, R.C. (1979). Methods of predicting complete elactation production from part lactation yields in Indian buffaloes. *Indian J. Anim. Sci.* 49: 735–38.

Kumar, R., Bhat, P.N. and Dwivedi, I.S.D. (1979). Inheritance of part lactationmilk yield in Indianbuffaloes. *Indian J. Anim. Sci.* **49**: 687–90.

Kumar, R., Bhat, P.N. and Garg, R.C. (1978). Inheritance of various segments of lactation curve in Indian buffaloes. *Indian J. Anim. Sci.* **48**: 652–55.

Kumar, R., Bhat, P.N. and Garg, R.C. (1979). Presistency off lactation in buffaloes. Indian J. Dairy Sic. 32: 318–20.

Kumar, S., Bhat, P.N. and Bhat, P.P (1991). Restriction endonuclease analysis of buffalo satellite DNA. Indian J. Anim. Genet. and Breed. 13, 4–7.

Kumar, S., Bhat, P.N., Bhat, P.P. and Rasool, T.J. (1993) Analysis of buffalo lymphocyte natigen (BuLA) system by Southern blot hybridization. 1. Polymorphism of DR genes. *Indian J. Anim. Genet. and Breed.* **15**: 21–25.

Kumar, S., Gupta, J., Kumar, N., Dikshit,K., Navani,N., Jain,P. and Nagarajan, M. (2006). Genetic variation and relationships among eight Indian riverine buffalo breeds. Mol. Ecol. **15**: 593–600.

Kumar, V., Joshi, H.C. and Kumar, M.(1987). Alteration in some clinico-haematological and biochemical values in buffaloes naturally infected with microfilariae of Setaria cervi. *Indian J. Vet. Med.* **7**: 5–9.

Kumar, C.R.,Moorthy, P.R.S., Rao, K.S., Naidu,K.V. (2002). Effect of year and season of birth on calf mortality in pure and corssbred cattle and buffaloes. *Indina J. Anim.Sci.* **72(10)**: 918–920

Kurar, C.K. and Mudgal,V.D. (1981) *Indian J. Anim. Sci.* **51**: 317.

Kushwaha, N.S., Mukherjee, D.P. and Bhattacharya, P. (1955). Seasonal variations in reaction time and semen quality of buffalo bulls. *Indian J. vet. Sci.* **25**: 317–28.

Kwatra, M.S.(1980). Mycotoxicosis as a possible cause of gangrenous syndrome in buffaloes- a preliminary report. *Indian J. Anim. Hlth.* **19**: 67–69.

Laipis, J.P., Wilcox, C.J. and Hauswirth,W.W. (1982). Nucleotide sequence variation in mito-chondrial DNA from bovine liver. *J. Dairy Sci.* **65**: 1655–1662.

Lakshmanan, V. and Pathak, N.N. (1979). Annual Scientific report of Livestock Poducts Tecnology Division, IVRI, Izatnagar. pp. 5.

Lakshmanan, V., Joshi, H.B. and Kondaiah, N. (1986). Annual Scientific report of Livestock Products Technology Division, IVRI, Izatnagar.

Lakshmanan, V., Kondaiah, N. and Anjaneyulu, A.S.R.(1984). Variety meats of buffalo. *Buffalo Bulletin* **3(1)**: 6–8 and 17.

Lakshmanan, V., Kondaiah, N., Joshi, H.B. and Sharma, N. (1985). Annual Scientific Report of Livestock Products Technology Division, IVRI, Izatnagar.

Lal, S.M. and Singh, I.P.(1973). Buffalo pox virus. Archges. Virus forsch. 40: 390–391.

Lall, D, Sikka, P., Khanna, S. and Chauhan, T.R. (2004). Effect of high plane of nutrition and mineral supplementaiton on blood mineral profile, tyroid hormones and fertility status of anoestrus buffaloes. Animal Nutrition and Feed Technology. 4(1):71–76

Lall, H.K. (1946). Tuberculin testing in buffaloes with a description of some differences in the macroscopic lesions of the cow and the buffalo. *Indian J. vet. Sci.* **16**: 24–6.

Lall, H.K. (1971). Veterinarian and the Law, 1st edn. Metropolitan Book Co. (Pvt.) Ltd., Delhi-6, India.

Lall, H.K. (1975). Study of economic characters in Murrah buffaloes. *Indian vet. J.* **52**: 337–44.

Lall, J.M. (1969). Tuberculosis among animals in India. *Vet. Bull.* **39**: 385–390.

Lall, J.M., Singh, G. and Sengupta, B.R. (1967). Incidence of tuberculosis among cattle and buffaloes in India. *Indian J. Anim. Sci.* **39**: 51–58.

Langar, P.N., Makkar, G.S. and Bakshi, M.P.S. (1984). Comparative studies on urea and fibre utilization in buffalo and cattle. *Indian J. Anim. Sci.* **54**: 413.

Langar, P.N., Malik, N.S. and Chopra, A.K. (1982). Nitrogen feeding trials with uromol–based concentrate diet on the milking buffaloes. *Indian J. Anim. Sci.* **52**: 53.

Langar, P.N., Sidhu, G.S. and Bhatia, I.S. (1968). Study of the microbial population in the rumen of buffaloes (Bos bubalis) and zebu (Bos indicus) on a feeding regimen deficient in carbohydrates. *Indian J. vet. Sci.* **38**: 333.

Latifova, E. (2000). Case study on buffalo recording. Breeding and management strategies of buffalo in Azerbaijan. Proc. workshop on animal recording and management strategies for buffaloes. ICAR technical series, 4: 45–50.

Lazarus, A.J. (1946). Buffalo as a dairy animal. *Indian Fmg.* 7: 247–50.

Lazary, S., Larson Glatt, P.A., Gerber, H., Kristensen, B. and Bakke, S. (1985). ELA in horses afflicted with sweet itch. Anim. Genet. 16: 93.

Le, H.D., Pham, S.L., Phan, S.L., Phan, D.L. and Nguyen, T.C. (1997). Some remarks on the clinical picture of young buffaloes and calves infected with Toxocara vitulorum and methods of treatment. Khoa Hoc Ky Thuat Thu Y. **4**: 54–58.

LEAD (Livestock, Environment and Development Initiative) (2004). Livestock and environment toolbox. FAO Rome, Italy. Accessed Feb.

Li, Y.Z. (1981). Control of coccidiosis in buffalo-calves. *Chinese J. Vet. Med.* **7**: 17–18.

Liggitt, H.D., DeMartini, J.C., Storz, J. and Coulter, G.R. (1980). Synovitis and bovine syncytial virus isolation an experimentally induced malignanat catarrhal fever. *J. Comp. Pathol.* **90**: 519–533.

Linford, E., Glover, R.A., Bishop, C. and Stewart, D.L. (1976). The relationship between semen evaluation methods and fertility in bulls. *J. of Reprod. and Fertility.* **47**: 283–291.

Littlewood R.W. 1936. Livestock of southern India. Superintendent, Government press, Madras. VIII: 23p

Liu Cheng Hua, Chang Shun Shu and Huang Hei Pein. (1985). The Chinese indigenous buffalo and its crossbreeding efficiency. *Buffalo J.* **1**: 9–18.

Liu Cheng Hua. (1978). The Improvement of Water Buffalo in China. China Buffalo Technical, China Buffalo Breeding Association.

Liu, Z., Zhao, J., Ma, L.H. and Yao, B.A. (1997). Babesia orientalis sp. soro- a parasite of buffaloes in China. Acta Vet. Zootech. Sinica 28: 84–89.

Lohan, I.S, Malik,R.K., Saini,M.S., Dhanda,O.P. and Baljit Singh (2000). Uterine and ovarian changes during the early postpartum period in Murrah buffaloes. *Buffalo Bulletin* **19 (1)**: 20–23.

Lohr, K.F., Pholpark, S., Srikitjakarn, L., Thaboran, P.,·Bettermann, G. and Staak, C. (1985). Trypanosoma evansi infection in buffaloes in North East Thailand. I. Field investigations. Trop. Anim. Hlth. Prod. 17: 121–125.

Loses, G.I. (1980). Diseases caused by Trypanosoma evansi. A Review. Vet. Res. Communications 4: 165–181.

Ludng, V.H., Vo. T.H., Vo. T.D., Nguyen, Q.T. and Tran, N.M.(1997). Prevalence of Fasciola gigantica infection and results of treatment of buffaloes and cattle in 11 provinces of southern Vietnam. Khoa Hoc Ky Thuat Thu Y. 4: 53–57.

Luktuke, S.N. (1957). (fide Bhattacharya, P. 1958. Indian Sci. Cong. Madras).

Luktuke, S.N. and Ahuja, L.D. (1961). Studies on ovulation in buffaloes. *J. Reprod. Fert.* **2**: 200–1.

Luktuke, S.N. and Roy, D.J. (1964). Studies on post-partum oestrus in Murrah buffaloes. *Indian J. vet. Sci.* **34**: 166–70.

MacCluer, J.W., Bailey, E., Weitkamp, L.R. and Balangero, J. (1988). ELA and fertility in American standardbreed horses. Anim. Genet., 19: 359–372.

MacGregor, M.R. (1941). The domestic buffalo. Vet. Rec. **53**: 443–50.

Madan, M.L., Singla, S.K. and Manik, R.S. (1994).*In vitro*production and transfer of embryos in buffaloes. *Theriogenology* **41**:139–143.

Madan, M.L., Singla, S.K., Ambrose, J.D., Prakash, B.S. and Jailkhani S. (1989). Embryo transfer technology in buffaloes. Endocrine responses and limitation. Annual Report of Embryo Biotechnology Centre, National Dairy Research Institute, Karnal, India.

Madhawi, D.L., Srinivasan, K.s., Kadgol, S.B. and Baliga, B.R. (1982). *J.Fd. Sci. and Technol.* **19**:214.

Magid, S. (1996). Buffalo population and production in Iraq. *Buffalo Newsletter* **6**: 6–7.

Mahadevan, P. (1992). Distribution, Ecology and Adaptation. In: Buffalo Production. World Animal Science C6. Tulloh, N.M. and Holmes, J.H.G. (editors), Elsevier, Amsterdam, pp. 1–12.

Mahanta, K.C. (1964b). Some technological aspects of mumiss production. *Ind. J. Dairy Sci.* **17**: 51–54.

Mahanta, K.C. (1964a). Handbook of Dairy Science. Kitabistan, Allahabad.

Mahapatra, S., Kar, B.C. and Mishra, P.R.(1996). Occurrence of mycotic mastitis in buffaloes of Orissa. *Indian Vet. J.* **73**: 1021–1023.

Mahmound, I.N. (1952). Some characteristics of the semen of Egyptian buffaloes. Bull. Fac. Agric. Found. I Univ. (Cairo), No. 15, 16 pp. (fide Anim. Breed. Abstr. 21: 260 1953).

Makaya, G., Cortadas, J. and Bernardi, G. (1978). Analysis of the bovine genome by density gradient centrifugation. *Eur. J. Biochem.* **84**: 179–88.

Makkar, G.S., Kakkar, V.K. and Ahuja, A.K. (1989). Uromin lick. A boon for drought areas. Indian Fmg, 5: 34–35.

Makkar, G.S., Kakkar, V.K., Bhullar, M.S. and Malik, N.S. (1984). Potato waste as a substitute of cereal grains in the rations of buffalo calves. *Indian J. Anim. Sci.* **54**: 1060.

Malik, N.S. and Chopra, A.K. (1977). Uromol as a source of dietary nitrogen for milch buffaloes. *Indian J. Anim. Sci.* **47**: 829.

Malik, N.S. and Chopra, A.K. (1978). Effect of feeding uromol along with formaldehyde treated groundnut meal in the concentrate mixtures on the digestibility of nutrients and nitrogen retention in buffalo calves. *Indian J. Anim. Sci.* **48**: 503.

Malik, N.S. and Chopra, A.K. (1980). Urinary nitrogen excretion in buffalo calves as affected by urea and uromol feeding at different levels of nitrogen intake. *Indian vet. med. J.* **4**: 181.

Malik, N.S. and Chopra, A.K. (1981). A modified 3–stage in vitro digestion of groundnut and mustard cake proteins protected by a quick oven–drying method of formaldehyde treatment. *Indian J. Anim. Sci.* **51:** 1168.

Malik, N.S. and Makkar, G.S. (1979). Processing and nutritive value of rice–bran-mixed uromol in ruminant diets. *Indian J. Anim. Sci.* **49**: 227.

Malik, N.S., Langar, P.N. and Chopra, A.K. (1978a). Uromol as a source of dietary nitrogen for ruminants: Metabolic and rumen fermentation studies on buffalo calves. *J. Agric. Sci. (Cambridge)*. **91**: 302.

Malik, N.S., Langar, P.N. and Chopra, A.K. (1978b). A note on the comparative growth studies of buffalo calves fed groundnut cake, urea-molasses and uromol-based concentrate mixtures. *Indian J. Anim. Sci.* **48**: 219.

Malik, N.S., Makkar, G.S. and Ichhponani, J.S. (1980). Effect of feeding wheat straw-based poultry litter on the digestibility of nutrients, growth rate and puberty age of buffalo heifers. *Indian J. Anim. Sci.* **50**: 938.

Mallick, K.P. and Dwivedi, S.K.(1981). A note on blood glucose level in clinical cases of bovine surra. *Indian Vet. J.* **58**: 162–163.

Malviya, H.C.(1972). Stephanofilarial infection in cattle and buffalo in Andaman Islands. *Indian J. Helminth.* **24**: 68–71.

Malviya, H.K. and Prasad, J.(1977). Ephemeral fever a clinical and epidemiological study in cross-bred cows and buffaloes. *Indian Vet. J.* **54**: 440–444.

Mamatelashvili, V.G.(1973). Review of studies on brucellosis in buffaloes. Shornik Trudov Gruzinskii Zootekhnich Esko-Veterinary Institut. pp. 55–59.

Mandal, G.P., Kurar, C.K. and Neelam Kewalramani (2002). Effect of methionine and lysine supplementation on dry matter intake, nutrient utilization and growth in buffalo caves. Animal Nutrition and Feed Technology. 2(1):39–47.

Mandal, P.C., Gill, B.S. and Grewal, B.S.(1981). Spontaneous pneumonia in a she-buffalo caused by Absidia corymbifera. *J. Res. P.A.U.* **18**: 452–456.

Manfredini, L. and Marangon, S.(1992). Psoroptic mange in buffaloes caused by Psoroptes natalensis. Obiettivi e Documenti Veterinari **13**: 65–68.

Mangrurkar, B.R. and Desai, R.N. (1981). Inter-relationship between heiferhood, body weight and first lactation performance in buffaloes. *Indian vet. J.* **58**: 199–202.

Manickam, A., Palaniswami, A. and Neelakantha, S.(1985). Mycotoxicosis in buffaloes. *Indian Vet. J.* **62**: 711–714.

Manji, B.P., Dwivedi, P. and Rao, J.R.(1998). Rocket immuno-electrophoresis analysis of ruminal and hepatic paramphistomid antigens. *Indian J. Vet. Res.* **7**: 1–7.

Mansjoer, M.(1961). Anthrax in man and animals in Indonesia. Communication Vet. 5: 61–67.

Maqbool, A. and Ifran, M.(1983). Effect of different fasciolicides against fascioliasis in buffaloes. *Pakistan Vet. J.* **3**: 70–72.

Maqbool, A., Arshed, M.J., Mahmood, F. and Hussain, A.(1994). Epidemiology and chemotherapy of fascioliasis in buffaloes. *Assiut Vet. Med. J.* **30**: 115–123.

Maqbool, A., Khan, M.A., Arshad, M.L., Aslam, M.S. and Mahmood, F.(1996). Epidemiology, chemotherapy and chemoprophylaxis of deg nala disease in buffaloes and cattle. *Indian J. Dairy Sci.* **49**: 643–647.

Maqbool, A., Raza, A., Saleem, A. and Mahmood, F.(1998). Effect of sodium hydroxide treated dry and mouldy rice straw on fungal growth, weight gain and blood picture in deg nala disease in buffaloes. *Indian Vet. Med. J.* **22**: 35–36.

Markandeya, N.M. and Bharkad, G.P. (2002). Effect of progesterone primed gonadotropin releasing hormone (GnRH) on induction of pubertal oestrus in buffaloes during different seasons. *Indian Vety. J.* B **79 (1):** 94–95.

Marmarian, Y. (2000). Buffalo breeding in Armenia. Proc. Workshop on Animal Recording and Management Strategies for Buffalo. ICAR Technical Series, **4**: 73–76.

Marwaha, C.L. (1974). Genetic, phenotypic and environmental relationships between body weights at different ages and milk production in buffaloes. M.Sc. Thesis, Punjab Agricultural University, Ludhiana.

Maske, D.K. and Ruprah, N.S.(1981). Note on the bionomics of psoroptic mange in buffaloes. *Indian J. Anim. Sci.* **51**: 494–497.

Matharu, J.S and Singh, Mehar (1980).Revivability of buffalo spermatozoa after deep freezing the semen using various extenders. Zantralblatt fur veterinarmeizin.A.27:385

Mc Cool, C.J. and Newton Tabret, D.A.(1979). The route of infection in tuberculosis in feral buffalo. *Aust. Vet. J.* **55**: 401–402.

Mc Nab, W., Muk, A.H., Duncan, J.R., Brooks, B.W., Van Dreumel, A.A., Martin, S.W., Nielsen, K.H., Sugden, E.A. and Turcotte, C.(1991). An evaluation of selected screening tests for bovine paratuberculosis. *Canadian J. Vet. Res.* **55**: 252–259.

Mc Williams, P.S., Searcy, G.P. and Bellamy, J.E.C.(1982). Bovine post-parturient haemoglobinuria: A review of literature. *Canadian Vet. J.* **56**: 309–312.

Meena, S.M. and Jat, R.P. (2003). Effect of some non-genetic facgtors on selective value of Surti buffaloes. *Indian J.Dairy Sci.* **56 (5)**: 323–326.

Meenakshi Gupta and Nangia, O.P. (1998). Effect of ammonia concentration on cellulase and transminases activity in the rumen of buffalo claves. Haryana Veterinarian. 37:15–19.

Mehmood, A., Anwar, M. and Javed, M.H. (1989). Superovulation with PMSG beginning on three different days of the cycle in Nili–Ravi buffaloes (Bubalus bubalis). Buffalo 5: 79–84.

Mehra, U.R., Nath, K., Ranjhan, S.K. and Chetal, U. (1976). A note on the difference between blood urea of buffalo and cow calves. *J. Agric. Sci.* **86:** 45–48.

Mendl, M., Paul, E.S. (2004). Consciousness, emotion and animal welfare: insights from cognitive science. Animal Welfare 13: S17–25.

Mephan, B. (1996). Ethical analysis of food biotechnologies: an evaluative framework. In: Food Ethics, pp. 101–19. Routledge, London.

Mercier, J.C. and Villote, J.L. (1997). The modification of milk protein composition through transgenics: progress and problems. In: Transgenic animals - Generation and Use. Ed. By L.M. Houdebine. 473–482. Harwood Academic Publishers GmBH Amsterdam.

Mercken, L., Simons, M.J., Swillens, S., Massaer, M. and Vassart. G. (1985). Primary structure of bovine thyroglobulin deduced from the sequence of its 8431 base complementary DNA. Nature, (Lond.), 316: 647–651.

Meredith, D., Elser, A.H., Wolf, B., Soma, L.R., Donawick, W.J. and Lazary, S. (1986). Equine leucocyte antigens relationship with sarcoid tumors and laminitis in two pure-breds. Immunogenet. 23: 221–225.

Miettinen, A., Husu, J. and Tuomi, J.I. (1990). Serum antibody response to Listeria monocytogenes, listerial excretion and clinical characteristics in experimentally infected goats. *J. Clin. Microbiol.* **28**: 340–343.

Minhas, K.S., Sidhu, J.S., Mudahar, G.S. and Singh, A.K. (2002). Flow behaviour characteristics of ice cream mix made with buffalo milk and various stabilizers. Plant Foods for Human Nutrition, **57 (1)**: 25–40.

Misra, A.K., Joshi, B.V., Agrawala, P.L., Kasiraj, R., Sivaiah, S., Rangareddi, N.S. and Sidiqui, M.U. (1990). Multiple ovulation and embryo transfer in Indian buffaloes (*Bubalus bubalis*). Theriogenology. 33, 1131–1142.

Misra, A.K., Joshi, B.V., Kaisraj, R., Sivaiah, S., and Rangareddi, N.S. (1991). Improved superovulatory regimen for buffalo. Theriogenology 35: 245.

Misra, A.K., Trivedi, K.R., Kasiraj, R., Rao, M.M., Rangareddy, N.S. and Pant, H.C. (2001). Effect of superovulation and embryo colelction on milk production in the buffalo (*Bubalus bubalis*).*Buffalo J.* **17**(**2**):263–267.

Misra, A.K., Yadav, M.C. and Motwani, K.T. (1988). Successful embryo transfer in a buffalo. (Bubalus bubalis). Proc. II World buffalo Congress, New Delhi, India **Vol. I**: 56.

Misra, B.S., Raizada, B.C. and Tewari, R.B.L. (1970). Gestation length, birth weight of calves and body weight changes during pregnancy in buffaloes bred during summer. *Indian J. Anim. Sci.* **40**: 103–109.

Misra, D.S.(1976). Bacterial buffalo mastitis. *Pantnagar J. Res.* **1**: 76–69.

Misra, S.P.(1969). Treatment of ear sore in buffaloes. Correspondence. *Indian Vet. J.* **46**: 632.

Mohan, R.N.(1968). Diseases and parasites of buffaloes. *Vet. Bull.* **38**: 567–576.

Mohanty, P.K. and Rai, A.(1990). Rapid detection of antibodies against buffalo pox virus by using counter immuno electophoresis. *Indian J. Virol.* **6**: 93–97.

Mokhtar, S.A. (1971). Study on some economic traits of Holstein Friesian cattle in U.A.R. M.Sc. Thesis, Fac. Agriculture, Ain Shams Univ., Cairo, Egypt.

Molina, H.A., Pena, F.de-la. Camer, G., Padilla, M.A. and De-la-Pena, F.(1994). An outbreak of haemorrhagic septicemia among carabos in Northern Samar. *Philippine J. Vet. Med.* **31**: 27–31.

Momongan, V.G. (1985) Reproduction in draught animals. In: Draught Animal Power for Production. ACIAR Proceedings Series No **10**: 123–129.

Momongan, V.G. Parker, B.A. and Sarabia, A.S. (1994). Crossbreeding of buffaloes in the Philippines. Proc. First Asian Buffalo Ass. Cong., BPRADEC, Bangkok. pp. 57–67.

Momongan, V.G., Palad, O.A., Singh, M., Sarabia, A.S., Chiong, R.D., Nva, Z.M., Obsima, A.R. and Del Barrio, A.N. (1984). Reproductive status and synchronization oestrus for predetermined time of insemination of Philippine carabaos (swamp buffalo) raised by small-holder farmers. In: The Use of Nuclear Techniques to Improve Domestic Buffalo Production in Asia. International Atomic Energy Agency, Vienna, pp 43–50.

Mondal, M.M.H.(1994). On the ectoparasites of buffaloes in Bangladesh. Proceedings 4th World Buffalo Congress, **Vol. 2**: 319–321.

Monga, D.P. and Kalra, D.S.(1971). Prevalence of mycotic mastitis among animals in Haryana. *Indian J. Anim. Sci.* **41**: 813–816.

Moran, J.B. (1978). Digestibility and water metabolism in Brahman cross, buffalo, Banteng and Shorthron steers fed roughage diets. *J. Augst. Inst. Agric. Sci.* **44**: 57–59.

Moran, J.B., Norton, B.W. and Nolan, J.V. (1979). The intake, digestibility and utilization of a low quality roughage by Brahman cross, buffalo, Banteng and Shorthron steers. *J. Augst. Inst. Agric. Sci.* **30**: 333–340.

Mouli, S.P.(1991). Gangrenous syndrome in bovines. *Indian Vet. J.* **68**: 1075–1077.

Muhammad, G., Athar, M., Rehman, F., Shakoor, A. and Khan, M.Z.(1995). Clinical and therapeutic aspects of Escherichia coli mastitis in water buffaloes. *Indian J. Dairy Sci.* **48**: 581–586.

Muhammad, G., Khan, M.Z., Athar, M. and Rehman, S.(1998). Clinico-epidemiological and therapeutic observations on pox outbreak in small holder dairy farms. *Buffalo J.* **14**: 259–267.

Muhammad, G., Lodhi, L.A., Athar, M. and Fazal-ur-Rehman.(1997). Evaluation of cephalexin in the treatment of clinical mastitis in buffalo. *Indian J. Dairy Sci.* **50**: 205–208.

Mukherjee, D.P. (1964). Melanizing activity of semen and its relation to live spermatozoa of bulls, goats and rams. *J. Reprod. Fertil.* **7**: 29–36.

Mukherjee, D.P. and Bhattacharya, P. (1953). Seasonal variations in semen quality and haemoglobin and cell volume contents of the blood in bulls. *Indian J. vet. Sci.* **22**: 73–91.

Mullick, D.N. (1960). Effect of humidity and exposure to sun on the pulse rate, respiration rate, rectal temperature and haemoglobin level in different sexes of cattle and buffalo. *J. agric. Sci. (Camb)* **54**: 391–4.

Multani, A.S. and Ahuja, A.K. (1985). Sulphur requirements of growing buffalo calves through different sources of sulphur. *Indian J. Anim. Nutr.* **2**: 17.

Mustafa, A.A., Ghalib, H.W. and Shigidi, M.T. (1978). Carrier rate of Pasturella multocida in cattle herd associated with an outbreak of haemorrhagic septicemia in Sudan. *British Vet. J.* **134**: 375–378.

Muthukar, M., Naveena, B.M, Sen, A.R and Babji, Y. (2005). Assessment of quality attributes and shelf life of buffalo haleem under refrigerated storage. *Buffalo J.* **21(1)**: 1–8

Naderfard, H.R. and Qanami, A. (1997). Buffalo breeding in Iran. Proc. Fifth Buffalo Congress, Caserta, 13 to 16 Oct 942–943.

Nagarcenkar, R and Sethi, R.K.(1981). Association of adaptive traits with performacne traits in buffaloes. *Indian J. anim. Sci.* **51**(**12**): 1121–1123.

Nagarcenkar, R. (1978). Riverine buffaloes of India and possibilities of genetic improvement vis-a-vis cattle. Procedings of FAO/SIDA/GOI/Seminar on Reproduction and Artificial Insemination of Buffaloes, NDRI, Karnal.

Nagpal, M.C., Gautam, O.P. and Gulati, R.L.(1968). Haemoglobinuria in buffaloes. *Indian Vet. J.* **45**: 1048–1058.

Nagpaul, P.K., Rao, M.V.N., Singh, C.B. and Batnagar, D.S. (1984). Draught efficiency of crossbred vis-a-vis zebu and buffalo bullocks in India. Wld Rev. Anim. Prod. **20**: 51–55.

Nagra, S.S. and Langar, P.N. (1984). Cotton textile effluent waste as possible source for chemical treatment of straws agaricultural wastes. 11: 91.

Naik, S.N., Sukumaran, P.K. and Sanghvi, L.D. (1969). Haemoglobin polymorphism in Indian zebu cattle. Heredity 24: 239–47.

Narayana Reddi, M.L. (1939). The buffalo in the Madras Presidency. *Madras Agri. J.* **27**: 50–58.

National Buffalo Project, (1988). Ministry of Jihad, Teheran, Iran.

Nautiyal, L.P. and Bhat, P.N. (1977). Growth curve in Indian buffaloes. *Indian J. Anim. Sci.* **47**: 517–20.

Nautiyal, L.P. and Bhat, P.N. (1979). Effect ofvarious factors on body weights in Indian buffaloes. *Indian J. Anim. Sci.* **49**: 979–83.

Naveena, B.M., Mendiratta,S.K. and Anjaneyulu, A.S.R.(2004). Tenderization of buffalo meat using plant proteases from CucumisTrigonus Roxb (Kachri) and Zingiber officinate roscoe (Ginger rhizome). *Meat Science* **68** (**3**): 363–369.

Navjot Kaur, Srivatava, A.K. and Bal, M.S.(2003). Subacute oral toxicity of chlorpyrifos in buffalo calves (Bubalus bubalis). *Indian J. Vet. Res.* **12** (**1**): 34–38.

Nayak, S. and Mishra,M. (1984). Dairy temperament of Red Sindhi, crossbred and Murrah buffaloes in relation to their milkingability and composition. *Ind. J. DairySci.* **37**: 20–23.

NDRI. (2006). Compendium of Lectures in Advanced Animal Breeding Technologies for improvement of livestock. 17 March 6 April, 2006.

Negi, S.S., Joshi, B.C. and Kehar, N.D. (1968). J. *of Research* **5**: 92.

Nelder, J.A. (1966). Inverse polynomial, a useful group of multifactor response functions. Biometrics 22: 128–41.

Nguyen, T.K., Thanb, P.D.L., Rruong, X.D. and Tran, T.L.(1996). Some physiological blood indices of buffaloes suffering from liverflukes. Khoa Hoc Ky Thuat Thu Y. 3: 82–85.

Nighot, P.K., Paliwal, O.P. and Ram Kumar.(1996). Chronic mastitis in buffaloes: granulomatous lesions. *Indian J. Vet. Path.* **20**: 23–26.

Nigm, A.A. (1996). Characterization of Egyptian buffalo. Proceedings of International Symposium on Buffalo Resources and Production Systems. Giza, Egypt, 17 Oct.

Nimasri, V., Bodhishri, L. and Yadi,Y. (1984). Personal communication

Oliver, A. (1938). A brief survey of some of the important breeds of cattle in India. Misc. Bull. No. 17, pp. 45, Imperial Council of Agricultural Research, New Delhi.

Pachauri, S.P. (1972). Cerebrospinal nematodiasis in buffalo- a case report. *Indian J. Anim. Res.* **6**: 17–19.

Pachauri, S.P. and Joshi, H.C. (1977). Note on clinical chronic bovine haematuria in buffalo. *Pantnagar J. Res.* **2**: 246–248.

Pal, M. (1997). Mycotic mastitis in a buffalo (Bubalus bubalis) caused by Candida tropicalis. *Buffalo J.* **13**: 91–94.

Pal, M. and Singh, D.K. (1983). Studies on dermatomycosis in dairy animals. Mycosen 26: 317–323.

Pal, S., Basu, S.B. and Sengar, O.P.S. (1971) Heritabilities and genetic correlations for milk constituents in Murrah buffaloes. *Indian J. Anim. Sci.* **41**: 1019–21.

Panda, B.K. and Misra, S.C. (1980). Observations on the etiology, clinical pathology and chemotherapy of immature amphistomiasis in buffalo-calves. *Indian J. Anim. Hlth.* **19**: 131–135.

Pande, B.P., Bhatia, B.B. and Arora, G.S. (1968). Studies on two fluke infections of the intestinal tract in buffalo calves (below two years of age) with further observations on histopathology. *Indian J. Vet. Sci.* **38**: 463–470.

Pande, B.P., Chauhan, P.P.S., Bhatia, B.B. and Arora, G.S.(1970). Studies on bubaline eyeworms with reference to the species composition of Thelazia and their pathogenic significance. *Indian J. Anim. Sci.* **40**: 330–345.

Pandey, M.B. and Shukla, P.C. (1979). Effect of digestible crude protin and urea feeding onmeat quality and quantity of Surti buffalo calves. *Ind. J. anim. Sci.* **49:** 696–698.

Pandey, M.D. and Roy, A. (1969). I. Variation in volume and composition of body fluids (intestinal, blood and urine) as a measure of adaptability in buffaloes to a hot environment. II. Variation in cardiorespiratory rates, rectal temperature, blood hematocrit and haemoglobin as measures of adaptability of buffaloes to a hot environment. *Br. vet. J.* **125**: 382–402 and 463–71.

Pandya,A.J. and Khan,M.M.H. (2006). Buffalo Milk-Production and utilization of buffalo milk. In Handbook of Milk of Non-Bovine Mammals. Park, Young W. and George F.W. Haenlein (Editors), Blackwell Publishers, USA, pp. 195–274.

Pansin, C., Srisa Kwattana, K., Parntone, S., Chethasing, S., Tasripoo K. and Kamonpatana, M. (2005). 4th Chulalong Korn University Veterinary Annual Conference. 60th Veterinary Anniversary building, Chulalong Korn University, 15th Feb.

Pant, H.C., Barot, L.R., Kasirraj, R, Misra, A.K., and Prabhakar, J.H. (2001). effect of clitoral stimulation after artificial insemination on cnception rate in the buffalo. *Bubalus Bubalis.* **7(3)**: 66–69.

Paranjape, V.L. and Das, A.M.(1986). Mastitis among buffalo population of Bombay– a bacteriological report. *Indian Vet. J.* **63**: 438–441.

Parkash, B. and Balakrishnan, C.R. (1993) Mendelian inheritance of NoRs in Indian water buffalo. *Indian J. Dairy Sci.* **46(5)**: 203–205.

Partoutomo, S., Soleh, M., Politedy, F., Day, A., Stevenson, P., Wilson, A.J., Copeman, D.B. and Owen, I. (1994). The epidemiology of Trypanosoma evansi and Trypanosoma theileri in cattle and buffalo in small holder farms in Java. Penyakit Hewan. 26: 41–46.

Patchimasiri, V., Tochantara, T., Cholayuth, V. and Muangyai, M.(1983). A survey of surra in Thai swamp buffaloes. *Thai. J. Vet. Med. Assoc.* **13**: 81–91.

Patel, P.R., Suthar, B.H., Soni, V.K., Dangaria, A.M. and Prajapati, C.B.(1993). Epidemiology, clinical findings and treatment of ephemeral fever in buffaloes. Australian Centre for International Agri. Res. Proceedings No. 44: 57–58.

Patel, H.G., Upadhyay,K.G. and Pandya, A.J. (1994). Buffalo milk Cheddar cheese -a review. *Indian* J. of Dairy Sci. **47**: 1–13.

Pathak, N.N. (1992). Behaviour and training of river buffaloes. In: Tulloh, N. M. and Holmes J, H, G., eds., Buffalo production, production-system approach, World Animal Science, Elsevier, Amsterdam, Netherlands, pp. 223–231.

Pathak, N.N. and Sharma, M.C. (1988). Buffalo health management. In: Buffalo Production and Health. I.C.A.R., New Delhi, pp. 156–184.

Pathak, N.N., Srivatsava, R.V.N., Jakhmola, R.C. and Kamra, D.N. (1980). Annual Scientific Report of Animal Nutrition Division, IVRI, Izatangar. Pp. 23–24.

Pathak, R.C. (1998). Buffalo diseases of mycoplasma origin undermining the animal productivity in *India. Indian J. Anim. Prod.* **30**: 153.

Pathak, N.N. and Ranjhan, S. K (1976). *Indian J. Anim. Sci.* **46**:323

Patil, N.A. and Appannavar, M.M. (1998). Pathological causes of mortality in surti buffalo calves. *Indian J. Anim. Prod.* **30**: 154.

Patil, N.A., Kumar, S.P., Mallikarjunappa, S., Ladshmaiah, K.R. and Bhat, A.R.S. (1992). Calf mortality in Surti buffaloes. *Indian Vet. J.* **69**: 1018–1022.

Patil, N.A., Mallikarjunappa, S. and Prasanna-Kumar, S. (1994). Efficacy of different drugs in the treatment of haemorrhagic septicemia in Surti buffalo calves. Livestock Advisor. 19: 3–5.

Patnaik, B. and Khan, M.H. (1980). Chemotherapeutic trials against stephanofilarial otitis externa in buffaloes. *Indian Vet. J.* **57**: 368–372.

Patnaik, M.M. (1965). Observations on the effect of sulphadimidine on Eimeria bovis in naturally infected buffalo calves. *Indian Vet. J.* **42**: 186–191.

Patnaik, M.M. and Pande, B.P. (1963). A note on parafilariasis in buffalo, Bob (Bubalus) *bubalis. J. Helminth.* **37**: 343–348.

Patro, B.N. and Bhat, P.N. (1979a) Effect of some non-genetic factors on production traits in Indian buffaloes. *Indian J. Anim. Sci.* **49**: 91–98.

Patro, B.N. and Bhat, P.N. (1979b) Inheritance of production traits in buffaloes. *Indian J. Anim. Sci.* **49**: 10–14.

Patro, G.K. and Mishra, M. 1986. Parlakhemundi breed of buffaloes. Indian J. Anim. Sci. 48 : 537–41.

Pattanaik, P., Kaushik, J.K., Grover, S. and Batish, V.K. (2001). Purification and characterization of a bacteriocin hike compund (Lichenin) produced anaerobically by Bacillus licheniformis isolated from water buffalo. *J. aspplied Microbiology*. **91(4)**: 636–645.

Pawan Singh and Singh, N. (2000) Effect of multiple showering and artificial aeration on sexual behaviour, quality and freezability of buffalo bull semn during hot-humid seaon. *Indian J. of Anim. Sci.,* **70 (9):** 914–916

Pawan Singh, Sengupta, B.P. and Tripathi, V.N.(2000) Effect of supplemental amino acid feeding on sexual behaviour, quality and freezability of buffalo bull semen. *Buffalo Journal.* **16 (1): 41–46.**

Peeva, T. (1996). Buffalo population and production in Bulgaria. Buffalo Newsletter, 4: 6–7.

Peeva, T., Vankov, K. and Vasilev, M. (1991). Persistency index of lactation milk production in buffalo cows. Proc. of the Third World Buffalo Congress, Agricultural Academy, Varna, Sofia, 13 to 18 May: 61.

Pei-Chien, W. (1978) The swamp bufalo and its improvement in the People's Republic of China. Proceedings of FAO/SIDA/GOI/Seminar on Reproduction and Artificial Insemination of Buffaloes, NDRI, Karnal.

Perera, B.M.A.O. and de Silva, L.N.A. (1985) Gestation length in indigenous (Lanka) and exotic (Murrah) buffaloes in Sri Lanka. *Buffalo J.* **1**: 83–87.

Pholpark, S., Pholpark, M. and Tantasuwan, D.(1985). Studies on the treatment of mange in buffaloes with asuntol 0.1% in the northeastern part of Thailand. Kasetsart Vet. 6: 24–28.

Pillai, S.R., Sherikar, A.J., Khot, J.B. and Jayarao,B.M.(1987). In : Advances in Meat research. Pp.33–46.

Pipalia, D.L., Ladani, D.D, Brahmkshtri, B.P. Rank, D.N, Joshi C.G, Vataliya, P.H. and Solanki, J.V. (2001). Kappa-casein genotyping of Indian buffalo breeds using PCR-RFLP *Buffalo J.* **17 (2)** 195–2002.

Pirzada, W.H. and Hussain, S.Z.(1998). Parturient haemoglobinuria in buffaloes- a review. Trop. Anim. Hlth. Prod. 30: 209–215.

Plucienniczak, A., Skowronski,J. and Jaworski.J.(1982).Nucleotide sequence of Bovine 1.715 satellite DNA and its relation to other bovine satellite sequence. *J.Mol. Biol.* **158**:305–315.

Polihronov, D., Vankar, K. and Aleksiev, A.I. (1967) The effect of calving season on the productivity of buffaloes and cows. Zhovt. Nask. 4, 17–27. (fide Anim. Breed. Abstr. 36, 2397).

Ponnappa, C.G., Nooruddin,M. and Raghavan, G.V. (1971).Rate of passage of food and its relation to digestibility of nutrients in Murrah buffaloes and Hariana cattle. *Indian J. Anim. Sci.* **41**:1026.

Pop, M. (1990). Lesions in the respiratory tract in malignant catarrhal fever in cattle and buffaloes. Zootehniesi Med. Vet. 44: 109–114.

Portngal, M.A.S.C., Giorgi, W. and Siqueira, P.A.de. (1972). Tuberculosis in buffalo herd in Sau Paulo State. Estado Sau Paulo Arquivosdo Instituto Biologico, Sau Paulo, Brazil.

Prabhakar, J.H., Kalsi, J.S., Jani, V.R. and Patel, S.H. (2002). Effect of live sperm count per inseminate on pregnancy rate in buffaloes. *Indian J. of Anim. Reproduction.* **23 (2)**: 170–172.

Prabhu, S.S. (1956) Influence of factors affecting sex drive on semen production of buffaloes. II. *Indian J. vet. Sci.*, **26**: 21–23.

Prabhu, S.S. and Bhattacharya, P. (1954) Influence of factors affecting sex drive on semen production of buffaloes. I. Physiological state of the 'teaser' cow. *Indian J. vet. Sci.*, **24**: 35–50.

Prajapati, K.B., Nagpaul, P.K., Chauhan, M.K., Kale, K.M. and Raina, V.S. (2000). The effect of exercise on seminal attributes of Mehsana buffalo bulls in different seasons. *Indian J. of Anim. Reprod.* **21(1):** 38–40

Prasad, C.B.(1967). Isolation and identification of microflora from the female genital tract of buffaloes. *Agra Univ. J. Res.* **16**: 187–188.

Prasad, J.(1979). A study on reticulo–ruminal movements and rumen pH in clinical indigestion in cattle and buffaloes. *Indian Vet. J.* **56**: 474–476.

Prasad, J., Ahluwalia, S.S. and Joshi, B.P.(1972). Clinico-biochemical studies in indigestion in cattle and buffaloes. *Indian J. Anim. Sci.* **42**: 911–914.

Prasad, V.L. and Raghavan, G.V. (1973) Effect of level of intake on reticulo-rumen cations of cattle and buffalo. *Indian J. Anim. Sci.* **43**: 693.

Puccini, V., Giangaspero, A. and Storto, E.Lo.(1985). Outbreak of psoroptic mange in buffaloes. Obiettivie Document Veterinari 7: 61–63.

Puri, J.P. and Garg, S.K. (2001). Effect of osmotic agent supplementaton in the diet of buffalo on some rumen functions and blood electrolytes. *Indina J. Anim. Sci.* 71(**10**):927–931.

Purohit, G.N.(1997). A clinical note on the use of insulin and dextrose in the treatment of ketosis. *Indian J. Anim. Sci.* **36**: 89.

Purohit, G.N., Vyas, S.K. and Yadav, S.B.S.(1997). Efficacy of ivermectin treatment for sarcoptic mange in Surti buffalo calves. *Indian Vet. J.* **74**: 689–690.

Qanemi, A. (1998). Buffalo population and production in Iran. *Buffalo News letter* **10**: 12–14.

Radadia, N.S, Pal, R.N., Sastry, N.S.R. and Juneja, I.J.(1980b). Effect of some summer managerial practices on lactating Murrah buffaloes. 2. Milk production and composition. *Haryana Agri. Univ. J. Res.* **10**:152.

Radadia, N.S., Pal, R.N., Juneja, I.J. and Sastry, N.S.R.,(1980a). Effect of some summer managerial practices on lacating Murrah buffaloes. 1. Feed and water intake. *Haryana Agric. Univ. J. Res.* **10**: 144.

Radostits, O. M., D.C. Blood and Gay, C.C.(1994). Veterinary Medicine. A textbook of the diseases of cattle, sheep, pigs, goats and horses. VIII ed. Bailliere Tindall and ELBS, London.

Ragab, M.T. and Abd-El-Salam, M.F. (1962). The effect of sex and month of calving on body weight and growthu rate of Egyptian cattle and buffaloes. *J. Anim. prod. UAR* **2**: 109.

Ragab, M.T. and Abdel-Salam, M.F. (1963). Heredity and repeatability of bodyweight and growth rate in Egyptian buffalo and cattle. *Egypt. J. Anim. Prod.* **3**: 15–26.

Ragab, M.T., Asker, A.A. and Hilmy (Miss), S.A. (1956). The relations between some fertility aspects and the milk yield or Egyptian cattle and buffaloes. *Indian J. Dairy Sci.* **9**: 53–60.

Ragab, M.T., Asker, A.A. and Kamal, T.H. (1958). The effect ofage and season of calving on the composition of Egyptian buffalo milk. *Indian J. Dairy Sci.* **11**: 18–28.

Raghu, H.M., Nandi, S., Ravindranatha, B.M. and Reddy, S.M. (2002). Effect of age of donor buffalo on the recovery rate and developmental potential of buffalo oocytes *in vitro. Indian J. Anim. Sci.,* **72(7):** 586–587.

Rahman, H., Baxi, K.K. and Sharma, S.N. (1981). Isolation of mycoplasma from a buffalo (Bubalus bubalis) with mastitis. Zbl. Vet. Med. 28: 285–286.

Rahman, M.A., Islam, M.S., Alam, M.G.S. and Shamsuddin, M. (1997). Seroprevalence of brucellosis in the buffalo (Bubalus bubalis) of a selected area in Bangladesh. *Buffalo J.* **13**: 209–214.

Rahman, M.M. and Prasad, S. (1985). Studies on the immune status of foot and mouth disease virus vaccinated buffaloes through viraemia index. *Bangladesh Vet. J.* **19**: 43–48.

Rahman, M.M., Prasad, S., Ahuja, K.L. and Sharma, R. (1988). Application of enzyme labelled immunosorbent assay for the detection of antibodies in foot and mouth disease virus vaccinated buffalo sera. *Buffalo Bull.* **7**: 8–11.

Raina, V.S., Gupta, A.K., and Singh Kiran (2002). Effect of antioxidant fortification on preservability of buffalo semen. *Asian Australasina J. of Anim. Sci.* **15** (**1**) 16–18

Raizada, B.C., Yadav, P. C., Tewari R.B.L. and Roy, A. (1968). Studies on the oestrous behaviour based on the cervical mucus crystallisation pattern in buffalo cows and its relationship to fertility. *Indian J. Vet. Sci.* **38**: 546–57.

Rajamohan, K. and Peter, C.T. (1975). Pathology of nasal schistosomiasis in buffaloes. *Kerala J. Vet. Sci.* **6**: 94–100.

Rajendran, M.P., Premkumar, E.S.J. and Ebenezer, D. (1998). The importance of brucellosis in buffalo milk and meat products. *Indian J. Anim. Prod.* **30**: 205–207.

Rakshe, P.T. (2003). Effect of age at first calving and subsequent period of breeding on the performance of buffaloes from the college of agriculture, Pune (MS). *Buffalo Bull.* **22** (**1**): 7–11

Ram, K., Singh, K. and Tomar, O.S. (1976). Economic efficiency of Brown Swiss crosses, zebu cattle and Murrah buffaloes for production of milk fat and milk solids. *J. Anim. Sci.* **46**: 269–73.

Ramamohana Rao, B. (1978). Preliminary studies on certain quantitative and ualitiative characters of buffalo meat. *The Ind. Vet. J.* **55**: 111–118.

Ramamurthy,M.K. and Narayanan,K.M. (1971). Fatty acid compositions of buffalo and cow milk fats by gas-liquid chromatography (GLC). Milchwissenschaft. 26:693–697.

Ramrao, W. Y., Tiwari, S. P. and Singh, P.(2005). Crop-livestock integrated farming system for augmenting socio-economic status of smallholder tribal of Chhattisgarh in central India. Livestock Research for Rural development. Vol.17(6) and Vol.18(7)) 2006., Article# 90 Retrieved May 17, 2006.

Rana, C.M., Dhami, A.J., Zalu, G.K., Vibhakar, K.V and Ladani, N.G. (2003) Preservation of gir (bos indicus) and Jafarabadi buffalo (Babalus Bubalis) semen at 5 degree C and ultra-low temperature using sephadex filteration techniques. *Indian J. Anim. Sci.*, **73(8)**, 858–863.

Randhawa, S.S., Joshi, B.P., Randhawa, C.S., Brar, R.S. and Trenti, F. (1994). Comparative evaluation of therapeutic trial in puerperal haemoglobinuria in dairy buffaloes. Proceedings 18th World Buiatrics Congress Vol. 2: 1541–1544.

Rangappa, K.S. (1965). "*Dahi Kusum*' or Indian Ice-cream. *Ind. Dairyman,* **17,** 13–15.

Ranjhan, S.K. (1977). Animal nutrition and feeding practices in India. 1st ed. Vikas Publishing House Pvt. Ghaziabad, India.

Ranjhan, S.K. (1983). Nutritional aspects of crop-livestock integration. Paper presented at Crop-livestock Research Workshop in the Asian Farming Systems Network. IRRL Los Banos, Laguna, April 15.

Ranjhan, S.K. and Patha, N.N (1979). Management and feeding of buffaloes. 1st ed.Vikas Publishing House Pvt. Ghaziabad, India.

Ranjhan, S.K. and Pathak, N.N. (1992). Nutritional and metabolic disorders of buffaloes. In: Buffalo Production. World Animal Science C6. Tulloh, N.M. and Holmes, J.H.G. (editors), Elsevier, Amsterdam, pp. 355–375.

Ranjhan, S.K.., Pathak, N.N. and Daniel, S.J. (1977). Annaul Scientific report of Animal Nutrition Division, IVRI. Izatnagar. Pp.23.

Ranjhan. S.K. and Pathak, N.N. (1983). In: Management and feeding of Buffaloes. Pp.232. 2nd revised Ed. Vikas Publishing House Pvt.Ld., New Delhi.

Rao, A.R. and Rao, S.V. (1968). A study on oestrus and conception rates in buffaloes in Andhra Pradesh. *Indian vet. J.* **45**: 846–52.

Rao, A.S.P., Luktuke, S.N and Bhattacharya, P. (1960) Studies on the viability of sepermatozoa in the reproductive tract of the buffalo cow. *Indian J. vet. Sci.* **30**: 246–64

Rao, A.V.N, Haranath, G.B., Sekhram, G.S. and Rao, J.R. (1986). Effect of thaw rate on motility, survival and acrosomal integrity of buffalo spermatozoa frozen in medium French straws. Animal Reproduction Science 22 (2): 123–125.

Rao, C.H., Rao, C.V. and Naidu, K.N. (1976) A note on the effects of off-season breeding in buffaloes. *Indian J. Dairy Sci.* **29**: 237–40.

Rao, K.S., Dinanath, K., Padmavati, P., Ramesh, A.J. and Chetty, M.S. (1993). Studies on the immunodiagnosis of Gigantocotyle explanatum. *Indian Vet. J. 70*: 903–905.

Rao, M.K., Jogi, R.V.K. and Nagarcenkar, R. (1985). Evaluation of influence of certain non-genetic factors on first lactation production in village buffaloes. Proc. Ist World Buffalo Cong., Cario, Egypt (ARE), 5: 1529–45.

Rao, R.A.(1979) Treatment og subestrus in buffaloes with cloprostenol. *Vet. Rec.* **105**: 168.

Rao, T.S., Devi, V.R., Babu, R.M. and Rao, A.V.N. (1999). Comparison of rapid plate agglutination, standard tube agglutination and dot-ELISA tests for the detection of antibodies to Brucella in bovines. *Indian Vet. J.* **76**: 255–256.

Rao, V.N.A., Palaniswami, K.S., Khan, G.A.R., Gnanabaranam, J.F. and Kamala, Krishnaswami.(1991). Hydrocyanic acid poisoning in cattle. *Indian Vet. J.* **68**: 887–888.

Rao, V.P., Prabhu, S.S.and Mishra, R.R. (1973) Studies on RI, E and J blood group antigens in Ongole and zebu-Jersey crossbred cattle. *Indian J. Anim. Prod.* **4**: 27–34.

Rao,M.K. and Nagarcenkar,R. (1977) Potentialities of the buffalo. World Review of Animal Production. 13:53–62.

Rasali, D.P. (1997). Present status of indigenous buffalo genetic resources in the western hills of Nepal. Proceedings of the Fourth Global Conference on Conservation of Domestic Animal Genetic Resources. Rare Breeds International: 168–170.

Rasali, D.P., Joshi, H.D., Patel, R.K and Harding, A.H. (1998b). Phenotypic clusters and karyotypes of indigenous buffaloes in the western hills of Nepal. Lumle Agricultural Research Centre, Technical Paper. 98/2: 24.

Rasali, D.P., Joshi, H.D., Shrestha, H. and Gautam, D.C. (1998a). Assessment of the infertility situation in cows and buffaloes in the western hills of Nepal. Lumle Agricultural Research Centre. Working Paper, 98/40: 16.

Rasool, G., Jabbar, M.A., Kazmi, S.E. and Ahmad, A. (1985). Incidence of subclinical mastitis in Nili-Ravi buffaloes and Sahiwal cows. *Pakistan Vet. J.* **5**: 76–78.

Rastagir, R. (1950). Animal Husbandry in the tropical nd subtropical regions of Iran. Proc. FAO meeting on Livestock Breeding in Tropical and subtropical conditions FAO, Rome pp 5.

Rastogi, M.K., Verma, I.S. and Paul, J. (1966). Nutritive appraisal of some Indian indigenous milk sweets. *Proc. 17th Int. Dairy Congr.*, **EF**,*pp.273–278.*

Rastogi, R., Youssef, F.G. and Gonzalex, F.O. (1978). Beef type water buffalo of Trinidad bufalypso. Wld Rev. Anim. Prod., 14, 49–56.

Rathi, S.S., Balain, D.S. and Acharya, R.M. (1971). Phenotypic and genetic parameters of some economic traits in Indian buffaloes. *Indian J. Anim. Prod.* **2**: 1–6.

Rathi, S.S., Balaine, D.S. and Kanaujia, A.S. (1973). Body weights and their relationship with economic traits in Indian buffaloes. *Indian J. Anim. Prod.* **4** : 1–8.

Ray, S.C. (1955). Pre-stratification Method of Ghee Making. ICAR Res. Series No.8. Ind. Coun. Agric. Res.: New Delhi.

Reddy, C.E. (1980). Genetic analysis of breeding records of buffaloes. Ph.D. Thesis, Kurukshetra University, Kurukshetra.

Reddy, C.E. and Taneja, V.K. (1982). Use of different types of records for estimation of heritability. Second World Congress on Genetics Applied to Livestock Production, Madrid, 4–8 October 1982, pp. 150–51.

Reddy, C.E. and Taneja, V.K. (1984). Factors influencing the first lactation traits of Nili-Ravi buffaloes. *Indian J. Dairy Sci.* **37**: 36–39.

Reddy, D.N.C., Gupta, D.C.L.R. and Radhakrishna, T.D.G. (1982). Deep freezing of buffalo semen and its fertility studies. *Indian Vet. J.* **59** (**7**): 574–575.

Reddy, K.M. and Mishra, R.R. (1980). Effect of age at first calving on lactation components and calving distribution in Murrah buffaloes. *Indian J. Dairy Sci.* **33**: 347–50.

Reddy, M. J., Reddy, M. S., Chetty, M. S., Kulkarni , R.D.and Reddy, G. M.(1990). Incidence of sarcocystosis in buffaloes in Hyderabad (A.P.). *J. Vet. Parasitol.* **4**: 59–61.

Rendel, J.M. and Robertson, A. (1950). Estimate of genetic gain in milk yield by selection in a closed herd of dairy cattle. *J. Genet.* **50**: 1–9.

Riet-Correa, F., Barros, S.S., Dame, M.C. and Peixoto, P.V.(1994). Heriditary suprabasilar acantholytic mechanobullous dermatosis in buffaloes. *Vet. Pathol.* **31**: 450–454.

Rouse, E.J. (1970). World Cattle, Vol. II. Univ. Oklahoma Press: Norman, Oklahoma.

Roy Choudhary, P.N., Mathur, B.S. and Deshmukh, S.N. (1971). Some environmental factors affecting calving interval and lactation length in Italian buffaloes. *Indian vet. J.* **98**: 711–716.

Roy, A. Raizada, B.G., Tewari, R.B.L., Pandey, M.D., Yadav, P.C. and Sen Gupta, B.P. (1968). Effect of management on the fertility of buffalo cows bred during summer. *Indian vet. J.* **98**: 711–16.

Roy, A., Pandey M.D. and Rawat, J.S. (1960). Composition of bovine semen. *Indian J. Dairy Sci.* **13**: 112–16.

Roy, A., Raizada, B.C., Tewari, R.B., Pandey, M.D., Yadav. P.C. and Sengupta, B.P. (1968). Effect of management on the fertility of buffalo cows bred during summer. *Indian J. vet. Sci.* **38**: 554–60.

Roy, D.J. and Bhat, P.N. (1973). Studies on deep freezing of buffalo semen using three different extenders. *Indian J. Anim. Sci.* **43 (12)**: 1097–1099.

Roy, P.K. and Nagpal, P.K., (1984). Studies on the effect of temperament score in different traits in dairy animals; *Indian J. of Dairy Sci.* **37 (1)** 74–75 .

Ruff, G. and Lazary, S. (1985). Investigation on the goat leucocyte antigen (GLA) system. Anim. Genet. **16 (1)**: 123.

Ruprah, N.S., Varma, R.K. and Chhabra, M.B. (1980). Evaluation of some acaricides against psoroptic mange in buffaloes. *Haryana Agri. Univ. J. Res.* **10**: 309–313.

Ryan, C.O., Harley, E.H., Bruford, M.W., Beaumont, M., Wayne, R.K. and Cherry, M.I. (1988). Microsatellite analysis of genetic diversity in fragmented south African buffalo populations. animal conservation 1: 85–94.

Saadat, M.N. (1997). In dairy goats and buffalo production. University of Teheran.

Sabry, M.Z., Dessonky, M.I., Hegazy, A.A. and Kamal, B.A. (1978). Haematological and histopathological studies on experimental infection of Egyptian cattle and buffaloes with contagious pleuropneumonia. *J. Egyptian Vet. Med. Assoc.* **38**: 93–106.

Sachindra, N.M., Sakhare, P.Z., Yashoda, K.P. and Rao, D.N. (2005). Microbial profile of buffalo sausage during processing and storage. *Food Control.* **16 (1)**: 31–35.

Sahni, K.L. and Roy, A. (1972). Deep freezing of buffalo semen. *Indian Vet. J.* **40**: 263–267

Saini, S.S., Sharma, J.K. and Kwatra, M.S. (1992). Assessment of some management factors responsible for prevalence of brucellosis among traditionally managed animal population of Punjab. *Indian J. Anim. Sci.* **62**: 832–834.

Salem, S.A.H., El-Mashed, I. and Hamada, F.K. (1996). Pathological and epidemiological studies on bovine rabies in Kaluobia Governorate. In: Procedings of third scientific Congress Egyptian Society for Calf Disease. 2: 303–311.

Salerno, A. (1960). Calving interval in buffaloes. Annali. Sper. Agr. 156, 191–206.

Salim Iqbal, Singh, R.V. and Setia, M.S. (1992). Influence of diets on the population of rumen microflora and fauna in buffalo calves. Buffalo J. (In Press).

Salim, S.A., Amin, M.M., El-Sherif, M.T. and El-Saifi, A.A.(1978). Indirect haemagglutination test as a means of diagnosis of tuberculosis in Egyptian buffaloes. *Bull. Anim. Hlth. Prod. In Africa* **26**: 315–319.

Samad, A. (1997). Host and environmental factors associated with phosphorus deficiency haemoglobinuria in buffaloes. *Buffalo J.* **13**: 385–395.

Samad, A., Singh, B. and Qureshi, M.I.(1979). Some biochemical and clinical aspects of haemoglobinuria in buffaloes. *Indian Vet. J.* **56**: 230–232.

Samizadeh-Yard, A., Karimi, A. and Ahmadi, M.(1981). Neoascariasis in buffalo calves of northwest of Iran. *J. Vet. Faculty Univ. Tehran.* **37**: 43–50.

Sandoe, P., Christiansen, S.B., Appleby, M.C. (2003). Farm animal welfare: the interaction of ethical questions and animal welfare science. Animal Welfare 12: 469–78.

Sane, C.R., Marathe, M.R., Pamaik, D.T., Salunke, B.K., Kaikini, A.S. and Desai, V.G.(1957). A case of milk fever in Nagpuri buffalo. Bombay Vet Coll. Mag. 6: 16–17.

Sane, D.D., Khanna, R.S., Bapai, L.D. and Bhat, P.N. (1972). Studies on Murrah buffalo (*Babalus bubalis*). 2. Genetic analysis of milk yield and peak yield. *Indian J. Anim. Prod.* **3**: 61–65.

Santa Rosa, C.A., Pestana de Castro, A.F. and Troise, C. (1961) Brucella agglutinin titres in buffaloes in Brazil (in Portuguese). Arq: Inst. Biol. S. Poulo, 28, 35–39. (fide Vet. Bull., **32: 736 1962)**

Sanyal, P.K. and Gupta, S.C.(1996). The efficacy and pharmacokinetics of long term low level intraruminal administration of triclabendazole in buffalo with induced fasciolosis. Vet. Res. Communications 20: 461–468.

Sanyal, P.K.(1996). Kinetic disposition and critical efficacy of triclabendazole against experimental bovine and bubaline fasciolosis. *J. Vet. Parasitol.* **10**: 147–152.

Sanyal, P.K., Ruprah, N.S. and Chhabra, M.B.(1985). Chemotherapeutic efficacy of sulphadimidine, amprolium, halofuginone and chloroquine phosphate in experimental Eimeria bareillyi coccidiosis of buffalo. Vet. Parasitol. 17: 117–122.

Sarabia, A.S. and Momongan, V.G. (1993). Evaluation of the physiological responses associated with work of the Philippine carabao and its crosses. Portion of a Ph. D. dissertation. Institute of Animal Science. at Los Banos, College, Laguna, Philippines (Unpublished).

Sarma, D.N. and Sharma, D.D. (1985). Effect of dietary protein levels ongrowth, body composition, cacass characteristics and meat quality of male buffalo calves. *Annual report of* NDRI, Karnal. Pp.64.

Sarma, K.M. (1984). Effect of dietary protein levels on growth and meat quality of male buffalo calves. Ph.D. thesis. NDRI, Karnal. Kurukshetra Univeristy, Kurukshetra.

Sarma, K.M. and Talapatra, S.K.(1963). Growth response in buffalo calves. *Indian J. of DairySci.* **16**: 236–244.

Sarma, P.A., Sharma, D.D., Basu, S.B. and Nagarcenkar, R.(1979).A note on Murarah buffalo for meat production. *Ind. J. Anim. Sci.* **49**: 570–572.

Sastry, N.S.R. (1982). Buffalo husbandry: constraints to successful bufflo farming and overcoming the same through management. Hohenheim, Stuttgart.

Sastry, N.S.R. (1983). Buffalo husbandry; constraints to successful buffalo farming and overcoming the same through management. Monograph, institute of animal management and breeding, university of Hohenheim, Germany, Discipline-Milk Production. pp 4–6.

Saxena, R. (1994). Economic value of some non-milk losses caused by foot and mouth disease in India. Working Paper, Institute of Rural Management, Anand, No. 62, pp. 15.

Saxena, S.P. and Mehra, K.N.(1973). A note on occurrence of dermatophyte infection in domestic animals. *Indian J. Anim. Sci.* **43**: 347–350.

Sebastian, L., Mudgal, V.D. and Nair, P.G., (1970). Comparative efficiency of milk production by Sahiwal cattle and Murrah buffaloes. *J. Anim. Sci.* **30**: 253.

Sehjpal.L., Mehta, A.K. and Rao, M.V.N. (1980). *Indian J. Anim. Sci.* **50**: 467.

Sekerden, O., Dogrul, F. and Erdem, H. (2000). Blood serumtransferrin types and genetic structure for transferrin types and growth performance of Anatolian buffalo. Kocatepe Enstitutu manda surusu sut ve dol verim ozellikleri. Ataturk University.

Sekerden, O., Kebapci, M. and Kopar, A. (1996a). Buffalo population and production in Turkey. *Buffalo Newsletter* **5**: 7–8.

Sekerden, O., Kebapci, M. and Kopar, A. (1996b). Blood transferrin types and genetic structure for transferrin types of buffalo population in Samsun province. *Buffalo Newsletter* **5**: 11–12.

Selim, M.K. and Tewfic, M.A.A.(1966). Incidence of Ascaris vitulorum in Egyptian buffaloes during the autumn and early winter in U.A.R. *Indian Vet. J.* **43**: 965–968.

Sen, A.R and Sharma, N. (2002). An enzymatic method designed to differentiate between fresh and frozen- thawed buffalo meat. *Buffalo Journal.* **18** (**3**): 339–347.

Sen, A.R. and Sharma,N. (2003). Quality changes in buffalo meat during storage in dry ice pack. *Indian Vety. J.* **80** (**2**): 166–168.

Sen, K.C., Ray, S.N. and Ranjhan, S.K. (1978). Nutritive value of Indian Feeds and feeding of cattle. ICAR Pub.25, New Delhi.

Sen, S.K. and Fletcher, T.B. (1962). Veterinary Entomology and Acarology for India. Indian Council of Agricultural Research, New Delhi.

Seneviratna, P. (1967). Economic losses due to parasitic diseases of livestock in livestock in Ceylon. *Ceylon Vet. J.* **15**: 5–8.

Sengar, O.P.S and Singh,S.N (1969). Paper presented in the seminar on "Ruminant Digestion and Metabolism" IVRI, Izatnagar, P.16

Sengar, S.S., Lakshmanan, V. and Joshi, D.C. (1985). Annual scientific report of animal nutrition division, IVRI, Izatnagar.pp.64.

Sengupta, B.P. and Sukhija, S.S. (1988). Current Status of Buffalo Frozen Semen Technology and Fertility: An Overview. FAO. Consultancy on Reproduction, Rome, Italy.

Sengupta, B.P., Bhela, S.L., Sukhija, S.S. and Tuli, R.K. (1982). On improving the freezability of buffalo spermatozoa effect of some experimental variables. 2nd International Conference on World Buffalo Production, Caserta, Italy IV, p–20–21.

Sengupta, B.P., Misra, M.S. and Roy, A. (1963). Climatic environment and reproductive behaviour of buffalaoes. Effect of different seasons on various seminal attributes. *Indian J. Dairy Sci.* **16**: 150–65.

Shabbir, A., Ali, F.A. and Ahmed, S. (1995). Babesiosis in cross-bred cattle (Bos indicus × Bos taurus) and buffaloes (Bubalus bubalis). *Punjab Univ. J. Zoology* **10**: 33–37.

Shafie, M.M. (1985). Physiological responses and adaptation of water buffalo. In M.K. Yousef, ed., Stress physiology in livestock, **Vol**. 2: Ungulates, Florida, USA, CRC.

Shafie, M.M. and Abou El–Khair, M.M. (1970). Activity of the sebaceous glands of bovines in hot climate (Egypt). J. Anim. Prod. U A R 10: 81–98.

Shaikh, H.U.D., Huq, M.M., Karim, M.I. and Khan, M.M.(1983). Parasites of zoonotic importance in domesticated ruminants. *Pakistan Vet. J.* **3**: 23–25.

Shalash, M.R. (1958) Physiology of reproduction in the buffalo cow. *Int. J. Fert.* **3**: 425–32.

Sharma, A. and Basu, S.B. (1984). Genetic architecture of Nili buffaloes. 1. Growth. *Indian Vet. J.* **61**: 227–32.

Sharma, A. and Basu, S.B. (1986). Optimization of breeding and managemental variables for maximizing lifetime milk production and profit in Murrah buffaloes. *Indian J. Anim. Sci.* **56**: 64–71.

Sharma, K.N. and Talapatra, S.K. (1963). *Indian J. Dairy Sci.***17**: 236

Sharma, K.N.S. and Singh, S. (1975). Estimation of life expectancy of cattle and buffalo. Symposium on "Introduction, Improvement and Evaluation of Germplasm for Milk Production" 28–29 Oct., 1975, NDRI, Karnal, Haryana, India, pp. 25–31.

Sharma, K.N.S., Jain, D.K. and Noble, D., (1975). Calf Mortality in pure and crossbred Zebu cattle and Murrah Buffaloes reared artificially from birth. Anim. Prod. 20: 207–211.

Sharma, M.C., Gupta, O.P., Dwivedi, S.K. and Das, S.C.(1980). Therapeutic efficacy of Seven 50 WP (carbaryl-1-nephthyl methyl carbonate) against bovine dermatophytosis (ringworm). *Indian Vet. J.* **57**: 509–510.

Sharma, M.C., Gupta, O.P., Dwivedi, S.K. and Das, S.C.(1981). Preliminary observations on thimerasol (merthiolate) in the treatment of bovine ringworm. *Indian Vet. J.* **58**: 241–242.

Sharma, M.C., Hung, N.N. and Vuc, N.V.(1982b). Leptospirosis as a cause of abortion in Murrah buffaloes. Buffalo Breeding Res. Cen. Song Be Annual. 5: 33–35.

Sharma, M.C., Pathak, N.N. and Hung, N.N.(1985b). Prevalence of haemoprotozoan infection in Murrah buffaloes in Vietnam. Indian J. Parasit. 9: 87–90.

Sharma, M.C., Pathak, N.N., Hung, N.N. and Vuc, N.V.(1982a). Incidence, pathogenesis and control of brucellosis. Buffalo Breeding Res. Cen. Song Be Annual. 5: 18–22.

Sharma, M.C., Pathak, N.N., Hung, N.N., Lien, N.H. and Vuc, N.V.(1982c). Studies on the prenatal calf loss in a herd of Murrah buffalo breed in Vietnam. *Vietnam J. Anim. Sci. Technol.* **1**: 57–62.

Sharma, M.C., Pathak, N.N., Hung, N.N., Lien, N.H. and Vuc, N.V.(1982e). Efficacy of nitrofurazone against coccidiosis in Murrah buffalo calves. *Vietnam J. Anim. Sci. Technol.* **1**: 67–73.

Sharma, M.C., Pathak, N.N., Hung, N.N., Nhi, D.L. and Vuc, N.V.(1985). Report on the outbreak of foot and mouth disease reared in Southern East Vietnam. In: Veterinary Viral Diseases. Antony, J. Dhella Porta (editor), Academic Press, Sydney, pp. 302–303.

Sharma, M.C., Pathak, N.N., Hung, N.N., Thuan, H.T. and Vuc, N.V.(1982f). Incidence of babesiasis in Murrah buffaloes reared in Vietnam. *Vietnam J. Anim. Sci Technol.* **1**: 46–50.

Sharma, M.C., Pathak, N.N., Hung, N.N., Vuc, N.V. and Thuong, N.V.C.(1983b). Abomasal milk impaction- an emerging disease of buffalo calves. *Indian J. Vet. Med.* **3**: 117–119.

Sharma, M.C., Pathak, N.N., Nhi, D.L., Hung, N.N. and Vuc, N.V.(1985a). Incidence of filariasis in Murrah buffaloes with particular reference to haematological changes and chemotherapy. Buffalo Bull. 4: 48–51.

Sharma, M.C., Pathak, N.N., Verma, R.P., Hung, N.N. and Vuc, N.V.(1983a). Presence of hair balls in the stomach of Murrah buffalo calves- a case report from S.R. Vietnam. Dairy Guide. 10: 45–46.

Sharma, M.C., Mishra, R.R. (1987). Livestock health and management, 1st edn. Khanna Publishers, Delhi, India.

Sharma, S.N. (1981). Veterinary Jurisprudence, 3rd edn. Oxford and IBH Publishing Co., New Delhi, India.

Sharma, S.N., Gahlot, A.K., Tanwar, R.K. (2003). Veterinary Jurisprudence, 5Th edn. NBS Publishers, India.

Sharma, M.C., Pathak, N.N., Verma, R.P., Hung, N.N. and Vuc, N.V.(1984). Studies on the anaplasmosis in Murrah buffaloes in Vietnam with reference to incidence, clinico-haematological changes and chemotherapy. *Indian Vet. J.* **61**: 890–894.

Sharma, M.C., Pathak, N.N., Verma, R.P., Hung, N.N., Vuc, N.V., Thuong, N.V. and An, D.T.(1982d). A note on the incidence of coccidiosis in a herd of Murrah buffalo (Bubalus bubalis). *Vietnam J. Anim. Sci. Technol.* **1**: 80–82.

Sharma, N., Gandemer, G., Goutefongea, R. and Kowale, B.N. (1986). Fatty acid composition of water buffalo meat. *Meat Science* B 16 **: 237–243.**

Sharma, P.K. and Sarker, A.B. (1968). Effect of sire, sex of calf and month of calving on gestation period and birth weight in Murrah buffaloes. *Indian vet. J.* **45** : 413–20.

Sharma, R.C. and Singh, B.P. (1978) Evaluation of the genetic potential of Bhadawari buffaloes. 1. Reproductive characters. *Indian vet. J.,* **55**: 595–600.

Sharma, R.C. and Singh, B.P. 1978a. Evaluation of the genetic potential of Bhadawari buffaloes. 3. Genetic and phenotypic correlations. Indian vet. J. 55 : 943–46.

Sharma, R.C. and Singh, B.P. 1978b. Evaluation of the genetic potential of Bhadawari buffaloes. 2. Reproductive characters. Indian vet. J. 55 : 751–55.

Sharma, R.K. (1982). Management of milk fever in dairy animals. Dairy Guide 4: 13–15.

Sharma, S.D., Rai, P. and Saxena, S.C.(1977). A survey of mycotic infections of udder in clinical and subclinical cases of mastitis in cattle and buffaloes. Indian Vet. J. 54: 284–287.

Sharma, S.K., Gupta, S.C., Wadhwa, M. and Bakshi, M.P.S. (2005). Effect of roughage to concentrate ratio on buffalo meat production.*Babalus Bubalis*, **11 (1)**: 44–51.

Sharma, S.K., Singh, G.R. and Murty, D.K. 1979. Allergic reaction in buffaloes after vaccination with foot and mouth disease vaccine. *Indian Vet. J.* **56**: 621.

Sharma, S.P.(1991). Treatment of clinical microfilariasis in buffaloes with ivermectin. Indian Vet. J. 68: 972–974.

Sharma, V.V. and Mardia, P.C. (1974) J. Agri. Sci. Cambridge, 83, 289.

Shashi Nayyar, Gill, V.K., Narinder Singh, Roy, K.S. and Rajvir Singh (2002) Levels of antioxident vitamins in anoestrus buffalo heifers supplemented with vitamin E and selenium. *Indian J. anim. Sci.* , **72 (5)**: 395–397.

Shelke, K.G., Bhokre, A.P., Aher, V.D. and Patil, V.T.(1996). Dilatation and right side displacement of abomasum in a she buffalo- a case report. *Indian J. Vet. Surg.* **18**: 53.

Sherikar, A.T., Waskar, V.S. and Paturkar, A.M.(1998). Incidence of sarcocystosis in buffalo meat. *Indian J. Anim. Prod.* **30**: 152.

Sherman, D.M., Gay, J.M., Bouley, D.S. and Nelson, G.H.(1990). Comparison of the complement fixation and agar gel immunodiffusion tests for diagnosis of subclinical bovine paratuberculosis. *Anim. J. Vet. Res.* **51**: 461–465.

Sherman, D.M., Markham, R.J.F. and Bates, F. (1984). Agar gel immunodiffusion test for diagnosis of clinical paratuberculosis in cattle. *J. Anim. Vet. Med. Assoc.* **185**: 179–182.

Shrestha, N.P. and Parker, B.A. (1992). 'Genetic evaluation of Philippine carabao (PC), Murrah (MB) buffalo and the F1 hybrids MB x PC and Nili–Ravi x PC'. Ph.D. Dissertation, Institute of Animal Science, U.P. at Los Banos, College, Laguna, Philippines, 327 pp. (unpublished).

Shukla, D.D. and Bhattacharya, P. (1949). Studies on the semen characteristics of Indian breeds of livestock. *Indian J. vet. Sci.* **19**: 161–70.

Shukla, K.P., Mithuji, G.F. and Buck, N.C. (1970). Breeding performance of Surti buffaloes during summer months. *Indian J. Anim. Sci.* **40**: 551–56.

Shukla, K.S., Ranjhan, S.K. and Netke, S.P. (1972). *J. of Dairy Res.* **39**: 421

Shukla, R.R. and Singh, G(1972). Studies on tuberculosis amongst Indian buffaloes. *Indian Vet. J.* **49**: 118–123.

Shukla, S.K., Dixit, V.P., Thapliyal, D.C. and Kumar, A.(1998). Bacteriological studies of mastitis in dairy cows. *Indian Vet. Med. J.* **22**: 261–264.

Sidky, A.R. (1953). The Buffalo of Egypt. 2. Breeding Efficiency in the Egyptian Buffalo. Age at First Calving. Government Press, Cairo.

Siebert, B.D. and MacFarlane, W.V., (1969). Body water content and water turnover of tropical *Bos taurus, Bos indicus, Bibos banteng and Bos bubalus bubalis. Aust. J. Agric. Res.* **20**: 613–622.

Sihag,Z.S., Chahal, S.M. and Punia, B.S.(2003). Influence of urea molasses mineral blocks (UMMB) containing different energy and protein sources on nitrogenous metabolites in the rumen of buffaloes.*Indian J. Anim. Nutri.* 2003; **20 (2)**: 124–130.

Sikka, L.C. (1950) A study of lactation as affected by heredity and environment. *J. Dairy Res.* **17**: 231.

Silva, T.D. and Silva, K.F.S.T.(1994). Total and differential cell counts in buffalo (Bubalus bubalis) milk. *Buffalo J.* **10**: 133–137.

Sindhu,J.S. and Singhal,O.P. (1988). Qualitative Aspects of Buffalo Milk Constituents for Products Technology. Proceedings of 2nd World Buffalo Congress. ICAR, New Delhi, India. **Vol. 2**: 263–287.

Singari, N.A., Bhardwaj, R.M., Chugh, S.K. and Bhardwaj, S.(1991). Status of erythrocytic glucose-6-phosphate dehydrogenase in phosphorus deficiency haemoglobinuria of buffaloes. *Indian Vet. J.* **68**: 226–230.

Singer, P. (1990). Animal liberation: a new ethics for our treatment of animals. Avon, New York.

Singh Sall, H., Flukiger, A. and Wieser, M.F. (1980). Deep freezing of buffalo semen. II-Deep freezing of buffalo semen using different dilutors. *Indian Vet. J.* **57 (2)**: 143–146.

Singh, A., Ahuja, S.P. and Abraham, R.(1998). Characterization of antigens of cellular membranes and pilli of Escherichia coli isolated from diarrhoeic buffalo calves. *Indian J. Anim. Prod.* **30**: 158.

Singh, A., Basu, S.B. and Bhatia, K.L. (1979). Milk fat and SNF percentages of Murrah buffaloes. *Indian J. Dairy Sci.* **32**: 446–49.

Singh, A., Kapoor, S. and Pandey, R.(1993). Molecular epidemiology of bovine rotavirus during 1987 as determined by electrophoresis of genomic RNA. *Indian J. Virol.* **9**: 1–8.

Singh, B. (1942). The blood groups of Indian cattle and buffaloes. *Indian J. vet. Sci.* **12**: 12–13.

Singh, B. and Joshi, S.J.(1991). Epidemiology, clinico-pathology and treatment of clinical Trypanosoma evansi infection in buffalo (Bubalus bubalis). *Indian Vet. J.* **68**: 975–979.

Singh, B., Gautam, O.P. and Sarup, S.(1974). Some biochemical and clinical aspects of milk fever (Parturient paresis) in buffaloes. *Indian Vet. J.* **51**: 612–614.

Singh, B.B. (1976). Evaluation of economic level of milk production in buffaloes. *Indian vet. J.* **53**: 829–33.

Singh, B.B. and Singh, B.P. (1967). Effect of calving season on lactation yield in Murrah buffaloes after eliminating the effect of lactation period, lactation order and non-orthgonality. *Indian J. Anim. Hlth* **6**: 251–55.

Singh, B.B. and Tomar, N.S. (1981). Evaluation of culling in Murrah buffaloes. *Indian vet. J.* **58**: 303–7.

Singh, B.K.(1982. Field trials with tetramisole (Nilverm) against ascariasis in buffalo-calves. Livestock Advisor. 7: 41–43.

Singh, C. (2003). Response of anestrus rural buffaloes (Bubalus bubalis) to the intravaginal progesterone implant during summer. *Indian J. Anim. Sci.* **73 (10)**: 1129–1130.

Singh, C.K. and Grewal, G.S.(1998). Comperative evaluation of rabies diagnostic sensitivity among seven laboratory techniques. *Indian J. Anim. Sci.* **68**: 1210–1213.

Singh, C.V. and Yadav, M.C. (1986). Estimation of genetic and nongenetic parameters affecting reproduction traits in Indian buffaloes. *Asian J. Dairy Res.* **5**: 171–74.

Singh, D and Basu, S.B.(1988). Genetics parametes of the first lactation traitsin Murah buffaloes. *Indian vet.J.*, **65**: 593–98

Singh, D. (1983). Genetic and financial consequences of different basis if selection in buffaloes. Ph.D. Thesis, Kurukshetra University, Kurukshetra.

Singh, G. and Tiwana, M.S. (1983). Estimates of genetic and phenotypic parameters of peak yield and some other economic traits in buffaloes. *Indian J. Dairy Sci.* **30**: 216–18.

Singh, G., Singh, B., Gupta, P.P. and Hothi, D.S.(1979). Epizootiological observations on malignant catarrhal fever and transmission of disease in buffalo calves (Bubalus bubalis). Acta Veterinaria Brno. 48: 95–103.

Singh, G., Taneja, V.K., Bajpai, L.D. and Bhat, P.N. (1971). Studies in Indian water buffaloes (*Babalus budalis*) Growth. *Indian J. Anim. Prod.* **2**: 13–19.

Singh, H.P. and Singh, R. (1977). Statistical studies of some economic traits of Indian buffaloes. *Indian vet. J.* **54**: 823–33.

Singh, I.P. and Singh, S.B. (1967). Isolation and characterization of the etiologic agent of buffalo pox. *J. Res. Ludhiana* **4**: 440–448.

Singh, J.,Nanda, A.S. and Adams, G.P. (2000). AH-Repro(gyn). The Reproductive pattern and efficiency of female buffaloes. *Anim. Repro. Sci.* **60/61**: 593–604

Singh, K.B., Singh, P.J. and Singh, C.K. (1999). Rabies in buffaloes clinical report. *Indian Vet. J.* **76**: 52–53.

Singh, N., Kumari, R. and Akbar, M.A. (1989). Biochemical changes in blood metabolites in buffaloes with indigestion. *Indian Vet. J.* **66**: 923–926.

Singh, O.P. (1974). Clinico-therapeutic studies of scour and ascariasis in Zebu and buffalo calves with a note on non-specific dermatitis in cow calves. *Agra Univ. J. Res. Sci.* **23**: 95–97.

Singh, R.B., Sharma, S.C. and Singh, S. (1958). Influence of the season of calving on inter-calving period in Murrah buffaloes and Haryna cows. *Indian J. Dairy Sci.* **11**: 154–60.

Singh, R.P. (1966). A study of production up to 10 years of age in buffaloes maintained at buffaloes, maintained at military farms. *Indian vet. J.* **40**: 149–54.

Singh, R.P. (1967). A study of body size and reproduction and relative efficiency of milk production in buffaloes maintained at military farms. *Indian Vet. J.* **40**:149–154.

Singh, R.V., Salim Iqbal and Setia, M.S. (1992). Effect of diets on rumen ciliate protozoa in buffalo calves (*Bubalus bubalis*). *Buffalo J.* (In Press).

Singh, R.V., Singh, R. and Setia, M.S. (1990). Inter-relationship of trace elements in certain body fluids of buffalo calves on different feeding regimens. In: Proc. of II World Buffalo Congress, New Delhi. 2, 345.

Singh, Ram Vir., Sachdeva, G.K., Garg, R.C., Kaushik, S.N., and Singh Kripal (2004). Conservation and Genetic improvement of important indigenous cattle breed of India. *In Proc. of National Symposium on Conservation and Propogation of indigenous breeds of Cattle Buffaloes. 26–28 Feb.*

Singh, S., Langar, P.N., Sidhu, G.S., Kochar,A.S. and Bhatia, A.S. (1968). Study of Rumen biochemical activity in the buffalo (*Bos bubalis)* and Zebu *(Bos indicus)* under non-urea and urea feeding regimes. *Indian J. Vet. Sci.* **38**: 674–681.

Singh, S.B. and Desai, R.N. (1962). Production characteristics of Bhadawari buffalo cows. *Indian Vet. J.* **39**: 332–43.

Singh, S.N. (1965). Protein requirements of Indian buffaloes. Annual report of the ICAR, Krishi Bhawan, New Delhi.

Singh, S.P. and Dutt, M. (1964). Study of reproductive efficiency in Murrah buffaloes. *Indian J. Dairy Sci.* **17**: 109–12.

Singh,R.V. and Nivasarkar,A.E. (2000). Production and reproduction potential of Bhadawari buffalo. *Buffalo J.* **16** (**2**):163–173.

Singla, S.K., Madan, M.L., Manik, R.S., Amros, J.D. and Chauhan, M.S. (1992). Fertilization and early embryo development pattern in superovuated buffaloes. XII International Congress on Animal Reproduction, The Hague, The Netherlands, Aug. 23–27, **Vol.** II, p 817.

Sinha, P.K. and Das, M.S. (1958). On a new stephanofilariasis in buffalo in Andaman Islands. Proc. 45th Indian Sci. Congress, Madras, Part III: 416.

Sinha, R.C., Sengupta, B.P. and Roy, A. (1966). Climatic environment and reproductive behaviour of buffaloes. IV. Comparative study of oxygen uptake and acerobic fructolysis by Murrah (B. bubalis) and Hariana (B. indicus) spermatozoa during different seasons. *Indian J. Dairy Sci.* **19**: 18–24.

Solbu, H., Spooner, R.L. and Lie, O. (1982). A possible influence of the bovine major histocompatibility complex (BoLA) on mastitis. In Second World Congress on Genetics Applied to Livestock Production, VII. pp. 368–371.

Spratt, D.M., Dyce, A.L. and Standfast, H.A.(1978). Onchocerca sweetae (Nematoda: Filarioidea): notes on the intermediate host. *J. Helminth.* **52**: 75–81.

Sreedharan, S. and Nagarcenkar, R. (1978). Relation of Body Size Measures to Production Efficiency in Murrah Buffaloes. pp. 5. Brief communication, 20th International Dairy Congress, Paris.

Sreetamulu, P. (1994). Epizootiology of nasal schistosomiasis in bovines in Andhra Pradesh. *Indian Vet. J.* **71**: 1043–1044.

Srinunthapanth, S., Triwanatham, N. and Khoonchan, P.(1985). Rabies in cattle and buffaloes. Case report. *Thai J. Vet. Med.* **15**: 279–285.

Srivastav, S.K., Sahni, K.L., Uma shanker, Saniwal, P.C., and Varshney, V.P. (1999). Seasonal variation in progesterone concentration during oestrus cycle in Murrah buffaloes. *Indina J. of Anim. Sci..* **69(9)**: 700–701.

Srivastava, A.K. and Sharma, D.N.(1980). Pattern of mortality in Zebu and buffalo calves. *Vet. Res. J.* **3**: 105–110.

Stahlheim, O.H.V.(1973). Stability of viable avirulent Leptospira pomona in pregnant cows. Am. J. Vet. Res. **34**: 173–174.

Stear, M.J., Newman, M.J., Nicholas, F.W., Brown, S.C. and Holroyd, R.G. (1984). Tick resistance and the major histocompatibility system. *Aust. J. exp. Biol. Med. Sci.* **62**: 47–52.

Sukhapensa, V.(1980). Anthelmintic activity of oxyclozanide and hexachlorophene against Fasciola spp. in buffaloes. *Thai J. Vet. Med. Assoc.* **10**: 91–101.

Sukhapensa, V.(1983). Anthelmintic activity of niclofolan against Fasciola gigantica in swamp buffaloes. *Thai J. Vet. Med. Assoc.* **13**: 12–17.

Sukhapensa, V.(1986). A study on endoparasites in swamp buffaloes. Kasetsart Vet. 7: 87–95.

Sukhapensa, V.(1992). Parasites of swamp buffaloes. In: Buffalo Production. World Animal Science C6. Tulloh, N.M. and Holmes, J.H.G. (edi.) Elsevier, Amsterdam, pp. 329–354.

Sukhija, S.S. (1984). Studies on certain morphological and biochemical changes in buffalo spermatozoa during freeze preservation. Ph.D. Thesis H.A. U. India.

Sundaresan, D., Eldridge,F.,E. and Atkson, F.W. (1954). Age at First Calving Used with Milk Yield During First Lactation to Predict Lifetime Production oflndian Cattle. *J. Dairy Sci.* **37**: 1273.

Sunil-Chandra, N.P. and Mahalingam, S. (1994a). Application of ELISA in the diagnosis of rotavirus infections in buffalo calves. *Buffalo J.* **10**: 237–248.

Sunil-Chandra, N.P. and Mahalingham, S.(1994b). Rotavirus-associated diarrhoea in buffalo calves in Sri Lanka. *Res. Vet. Sci.* **56**: 393–396.

Sunil-Chandra, N.P., Mahalingam, S. and Chandra, N.P.S.(1994). Rotavirus associated diarrhoea in buffalo calves in Sri Lanka. Proceedings 4th World Buffalo Congress, Brazil, 2: 322–324.

Syndenham, F.W., Thiel, P.G. and Vleggaar, R. (1996). Physicochemical data for some selected Fusarium toxins. *J. AOAC International.* **79**: 1365–1379

Tanaka, K., Matsuda, V., Masang kay, J.S., Solis, C.D., Anunciado, R.V.P and Nami Kawa, T. (1999). Characterization and chromosomal distribution of satellite DNA sequences of the water buffalo (*Bubalus bubalis*).

Taneja, V.K., Nanda, S.K., Dutta, T.K. and Bhat, P.N. (1990) Embryo transfer in buffaloes. Present status and future research needs. Proceedings II World Buffalo Congress, New Delhi, ICAR Invited papers and Special Lecture, Vol II, Part II, p. 603.

Taparia, A.L. and Sharma,V.V. (1980). *J. of Anim. Sci. Cambridge*, **94**: 167.

Techakumphu, M., Lohachit, C., Chantaraprateep, P., Prateep, P. and Kabayashi, G. (1989) Preliminary report on cryopreservation of Thai swamp buffalo embryos. Manual and Automatic methods. *Buffalo Bull.* **8** (**2**), 29–36.

Tenora, F., Kotrla, B. and Blazek, K. (1974). Finding of the trematode Gigantocotyle siamense (Stiles et Goldberger, 1910) in buffalo in Afghanistan. Acta. Veterinaria Brno. 43: 111–116.

Thakuri, K.C., Mahato, S.N. and Thakur, R.P. (1992). Diseases of cattle and buffaloes in the Koshi hills of Nepal. A retrospective study. Vet. Rev. (Kathmandu). 7: 41–46.

Tham, K.M. (1997). Molecular and clinico-pathological diagnosis of malignant catarrhal fever in cattle, deer and buffalo in New Zealand. Vet. Rec. 141: 303–306.

Thapliyal, D.C. (1998). Food-born infections and intoxications with buffalo milk and milk products. *Indian J. Anim. Prod.* **30**: 200–201.

Thungtanawat, P., Arraprasert, D., Kielsommat, S., Lohachit, E., Chantaraprateep, P. and Bhodhipaksha, P. (1981). Induction of superovulation in buffaloes and non-surgical embryo collection. Annual Case Conference Report, Fac. Vet. Sci., Chulalonkorn Univ., Bangkok, Thailand, p. 16.

Tomar, N.S. and Tomar, S.P.S. (1960). A study of calving season in Murrah buffaloes. *Indian vet. J.* **37**: 445–52.

Tomar, S.P.S. (1969). Inheritance of economic traits and construction of selection indexes in buffaloes. Ph.D. Thesis, G.B. Pant University of Agriculture and Technology, Pantnagar, U.P.

Tomar, S.P.S. and Desai, R.N. (1965). Study of growth rate in buffaloes maintaiined on military farms. Genetic, phenotypic, environmental correlations and regression. *Indian Vet. J.* **42** : 947–57.

Tomar, S.P.S. and Desai, R.N. (1967). Factors influencing the inheritance of birth weight of buffalo calves on military farms. *Indian vet. J.* **44**: 694–701.

Tomar, S.P.S. and Desai, R.N. (1968). Genetic relationship of age at first calving with first lactation yields in buffaloes on military farms in North. *Indian J. Anim. Hlth.* **7**: 189–92.

Tomar, S.P.S. and Desai, R.N. (1969) Genetic relationship of body weight at first calving with milk production in buffaloes at military farms in the North. *Indian J. Anim. Hlth.* **8**: 25–29.

Tomar, S.S. and Basu, S.B. (1981) Genetic study on lifetime reproductive traits of Murrah buffaloes. *Indian J. Dairy Sci.* **34**: 195–98.

Tomar, S.S. and Tripathi, V.N. (1987). Disease spectrum in a herd of Murrah buffaloes. *Indian Vet. J.* **64**: 683–688.

Tongson, M.S. (1971). Neoasacris vitulorum larvae in milk of Murrah buffalo. *Philippine J. Vet. Med.* **10**: 60–63.

Torky, H.A. and Hammad, H.A.S. (1981). Trichophyton in farm animals and trials for treatment. Bull. Anim. Hlth. Prod. In Africa **29**: 143–147.

Totey, S.M., Dalivi, M., Apparao, K.B.C., Pawske, C.H., Taneja, M. and Chillav, R.S. (1996). Differential cleavage and developmental rates and their correlation of cell number and sex ratios in buffalo embryo generated *in vitro*. Theriogenology **45**: 521–533.

Tsai, S.J., Hutchinson, L.J. and Zarkower, D. (1989). Comparison of dot immunobinding assay, enzyme-linked immunomsorbent assay and immuno diffusion serodiagnosis of paratuberculosis. *Canadian J. Vet. Res.* **53**: 405–409.

Tuli, R.K., Singh, M. and Mataroo, J.S. (1981). Effect of different equilibration times and extends on deep freezing of buffalo semen. Theriogenology 16: 99/0104.

Tulloch, D.G., (1974). The feral swamp buffaloes of Australia's northern territory. In: W.R. Cockrill (Editor), The Husbandry and Health of the Domestic buffalo. FAO/Rome, pp–493–505.

Tunio, A.N. (1999). Personal communication.

Turabov, T. (1991). Selection of Caucasian breed of buffalo. PhD Thesis. Academy of Agriculture, Baku, Azerbaijan.

Turabov, T. (1997a). The modern condition of buffalo production in Azerbaijan and the ways for further improvement. Proc. of the Fifth Buffalo Congress, Caserta, 13 to 16 Oct.: 884–888.

Turabov, T. (1997b). Buffalo population and production in Azerbaijan. *Buffalo Newsletter* **8**: 9–10.

Udupa, K.G., Reddy, P.M.T. and Prakash, N. (1995). Treatment of thelaziasis in a buffalo- a case report. *Buffalo Bull.* **14**: 68.

Umrikar, U.D. and Deshpande, K.S. (1985). Genetic studies on lactation length and dry period in Murrah buffaloes. *Indian J. Anim. Sci.* **55**: 888–92.

Upadhyay,K.G., Patel,G.C., Vaghela,M.N. and Sunder,M.R. (1986) Manufacture of Mozzarella cheese from buffalo milk. *Indian Dairyman*. **38**: 479–483, 486.

Uppal, S.K., Sigh, K.B. and Bansal, B.K. (1996). Efficacy of teat dipping and dry therapy in the control of subclinical mastitis in buffaloes. *Buffalo J.* **12**: 347–352.

Uppal, S.K., Singh, K.B., Bansal, B.K. and Jand, S.K. (1998). Antibiogram of bacteria isolated from sub clinical and clinical cases of mastitis in buffaloes. *Buffalo J.* **14**: 253–258.

Uppal, S.K., Singh, K.B., Bansal, B.K., Rajesh Mohan and Mphan, R. (1995). Mastitis treatment with ampicillin and cloxacillin combination in lactating buffaloes. *Indian Vet. J.* **72**: 1001–1002.

Uppal, S.K., Singh, K.B., Roy, K.S., Nauriyal, D.C. and Bansal, B.K. (1994). Natural defence mechanism against mastitis. A comperative histomorphology of buffalo and cow teat canal. *Buffalo J.* **10**: 125–131.

Vagsholm, I., Nesse, L.L. and Gudding, R. (1991). Economic analysis of vaccination applied to ovine listeriosis. *Vet. Rec.* **128**: 183–185.

Varshney, T.R. and Singh, Y.P. (1975). Anthelmintic efficacy of trodax (M&B) in sheep. *Indian Vet. J.* **52**: 793–799.

Vassart,G, Brocas, H. Christophe,D.,De Martynoff,G,Mercken, L, Pohl, V. and Van Heuverswijn, B (1983). Structure and expression of the thyroglobulin gene. In hormone and cell regulation 7: 335–346.

Venitkul, M. (Ed.) (1989). Progress on embryo transfer in buffalo (Thai.). *Cattle Monthly* **2** (**18**): 78–81.

Venkatachar, M.C. and Sampath, S.R. (1978). Studies on the growth rate of Surti buffalo calves. Annual Report, NDRI, Karnal, pp. 141

Venkataswami, V. and Vedanayagam, A.R. (1962) Biometrics of spermatazoa of cattle and buffalo. *Indian vet. J.* **39**: 287–91.

Venkateswarlu, M., Choudhuri, P.C. and Reddy, J.S. (1998). Biochemical changes in rumen liquor and blood in clinical cases of alkaline indigestion in buffaloes. *Indian Vet. J.* **75**: 438–440.

Venkayya, D. and Anantkrishnan, C.P. (1957). Effect of season of calving, weight at freshening and length of service period of milk yield of Murrah buffaloes. *Indian J. Dairy Sci.* **10**: 123–30.

Verma, B.B. and Gautam, O.P. (1978). Studies on experimental surra (Trypanosoma evansi infection) in buffaloes and cow calves. *Indian Vet. J.* **56**: 648–653.

Verma, K.A.K., Kumar, M. and Sharma, S.P. (1994). Therapeutic studies on experimental Setaria cervi infection in rabbits. *International J. Anim. Sci.* **9**: 231–232.

Verma, P.C. and Kalra, D.S. (1974). Mortality in buffalo calves (Bos bubalis). *Indian J. Anim. Sci.* **44**: 163–168.

Verma, R. (1998). Haemorrhagic septicaemia in buffaloes current trends and prophylaxis. *Indian J. Anim. Prod.* **30**: 159.

Villaegas, V. (1928). The trend of sexual reproductive seasons amongst horses, cattle, water buffaloes, sheep and goats under Los Banos conditions: A preliminary report. Philipp. Agric. **17**: 477–85.

Villegas, V. (1930). Observations on the breeding activities of carabao. Philipp. Agric. **19**: 3–9.

Virk, A.S., Malik, N.S. and Chopra, A.K. (1981) Performance of growing and lactating buffaloes on diets containing dried brewer's grain. *Indian J. Anim. Sci.* **51**: 17.

Vlakhov, K., Karaivonov, K.H., Petrov, M., Kacheva, P., Alexiev, A., Polikronov, A. and Danev A. (1985). Studies on superovulation and embryo transfer in water buffaloes. First World Buffalo Congress, Cairo, Egypt, Vol. III, p. 510.

Wahby, A.M. and Hilmy, M. (1946). Prevalence of bovine mastitis in Egyptian dairy cattle. *J. comp. Path.*, **56**: 246–53.

Wahid, A. (1976). Pakistani Buffaloes. Monograph 7, University of Karachi, Pakistan.

Wakana,S., Watanabe,T., Hayashi,Y. and tomita, T., (1986). Avariant in the restriction endonuclease cleavage pattern of mitochondrial DNA in the domestic fowls (*Gallus galus domesticus*), *Anim. Genet.*, **17**:159–168.

Walker, A.R. (1986). The Toda of south India. A new look. Hindustan Publishing Corporation, Delhi, XVI+371.

Wang, X., Huang, K.G.J., Shen, X, Lin, C.Yi., Zhang, G. D.I. (1997). Chemical composition and microstructure of uroliths and urinary sediment crystals associated with the feeding of high-level cottonseed meal diet to water buffalo calves. *Res. Vet Sci.* **62**: 275–280.

Ware, F. (1942). A further survery of some important breeds of cattle and buffaloes in India. Misc. Bull. No. 54, pp. 16. Imperial Councial of Agricultural Research, New Delhi.

Warner, C.M. (1986). Genetic manipulation of the major histocompatibility complex. *J. Anim. Sci.* **63:** 279–287.

Warner, J.N. (1951). "Khoa, Kheer, Sar, Malai, Rabri, Chena and Sweets" In "Dairying in India" Animal Husbandry Manual, ICAR, New Delhi, p.223.

Watanabe, T., Hayashi, Y., Semba, R., and Osgasawara, N. (1985) Bovine mitochondrial DNA polymorphism in restriction endonuclease cleavage pattern and the locations of the polymorphic sites. Biochem. Genet., 27: 431–438.

Webster, C.C. and Wilson, P.N. (1980). Agriculture in the tropics, (ELBS, 2nd edn.) Longman, London, UK, pp. 390–400.

Webster, J. (1994). Animal Welfare: A cool eye towards eden. Blackwell Science, Oxford.

Webster, J. (2005). Animal welfare: Limping towards eden, 2nd edn. Blackwell Publishing Ltd.

Wilesmith, J.W. (1982). Jone's disease: a retrospective study of vaccinated herds in Great Britain. *British Vet. J.* **138**: 321–331.

Williamson, G. (1949). Iraqi livestock. *Emp. J. exp. Agric.***17**: 48–59.

Wiyono, A., Baxter, S.T.F., Saepulloh, M., Damayanti, R., Daniels, P. and Reid, H.W. (1994). PCR detection of ovine herpesvirus-2 DNA in Indonesian ruminants- normal sheep and clinical cases of malignant catarrhal fever. Vet. Microbiol. **42**: 45–52.

Wood, P.D.P. (1967). Algebaic model of the lactation curve in cattle. Nature (Lond.), 216, 162.

Wood, P.R., Corner, L.A., Rothel, J.S., Baldock, C., Jones, S.L., Cousins, D.B., McCormick, B.S., Frances, B.R., Creeper, J. and Tweddle, N.E. (1991). Field comparison of the interferon gamma assay and the intra dermal tuberculin test for the diagnosis of bovine tuberculosis. *Aust. Vet. J.* **68**: 286–290.

Wu, J.J., Ni, Y.H., Wu, J.Z. and Gao, F. (1991). Aetiological surveys on fescue foot in farm cattle: Toxicity test of metabolites of Fusarium sp. *Chinese J. Vet. Sci. Tech.* **21**: 17–20.

Yadav, B. R., Kumar,P., Tomer,O. S. and Balain, D. S. (1990). Monosomy X and gonadal dysgenesis in a buffalo heifer. Theriogenology, **34**: 99–105.

Yadav, B.R. and Balakrishnan, C.R. (1984). Chromosome analysis of isosexual and heterosexual multiple births in cattle and buffaloes. *Indian vet. J.* **61**: 126–30.

Yadav, B.R. and Balakrishnan, C.R. (1985). Parallelism of XX/XY chimaerism in cattle and buffalo multiple births. *Indian J. Dairy Sci.* **38**: 62–65.

Yadav, B.S. and Singh, L.N. (1974). Chemical composition of buffalo meat avaialble from the local slaughter house, *Ind. J. Anim. Sci.* **44 (10)**: 746–751.

Yadav, K.R., Paliwal, V.K., Mandal, A.B and Krishna, G. (2000). Effect of feeding formaldehyde treated clusterbean meal on nutrient utilization and growth in buffalo calves. *Indian J. of Anim. Nutr.* **17(3)**:206–210.

Yang, J.Q. (1983). The epidemiology and clinical diagnosis of rabies in buffaloes. *Chinese J. Vet. Med.* **9**: 11–12.

Yaqub, T., Das, P., Cheema, R.N., Muhammad, K. and Butt, I.A. (1997). Therapeutic trials in an outbreak of haemorrhagic septicemia in buffaloes. *Pakistan Vet. J.* **17**: 199–201.

Yashoda, K.P., Sachindra, N.M., Sakhare, P.Z. and Rao, D.N. (2000). Microbiological quality of hygienically processed buffalo carcasses. *Food Control, 11(3):* 217–224.

Youkhana, S.O. (1997). Pathological study on mastitis in buffaloes. *Iraqi J. Vet. Sci.* **10**: 107–112.

Zaki, R (1948) *Brucella abortus* infection among buffaloes in Egypt. *J. Comp. Path.* **58**: 73–79

Zakrzewski, H., Cybinski, D.H. and Walker, P.J. (1992). A blocking ELISA for the detection of specific antibodies to bovine ephemeral fever virus. *J. Immunological Methods.* **151**: 289–297.

INDEX

C

D

E

F

G

H

I

N

O

P

Q

R

S